CONTENTS

Acknowledgments ix

Contributors xiii

Foreword xv
JOHN L. HARPER

Introduction 3
RODOLFO DIRZO AND JOSÉ SARUKHÁN

SECTION I. NEW AND CONTRASTING APPROACHES TO THE STUDY OF PLANT POPULATIONS 9

The Morphological Basis of Plant Population Ecology

1. Plant Metamerism 15
JAMES WHITE

2. Dynamic Morphology: A Contribution to Plant Population Ecology 48
ADRIAN D. BELL

Contrasting Levels in the Study of Plant Populations

3. Dynamics of Plant Populations from a Synecological Viewpoint 66
EDDY VAN DER MAAREL

4. The Analysis of Demographic Variability at the Individual Level and Its Populational Consequences 83
JOSÉ SARUKHÁN, MIGUEL MARTÍNEZ-RAMOS, AND DANIEL PIÑERO

Populational Consequences of Biotic Interactions

5. Local-Scale Differentiation as a Result of Competitive Interactions 107
ROY TURKINGTON AND LONNIE W. AARSSEN

6. Some Evolutionary Aspects of Plant–Plant Interactions 128
SUBODH JAIN

7. Herbivory: A Phytocentric Overview 141
RODOLFO DIRZO

Approaches to the Study of Plant Life-Histories

8. Using Intraspecific Variation to Study the Ecological Significance and Evolution of Plant Life-Histories 166
D. LAWRENCE VENABLE

9. Life-History Variation, Natural Selection, and Maternal Effects in Plant Populations 188
BARBARA A. SCHAAL

SECTION II THE INTERFACE BETWEEN ECOLOGY AND GENETICS 207

Consequences of the Genetic Structure of Populations

10. Ecological Significance of Genetic Variation Between Populations 213
A. D. BRADSHAW

11. Genetic Variation Within Populations 229
JANIS ANTONOVICS

12. Immigration in Plants: An Exercise in the Subjunctive 242
DONALD A. LEVIN

Consequences of Plant Breeding Systems

13. Mating Patterns in Plants 261
MARY F. WILLSON

14. Gender Allocations in Outcrossing Cosexual Plants 277
DAVID G. LLOYD

SECTION III PLANTS AS INTEGRATED ECOPHYSIOLOGICAL UNITS 301

15 The Study of Plant Function: The Plant as a Balanced System 305
HAROLD A. MOONEY AND NONA R. CHIARIELLO

16 Demographic Consequences of Plant Physiological Traits: Some Case Studies 324
FAKHRI A. BAZZAZ

17 The Phenotype: Its Development, Physiological Constraints and Environmental Signals 347
ROBERT L. JEFFERIES

PERSPECTIVES ON PLANT POPULATION ECOLOGY

Edited by Rodolfo Dirzo
and José Sarukhán
UNIVERSIDAD NACIONAL AUTÓNOMA DE MÉXICO

SINAUER ASSOCIATES INC. • PUBLISHERS
Sunderland, Massachusetts 01375

PERSPECTIVES ON PLANT POPULATION BIOLOGY

For information address
Sinauer Associates Inc.
Sunderland, MA 01375
U.S.A.

Printed in U.S.A.

Library of Congress Cataloging in Publication Data

Main entry under title:

Perspectives on plant population ecology.

Bibliography: p.
Includes index.
1. Plant populations. 2. Botany—Ecology.
I. Dirzo, Rodolfo. II. Sarukhán, Jose.
QK910.P47 1984 581.5'248 83-20182
ISBN 0-87893-142-2
ISBN 0-87893-143-0 (pbk.)

9 8 7 6 5 4 3 2 1

Perspectives on Plant Population Ecology

SECTION IV AGRONOMIC IMPLICATIONS OF PLANT DEMOGRAPHY **359**

18 Population Ecology and Weed Science 363
A. MARTIN MORTIMER

19 Plant Demography in an Agricultural Context 389
R. W. SNAYDON

Literature Cited 409

Index 469

ACKNOWLEDGMENTS

This book is the outcome of a scientific meeting of plant population biologists who gathered at Oaxtepec, Morelos, Mexico, at the initiative of the Instituto de Biología (Universidad Nacional Autónoma de México), to commemorate Charles Darwin on the centennial of his death. The book is intended as a permanent record of the information and ideas presented and discussed at the meeting, and as homage to the memory of the biologist who, in the opinion of all the participants at the Symposium, has been the major intellectual driving force behind modern plant population biology and a source of inspiration for many ecologists interested in asking themselves questions similar to those which fueled Darwinian thought.

From the onset of the organization of the meeting in early 1981, we faced the seemingly impossible task of covering a very ample field of knowledge, represented by many excellent ecologists, while at the same time creating an event with a focus on fresh ideas. We were fortunate in achieving this goal and in being able to assemble not only the excellent roster of speakers whose papers form the chapters of this book, but also in attracting many other senior and junior ecologists, students from thirteen countries, and an enthusiastic and numerous group of Mexican biology students. The excellent poster presentations (46 in all) that complemented the talks were clear proof of the impressive quality of the attendants. These did much to increase the level of interaction and quality of the discussions during the meeting and this book has benefited considerably from such interaction. The many efforts directed towards the realization of a Darwin Centennial meeting have been more than amply rewarded by the results. The meeting remains in our memory as a worthwhile and satisfactory event.

We want to express our gratitude to the following institutions for their economic support which made the Oaxtepec meeting possible: Instituto de Biología (UNAM); Dirección General de Asuntos del Personal Académico (UNAM); Academia de la Investigación Científica, A.C. (Mexico); The British Council (Mexico) and The Royal Society. Very efficient logistic support was provided during the meeting by several persons, especially Ana Mendoza, Tere Medina and Gerardo Coronas. Their efforts made the meeting a most pleasurable experience for all participants.

Support for the preparation of this volume was provided by the Instituto de Biología (UNAM); we particularly thank Daniel Piñero for help in the editorial review of one of the chapters. Tere Medina diligently assisted us in the extensive retyping of manuscripts and Marlenne de la Cruz patiently helped us in the literature search. Finally, we thank our Editor, Andrew Sinauer, who extended to us a most open and valuable collaboration, considerably easing our task.

RODOLFO DIRZO
JOSÉ SARUKHÁN

CHAPTER ACKNOWLEDGMENTS

CHAPTER 1/James White
I am grateful to José Sarukhán for inviting me to prepare this paper and for his hospitality in Mexico. I thank David Agnew, Arne Dunberg, John Harvey and Maxine Watson for permission to consult and quote from unpublished theses or manuscripts.

I acknowledge the support of a Fellowship from The Royal Society of London and the facilities of the Botany School, Cambridge in 1982 during the preparation of this chapter.

CHAPTER 2/Adrian D. Bell
Sincere thanks are due to José Sarukhán and Rodolfo Dirzo, indeed to all at the Institute of Biology, UNAM, for the opportunity to deliver this paper on such a prestigious and stimulating occasion. Acknowledgment is due to the organizers of the Symposium at Yale, 1982: "Population Biology and Evolution in Clonal Organisms" for catalyzing thoughts concerning "foraging behavior" as applied to plants.

Financial support was received from The Royal Society, The British Council, and the University of Wales, U.K.

I am indebted to Jim White (University College, Dublin) for the crucial adjective, Dynamic.

CHAPTER 4/José Sarukhán, Miguel Martínez-Ramos, and Daniel Piñero
Many people have helped us since 1974, when we started our studies on populations of *Astrocaryum mexicanum*. We especially want to acknowledge the help of R. Dirzo and A. Mendoza in the field, as well as the staff at the Estación de Biología Tropical "Los Tuxtlas" where we are continuing our studies. These have been possible in part due to a CONACyT Research Grant.

We thank Francisco Molina and Alejandro Castellanos for allowing us to use some of their unpublished data on *Simmondsia chinensis*.

CHAPTER 7/Rodolfo Dirzo
I am grateful to Victor Jaramillo for critically reading this chapter and offering useful suggestions.

CHAPTER 8/D. Lawrence Venable
I thank D. Schemske and H. Dingle for reading an early version of this chapter and offering useful suggestions.

I would like to thank Eduardo Morales for allowing me to use some unpublished data from his thesis. Special thanks go to José Sarukhán for his kindness and help while I was working at UNAM. The studies in Mexico were partly supported by Grant No. PCENAL-80078 PNIE of the Consejo Nacional de Ciencia y Tecnología.

CHAPTER 9/Barbara A. Schaal
I am grateful to Joe Leverich for his help throughout this study, from the initial field collections to comments on the final manuscript. Randy Smith, Alice Sterkel, Sheri Baumgarte, and Leeju Wu spent many hours in the greenhouse counting thousands of flowers. Without their help this study would not have been possible. My initial interest in maternal effects was stimulated by Leah Leverich. This study was supported by NSF grant DEB 8141023.

CHAPTER 11/Janis Antonovics
The experimental work reported here was carried out in collaboration with Dr. Norman Ellstrand and Dr. Annie Schmitt: without their stimulus and effort these studies would not have been possible.

CHAPTER 13/Mary F. Willson
I am grateful to F. A. Bazzaz, D. G. Lloyd, and L. Gilbert, each of whom offered useful suggestions on the thoughts developed here. M. Lynch and F. A. Bazzaz commented helpfully on the manuscript. D. A. Levin made parts of this assignment easy, by having reviewed many aspects of gene flow and population structure.

CHAPTER 14/David G. Lloyd
I am grateful to Kamal Bawa, Ric Charnov, Dan Schoen and Colin Webb for their comments on a draft of the chapter, and to Dan Schoen and Mary Willson for providing unpublished data.

CHAPTER 15/Harold A. Mooney and Nona R. Chiariello
This chapter grew out of studies supported by the National Science Foundation: "Studies on the Limitation to Energy Capture in Natural Ecosystems."

CHAPTER 17/Robert L. Jefferies
I thank Susan Cargill for her criticism of the manuscript.

CONTRIBUTORS

Lonnie W. Aarssen, Department of Botany, University of British Columbia, Vancouver, British Columbia, Canada

Janis Antonovics, Department of Botany, Duke University, Durham, North Carolina, U.S.A.

Fakhri A. Bazzaz, Department of Botany, University of Illinois, Urbana, Illinois, U.S.A.

Adrian D. Bell, School of Plant Biology, University College of North Wales, Bangor, Wales

A. D. Bradshaw, Department of Botany, University of Liverpool, Liverpool, England

Nona R. Chiariello, Department of Biology, University of Utah, Salt Lake City, Utah, U.S.A.

Rodolfo Dirzo, Institute of Biology, Universidad Nacional Autónoma de México, Mexico City, Mexico

Subodh Jain, Department of Agronomy and Range Science, University of California, Davis, California, U.S.A.

Robert L. Jefferies, Department of Botany, University of Toronto, Toronto, Ontario, Canada

Donald A. Levin, Department of Botany, University of Texas, Austin, Texas, U.S.A.

David G. Lloyd, Department of Botany, University of Canterbury, Christchurch, New Zealand

Miguel Martínez-Ramos, Institute of Biology, Universidad Nacional Autónoma de México, Mexico City, Mexico

Harold A. Mooney, Department of Biological Sciences, Stanford University, Stanford, California, U.S.A.

A. Martin Mortimer, Department of Botany, University of Liverpool, Liverpool, England

Daniel Piñero, Institute of Biology, Universidad Nacional Autónoma de México, Mexico City, Mexico

José Sarukhán, Institute of Biology, Universidad Nacional Autónoma de México, Mexico City, Mexico

Barbara A. Schaal, Department of Biology, Washington University, St. Louis, Missouri, U.S.A.

R. W. Snaydon, Agricultural Botany Department, University of Reading, Reading, England

Roy Turkington, Department of Botany, University of British Columbia, Vancouver, British Columbia, Canada

Eddy van der Maarel, Institute of Ecological Botany, University of Uppsala, Uppsala, Sweden

D. Lawrence Venable, Department of Ecology and Evolutionary Biology, University of Arizona, Tucson, Arizona, U.S.A.

James White, Department of Botany, University College Dublin, Dublin, Ireland

Mary F. Willson, Department of Ecology, Ethology and Evolution, University of Illinois, Urbana, Illinois, U.S.A

FOREWORD

John L. Harper

For a long time the literature of population biology was dominated by, or even the exclusive concern of, zoologists, particularly of zoologists concerned with animals that run, jump, fly or swim. The science developed with a powerful skeleton of theory that has supported and stimulated the development of an experimental science of animal population biology. The study of plant populations was slower to develop, though there has been a recent tendency for a plant biologist to be included in many collections as a token representative of his kingdom. The present volume seems to be the first specifically given over to the population biology of plants. It is important, therefore, to ask what, if anything, it is about plants that makes them appropriate for separate study and discussion. After all, the fundamental properties of birth and death and potentially exponential increase in numbers leading to potentially competitive interactions characterize all living organisms; plants are no exception. That most plants are photosynthetic and depend on "raw," not prepackaged, resources, differentiates them sharply from heterotrophic organisms and means that their resource demands are at the same time more similar (all green plants require essentially the same range of resources) but also more diverse [all plants require a large number of different resources, most of which must be obtained independently of each other (nitrogen, phosphorus, potassium, light, water, nutrients, etc.)]. There are, however, other differences between animals and plants which appear at first sight to be more profound. These concern the manner in which an individual zygote grows in the process of leaving progeny zygotes. In all higher plants this process of growth is modular, proceeding by the iterative development of repeated units of structure. The form that results is almost always branched and has at least the theoretical potential for exponential increase in size (and of fecundity) and for infinite life. Much of the variation in the branching architecture and much of the way in which an individual enters into the environment of others depends on this architecture. Moreover, the modular or metameric form of growth is associated with a relatively unsegregated germ plasm. This point is seldom made, yet in view of the importance usually attached in evolutionary theory to Weismann's theory of the segregation of the germ plasm, it needs perhaps to be emphasized. There are possible consequences in the role that somatic mutation might play in the evolution of modular organisms that are denied to organisms that have segregated germ plasms.

Once it has been recognized that this set of special properties confers peculiar populational consequences on the plant kingdom, it becomes apparent that there are some animals with the same features. They happen to be those forms that are most often neglected by the population zoologist. The hydroids, bryozoans, colonial ascidians, corals and aphids all have life cycles in which the progression from zygote to zygote involves the iteration of modular units (metamers) and sometimes also a branching habit, a characteristic architecture, potentially exponential growth of the individual, potentially exponential increase in fecundity and discontinuity of the germ plasm. We could add to this set of characters of "modular organisms" that the phenomenon of senescence appears in many cases to be concentrated at the level of the repeated module rather than at the level of the genetic individual. There is, therefore, beginning to emerge a science of the population biology of modular organisms that brings together both botanists and zoologists. Such a meeting of minds has also been encouraged by the very rapid growth of interest in the ways in which the activity of herbivorous animals affects the behavior of plant populations, and there is an appropriate place for papers on plant-animal interactions in this book. I suspect we may come to recognize that the plant kingdom is very largely what the animal kingdom made it, that the animal kingdom herbivores, pollinators and dispersers of seeds have been a major, if not the major, driving force in the evolution of the diversity that we now know in the plant kingdom.

Of course, it is in the nature of the growth form of many plants, particularly trees, that their life is limited and that the risk of death increases after they have achieved a particular size. However, it is extremely difficult to find evidence of any phenomenon that might be called a true senescence at the level of the meristem. The tree may become liable to fungal attack or more liable to be felled by a typhoon as it becomes older, but if shoots are repeatedly rooted artificially the genetic individual would seem to have an infinite life. Amongst plants that do not accumulate an ageing body, such as the duckweeds and many plants of clonal growth, the potential for an infinite life seems to be real. A different set of populational phenomena, however, become apparent at the level of the iterated unit of construction. An individual leaf passes through a juvenile phase to maturity, senescence and death, and at any point in time a plant may be composed of parts that have quite distinct differences in age. The whole plant, the whole genetic individual, comes to have an age structure. This would be true also of modular animals but in no real sense is it true of unitary animals: it would be rather shattering to discover that an animal had legs of different ages! A plant is a population of parts that are born and die at different times. The whole may only be said to be dead when its last remaining living part dies. The phenomenon is easily seen in a dense forest stand where individuals that are doomed soon to die are composed mainly of dead tissue, aborted or dead buds, and may carry only one or two lingering live shoots. We can distinguish in plant populations two phenomena which we can call growth: the growth in the number of genetic individuals as zygotes develop and produce more daughter zygotes, but also the process by which the product of an individual zygote increases in size, the increase in the number of its parts exceeding the rate of death of its parts. In modular organisms demography is a science to be applied both *of* organisms and *within* organisms.

The present volume deals with the general topic of the ecology of plant populations. Population biology has been a relatively recent invader into (some have called it an insidious influence on) plant ecology. I believe that in essence what population biology has to add to ecology is a focus for the links that tie it to genetics, to evolutionary theory, to conservation and to management. Overwhelmingly, plant ecology has, in the past, been concerned with the description of vegetation and with the correlation of vegetation with physical factors of the environment, soil, climate, etc. It has given us a strong predictive science based on correlation, but rarely on causation. It has enabled us to make predictions about what species are likely to be found where and what types of plant form and physiology are likely to be found in particular environments. However, by its nature it has been a correlative science, seldom able to provide causal answers or causal explanations.

There is an analogy which seems appropriate for describing the variety of concerns of ecologists in understanding and "explaining" the diversity of forms found in nature. The analogy is with a watch. If we sample from members of a human community the wristwatches that are worn, we may be able to classify these or to arrange them in order. They may have hands that move or figures that flash, be made of steel, silver or gold, the strap may be of leather or metal. We would most probably be able to establish some correlation between the nature of the watch and the type of individual from whom it was taken, indicating his age or his wealth. If he were to lose the watch, the information from the classification or ordination and perhaps from the correlations would be what he needed to report the loss to the police or to advertise in the "lost and found" columns of a newspaper. These watches possess obvious holistic qualities. They are assemblages of parts that form integrated wholes and have a recognized activity (recording and reporting the passage of time) that is lacked by the kit of parts from which the watch could be assembled. The qualities of the whole watch can be described and compared—how far they conform in the way they record time, whether they tick, whether they are systems powered by daily winding or from the daily natural movements of the wearer or by annual insertion of a battery. All of these ways in which we can describe and analyze the form and behavior of watches have exact parallels in the ways in which the ecologist may look at communities or ecosystems. None of these ways is in the least appropriate for repairing the watch if it goes wrong or accounting for how the watch comes to behave in the way that it does (how it has evolved). The watch maker and repairer needs to know causes and effects—which cogs drive which cogs, how power is transmitted, which parts wear out most quickly. It is this same level of concern that, in ecology, is the concern of the population biologist. Moreover, if the ecologist is to discover not only how the ecosystem works, but also how it has come to be what it is and does what it does, there must be an evolutionary dimension to his study, and this has to involve concentrated observation at the fine level of the individual organism on which natural selection acts. Study of the integrated community tells us virtually nothing about evolutionary processes—yet all "ultimate" biological explanation has to be in evolutionary terms.

It is of the greatest significance that this volume celebrates a Darwinian centennial—it is from Darwin's *Origin of Species* that the greatest stimulus to developing a causal ecology derives.

> It is interesting to contemplate a tangled bank, clothed with many plants of many kinds, with birds singing on the bushes, with various insects flitting about, and with worms crawling through the damp earth, and to reflect that these elaborately constructed forms, so different from each other and dependent on each other in so complex a manner, have all been produced by laws acting around us. . . . These laws, taken in the largest sense, being Growth with Reproduction; Inheritance which is almost implied by reproduction; Variability from the indirect and direct action of the conditions of life, and from use and disuse: a Ratio of Increase so high as to lead to a Struggle for Life, and as a consequence to Natural Selection, entailing Divergence of Character and the Extinction of less improved forms.

If population biologists feel in need of a manifesto or a creed, Darwin's words provide it.

There is special delight that the Conference that generated this volume was organized and held in Mexico. Here is both the "tangled bank" in all its variety and profusion and also a vigorous group of population biologists—especially botanists—working in the tangled bank. I was deeply sorry than an illness prevented my attendance: for me and for many others this conference was a career highlight. I hope that Darwin's ghost has been able to be present; both the conference and the venue would give him great delight.

Perspectives on
Plant Population Ecology

INTRODUCTION

Numerous recent books, papers and talks have pointed out repeatedly that only through the marriage of population ecology and population genetics will it be possible to understand evolutionary processes in populations. But it is equally true that such a marriage will bear unfit theory unless the infinite variegation of the environment, both physical and biotic, is taken into account.

This interaction between population ecology and population genetics, together with the effects of the environment on the former, is particularly important in plant populations. Several basic factors account for this.

In most cases, plants grow fixed in space and are thus unable to average out their environments, except over time and as result of growth. At the individual level, their environmental exploration occurs through normally subtle probing via their shoot and root systems. The energy expenditure in such probing may be risky, and its cost–benefit relations are largely unknown.

The virtually all-or-nothing situation for growth and reproduction (due to rigid metabolic thresholds) predominant among animals would also be predominant in plants if it were not for their great ability to respond differentially to the infinite range of situations encountered by each individual. Plasticity, then, is a response to the inability to escape from the inadequacies of a given environmental spot.

Since it is impossible for individuals to actively explore their environment, we might expect plants to be active wanderers at the progeny-population level. Ironically, plants appear, according to an ever-growing body of evidence, to be utterly conservative in their search for environments: progenies are, on the whole, mostly distributed around their mothers; vegetative propagules usually place themselves as new physiologic units just next to the plants from which they originate, at least on a small time scale; individual leaves, and root and shoot tips, explore their surroundings through very restricted movements whereby, for example, competing plants (or plant parts) or obstacles can be surmounted, or suitable environmental spots can be exploited. This modest exploration, within a region of only a few centimeters or meters, is suggestive of an environmental heterogeneity much greater than we may at first suspect. The typical studies on pattern analysis usually detect one or two scales of pattern as a result of the way plants sense the environment; very likely, if more sensitive analyses and equipment were available, more scales of pattern could be detected, particularly toward the smaller range of space.

A great capacity to "sit-and-wait" (through seed dormancy and a state of suspended growth in seedlings) is the only option that enables plants to "use" new environments that become available only through the passage of time and by the modification of their present environment.

The reactions of a spatially fixed organism to its proximal environment and to other interacting organisms (such as competitors, predators, or parasites) must by necessity be far more versatile than that of an organism able to run, hide or chase. In stark contrast to animals, plants feed on their physical environment: *light* is their energy source, *minerals* are their building bricks.

Plants make use of an astonishingly narrow band in the already small luminous part of the energy spectrum. The constraints of the photosynthetic process necessitate that other factors of the physical environment must be considered along with light in any study of the way an individual fixes CO_2. Temperature and water availability (both in the atmosphere and the soil) need to be studied concomitantly.

Competitive interactions are one obvious consequence of this situation. These interactions have been thoroughly studied by experimental plant ecologists and agronomists, who generally use crop plants and their associates (weeds) in extremely simplified or model systems. It is in these systems too that the detailed relationship between light and plant growth has been investigated. Somehow, this level of research has not transcended to studies with wild species growing in natural (non-experimental) systems, particularly highly structured systems such as forests, where light is not evenly available either in space or time.

Of the many concepts and ideas from animal population ecology, it is surprising that the conceptual framework of trophic relations—which plays such a pivotal role in zoological thought—has not been adopted as easily as other (seemingly less relevant) concepts to the development of plant population ecology. Perhaps the reductionist approach and the very ideal conditions under which most of our knowledge of plant physiology, especially CO_2 fixation and photosynthesis, has been derived was a deterrent to ready exploration of the energy/capture relations of plants in natural ecosystems.

That plants are fixed in a two-dimensional space, and that the energetic resource is normally perpendicularly incident onto the plane of that space and is exhausted by intervening canopies, are plausible causes of the modular form of growth of plants. Modular growth of their aerial tissue is the only way left to plants to explore new environments and to become more competitive for the energetic resource. Similarly, modular growth of roots allows continuous probing for mineral resources and water in the soil.

Because of their metameric construction, plants have the potential to grow indefinitely and, hence, to live forever. Continuous growth is the only way open to plants for short-term search of new environments, and also for achieving better competitive status with their neighbors. But the potential to continually place new photosynthetic tissue and new root tips in unexplored parts of the environment exacts a concomitant structural cost which partly offsets the gains achieved by discovering new and favorable environments.

Also, the ability of long-lived plants to integrate the environmental characteristics of the spot where they live through their ontogeny is a result of their space fixation. This is a situation not encountered in most animals, and one which has been largely neglected as a potential research tool on the relationships between plants and the relevant factors of their environment, and the populational consequences of such relationships.

The time-integration of the environment occurs through the different growth responses of the individual to varying environments. The recognition of permanent "reading devices" (growth rings, leaf scars, internodal lengths, etc.) in plants may constitute a useful way to trace individual behavior back in time and to explore its relations with the environment. The application of these analyses at the population level has been scant and may be rewarding in understanding long-term population trends and individual behavior.

Fixation in space is also responsible for the very special interactions that characterize plants in their fast responses to interacting organisms. High plasticity in response to competitors, and the rapid response to predators by the profuse design of a variety of secondary compounds, are two clear examples of adaptation to the inability to escape in space. Similarly, floral biology (and consequently breeding systems and progeny dispersal) show an enormous array of adaptations to the interacting organisms—pollinators and dispersal agents. The preadaptation of genotypes to specific, fine-scaled environments, which is increasingly becoming evident through recent studies, is good evidence of the great capacity of individual plants to fit snugly into very specific, constrained environments determined by certain neighbors and a given set of physical factors.

Plant population ecology has been concerned mostly with average performances in populations and with the final outcomes of the processes by which plants capture energy and nutrients: ecologists record whether, on average, plants survive, grow and reproduce as an expression of their success for CO_2 fixation and photosynthetic efficiency. But usually, ecologists are unable to explain which morphological or physiological trait or which environmental condition has been responsible for a given outcome. So far, plant population ecology has been able only to describe results—that one individual performs better or worse than another; it cannot explain why.

At the present time it has become obvious that knowledge of within-population demographic variability at the individual level is essential to understanding the causes of the average patterns observed. We have gained detail in the description of demographic behavior by this approach, but unless we can explain the source of such individual variability, we will not be able to transcend description, no matter how detailed the description may be. The explanation of the source of variability at the individual level requires the study of its physiology and genetics. Knowing to what degree an individual's differential survival, growth and reproduction is the result of the environment in which it happened to occur, and to what degree it is determined by genetic traits, is essential to define whether natural selection is acting or not in a population.

The triadic approach to plant population biology—demography, genetics and physiology—conceived at the level of the individual appears to be the best route to understanding selective processes in populations and to explaining adaptive traits of individuals within populations. Such an approach will undoubtedly prove useful for the understanding not only of natural populations but also of those managed by man. Understanding individual variability may help agricultural systems to better manipulate the plants concerned. It will be very difficult to improve average (population level) performances in a crop if one ignores the variance around the mean behavior and one cannot benefit from the knowledge of *the causes* of extreme performances. This applies equally to the management of plants associated with crops (the so-called weeds), whether these are injurious or not.

The chapters in this volume attempt to provide not only a perspective in plant population genetics, plant population ecology and whole plant physiology within a population context, but also to state the front-line of research on these three major subjects. The perspective exercises do not remain isolated from each other. It is clear that the three fields point towards the need of merging with each other in a novel and very powerful approach for the study of many fundamental aspects of plant population biology. No doubt, this becomes an interdisciplinary, team-level approach, with all its great advantages, its demands and its limitations.

Characteristically, Darwin attempted a holistic approach to most of his studies; we hope that the holistic approach that permeates this book will prove to be a useful one, and that the perspectives in plant population ecology presented here will be considered by plant scientists in the future as an appropriate means of having commemorated Charles Darwin on the centennial of his death.

SECTION I
NEW AND CONTRASTING APPROACHES TO THE STUDY OF PLANT POPULATIONS

The Morphological Basis of Plant Population Ecology
Contrasting Levels in the Study of Plant Populations
Populational Consequences of Biotic Interactions
Approaches to the Study of Plant Life-Histories

INTRODUCTION TO SECTION I

The variety of themes dealt with in this section is an appropriate reflection of the multiplicity of areas in which a populational approach is relevant, and frequently essential, to the proper understanding of the biology of plants. Conversely, it could be argued that this variety of themes is indicative of the many botanical (or even zoological) aspects that must be taken into account for a proper development of plant population ecology. A modern science of plant population ecology needs the foundations of other related (and some seemingly unrelated) areas to become a source of better understanding of phenomena in plant biology.

The first part of this section, "The Morphological Basis of Plant Population Ecology," deals with a very fundamental level of organization in plants: subunitary plant parts (metamers) whose arrangement and disposition determine morphology, an essential attribute of plants —an attribute whose populational relevance has been much neglected, and that White (Chapter 1) scholarly rescues for us.

White first presents a splendid historical account of metamerism as a morphological concept; he goes back to the very roots of the concept of morphology in human thought, and analyzes in detail the late eighteenth–early nineteenth century morphologists up to Darwin, and then those of our own time. As he rightly states, Darwin (very much influenced by Owen) occupies a pivotal position in the history of morphology. By dissecting the concept of a metamer (and metamerism),

White analyzes some of its contemporary applications (plant morphology, plant competition, plant productivity, and plant–animal interactions). There is great potential for the application of plant metamerism to many topics which are of major concern for plant population ecologists. This new "ecological morphology," in which demographic, life-history and morphological concepts intermingle, offers a very promising field of interactions in plant ecology.

A novel and very important extension of the discussion on morphology (emphasizing the relevance of morphological factors to population ecology) has been made by applying one particular aspect of morphology—clonal branching patterns—to population processes (Bell, Chapter 2). In describing the general features of clonal branching patterns in plants, Bell shows that the process of progressively larger and apparently more complex growth in clonal plants can be described by relatively simple and repetitive branching rules. Computer technology aids considerably in showing the importance of plant morphology to the study of how plants with different growth rules explore their environment. Computer simulation allows rapid assessments of, for example, long-term spread and spatial disposition of expanding clones; and analyses of comparative efficiency in the invasion of space by clones with different growth tactics. One can create hypothetical situations in which the effects of different growth patterns or environmental features (such as neighbors) can be assessed. This approach can even be useful in asking questions of evolutionary significance: A bank of rules (genes) of growth can be created, mutations included, replacement rules (selection) introduced and varied, etc. Thus classical plant morphology moves to a position of more incisive exploration of its relations with environmental factors, and to the realm of prediction of the behavior of plant growth and its populational consequences.

At higher levels of organization in plants—individuals and communities—Part 2 of this Section shows two contrasting approaches to the study of plant populations. Sarukhán et al. (Chapter 4) make the point that an appropriate understanding of the dynamics of plant populations can only be conceived through the analyses of the considerable demographic variability displayed at the level of the individual plant. This chapter emphasizes the importance of individual demographic variance in explaining the components of fitness—survivorship, growth and reproduction. In keeping with the general intention of this book, Sarukhán et al. also contend that studies on intrapopulational variation within a demographic context constitute one of the most promising points of interaction of plant population genetics with demography and physiological ecology.

In contrast, van der Maarel (Chapter 3) explores a higher level of organization in plants, the community, in relation to population processes. He discusses several ecological phenomena usually studied and conceived as exclusively synecological, such as species performance along gradients; vegetation dynamics in time and space; and community diversity, and argues that they are better understood in the light of population processes such as niche replacement (competitive exclusion); plant population dynamics (establishment, growth, and disappearance); and competitive interactions, respectively. A major point emerging from this contribution is the view that synecological processes are the result of very dynamic populational phenomena; this hopefully will help vegetation scientists to incorporate a functional perspective into their usual geographical approach.

In the third part of this Section, "Populational Consequences of Biotic Interactions," the emphasis shifts to the role of biotic interactions (neighboring plants and herbivores) on a number of proximal (ecological) and ultimate (evolutionary) features of plant populations. Turkington and Aarssen (Chapter 5) review comprehensively the potential of plant competition as a mechanism leading to local-scale population differentiation. A major outcome of this chapter is the exciting notion that plant–plant interactions produce the necessary conditions for specialized genotypes vis a vis biotic ecotypes among individuals living in close proximity, leading to biotic specialization at the genotype level (in this particular case in relation to competition, but presumably extendable to other environmental factors). This notion calls for a view of community structure and microevolutionary processes that centers on the individual plant.

Chapter 6 is a more general elaboration of the discussion of neighborhoods in which Jain discusses the impact of both competitive and reproductive plant–plant interactions within highly structured populations composed of smaller spatial units (neighborhoods) that display a number of ecological and genetic relationships. In particular, he presents a number of examples in order to explore the consequences of frequency-dependent selection in plant neighborhoods. His examples include:

(1) Selection in mixtures and composite cross populations (discussions along the same lines as Turkington and Aarssen) which sometimes lead to situations of mutual advantage among competing genotypes so that yield may be increased and genetic diversity may be retained and even enhanced. The mechanisms responsible for this are widely different (e.g., heterozygote advantage; host–pathogen mediating effects; frequency-dependent selection) and, in most cases, are very poorly understood.

(2) Natural selection in gynodioecious populations. Jain indicates that models of gynodioecy and other breeding systems generally depend on mating patterns occurring between neighboring plants and on sib competition arising within progenies of different mother plants dispersed within rather short distances. Clearly, plant–plant interactions must play a fundamental role; the examples he discusses strongly suggest that this is so.

An important conclusion Jain reaches is that any plant population with fitness variance might have a heritable component due to intergenotypic interactions, and that this can be better expressed in relation to the neighborhood concept. This offers a clear contrast to Darwin's classical theory of natural selection by individuals rather than population units and offers, too, a challenging and exciting field to work on.

A complementary chapter to this part of Section I is provided by Dirzo's phytocentric overview of herbivory. He proposes that a science of plant population ecology needs to incorporate animals, in their role of herbivores, as a potentially important ecological factor in plant populations, and also as an important selective force responsible for a number of plant attributes (e.g., phenological, biochemical) of evolutionary relevance. Accordingly, he suggests that the role of herbivores in plant populations can be studied and interpreted on the one hand at a proximal (ecological) level, and on the other at an ultimate (evolutionary) level. A major part of the chapter is dedicated to the analysis of the proximal interaction: He discusses a number of variables—including the modular structure of plants—that determine the impact of herbivory on the individual plant. An interesting feature of this discussion is that often the real impact of herbivory can only be observed in the context of neighboring (competing) plants and, sometimes, in the context of other physical features of the plant's environment. The fact that herbivory can be important in tipping the balance of competitive interactions between plants is extended to explore the conditions necessary for herbivory to act as a force to reduce, maintain or increase diversity in plant communities. The analysis of the ultimate (evolutionary) interactions shows several examples where it has been argued that the plant–herbivore patterns we observe today are the result of past evolutionary interactions between the ancestors of the present interactants. Plant polymorphisms and their interaction with herbivores seems to offer a suitable system to evaluate the evolutionary potential of the plant–herbivore interface. This area emerges as a fertile field for the interaction of plant (and module) demographers, physiological ecologists and geneticists.

The final segment of this section is a refreshing view of a well-established topic in population ecology: the study of plant life-histories. Venable offers a very detailed analysis of the strengths and weaknesses of the different ways of investigating plant life-histories (mathematical modeling, comparative surveys, detailed comparisons of a few species, and the study of intraspecific variations), of which intraspecific variations in life-history traits appear to offer the most convenient and sensitive system of study. The chapter is complemented with a description of the author's own work. Venable shows that species with somatic seed polymorphism are a rather convenient system to work with in the field, since highly controlled selection experiments are possible. Because the variation is not only intrapopulational but is contained within the progeny of single individuals, the genetic background is quite constant and the variation (i.e., seed morphs) is manipulable.

Schaal (Chapter 9) presents a detailed account of her own work on variations of life-history features with the highly variable legume *Lupinus texensis*. This chapter illustrates a very different approach to the study of life-history variations: to reduce the environmental component of variation and to expose the inherent variation by working in controlled greenhouses. As Schaal shows, a great deal of realism and environmental control is achieved, but caution is necessary in any inferences to field situations. She was able to detect significant differences in survivorship patterns, viability components of fitness, and reproductive values; since these differences were shown in the uniform greenhouse environment, it is inferred that most of this variation is genetically determined.

Schaal then explores the role of maternal effects as a source of variation in offspring fitness, in particular via seed size. Her data show that seed size strongly influences fitness and suggest that seed size is in large part a maternal characteristic; thus, the fitness of an individual plant may be, at least in part, a function not of that very individual but of the maternal plant's genotype (genotypic maternal effect), and of the environment the maternal plant might have experienced (environmental maternal effect). Finally, she documents that the breeding system may be another source of variation in life-history attributes. A comparison of selfed and outbred progeny showed a strong inbreeding depression (and hence a genotypic effect) on several components of fitness.

This chapter illustrates the complexity of the relationships among some of the different sources of variability and their effects in influencing life-history features. The genetic aspects considered in this and the previous chapters prepare the way for the following section of the book, on the interface between ecology and genetics.

In general, Section I offers a balanced view of the numerous avenues plant population studies have explored and the numerous fields for which a populational approach can provide a fertile area of inquiry and understanding of nature. As most chapters in this section show, Darwin has indeed been a major source of inspiration for the study of plant populations.

SECTION I
NEW AND CONTRASTING APPROACHES
The Morphological Basis of Plant Population Ecology

CHAPTER 1

PLANT METAMERISM

James White

Among plants, form may be held to include something corresponding to behaviour in the zoological field . . . for most though not for all plants the only available forms of *action* are either growth, or discarding of parts, both of which involve a change in the size and form of the organism.

—Agnes Arber
The Natural Philosophy of Plant Form (1950:3)

INTRODUCTION

Darwin devoted only a few pages of *On the Origin of Species* to morphology, but indicated vividly his belief in its importance:

This is the most interesting department of natural history and may be said to be its very soul.

He was intimately familiar with this science, having recently spent some eight years (1846–1854) studying barnacles. The significance of morphology for the "long argument" of *Origin* was manifest: the similarity of structural patterns in members of the same class could be interpreted by descent by means of natural selection of successive slight modifications from an ancient progenitor or archetype. In this essay I shall explore a particular morphological topic, metamerism, and its application to plants. The etymology of the word (shared parts) is straightforward and leads to a formal definition: the serial repetition within or along an organism of unit structures—the metamers or segments—which are either identical or homologous in structure. Darwin was quite familiar with the concept as it applied both to animals and to plants (*Origin*, pp. 435–437):

> There is another and equally curious branch of [morphology] namely the comparison not of the same part in different members of a class, but of the different parts or organs in the same individual . . . Why should similar bones have been created in the formation of the wing and leg of a bat, used as they are for totally different purposes? Why should one crustacean, which has an extremely complex mouth formed of many parts, consequently always have fewer legs; or conversely, those with many legs have simpler mouths? Why should the sepals, petals, stamens and pistils in any individual flower, though fitted for such widely different purposes, be all constructed on the same pattern?
>
> On the theory of natural selection we can satisfactorily answer these questions. In the vertebrata, we see a series of internal vertebrata bearing certain processes and appendages; and in flowering plants, we see a series of successive whorls of leaves. An indefinite repetition of the same part or organ is the common characteristic (as Owen has observed) of all low or little-modified forms; therefore we may readily believe that the unknown progenitor of the vertebrata possessed many vertebrae; the unknown progenitor of the articulata, many segments; and the unknown progenitor of flowering plants, many spiral whorls of leaves. We have formerly seen that parts many times repeated are eminently liable to vary in number and structure; consequently it is quite probable that natural selection, during a long-continued course of modification, should have seized on a certain number of primordially similar elements, many times repeated, and have adapted them to the most diverse purposes.

I have quoted Darwin at length, not just because it is his death we are remembering, nor because he may always be read and quoted with profit, but because he occupies a pivotal position in the history of morphology. The impact of *Origin* on morphology since 1859 has been ambivalent. Most of the concepts and techniques of morphology had been enunciated in the previous half-century by Darwin's predecessors and contemporaries, and to them he owed a huge intellectual debt: Owen, for example, mentioned in the foregoing passage, the spectacular doyen of English comparative anatomists, is one of the most cited authors in *Origin*. Despite Darwin's advocacy of the significance of morphology for natural history it is doubtful if, even today, his theory of natural selection has much influence on morphology (Bock, 1980; Lauder, 1981). Many of its contemporary concepts can be traced, almost seamlessly intact, back along an intellectual tradition to their pre-Darwinian originators. Ghiselin (1980) has argued persuasively that morphology failed to assimilate Darwinism and contributed virtually nothing to the synthetic theory of evolution, a view with which Coleman (1980) concurred. On the other hand, several morphologists accepted Darwin's theory of natural selection wholeheartedly, notably Gegenbaur and Haeckel in Germany (Coleman, 1976), even if an extraordinary "feat of legerdemain" (Arber, 1950:64) was required to reinterpret morphological series (unspecialized–specialized, "primitive-derived") as phylogenetically ancestral or descendant. Many morphologists inveighed against this practice (Russell, 1916:24): Thompson, for example, rejected Darwinian natural selection in his seminal monograph *On Growth and Form* (1917). More recently Arber (1950:63) expressed the reaction forcefully:

> In the Darwinian reorientation of biology, the attention of most botanists was directed from pure morphology to the use of form data in support of speculations about evolution. This was particularly so where flowering plants were concerned, since the most direct kind of evidence, that of the geological record, was rarely available. To evolutionary schemes, the type concept fell an immediate victim. . . . Gradually, however, the facile Darwinian view, so easy to understand, and therefore so fatally easy to accept, lost its hold.

I find it difficult not to sympathize with this attitude a little, when I recollect the striving by many nineteenth century morphologists to achieve a rational understanding of biological form and organization. The dissatisfaction with the Darwinian reorientation of biology vis-à-vis morphology continues to be expressed (Goodwin, 1982).

But there are winds of change perceptible nowadays. For some time a handful of animal morphologists, notably Böker (1935–7), have been advocating a proper synthesis of morphology and evolutionary theory and a resuscitation of morphology with a strong functional and ecological bias (Bock, 1977, 1980; Riedl, 1977). Plant ecologists, too, have independently begun to pay careful attention to the construction of plants: less to the well-established botanical practice of growth-form and life-form analysis (e.g., du Rietz, 1931) than to the dynamical details of growth and form in a functional-ecological context. One may already detect in the published literature and through the informal contacts of the "invisible college" the beginnings of a new synthesis of morphology and ecology in an evolutionary context, by biologists who accept wholeheartedly the Darwinian perspective.

In the first part of this chapter I shall trace the history of the concept of metamerism, an idea in continuous use since the 1790s, still well known to zoologists but almost forgotten by botanists until its recent revival (Hallé and Oldeman, 1970; Harper and White, 1974; Harper, 1977: Chapter 1; White, 1979). In the second part I shall cast an eye over some contemporary applications of plant metamerism. It is altogether appropriate to do so, I believe, under the rubric of "perspectives on plant population ecology" because a *rapprochement* between the morphology and ecology of plants is being facilitated through the modern science of plant demography, by the choice and census of suitable plant organs.

THE HISTORY OF METAMERISM AS A MORPHOLOGICAL CONCEPT

Goethe to Darwin

The repetition of similar or dissimilar elements in the construction of organisms is perhaps one of the oldest concepts of morphology. The idea was clearly expressed by Aristotle and by his pupil Theophrastus who was the greatest botanist of antiquity.

> Every tree has many starting points for sprouting and fruiting. This is of the essence of a plant, that it lives from a multitude of parts, which is why it can also sprout from them . . . the sprouts are neither all of similar or equal size nor of simultaneous production . . . the position of the parts will make a difference in their sprouting.
> Theophrastus, *De Causis Plantarum* (Einarson and Link, 1976).

I have briefly referred elsewhere (White, 1979) to later botanists who held similar views. It was among early nineteenth century morphologists that the concept of metamerism became firmly established and it is only from them that we can trace its almost unbroken intellectual tradition to our own time. As we commemorate the centenary of Darwin's death, I recall also that it is 150 years since the death of Goethe (1749–1832) and of Cuvier (1769–1832) in whose writings one may find the most influential sources of the idea that plants and animals are composed of serially repeated parts, arranged in harmonious, correlated combinations. Goethe, by common consent (Russell, 1916; Mayr, 1982) the seminal source, expressed the idea strikingly in his essay of 1790, *Versuch die Metamorphose der Pflanzen zu erklären* [available in an English translation by Arber (1946)]. An important element in his thinking was the concept of archetype (*Urbild*), a unity of plan or of type which characterized each major group of animals and plants: groups were united by sharing common features of design. He asserted that vertebrates, for example, were formed according to an archetype which varied only more or less in its basically constant parts, akin to his conception of the shoot of a plant archetype (*Urpflanze*) being composed of serially repeated structures belonging to the organ category of leaf. Such notions ushered in an active period of idealistic morphology, whose roots lay deep in the Romantic Movement (e.g., Lenoir, 1978), and which flourished in Continental Europe for several decades. The idea of unity of plan was espoused by Cuvier, who believed that the prodigal variety of animal forms was based on relatively few possible combinations of the principal parts or organs in order to ensure the functional integrity of the organism (Russell, 1916; Coleman, 1964). His principle of correlation of parts was the cornerstone of his work: it is fascinating to observe how even today this principle underpins the morphological investigations of many French botanists (e.g., Nozeran et al., 1971). Through the work of Geoffroy Saint-Hilaire (1772–1844), a method was established to determine homologies between organisms based on the positions, relationships, and interdependencies of parts (the principle of connections). This was especially necessary when parts were modified by structural transformations.

Such ideas had a profound impact on morphologists, including Darwin. Ospovat (1981) has dealt in detail with Darwin's debt to them. Of course, he used such insights on unity of plan as one of the strongest lines of evidence for a theory of descent in *Origin*: this was the reason for his assessment of morphology as the soul of natural history. Mayr (1982), echoing Russell (1916), stated that "the idealistic morphology concept of structural variation ultimately provided a perfect stepping stone to the theory of common descent."

The German school of idealistic or transcendental morphology established by Goethe's writings was primarily responsible for the theory of the repetition or multiplication of parts within the individual organism (Russell, 1916: Chapter 7): this was the concept of metamerism. I have not located the earliest use of this term, but the concept was certainly familiar decades earlier than the *Oxford English Dictionary* indicates (the 1870s) for the first use of the term in English. For example, von Baer, the founder of embryology and the study of development, gave it detailed attention in the second volume of his *Über Entwickelungsgeschichte der Thiere* of 1837 (Russell, 1916: Chapter 9). Goethe himself regarded his archetypal plant as a supersensible conception, not representable pictorially, and he never made an illustration of its metameric construction. But several other botanists attempted it, and some of their efforts are reproduced by Schmid (1930), including a bizzare depiction by Turpin; I might add that these efforts persist (Ritterbusch, 1977). Schmid does not include an illustration which in my opinion is the finest in this strange genre, that by Richard Owen.

Owen (1804–1892) was the last great idealistic morphologist of the period before *Origin*. He had great influence on Darwin and is the third most cited author in *Origin* (Barrett et al., 1981). In the spring of 1846, Darwin discussed the phenomenon of vegetative repetition with Owen (Ospovat, 1981), a matter which engaged Owen particularly at the time. Following his celebrated monograph *On the Archetype and Homologies of the Vertebrate Skeleton* (1848) in which he sought to produce a comprehensive and consistent theory of morphology, Owen in 1849 published the short book *On Parthenogenesis*. Its interest in the present context is not for his theory of parthenogenesis, which did not

Plate I.

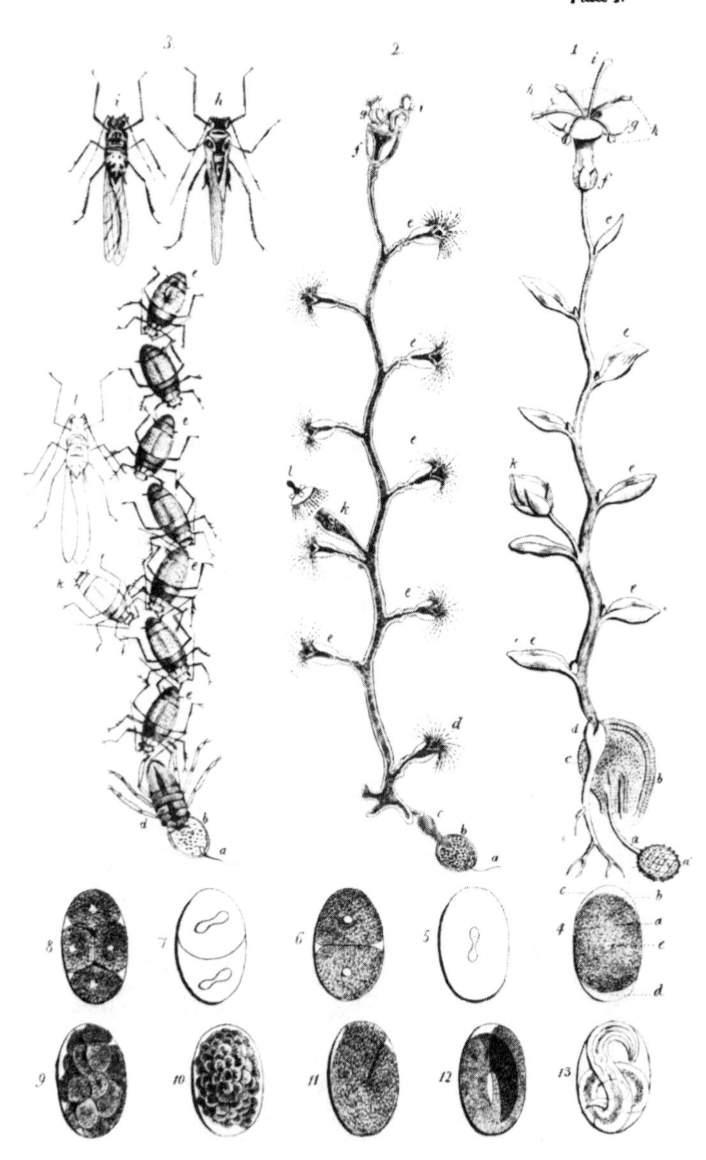

"*Natura infinita est, sed qui symbola animadverterit omnia intelliget, licet non omnino*" Goethe.

R. Mendel. J. Erxleben, on stone. Day & Son, Lithrs to the Queen.

◀**FIGURE 1.** Frontispiece of Richard Owen's *On Parthenogenesis* (1849) in which he analogized the "internal gemmation" (parthenogenesis) of aphids to the "external gemmation" of a hydroid and a plant. The drawing of the plant is based on Owen's realization of Goethe's conception of a metamerically constructed plant archetype. [The lower figures (8–13) depict Owen's understanding of embryological details in fertilized ova and are not relevant to my discussion.]

stand the test of later research, but for the frontispiece (Figure 1) and associated text. In his earlier (1848) treatise, he had formally distinguished different types of homology, among them "serial homology," the repetition of similar parts within one organism. In seeking to understand the parthenogenesis of aphids, he tried to analogize it with other models of embryological development and eventually turned to a botanical example to resolve his dilemma since "the botanist [had] arrived earlier than the zoophytologist at an intimate philosophical comprehension of the nature of his composite subjects." (Owen, 1849:53). He sketched his idea of an *Urpflanze* following Goethe's "true poetic insight into [its] essential nature." He believed that the "most familiar, if not the commonest form of the individual plant, is the leaf." Leaves with their associated vascular supplies constituted the trunk and branches of the compound whole, he maintained.

A generation of leaves, the breathing parts of the individual phytons, perishes each year; but not before each individual has made preparation for its successor by developing a bud at the axil of the leaf, which remains when the leaf falls and supplies its place in the following year. Here the individuals are successively propagated by gemmation . . . but continue associated together like the compound polypes.

He could not accept that organic individuality lay in the whole grouping of plant parts and cited the opinions of Steenstrup and of Forbes to bolster his view:

It is certainly the great triumph of morphology that it is able to show how the plant or tree (that colony of individuals arranged in accordance with a simple vegetative principle or fundamental law) unfolds itself, through a frequently long succession of generations into individuals becoming more and more perfect. (Steenstrup)

Every botanist knows that [a plant] is a combination of individuals, and if so, each series of buds must certainly be strictly regarded as generations. (Forbes)

Owen's analogy for the parthenogenesis of aphids was, accordingly, as follows: whereas the plant and zoophyte showed "ordinary or external gemmation" of parts, the aphid showed "internal gemmation." "The aphids generated from virgin parents, by this process of internal gemmation, are as countless as the leaves of a tree, to which they are so closely analogous."

By 1859 and the publication of *Origin*, the concept of serial homology of unit parts was firmly established and found its expression in the passage I quoted in the introduction. *Origin* had little effect on the concept as a heuristic device in morphological analysis; but it had a larger effect, as it redirected the attention of many biologists from questions of pattern to those of process (Lauder, 1981), to develop a science "dominated by content (historical explanations, contingencies) rather than by form (universals, laws)" (Goodwin, 1982).

After Darwin

Russell (1916) has documented well the conversion of morphology to evolutionary ideas, particularly by Haeckel and Gegenbauer in Germany. Haeckel was much concerned with the problem of organic individuality (Rinard, 1981), which is scarcely surprising since one of his most influential teachers at Berlin had been Alexander Braun, a botanist who had a special affinity for Goethe (shared by Haeckel) and who wrote on plant individuality (see White, 1979; Rinard, 1981). In his major work, *Generelle Morphologie* (1866), Haeckel developed a complex multilevel scheme for understanding the organic individual: above the levels of cells, tissues, and homologous parts were what he called metameres, repeated segments along a main axis, such as the internodes of seed plants or the segments of vertebrates. Beyond this level were persons (individuals in the ordinary sense) and colonies, six levels in all. Botanists at this time were not idle either, and Cusset (1982) has provided a comprehensive synthesis of the various morphological traditions which took a not-dissimilar reductionist view of plant construction.

Among animal groups, repetitive segmentation is more evident in some than in others: annelids and arthropods show it more clearly than vertebrates. The debate still continues today about its significance (e.g., Chaudonneret, 1979; Gasc, 1979). After several decades of research, Lankester (1904) attempted to summarize opinion on the construction of arthropods in an essay which included what, he stated, "may be called the laws of metamerism," though he did so with a caveat that "these are not so fully ascertained or formulated as might be expected" and are really "statements of a more or less general proposition." Thirteen "laws" were enunciated. I do not wish to repeat them here at length, but I believe that a few may be cited for their relevance to our contemporary search for an understanding of the principles of plant construction. I shall paraphrase the laws somewhat for brevity:

Law 1. *Metamer formation is either indefinite or definite within an organism.* Determinate or indeterminate growth comes to mind as a botanical example.

Law 2. *Metamers may be all alike or may differ from one another greatly by modifications of the various constituent parts of each: the constituent or subordinate parts are called meromes.* The notion that metamers were either similar or differently specialized had been suggested by Owen (1849). Lankester had in mind for arthropods the various tegumentary plates, coelomic features, muscles, nerves, alimentary tract, etc. which composed each metamer. A botanical analogy of a metamer in this sense (similar also to Haeckel's) would be a phytomer with its constituent leaf, axillary meristem, node, and internodal segment. As far as I can judge, the use of the word *metamer* by zoologists has traditionally had this composite property: it is a repeated segment, assembled from various elements.

Law 3. *Some regions may show a like modification of metamers differing in their modification from that in regions before and behind them. Such a region is called a tagma (plural: tagmata). Some organisms may have more than one tagma.* A botanical example of a tagma in this sense is the flower vis-à-vis the vegetative axis.

Law 4. *The parts of a metamer may be separately and dissimilarly affected by changes in form.*

Law 5. *Some parts within a metamer may be repeated while others may not show any repetition.* Multiple meristems in the leaf axil of some plants come to mind as one botanical example.

Laws 6, 7, and 8 are concerned with transformation and transposition of metamers in arthropods. They prefigure the ideas of Corner (1958) and of Stebbins (1970) on "transference of function" in plants. However, the difficulty with such rules is that they appear to be quite ad hoc and seem to be called for once other concepts or statements have been hypostatized to an almost axiomatic rank.

Laws 9, 10, 11, and 12 are concerned with proliferation and atrophy of metamers in arthropods.

Law 13. *The homologous meromes (parts) of two or more adjacent metamers tend to fuse with one another; this is often preceded by extensive atrophy of the metamers concerned.* This law echoes de Candolle's theory of the fusion of floral parts, elaborated in 1813 and still the conceptual basis of angiosperm systematics. Floral metamers typically lack the complexity of their vegetative counterparts formed by the apical meristem in the earlier ontogeny of the shoot: for example, the nodal plexus and axillary meristem are rudimentary or undeveloped.

Many of these principles of arthropod construction are still accepted, as a glance at a recent summary by Cisne (1979) reveals. I find it instructive to apply them mentally to plants, whose metameric construction is as self-evident as that of arthropods. The array of diverse forms in the Arthropoda (with over a million modern species)—crabs, spiders, insects, centipedes, etc.—shows the immense range of structures that may be developed by evolutionary specialization of the basic segmental *Bauplan*: the number of metamers in the adult body varies from <10 to >100 (some Carboniferous millipedes were 2 m long); different sets of metamers are enhanced in size and complexity or reduced to the point of vanishing. There is a striking analogy in the theory of pattern formation in arthropods and the theory of axis and floral construction in angiosperms expressed by de Candolle in his *Organographie Végétale* of 1827. Angiosperms have a considerably shorter evolutionary history than the arthropods, but there is copious evidence all about us that they too manifest the prodigious variety of construction which a repetitious segmental design or *Bauplan* allows.

ADAPTIVE ADVANTAGES OF METAMERISM

The adaptive advantages of metamerism have been quite inadequately explored, perhaps because of the mutual aloofness that has existed between morphologists and evolutionary biologists for so long (Riedl, 1977; Stebbins, 1980; Lauder, 1981): Riedl believes that morphology has "virtually been cut off from modern biology." As the ecological consequences of morphological patterns become better explored, we may expect an improved insight into their evolutionary development. For plants we are only on the threshold of understanding, but the prospect is exciting, as Niklas has demonstrated (Niklas, 1978; 1982; Niklas and O'Rourke, 1982).

R. B. Clark (1964) has pointed the way for animal morphologists, and his work has a certain interest for botanists since he strives to understand the ecological and evolutionary benefits of metameric construction. Segmented animals (articulates and chordates) have enjoyed enormous evolutionary success compared with nonsegmented groups. "Whatever the initial advantages of metamery, it predisposed its possessors to adaptive evolution on the grand scale" (Mackie, 1963); Clark attributes their success to structural plasticity. Metamerism is considered by Clark (1964, 1980) to have arisen independently in most groups of animals which show it—turbellarians, nemerteans, mollusks, annelids, pogonophorans (Southward, 1980), chordates, cestodes, etc. In a brilliant functional and mechanical analysis, he established that its primary evolutionary advantage for animals lies in locomotion (Clark, 1964). No similar study of the evolutionary significance of metamerism in plants has ever been attempted.

In plants, as in animals, there are many species which lack metameric organization (many algae, some hepatics, fern gametophytes): they are typically confined to water or to continuously moist habitats in a subaerial environment. Did the aerial transmigration confer a relative selective advantage on metamerically organized vegetative structures? It is difficult to envisage (for mechanical reasons) the growth of large thalloid structures such as we know among the algae in an aerial environment. Metameric organization, once established and relatively advantageous, presumably became phylogenetically conserved, even in secondarily aquatic plants. As in animals, metamerism may serve an important locomotory function for a genet, allowing coloniality and expansion in space through the relative or complete autonomy of metamers or clusters of metamers: *Trifolium repens* provides an instructive example, since each metamer (stolon node plus associated meristem, leaf, and internode) is capable of independent existence if severed from the remainder in an environment suitable for establishing roots.

There are some fundamentally different types of metameric organization which lead to very different life-history attributes. In many metameric animals the formation of the major organ categories is confined to early life. Others share with plants a continuous embryogeny of the genet during life. [Bell (1982) has provided a comprehensive review of this phenomenon in animals.] In plants the genet may develop aerially either from one or from a small set of meristems and remain noncolonial, or it may develop from a continuously increasing number of meristems and become either colonial or noncolonial: examples of each category occur among both herbaceous and woody plants. The evolutionary advantages of metamerically organized colonial genets, which are so common among angiosperms, remain to be considered by biologists, since many of the current concepts of life-history theory are derived from noncolonial organisms whose embryogeny is restricted to early life. Fitness, reproductive allocation, age-specific fecundity, senescence, and such concepts tend to become operationally intractable when applied to colonial genets (e.g., Hughes and Jackson, 1980). How do we assess the relative fitness (and on what time scale and in what ecological arena) of genets, one of which creates and propagates most meristems vegetatively while the other creates most meristems in seeds sexually, the one being a 100% copy of the genet, the other as little as 50% like its parental genet, depending on the breeding system and population structure? Vegetatively dispersed meristems are commonly the fitter in the short term (e.g., Sarukhán and Harper, 1973); but the "short term" for a genet may be hundreds or even thousands of years [some examples quoted by Harper and White (1974) and Vasek (1980)]. The usual meanings of r- and K-selection are nebulous when applied to genets of colonial organisms.

Metameric or modular fragmentation in plants is commonplace. It is notable among aquatic plants (Sculthorpe, 1967) where there may be special anatomical features to ensure it, for example, the segmentation region in the internode just above each lateral bud in *Anacharis densa* (Jacobs, 1946). It occurs widely among species which are regarded as weeds of cultivation (Leakey, 1981; Håkansson, 1982) and though less evident, is also common in rhizomatous herbs generally (Bell and Tomlinson, 1980). Cook (1979b) suggested that the selective advantage of coloniality in plants could be viewed as a spreading of the risk of genet extinction, particularly if ramets were physiologically independent. His idea may be generalized, as Highsmith (1982) did recently for corals, to the statement that "the probability of genet mortality decreases as the rate of formation and physiological independence of ramets increases and is equal to the product of probabilities of mortality for the independent ramets." Insofar as metamerism in plants favors the formation of independent fragments of the genet, it may be an important component of genet fitness. This leads one to doubt that vegetative *growth* of an intact genet is equivalent to asexual *reproduction* by vegetative disarticulation of a genet, at least in fitness terms. Harper (1977:27) considered the two processes to be indistinguishable for a clone of *Trifolium repens*, a view echoed by Janzen (1977). The rigid distinction (Harper 1977, 1981) between reproduction (restricted to products arising from a single cell) and growth seems inappropriate to clonal plants, which may fragment at a metameric or modular level into independent units; this is more than growth and should be considered asexual reproduction. Charlesworth (1980) has argued that the behaviors regarded as growth by Harper and by Janzen are not equivalent with respect to a feature of evolutionary importance: the cost of meiosis. Clearly the theoretical and practical consequence of metameric fragmentation remain to be more completely understood. For intact genets that are relatively or completely sessile, recovery from predation is also a significant necessity for survival: metameric construction through continued meristem embryogeny enables the genet to reconstitute itself effectively.

CONTEMPORARY APPLICATIONS OF PLANT METAMERISM

In this section I shall attempt to indicate a few of the topics in plant biology to which the concept of metamerism is applicable. These are wide-ranging, their variety is increasing, and I can only hint at their scope here. I have previously collated some of them (White, 1979), and shall confine myself here to different or more recent examples.

Plant morphology

Crop growth dynamics. Dynamic models of plant construction using subunitary parts are by no means recent in origin and many lie outside the usual bounds of what I may for convenience refer to as "academic plant morphology." An intimate understanding of the demography of shoots and buds forms the economic basis for the management of tea plantations (e.g., Herd and Squire, 1976; Tanton, 1981). The demography of nodes, buds, branches, or flowers is central to the yield of cotton (Munro and Farbrother, 1969), Brussels sprouts (Hodgkin, 1981), cucumbers (Lint et al., 1982), raspberries (Jennings and Dale, 1982), beans (Lovett-Doust and Eaton, 1982), soybeans (Hansen and Shibles, 1978), and many others; the list is very extensive (and poorly appreciated by botanists). Multiple repetitions of a chosen unit typically arise by plant growth: their demography constitutes the dynamic model of cropping and yield. As I have suggested previously (White, 1979), there is no basis for believing that agronomists are working out a research program on plant metamerism or that they are necessarily aware of the intellectual lineage of morphologists who have for nearly two centuries advocated a reductionist approach to plant construction. Instead, by a Whewellian "consilience of induction" (a methodology employed by Darwin in the later chapters of *Origin*), they have come to plant metamerism by sharing a common pragmatic interest. This is particularly evident among agronomists studying grasses for pasture and cereal production: their focus is the tiller or the phytomer, and the dynamics of these parts have been discussed in a prodigious literature. To take but one example, Masle-Maynard and Sebillotte (1981) have recently provided a sophisticated model of wheat growth which depends on the detailed morphology of tillers, their age-structure, and their age-dependent characteristics. As with the study of animals, where the economic incentives of fishery management or pest control played such a large role in the development of demographic theory, so too one repeatedly finds that many seeds of theoretical plant demography have been well drilled in agronomy, horticulture, or forestry: this is as true for the dynamics of plant structure as for competition theory.

Module and metamer. The recent advent of a dynamical approach to plant morphology among botanists was heralded by the celebrated monograph of Hallé and Oldeman (1970), which collated in a simple, systematic, and comprehensive manner the diverse researches on tree morphology of the previous half-century. Their ideas were rapidly assimilated and promulgated by plant demographers (Harper and White, 1974; Harper, 1977: Chapter 1, Harper and Bell, 1979; White, 1979) and plant morphologists (Tomlinson and Zimmermann, 1978) and have received widespread publicity through their extended monograph in English (Hallé et al., 1978). A pivotal concept of the Hallé-Oldeman analysis is the *module* (a translation of *l'article* which was first used by Harper and White, 1974). The concept was Prévost's (1967, 1978) and was carefully defined by her and by Hallé and Oldeman. *Module* is a morphological term and refers to the developmental products of a single apical meristem, from birth to death. It is a monopodial axis with its attendant node(s), leaf (or leaves) and axillary meristem(s) terminated by an inflorescence (which invariably leads to loss of further vegetative extension by the apical meristem), by a vegetative structure (such as a spine or tendril), or by parenchymatization (loss of mitotic capacity). A module may, thus, be extremely short or meters long: in *Philodendron selloum*, it consists of a condensed axis with a prophyll, a foliage leaf, and two inflorescences, the apparently monopodial axis of the plant being, in fact, sympodial (Hallé et al., 1978:135); in the Marantaceae, it is a little more complex since a small number of internodes are formed (Tomlinson, 1970: Figure 9); in many palms assignable to the models of Holttum and Corner, a single apical meristem is responsible for the whole axis construction which may be tens of meters long (Hallé et al., 1978). A module is a developmental unit, the product of a single apical meristem. The term has now come to be used by several plant ecologists indiscriminately for any part of a plant. Harper's (1981) definition is "deliberately wide" and disregards the care taken in the definition by the morphologists who conceived it. I believe that it is desirable to use morphological categories as precisely as possible, so that the fertile interchange of ideas between morphologists, demographers, and ecologists begun in the past decade may be maintained and fostered. Ad hoc definitions conflate categories and can lead to confusion in an ordered morphological theory (Riedl, 1978). Insofar as students of colonial animals attempt to build bridges between plant and animal morphology (witness Rosen, 1979;

an echo of Owen's attempt in 1849) it may be profitable to be as precise as possible on the limits and definitions of structural units, though I acknowledge that the terminology can become scholastically formidable, as Riedl (1978) so well demonstrates.

I have attempted previously (White, 1979) to provide a vocabulary for subunitary plant parts and have suggested that the best generic term was *metamer*. I indicated that it was already in use by botanists (Nozeran et al., 1971; Smirnova, 1970; also Gatsuk, 1974a,b; Guédès, 1979; Shafranova, 1980). I have since realized its widespread currency among animal morphologists, as I have partly documented earlier in this chapter. It is a term that is not usually employed to denote an organ (such as a leaf) but rather an intimate assemblage of organs which are collectively repeated in the organism (e.g., Lankester, 1904; Clark, 1964). Since organs already have botanically unambiguous names (though there are problems with these, too, as Sattler (1974) indicated), I favor the use of the term *metamer*, by zoological analogy, to denote the node–leaf (leaves)–axillary meristem(s)–(roots)–internode complex, for which a generally accepted morphological term has been lacking. The term *phytomer* as used, for example, by Evans (1958) is exactly comparable, but seems to have been virtually restricted to grasses (e.g., Hyder, 1972). *Phyton* comes close as a term, but its definition has been ambiguous (see White, 1979, for sources) and seems generally to have excluded the axillary meristem. Not all plants by any means are metamerically constructed (many algae and hepatics, and all fern gameophytes, for example), but I have confined my attention in this chapter to those, notably the angiosperms and gymnosperms, which show repeated or iterated metamers in their construction. The term may be virtually synonymous with *module* in some plants (*Philodendron selloum* comes close), but generally it is not. A good example of my concept of metamer is provided in an illustration by Kurihara et al. (1978), which I reproduce in Figure 2. Gatsuk (1974a) also employs the concept in a similar manner.

Metamers may differ in their exact morphological expression within the plant: heteroblasty, for example, is a common manifestation (e.g., Wardlaw, 1968:209). If we accept the Goethe–de Candollean theory of the flower (e.g., Eyde, 1975), vegetative metamers are transformed in various ways in the reproductive regions of the plant. Elements within the metamer may show duplication or difference in potential: axillary meristems provide many examples of both, as Hallé et al. (1978) have shown for trees; *Coffea arabica* has been especially well studied in this respect (de Reffye, 1981, 1982). Furthermore, metamers may be grouped into clusters which themselves have a

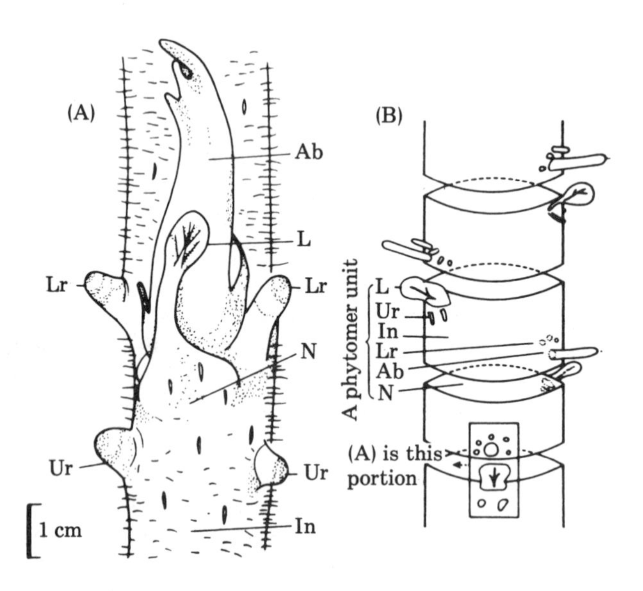

FIGURE 2. Metameric organization of the stem of potato, showing (A) the component parts of a metamer and (B) the schematic pattern of metamer accumulation. Ab, branch on the above-ground main stem (A) or an axillary-bud or stolon initial (= meristem) (B). N, node; L, leaf; In, internode; Lr, lower root; Ur, upper root. (From Kurihara et al., 1978.)

structural integrity (*tagmata* in the terminology of zoologists): short shoots and long shoots, plagiotropic and orthotropic branches, flowers or inflorescences come to mind. *A set of metamers which is the product of one apical meristem constitutes a module.* "The extent to which subcategories of construction can be recognized is too little recognized even in descriptive morphology," argued Tomlinson (1982) (an opinion with which I concur), and he gave as an example "the more complex species of *Lycopodium* (which) have 8–10 discrete kinds of axis, indicating a very highly organized and integrated construction." [Callaghan (1980) has shown the importance of segmental axis construction in *Lycopodium annotinum* for understanding its productivity and nutrient allocation, an example of the original application of metameric concepts to plant productivity which he has pioneered in recent years (Callaghan and Collins, 1981).]

Not only may plant metamers vary in morphological expression, but they may differ genetically in certain cases. For example, Nielsen (1968) has shown that clonal progenies developed from individual nodal segments on a bromegrass rhizome are morphologically variable. This may be due to intraplant variation in chromosome numbers or to the chimeral nature of some plants (Neilson-Jones, 1969). The ecological and evolutionary implications of genetically distinct metamers within a plant are still poorly understood, although Tuomi et al. (1983), Whitham (1983), and Whitham and Slobodchikoff (1981) have recently explored them speculatively in relation to plant–herbivore interactions.

Statics and dynamics of plant form. Goethe made a distinction between *Gestalt* or fixed form and *Bildung* or form change. *Gestalt* was a momentary phase of *Bildung* and "could be considered apart and in itself only by an abstraction fatal to all understanding of the living thing" (Russell, 1916:49). The science which should discover the inner meaning of organic *Bildung* was named by him in 1807 "morphology." Despite Goethe's prescription, plant morphology has been largely preoccupied with *Gestalt* rather than with *Bildung*. The modern resuscitation of his distinction after 160 years in exactly homologous terms—*architecture* and *architectural model*, the one static, the other dynamic (Hallé and Oldeman, 1970)—has reinvigorated plant morphology in an age when the conceptual tools of demography are available to explore plant *Bildung* in some detail.

The static depiction of plant form owes much to ecologists who sought repeatedly to provide systematic classifications: much of this literature was reviewed by du Rietz (1931), himself the most distinguished Swedish exponent of this tradition. Nordic botanists, indeed, seem to have had a predilection for ecomorphology. The important Finnish school (e.g., Kujala, 1926) survived in the work of Oinonen [see Harper and White (1974) for references to his work] and through the influence of Linkola (for example his 1935 paper) inspired T. A. Rabotnov (personal communication), the founder of the Russian School of morphological demography (e.g., Rabotnov, 1950, 1978b). Plant morphology in Russia is particularly brilliant, since it combines on the one hand a traditional level of detailed structural analysis (*Gestalt*) and on the other a fine appreciation of the dynamics of morphological expression (*Bildung*) in natural environments: the very extensive publications on this topic have been conveniently listed by Rabotnov (1980, 1981), but, hitherto, little has been published outside the Soviet Union (see Gatsuk et al., 1980), and the work still remains poorly known to many botanists.

It is far from my intention to review the various schools of plant morphology in this chapter, since I only wish to make the point that repeatedly one can find detailed diagrams of the diverse ways in which plants are metamerically constructed. The research of the school associated with Meusel in Halle (DDR) is particularly instructive (e.g., Gluch, 1967; Hagemann, 1983; Kästner, 1981; Meusel and Mörchen, 1977, among many) since they typically explore the varieties of axis

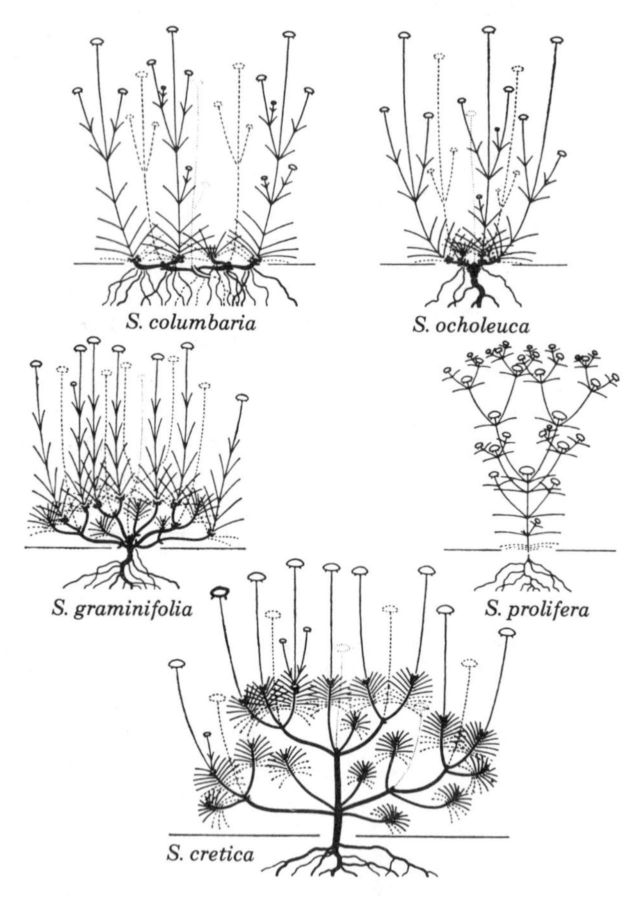

FIGURE 3. Schematic growth forms of *Scabiosa* species from Mediterranean (M), sub-Mediterranean (SM), and Central Europe (C) in a late-autumn stage. Thick lines represent persistent (lignified) parts; thin lines, annual parts; dotted lines, dead parts. *S. columbaria*: semirosette rhizomatous perennial with winter dormancy (SM, C); *S. ochroleuca*: semirosette/geophytic perennial with winter dormancy (C-steppe); *S. graminifolia*: semishrub (M, SM); *S. prolifera*: summer annual (M); *S. cretica*: evergreen cushion shrub (M). Root morphology is not shown in detail. (From Meusel, 1970.)

construction of species within a genus and speculate on their ecological significance. It is tempting to see, for example, in the species of *Scabiosa* subtle variations in the disposition and persistence of metamers of the same fundamental construction (Figure 3), although Meusel has not published an analysis of their dynamics.

In the past ten years the growth dynamics of plants have been actively investigated by morphologists and plant ecologists, undoubtedly under the influence of the research of Hallé and Oldeman (1970). Models of plant growth based on modules (in the strict sense), metamers, or meristems are becoming increasingly familiar in the botanical literature, and we may expect many more. Bell (1974) provided the first detailed analysis of branching and vegetative dispersal (in *Medeola virginiana*) using metameric concepts and has extended his work to a variety of branching systems (Bell and Tomlinson, 1980). Formal (and realistic) growth rules have now been determined for several branching systems (Bell et al., 1979, Chapter 2; Fisher and Honda, 1979; Honda et al., 1981); a convenient guide has been provided by Waller and Steingraeber (in press) to this rapidly developing literature.

Meristems. Although a variety of subunitary structures may be used as a starting point for exploring the dynamics of plant construction, I believe a strong case can be made for advocating the centrality of apical meristems in plant growth models. They are a unique feature of plants and are defined as embryonic regions at or near the extremity of axes which permit continual embryological development of the organism. (Embryogeny in most animals is restricted to early life). Trewavas (1981) has documented their pivotal role in plant development from a physiologist's viewpoint. Bell has best expressed their significance for dynamic plant morphology (Bell et al., 1979). As the single shoot meristem of the seed begins to grow, it typically forms new meristems in the axils of leaves. Each individual meristem may live or die, remain dormant or grow into a shoot. Its potential to grow into a specialized organ or to form a plagiotropic or an orthotropic shoot may depend on its position *vis-à-vis* other meristems. Meristem development may be tightly programmed (e.g., Nozeran et al., 1971) or opportunistic. Although meristem potential, position, and fate determine plant morphology, the demography of meristem populations has scarcely been explored, and we may anticipate a considerable expansion of research on this topic soon. De Reffye (1981, 1982) has constructed very detailed stochastic growth models for *Coffea arabica* by monitoring the dormancy, growth, potential, and fate of meristems; when allied to other growth parameters such as internode length, branch angle, and plastochron index, the simulations are quite realistic.

In extratropical woody plants, the only convenient denumerable apical meristem is contained in the bud, defined by Romberger (1963) as "an unextended, partly developed shoot having at its summit the apical meristem which produced it": it may therefore have axillary apical meristems. The bud is in fact an unextended module. Maillette (1981, 1982a,b) in a seminal series of papers has attempted to model tree morphology through the demography of buds, and a similar approach has been adopted for *Fuchsia* cultivars by Porter (1983), for

Fagus sylvatica by Agnew (1981) and for *Picea abies* by A. Dunberg (personal communication).

The relative allocation of the meristem "bank" or reserve of a plant to vegetative growth or sexual reproduction (the latter precludes the former) certainly affects plant morphology (e.g., Maillette, 1982a,b), but so far we have only a vague understanding of its effect on plant life-history. A. Dunberg (personal communication) and Edelin (1977) have shown that female strobili on *Picea abies* terminate first- or second-order branches of the main axis, consequently inhibiting further extension growth of these branches. This has significant implications for tree morphology and for sylvicultural productivity: Dunberg has suggested that the consumption of meristems is likely to be a more fundamental problem than nutritional requirements for some woody perennials. Similar work on the balance between vegetative and reproductive meristems has been carried out on some other trees of economic importance, such as *Abies balsamea* (Powell, 1977): in this species the female strobili are, however, lateral and have less influence on vegetative extension than in *Picea abies*.

Watson (in press) has argued persuasively that some plants may be growth-limited by the developmental capacities of their meristem population. This seems true of the aquatic plant *Eichhornia crassipes* which she has investigated: it has only a small number of meristems per shoot which are regularly employed during normal development, which facilitates its analysis. Commitment of a meristem to one developmental pathway (flowering, for example) precludes its commitment to another: inflorescence production in *E. crassipes* was found to be inversely correlated with clonal growth (production of ramets sympodially). Watson has therefore suggested that meristems are a more appropriate "currency" of resource allocation in plants than fixed carbon (dry weight) or nutrients and that their relative allocation has important life-history consequences. This view has been independently advocated by Tuomi et al. (1982), who believe that there is some trade-off between seed production and somatic investment in plants, the trade-off being expressed at a metameric level.

Genetics of plant form. Plant architecture has repeatedly been modified in breeding programs for improved crop yields. This may be an unwitting outcome of selection, but plant breeders are increasingly concerned with the systematic alteration of plant shape, insofar as this can be correlated with yield. The philosophy has been well expressed by Donald (1968) as the breeding of crop *ideotypes* (an expression redolent of the Goethean *archetype*), meaning a plant model expected to yield an increased amount of useful product when developed as a cultivar. The concept has achieved wide use in recent years and is well demonstrated in the development of dwarf cereal cultivars. Coyne

(1980) has partially reviewed the extensive horticultural literature. Many architectural ideotypes have been developed in *Pisum sativum* and their genetic bases elucidated (Hedley and Ambrose, 1981; Snoad, 1981). In the present context of plant metamerism, what can one glean about the genetic control of metamer number, size, and development from this literature? Do those changes in plant shape whose genetic basis is understood involve alterations in metamer number and/or size? The pertinent literature is large and this is not the place to review it. Since breeders do not normally discuss this issue explicitly, I have so far found it difficult to make any generalizations and shall mention only a few examples.

Rice breeders have succeeded in reducing plant height, typically not by lowering node number but by reducing internode length (e.g., Chandler, 1969: Figure 1): modern rice cultivars do not increase internode length as much as do older, less productive cultivars in response to nitrogen. How many genes determine such reactions in crop plants? The answer at present is unclear. A single major gene may control overall plant shape in some inbred lines of *Nicotiana rustica*, but the control of the numbers of nodes and branches, branch length, and plant height, which are the components of plant shape, is genetically complex (Caligari and Hanks, 1978). On the other hand, the genetic basis of the difference between determinate and indeterminate cultivars of *Glycine max* (soybean) is well known: two major genes control the number of nodes produced by the apical meristem (Bernard, 1972). Flowering and maturity are controlled by a further three major genes, which also influence the numbers of main stem nodes, branches per plant, and fruit pods per node (Hartnung et al., 1981). Hodgkin (1981) has provided good evidence for the existence of major genes influencing node number in *Brassica oleracea* var. *gemmifera* (Brussels sprouts): the number of nodes at which the enlarged axillary buds (sprouts) are produced can be manipulated by selection. Gönen and Wricke (1978), in their attempt to breed determinate cultivars for mechanical harvesting, have shown that node number in *Cucumis sativus* (cucumber) is controlled by perhaps two genes. Branching patterns in *Helianthus annuus* may be governed by a few dominant genes in some cultivars but is under more complex control in others (Hockett and Knowles, 1970).

One could cite several other examples. L. Gottlieb (in an unpublished lecture on macroevolution and plant genetics to the American Society of Plant Taxonomists, August 1982, and personal communication) believes that there are dozens of examples of structural characters governing plant shape which seem to be controlled by one or two major genes; this literature has not been systematically collated as far as I know.

Despite the fact that most ecotypic variation observed by ecologists in natural plant populations is morphological (or more commonly reported than physiological or biochemical variation, of which there are, of course, several examples), the genetics of environmentally related (ecotypic) morphological changes are very poorly known (Stebbins, 1950; Briggs and Walters, 1969). Morphological differences with *some* genetic basis continue to be documented (e.g., Mahmoud et al., 1975; Warwick and Briggs, 1979; Baker and Dalby, 1980; Hume and Cavers, 1982—examples from a copious literature), but their precise genetic control remains unclear. More is known about *Trifolium repens* in this respect than most species (e.g., Burdon, 1983) because of its interest to agronomists. A single gene apparently accounts for the difference in prostrate and erect forms of *Viola tricolor* (Clausen, 1926). Since genecologists have provided almost no information on the genetics of plant architecture for wild populations, as with Darwin in *Origin* one can at present only analogize natural with cultivated plants. My speculative analogy is that the metameric structure of plants is consonant with a view that rather few major genes control the structural form of plants. The *detailed* anatomy, physiology, or phenotypic expression of each metamer may be under complex genetic influence, as are many yield characters involving size or quantity of product. Clausen et al. (1940) suggested that morphological characters in *Potentilla glandulosa* ecotypes involved complex combinations of genes and Grant's (1981:98) recent assessment generalized this view. However, the various attributes of morphology are conflated in these analyses, which do not distinguish numbers of parts from their sizes, particularly in vegetative tissues. For example, how many major genes control the structural differences in the five *Scabiosa* species shown in Figure 3? My (Darwinian) analogy with the agronomic literature is that the numbers and spatial dispositions of metamers are under relatively simple genetic control. The variety of growth patterns that may be produced by very slight alterations in the rules of growth of simulated (though realistic) plant structures (Bell, Chapter 2; Bell et al., 1979; Fisher and Honda, 1979; Honda et al., 1981; Agnew, 1981) suggests to me that small or few gene changes may cause major overall shape changes phenotypically. Small alterations in developmental ontogeny may have manifold effects at maturity, as Stebbins (1950: Chapter 13) argued. I believe that is particularly true of organisms with multiple metameric construction, in which small changes can be accentuated by cumulative growth of metamers. This is not to say that overall shape and form is not constrained by metamer interactions, subject to Sinnott's law of developmental allometry (Hamid and Grafius, 1978) or other manifestations of internal correlation (Nozeran, 1978a,b) and meristem dependence (Tomlinson, 1974).

Stebbins (1950:479) attempted to formulate a general principle of morphological phylogeny which had a metameric basis:

> Reduction in size and complexity of individual structures is often accompanied by an increase in the number of these structures produced, while reduction in number may be balanced by increase in size and complexity of individual structures. . . . In a plant which normally differentiates organs serially, one after the other, the way of producing a single complex structure with the least modification of ontogeny is by compressing together and modifying several relatively simple organs.

His examples were typical of the traditional focus of botanists on floral structures. Systematic examination of vegetative architecture of ecotypes or of closely related species have never been undertaken in these terms to my knowledge. Stebbins' rule can be demonstrated with floral metamers, if one accepts the probable homologies usually predicated. But does it apply to vegetative metamers? The genetic basis of plant form remains a virtually unexplored area of morphology, yet one which is essential to a comprehensive understanding of its ecological significance and evolutionary history.

The expression of a metameric character may be changed by environmental conditions (including competition and predation). The amount by which it is changed from some chosen reference value is a measure of its plasticity. One may expect that there will be a "hierarchy" of plasticities in the construction and elaboration of the parts of plants, an ordering that will have significant effects on ultimate size and shape of the whole plant in any given set of circumstances. Stebbins (1950:492) suggested that characters formed by long periods of meristematic activity, being more subject to environmental influences, are likely to be more plastic than those formed rapidly in ontogenesis. The evidence remains equivocal as far as I know (witness the effects of environment on internode length in different rice cultivars, mentioned earlier). Bonaparte and Brawn (1975) have ranked phenotypic characters in *Zea mays* hybrids by their plasticity. Koblet (1979) has explored the plasticity of meadow plants by comparing their growth in pure culture, in mixtures, and in permanent meadows: species differed greatly in their ability to expand and exploit an enlarged space or to survive by adjusting their morphology in a narrowly confined space within a sward. (The order of decreasing plasticity was *Dactylis glomerata* > *Taraxacum officinale* > *Ranunculus friesianus* > *Anthriscus sylvestris*). The morphological responses of plants to ous ecological circumstances remain so poorly known that precise lytical formulations of their metameric plasticities cannot be made ond the very gross level of changes in productivity that has typically characterized this research. Recent work on *Trifolium repens* (which bids fair to become the paradigmatic plant in demographic morphology) has given much new insight into the detailed structural and demographic changes that a plant may undergo in a variety of artificial and natural habitats (Harvey, 1979; Turkington, 1983); moreover, the genetic basis of many of its morphological features are well understood from research by breeders and agronomists (Burdon, 1983).

Plant competition

Between plants. The typical response of plants to the stimulus of interference by neighbors is reduction in size. In the huge literature of plant competition, the response is almost invariably expressed in terms of biomass. It is often clear that the morphology of the interactants is altered: this is strikingly seen in branch production by trees along density gradients and is casually noted among crop-weed interactions, as by New (1961), who recorded that *Spergula arvensis* plants were almost completely unbranched among oats and luxuriantly branched in an adjacent field of cabbage and peas. Rarely is the outcome of competition documented in precise morphological terms. Yet commonly changes in morphology are just those that underlie the differences in yield, particularly of reproductive tissues of economic importance. A fine example of the sort of detailed morphological demography that I believe will become increasingly necessary to understand the processes of plant interference adequately is by Darwinkel (1978), from whose paper I have reproduced some results (Figure 4). The effect of density on the grain production of winter wheat was analyzed in terms of the demography of tillers (modules in the formal sense).

Explicit examples of the effects of competition on the metameric construction of plants appear to be scanty. I shall mention only a handful (see also White, 1979); the agronomic literature usually provides them. Verheij (1970) has shown that density influences plant habit in Brussels sprouts through its effect on node number, internode length, and associated allometric changes in stem diameter. In *Hibiscus cannabinus* (kenaf), Muchow (1979) has shown that as initial planting density was increased from 13.6 to 90.6 plants $\cdot$ m^{-2}, after 98 days the number of nodes per plant decreased monotonically by 32% while internode length increased monotonically by 26%; the number of branches per plant declined nearly 11-fold. Plant growth form, survival, and yield are all integrated in Muchow's analysis. Kobayashi

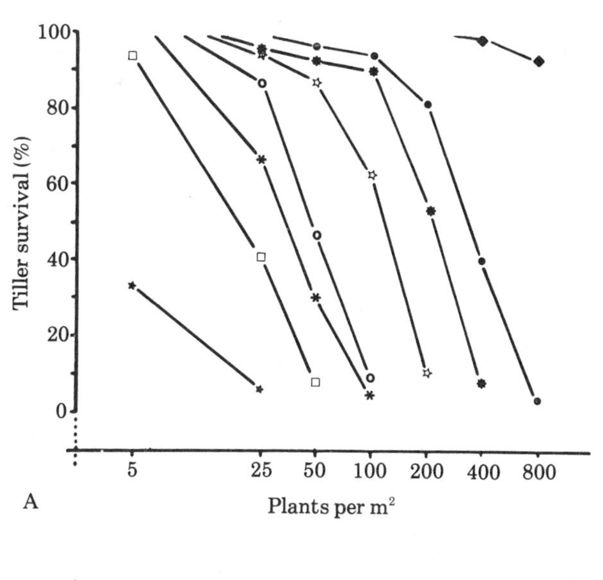

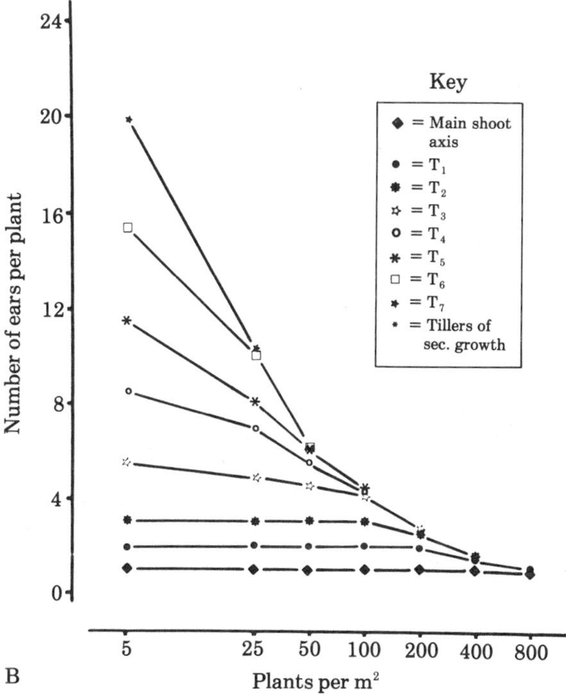

FIGURE 4. Response of winter wheat to density. A. Survival of tillers of different age classes. B. Mean number of reproductive structures (ears) per plant, represented cumulatively for each density. Ms, main shoot axis; T_1–T_7, first to seventh age class of tillers, tagged successively on emergence at 10 to 20-day time intervals. (From Darwinkel, 1978.)

(1975) also showed that increasing density led to a decrease of nodes per plant in *Helianthus annuus* (sunflower), in a study that attempted a quite original metameric analysis of plant productivity. Independent ramets of *Trifolium repens* respond to increasing density by a reduction in node numbers, but this is strictly hierarchical, with main stolons being relatively unaffected compared with the severe reductions in node development of first- and second-order branches (Harvey, 1979); this was due to the inhibition at increasing density of axillary meristems on the main stolon axis. Harvey also demonstrated that main stolon growth was maintained under conditions of nutrient impoverishment and suggested that its extension, being least inhibited, enabled the genet to colonize new areas, discover new resources, and escape competitive depletion. Turkington (1983) has extended the metameric analysis of growth responses of *Trifolium repens* to interspecific competition. Smith (1983) has investigated the vegetative plasticity of a forest herb in its natural habitat at various densities by a non-destructive census of nodes, leaves, branches and vegetative buds: he found that morphological processes that were initiated later and that last longer during the growing season were those most affected by density. De Reffye (1981) has shown how density affects the demography of meristems in *Coffea arabica*, which brings the focus of analysis of competition to the structure which above all others influences plant shape. I believe that this level of investigation is the harbinger of much greater insight into the competitive interactions of plants.

Within plants. Metamer or organ formation is clearly influenced by interference between genets, and this is expressed within an individual genet by interactions between metamers or organs. The nature of these interactions is by no means well understood though it has been extensively studied, notably by French investigators of plant morphogenesis (e.g., Barthou, 1979; Melin, 1977; Nozeran et al., 1971; Nozeran, 1978a,b; Pfirsch, 1972; 1978).

One view, which at least polarizes the debate, is that metamers are relatively autonomous in their carbon economy. It was originally propounded by Adams (1967) (though he acknowledged the germ of the idea to earlier workers), who conceived the (aerial) structure of *Phaseolus vulgaris* (bean) as a succession of cumulated "nutritional units." The nutritional unit consists of a leaf on the main axis together with the flower-bearing raceme in its axil and a second smaller leaf borne on the pendUncle below the flowers, all of which are served by the same vascular supply from the main stem. He suggested that the primary competition for nutrients took place in a "peck order" among the developing yield components within this unit: rapidly developing young pods > unopened flowers > open flowers > young fertilized ovules in developing pods > freshly pollinated embryos and very young pods. The nutritional units, which are clearly akin to metamers in my terminology, were not regarded as being totally independent, for obvious reasons. Adams' analysis was based on correlations between yield components but has been substantiated by Olufajo et al. (1982), who demonstrated that labeled photosynthates were mainly retained within each nodal unit, the branch pods acting as the main sinks.

Watson and Casper (in press) also argue the case for the relative carbon autonomy of metamers or modules, particularly in plants where translocation is restricted by vascular anatomy. However, the evidence remains ambivalent in my opinion, since physiologists have rarely analyzed plants in metameric terms. Reciprocal movement of radiocarbon between plant parts has been demonstrated frequently by Sagar and co-workers (e.g., Ismail and Sagar, 1981, who give earlier references). Chacko et al. (1982) have demonstrated that even the 30 leaves on the shoot supporting a single mango fruit were insufficient to supply it with assimilates to bring it to normal size and that stored reserves were called upon. Harvey (1979) argued from the results of resource depletion and defoliation studies on the growth of *Trifolium repens*, where the dominance hierarchy among stolons is main > first-order branch > second-order branch, that there was a common pool of carbohydrate resources from which all sinks could draw. He suggested that strict source–sink linkages would be disadvantageous to a species such as *T. repens*, which must normally tolerate grazing and trampling by animals, the typical effect of which is to damage or kill metamers or organs rather than whole genets. Newell (1982) has provided a succinct review of this topic for herbaceous plants and has shown that the degree of physiological integration varies among stoloniferous species of *Viola*.

Some metamers or modules, while possibly being self-sufficient in their own carbon economy, may also supply others. Many trees have a dimorphism of long and short shoots (Büsgen and Münch, 1929): the former may be regarded as exploring new spatial environments, carrying buds on long internodes, the latter as exploiting locations already achieved, for as long as they have enough light in the progressively developing crown (few live longer than a decade or two in many trees). Münch (quoted in White, 1979) regarded the form of trees as the outcome of "egotistical" interactions among shoots. However, Renard (1971) has shown that whereas short shoots account for only 37% of the biomass of shoot extension growth annually on 120-year-old *Fagus sylvatica* (beech) trees, they carry 77% of the leaf area; the remaining leaf area is carried by the new long shoots (63% of the biomass of annual extension growth). On young saplings up to 25 years old, the relative distribution is more equitable: long shoots account for 70% of the annual extension growth and support 60% of the leaf area (Agnew, 1981). There is probably an ontogenetic trend in the relative distribution of leaves between long and short shoots, but this remains to be investigated for trees with this type of branch dimorphism. Short shoots (which are modules) scarcely thicken by secondary growth, extend by only a few millimeters per year, do not generally elaborate buds or branches, and are short-lived [10–15 years in beech saplings, according to Agnew (1981)]. Arguably they show "altruistic" behavior toward other modules in the genet. (I use the term colloquially in the absence of a suitable term not already expropriated by inclusive fitness theorists: in their sense, it would be inappropriate for structures within a genet.) Presumably short shoots export most of the locally garnered photosynthate to build up the enlarging trunk and branch axes. The demographic and physiological dynamics of tree shape offer many challenges, but appropriate questions have scarcely been formulated. Even the mechanics of internal water movement and its relative distribution between parts are only beginning to be understood (Zimmerman, 1978). Metameric analysis of the type pioneered by de Reffye and Maillette seems to provide plant demographers with a suitable approach which can be complemented by physiologists, and we may expect considerable progress in this area of tree biology.

Plant productivity

Tomlinson (1974) appears to have been the first to advocate that plant productivity, at least of seagrasses, could only be adequately understood by recording the dynamics of meristems. Jefferies (Chapter 17) has echoed this by emphasizing the importance of linking productivity studies with demography. Only recently has the significance of the dynamics of plant organs been fully appreciated by production ecologists; the mortality and disappearance of subunits such as leaves and branches can lead to substantial underestimates of primary productivity if they go unrecorded. Attempts have been made to circumvent the problem, among the most interesting being Carpenter's extension of cohort production methods. Carpenter (1980) considers the plant in terms of subunits, hierarchially structured, and estimates productivity by following the fate of cohorts of these subunits through time. Loss of subunits is monitored by censusing abscission scars or changes in the structure of marked shoots. The method has much promise as an accurate means of assessing productivity, but it has only been applied to a few aquatic plants so far and at a rather gross level of morphological discrimination.

However, undoubtedly the most brilliant application of plant metamerism to primary productivity is by ecologists working in boreal regions. I have previously mentioned the researches of Callaghan and his colleagues, studies that are the seminal examples for herbaceous plants and mosses; similar work has been reported by others in *Holarctic Ecology* 5(2). A clear statement of the new outlook was provided by Flower-Ellis (1980): "It is intended to combine two main approaches to the problem of estimating the production of plants: plant demography and sampling methods of the production ecologist. From their synthesis a long-term perspective of the production of a site may be obtained." I have chosen an example from his work (Figure 5) which illustrates splendidly the combiantion: numbers of organs (leaves and shoots of different ages, flowers and capsules) have been carefully censused and their production and biomass attributes recorded. The method has also been applied to trees (Flower-Ellis et al., 1976; Flower-Ellis and Persson, 1980), with an emphasis on the demography of leaves—their survivorship, age structure, and age-dependent photosynthetic capacity. This level of demographic–productivity analysis has been extended to investigate the effects of defoliation by insects on *Pinus sylvestris* (Larsson and Tenow, 1980; Ericsson et al., 1980a). "Studies of photosynthesis are primarily concerned with individual shoots of defined age and position within the crown; studies on the effects of consumption by insects . . . are similarly specific. Hence the degree of resolution in the structural description (of trees) must be commensurate with these" (Flower-Ellis and Persson, 1980).

I have argued already that plant shape and size are the product of accumulated metamers, or, to be more precise, the outcome of the demography of metamers, since formation and loss (birth and death), survivorship, and age structure are all involved (White, 1980). The biomass accumulated by most plants may be viewed also as the product of the cumulative life time of leaves, at least in those plants (such as trees) where stems contribute relatively little to carbon gain. The idea is not new as it was expressed decades ago by MacDougal (1936, 1938) but has remained dormant until recently. MacDougal recorded annually the leaf numbers on specimens of *Pinus radiata* and expressed the wood production as the cumulated product of leaf-years. Tree No. 20 in his series, for example, when felled and dissected after 18 years had a fresh weight of 563 kg (roots, trunk, branches) and was, he stated, "the net product of 10,722,400 leaf-years of photosynthetic activity, an average of 57 mm^3 fresh material for a year of activity of a single leaf." Tree No. 17 at 25 years of age was estimated to be the product of 42.5×10^6 leaf-years. I have suggested previously (White, 1979:128) that his work is a rich source of undigested raw data: a combination of demography and productivity, it remains unique in its

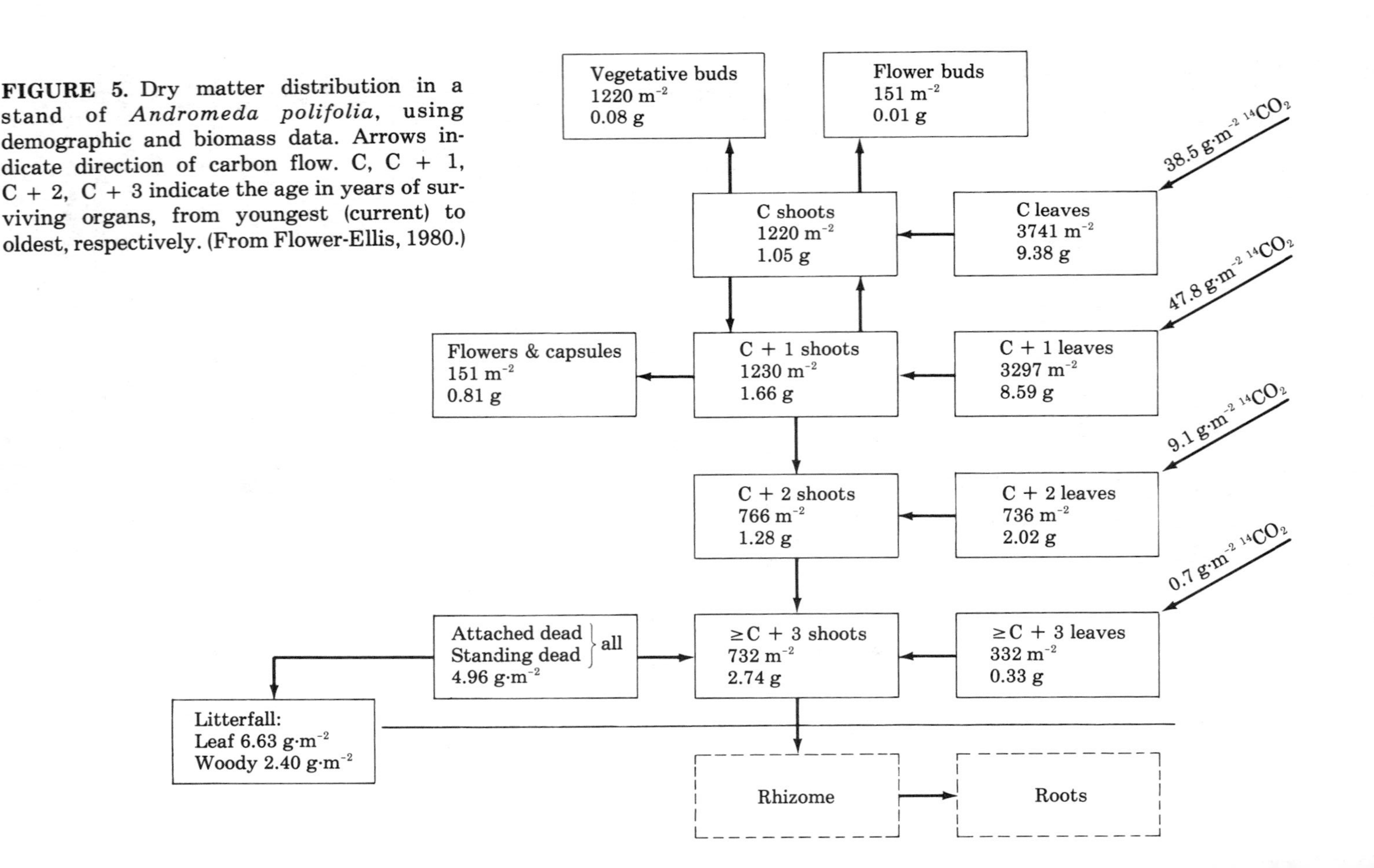

FIGURE 5. Dry matter distribution in a stand of *Andromeda polifolia*, using demographic and biomass data. Arrows indicate direction of carbon flow. C, C + 1, C + 2, C + 3 indicate the age in years of surviving organs, from youngest (current) to oldest, respectively. (From Flower-Ellis, 1980.)

detail as far as I know. A similar rationale lay behind the painstaking investigations of Bürger in Switzerland, who attempted to calculate the mass of leaves required to produce 1 m^3 of timber in several tree species of commercial importance (see White, 1979). The present interest in the demography of leaves (partly reviewed by White, 1979; also Chabot and Hicks, 1982; Ford, 1982; Whitney, 1982; Kikuzawa, 1983) should lead to a much better understanding of the dynamics of plant productivity.

Plant–animal interactions

There is another topic to which metameric concepts are so relevant that I should say a little about it, however briefly: plant–animal interactions. Animals typically feed on particular plant parts and do not normally harvest or destroy genets wholesale. The number and growth potential of the remaining metamers or meristems determine the outcome of the interaction for survival of the genet, the more so in plants with a limited meristem population. Plant–animal interactions remain to be quantified from the plants' point of view (but see Dirzo, Chapter 7), and this will require more detailed attention to morphology than is typically given. Lubchenco and Gaines (1981) have sown the seed of a formal analysis by considering the plant as being composed of small pieces and the expected change in total genet fitness as the cumulative product of (a) the probabilities of each piece being encountered and consumed and (b) the loss of genet fitness resulting from the consumption of each part. (However, I believe that the effects of consumption of parts on genet fitness are ambiguous and may be either positive or negative, depending on plant morphology; space precludes a review of the evidence here.) Can we rank plant parts by their contribution to genet fitness when removed or destroyed by predators? The loss of a leaf seems trifling to a tree when compared with lethal predation or parasitism of its vascular supply, xylem (e.g., elm disease), or phloem (e.g., coconut yellows). Williamson (1982) implies, given the relative autonomy of some plant parts and the (presumed) small effect of their loss on genet fitness, that mimicry is less likely to have evolved in plants than in animals since its selective advantage is much less in plants than in those (noncolonial) animals where loss of a part entails death of the whole. Botanists need to put the detailed morphology of plants and the demography of their parts more conspicuously into the natural history of plant–animal interactions. This should not be too difficult as the copious literature on grazing already indicates.

CONCLUSION

I have attempted in this chapter to revivify the concept of plant metamerism, to decant a morphological idea of venerable vintage into a variety of new vessels. I do not wish to pretend that the draught is intoxicating, but I find it pleasurably stimulating and I have taken the opportunity provided by cordial Mexican hosts to share it.

I believe that the concept of metamerism has immense heuristic potential for a wide range of studies with which plant population biologists may concern themselves. Demographers, as they turn their attention to plant morphology with its long tradition, rich in theory (Cusset, 1982), may wonder at its dearth of interest in evolutionary theory, its paucity of evolutionary interpretations of growth patterns, the questions that remain unasked or inadequately and unconvincingly answered. The new "ecological morphology" (Bock, 1980), illuminated by demographic and life-history concepts unfamiliar to older morphologists, holds the promise of an intellectual revolution in the understanding of the construction of vegetation. We are only in the early stages of the synthesis.

As Darwin noted in *Origin*, a conspicuous feature of organic evolution has been the achievement of increased complexity by means of replication of parts followed by their differentiation and specialization. Lauder (1981) believes that hypotheses about the consequences of metamerism may be among the most general to emerge from morphology, since the most basic aspects of biological organization are the number of structural elements and the number of connections between them. Simon (1962) has argued that the evolution of statistically improbable assemblies proceeds more rapidly if there is a succession of intermediate, stable subassemblies from which they may be serially derived. This view was echoed by Riedl (1977), who advocated the analysis of morphological patterns in terms of the hierarchical and symmetrical ordering of functionally interdependent standardized structures. Patterns of complexity can be generated by the interaction of a small number of simple rules operating on standardized, functionally integrated parts (Braverman and Schrandt, 1966). The metameric and modular organization of seed plants has certainly allowed an immense diversity of shape and size: their size range spans 11 orders of magnitude of biomass, from *Lemna* to *Sequoia*. But the evolution and selective advantages of their metameric construction have not hitherto received adequate attention from botanists.

I have mentioned earlier the mutual distrust of morphologists and evolutionists: this was especially notable among influential botanists like Arber, Bower, and Bailey, who regarded adaptation and selection as teleology and were strongly opposed to truly Darwinian concepts: Bailey saw no benefit in evolutionary explanation (Stebbins, 1980). Indeed, among many botanists a belief in "soft" selection (the inheritance of acquired modifications) has only relatively recently given way to an acceptance of "hard" (Darwinian) selection (Stebbins, 1980). Of course, in plants the germ plasm is not separate from somatic tissues: the same apical meristem that forms a vegetative axis gives rise later by subtle, environmentally induced stimuli to reproductive tissues ["somatic embryogenesis" to use the terminology of Buss (1983)]. "Until the molecular revolution which demonstrated that DNA replication is independent of the environment there was no theoretical reason for denying the inheritance of acquired modification" (Stebbins, 1980). However, the evolutionary and morphological consequences of somatic embryogenesis (in plants, fungi and some animals) are still far from clear (Buss, 1983) and have exciting intellectual prospects. Perhaps the acceptance of Darwinian selection, *and all that it entailed*, was easier for zoologists studying those animals which have sharply demarcated somatic and reproductive tissues. Animal morphologists have been particularly aware of the interacting networks of constraints among parts which impose developmental and architectural limits to the evolution of structural diversity: this is well expressed by Raup (1972) and by Gould and Lewontin (1979), following Seilacher's researches. Botanists seem to have investigated the structural integrity of plants much less intensively, despite the advocacy of Nozeran et al. (1971): the potential opportunism of vegetative growth from multiple meristems perhaps gives a misleading impression of great flexibility of morphological expression in plants. In fact many of the contemporary zoological concepts of basic *Bauplan* in animal groups are congruent with those of "pre-Darwinian" morphologists, and were, as I have indicated earlier, once shared by botanists and zoologists alike in elaborating generalizations about fundamental structural patterns in organisms. Plant morphologists have since, by and large, remained outside the evolutionary fold and their interpretations of vegetative structure have developed with too little recognition of the organizing principles of evolution and population biology. A century is time enough without Darwin.

CHAPTER 2

DYNAMIC MORPHOLOGY: A Contribution to Plant Population Ecology

Adrian D. Bell

INTRODUCTION

Plant morphology is concerned with the external features of plants and, as such, must be recognized as one of the principal factors influencing survival. It is one of the cornerstones of population ecology. Darwin, in his preoccupation with survival, was fascinated by plant morphology and was a great observer of form and function. His series of books on climbing plants (Darwin, 1885), insectivorous plants (Darwin, 1875), and flower form (Darwin, 1884) is well known, although he denies modestly any particular affinity with plants in the latter, where he states, "The subject of the present volume ought to have been treated by a professed botanist, to which distinction I can lay no claim."

This chapter is intended to emphasize the relevance of morphological factors to population ecology and to demonstrate one area at least in which a knowledge of plant morphology can be applied to population ecology only now that computer power is available—a possibility denied, perhaps thankfully, to Darwin.

Plant morphology impinges on the dynamics of plant populations in numerous ways, some more obvious than others. Breeding systems are commonly determined by pollination mechanisms, and the morphological complexity of some flowers provided a major stimulation for Darwin. Subsequent seed or fruit dispersal is often a dramatic morphological event; form and function go together, teleological pitfalls cannot detract from the fact that exploding seed pods disperse seed.

However, plant morphology plays a much deeper part in population ecology in a manner not sufficiently recognized in the past. A study of the morphology of plants must consider their form and their shape. And a plant's shape represents its ability to physically fit and expand, in spatial terms, into the environment and therefore its ability to collect light, water, and nutrients in the face of competitors.

Indeed, an elementary distinction to be made between an animal and a plant is that the former searches for food whereas the latter accumulates food and energy at a fixed position. Darwin, not withstanding his disclaimers, studied both.

Thus, one major factor in the study of the population ecology of animals is the important concept of foraging behavior, whether applied to individuals or to groups. There is a recognition of optimal foraging—an economy of effort in the collection of resources brought about by an organized and efficient foraging pattern. However, is not the foraging behavior of a plant represented by its morphology as exemplified by its pattern of branching, and is not the morphologist, in contemplating the framework of a plant, the counterpart of the behavioral zoologist? Indeed, a common interest is found in the case of one ubiquitous growth form— the clonal organism, animal or plant—and the morphologist's contribution to population ecology can be demonstrated in a discussion of this mode of development.

CLONAL ORGANISMS: MORPHOLOGY AND POPULATION ECOLOGY

Clonal organisms have the distinction of feeding simultaneously at many different locations. In a substantial number of cases, the sites at which feeding takes place remain interconnected by a communication network. A zoological example of such a clonal organism is provided by the hydroid *Podocoryne*, feeding hydranths being located on a horizontal system of stolons (Braverman and Schrandt, 1966); botanical examples are represented by any rhizomatous or stoloniferous plant, that is, plants having a horizontal branching system at or below ground level. In this chapter a distinction will be made between the two constructional components of a clonal branching system: (1) feeding *sites* which are separated by various lengths of (2) *spacers*. Thus, in a typical rhizomatous plant, *sites* are represented by vertical shoots bearing green leaves and feeding roots and are separated by intervals of often rootless stem (*spacers*) bearing nonphotosynthesizing scale leaves.

An animal will demonstrate, in many instances, a behavior pattern when feeding and will visit the environment in an organized manner that will result in a recognizable foraging pattern. The fossilized meander tracks of ancient sediment feeders provide an intriguing example (Figure 1) (Raup and Seilacher, 1969). Similarly, a clonal organism will exhibit an organized location of the feeding sites. The equivalent of the foraging behavior of a mobile animal will be found in the geometry of clonal branching patterns—the juxtaposition of site and spacer. These organized branching patterns will spread through

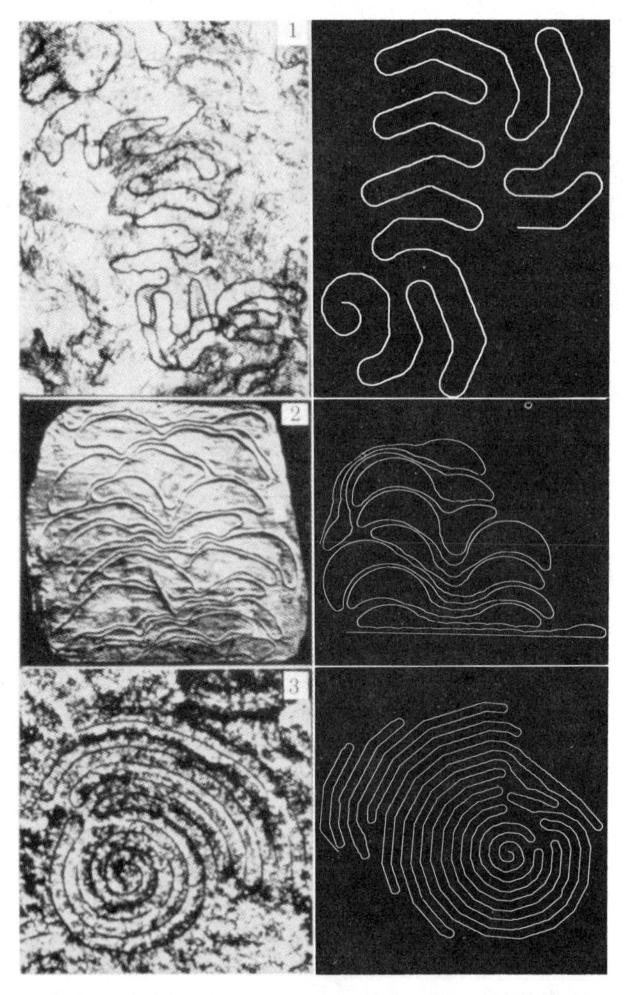

FIGURE 1. A. Meander patterns of sediment feeders: (1) *Dictyodora*, Ordovician flysch; (2) Cretaceous flysch; (3) *Paraenis*, modern beach worm. (From Raup and Seilacher, 1969.) B. The simulated meander patterns of sediment feeders. Compare with A.

the environment as the organism grows. The old end may decay in time and the clone fragment as it enlarges (see Appendix Bibliography in Bell and Tomlinson, 1980).

Thus, feeding strategies can be summarized as follows:

1. Single organism mobile along a foraging route ("an animal").
2. Single organism feeding at one site ("a plant").
3. Clonal organism feeding at many sites reached by growth of a branching system. The branching system may rapidly fragment.

A final mode of feeding is demonstrated by some social insects. The whole community forages along an organized branching pattern, each individual feeder being a part of this unified whole. Such systems have been compared with the plant (strategy 2) (Hölldobler and Moglich, 1980) (Figure 2A).

This chapter will consider the "foraging behavior" of clonal plants as represented by their patterns of branching, thus emphasizing the impact of form on survival.

ORGANIZATION IN CLONAL BRANCHING PATTERNS

The persistence and success of a plant depends upon its morphological potential in its particular ecological circumstances. The outcome can be monitored in terms of the birth and death of morphological units (Harper and Bell, 1979; Maillette, 1982a,b). The impact of morphology on plant population ecology is seen most clearly among clonal plants, that is, plants that spread and multiply by vegetative means. As the genet expands, ramets, which may or may not remain interconnected, are located throughout the environment. The manner in which this takes place depends to a large extent upon the organization of the form or architecture of the plant. Hallé et al. (1978) show convincingly that a plant will have an intrinsic blueprint (or model) of branching which is expressed in a flexible manner in a heterogeneous environment. Dormant meristems can repeat the initial pattern of branching in response to damage or in response to enhanced growth conditions (reiteration of the model). A clonal plant grows progressively larger and apparently complex, although the developmental details of its branching processes may be relatively simple and repetitive and are open to study. It is important to stress that the concept of morphological form as applied to population ecology must be considered as an ongoing developmental process and not as a static situation. An animal moves about modifying instinctive foraging behavior according to ex-

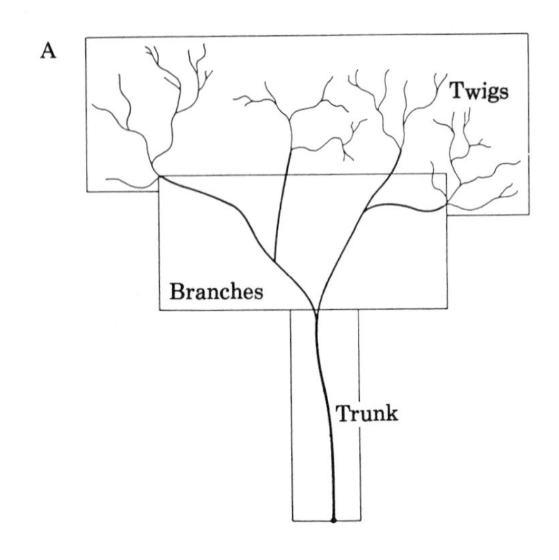

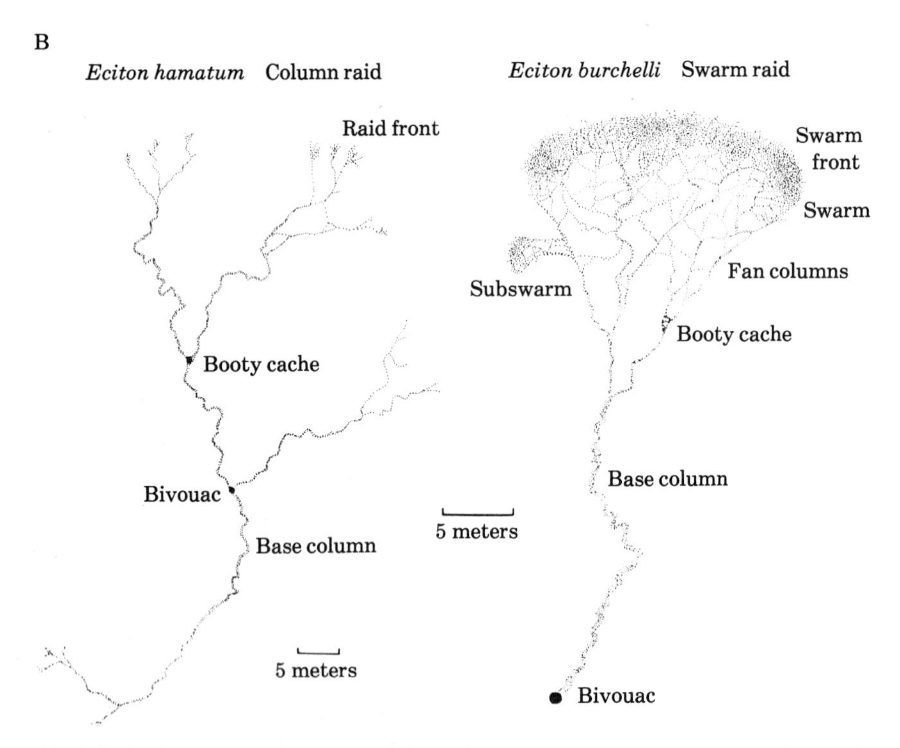

FIGURE 2. The trunk route system. A. The foraging system of *Pheidole militicide*, an ant colony. (From Hölldobler and Moglich, 1980.) B. Linear and multibranched foraging systems in army ants. (From Wilson, 1971.)

perience; a plant pervades space slowly, its intrinsic pattern of spread being augmented in response to environmental factors.

A study of the branching patterns of rhizomatous and stoloniferous plants leads to some massive, but simple, hypotheses. The branching process can be considered as a system locating functional sites at discrete intervals by the interposition of spacers. Sites will be expected to be organized such that the maximum number occupy an area with a maximum economy of spacer material. The pattern would be expected to be displayed such that interference between parts of the same clone are minimized. Not many plants grow in a homogeneous environment. In a heterogeneous environment, the expanding clone would encounter an alien background. The features of this background will either be *static* [and then "passive" (rocks) or "active" (rooted plants of strategy 2)] or *mobile* (other clonal organisms branching in a more or less organized manner). The innate branching pattern of the clonal organism might be expected to retain an ability to respond to this environmental disruption by, for example, the process of reiteration.

Examples of organized pattern in rhizomatous plants tending to substantiate these hypotheses occur throughout the taxa of the plant kingdom. Moreover, the same detail of pattern occurs time and again in totally unrelated plant groups. One comparison, that of the "gingers" (Zingiberales) and the bamboos (Gramineae) will serve to emphasize such convergent evolution (Bell and Tomlinson, 1980). Figure 3A shows diagrammatically the plan view of rhizome systems found among various gingers. The rhizomatous stems represent the "spacers" in the pattern. The feeding sites are identified by dots. An almost precise match of these patterns is illustrated by McClure (1966) for the bamboos (Figure 3B) and is based to some extent on the work of Tachenouchi (1926). Both ranges of patterns include hexagonally based details with or without additional linear components. These two aspects alone can be selected to allow a complete simplification of clonal branching in plants: (1) linear systems giving a sparse distribution of sites and branching only occasionally; (2) multibranched systems giving an aggregated array of sites.

These represent the extremes of behavior in terms of the invasion of new territory and are analogous to the two tactics of army ants described by Wilson (1971) as "column raiders" and "swarm raiders" (Figure 2B). A similar tactical analogy is made by Lovett Doust (1981b), specifically relating to clonal plants (guerrillas, sparse linear patterns; and phalanx, multibranched compact patterns).

However, interpretation of a particular pattern should consider the

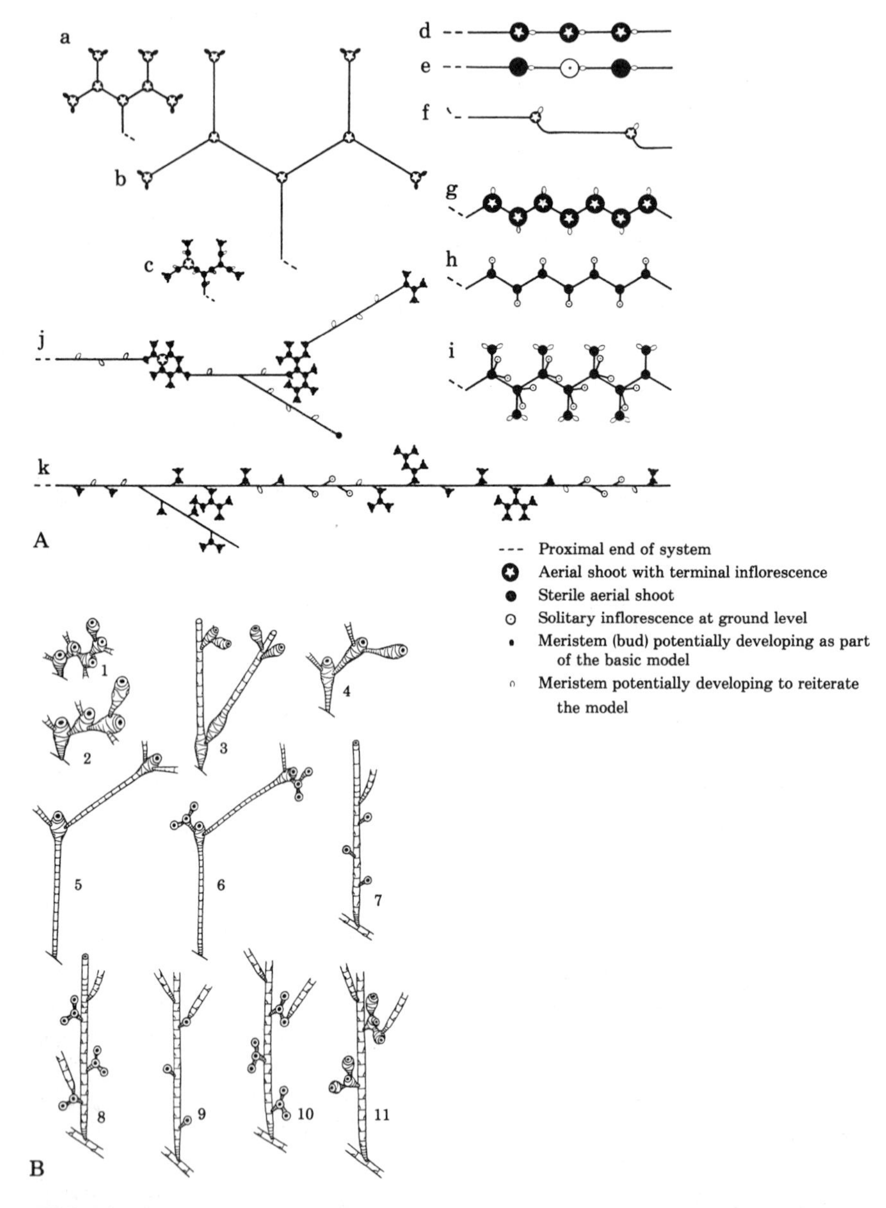

FIGURE 3. A. Clonal branching patterns within the "gingers" (Zingiberales). Each letter corresponds to branching patterns typical of one or more species (see original reference for details). (From Bell and Tomlinson, 1980.) B. Clonal branching patterns within the bamboos (Gramineae). Each number corresponds to a different bamboo species. (From McClure, 1966.)

scale of the system. An apparently linear portion of rhizome may represent just one small component of a much larger and multi-branched overall plan. This aspect can be judged in relation to the ratio of spacer length to site radius, long spacers coupled with small sites being indicative of a sparse linear system. The ratio of spacer length to site radius might be expected to be related to the grain of the environment. Long spacers result in rapid exploration; short, branched spacers result in localized exploitation. Some clonal branching patterns combine both these features (Figure 3). There is a second aspect of scale to consider. A multibranched "local" system may demonstrate a very long-term meander, whole portions of the clone demonstrating a large-scale mobility (Bell, 1976). Tracking such extended mobility might reveal meander patterns of clone fragments, comparable to those of the foraging worms. This has not been investigated.

One concern of the population ecologist will be with the flux of numbers and densities of ramets and genets. Attention to morphological detail can enable assessments of past productivity (Tomlinson and Soderholm, 1975), of dry matter turnover without recourse to destructive sampling (Tietema and Vroman, 1978), of potential productivity (Tomlinson, 1974), and of the cost of reproduction (Sohn and Policansky, 1977). In addition, it can provide perhaps the only way of studying the spatial dynamics of ramet and genet. If the developmental morphology of a clonal plant is understood, there is a chance that the manner in which it is able to fit into the environment will become apparent. The same attention to morphological detail is also valuable in the study of static plants, the birth and death of buds regulating the metapopulation of structural units of which the plant is built (White, 1979).

The use of computers has become an accepted part of almost any quantitative population study, but qualitative morphological considerations may seem somewhat unamenable to computer activity. This is not necessarily so. For example, Barkham and Hance (1982) have modeled the spatial population dynamics of wild daffodil (*Narcissus pseudonarcissus*) over periods representing up to 1000 years. *Narcissus pseudonarcissus* is a clump-forming species; successful establishment of ramets (bulbs) or seeds depends on the juxtaposition of existing shoots within the clones. These authors used a simple square lattice grid layout to locate their plants, each occupied square being surrounded by eight potential daughter sites. This may or may not be a morphological possibility, but it does lead to a spatial understanding of the population ecology of that species and can predict genet behavior under a variety of environmental conditions.

Many clone-forming plants have a fairly precise pattern of branching; hence, space is occupied in an economical and efficient manner in terms of the amount of rhizome or stolon material needed to reach an unoccupied site. An alternative to a rigid pattern would be a "deliberately" random arrangement of ramets. This appears to be the case for a number of *Viola* spp. as reported by Schellner et al. (1982). The stolons of these species are deployed in random directions and have random lengths, shoot locations being scattered accordingly. This phenomenon is rather reminiscent of the foraging behavior of certain leaf-cutting ants (*Atta* spp.) (Cherrett, 1968). Very variable vegetative morphology of this nature necessitated detailed analysis of ramet distribution in space and time rather than a study of architectural events. Nevertheless, the morphologies of the plants have a direct bearing on their population ecology.

Narcissus and *Viola* demonstrate again the two contrasting forms of clonal growth: "clumps" and "runners." There are numerous recognized terms for these modes of growth (caespitose, tufted, spreading, leptomorph, and pachymorph, for example), and the different forms lead to either dense or lax stands of aerial shoots. A ramet of a clone may collide with "individuals" of the same clone, another clone of the same species, or another clone of a different species.

It is under these differing conditions that the morphology of the plant is put to the test, although intermingling clones are not inevitably competing with one another. There are advantages for both if they are, for example, dependent upon the foraging behavior of common pollinators (Thomson, 1981).

SIMULATED MORPHOLOGY

The importance of considering plant morphology when studying the ecology of a species and the application of computer technology to morphological problems can be demonstrated by a simple theoretical game using two hypothetical plants: a clumper and a runner. A clumping plant will stay in one place with time and may form a reasonably unpenetrable fairy ring. A runner will sample a wide area in a transitory manner. Are these merely alternative routes to the same end? Simple computer exercises can compare the relative attributes of contrasting tactics.

Thus, an organized geometry of branching in a clonal organism or organized meandering in a marine worm allows graphic simulation of foraging patterns. Such a simulation for the worm tracks of Figure 1A is demonstrated in Figure 1B. It has been suggested that the genetic control of such organization need not be any more complex than the simplest set of "rules" that can be compiled to mimic a particular design (Gould, 1970; Raup and Seilacher, 1969). Certainly minimal changes of geometric rules can substantially change the architecture of an organism. The casual transformation of *Obelia* (a hydrozoan) into *Alstonia* (a tropical tree) (Harper and Bell, 1979) demonstrates this and emphasizes an underlying similarity in the manner in which it is possible to conceptualize the construction of sessile animals and plants. The simulation program (RHIZOM) used in that exercise is recognized as stationary and stochastic by Waller and Steingraeber (in press). Thus, in its published form (Bell et al., 1979), its rules of branching, although incorporating an element of chance, cannot change during the simulated growth of the organism. An updated version of the same program (SEED3) can operate in either a stationary or nonstationary manner, that is, interaction between components is possible. The identifiable order in rhizome branching patterns allows the simulation of the model for any particular plant. An example is shown in Figure 4A, in which four of the bamboo branching types of McClure and Tachenouchi are simulated. Growth in Figure 4A is "stationary" in the sense that there is no interaction between parts of the same clone or parts of invading clones. Such a simulation is transformed into a nonstationary development if there is interference between components of the clones. The clonal growth of bamboos illustrated in Figure 4A is replicated in Figure 4B. In this simulation, overcrowding of sites within a clone depresses its growth, and proximity to another clone stops growth at that point.

Such simulation procedures form powerful tools. They allow rapid assessment of the long-term spread and spatial disposition of an expanding clone and also permit an analysis of the efficiency of the invasion of space (Bell, 1979; Honda et al., 1981). Thus, Bell (1979) describes the hexagonally based clonal branching in a ginger (*Alpinia*) (top left in Figure 3A) and shows that the inherent economics of a hexagonal tessalation (e.g., Woldenberg, 1968) slips up in this system as sites find themselves sitting on preexisting sites. In *Alpinia* a subtle departure from the 120° of a regular hexagon circumvents this problem, coupled with the possibly programmed failure to grow of some spacers. *Alpinia* would be superbly efficient if it followed the pattern of growth generated by Ulam (1966). Ulam's pattern is repeated here (Figure 5) in a computer simulation reaching generation 44 (compare the top half of Figure 105 in Stevens, 1974). His expectation of repeated failures on six fronts followed by resumption of growth at the corners is seen to hold true. A clone operating this system would be exceedingly economical but would have to anticipate the proximity of neighboring branches or grow to some very complicated rules—changing the angle is simpler. Pattern simulations can also be of value as an

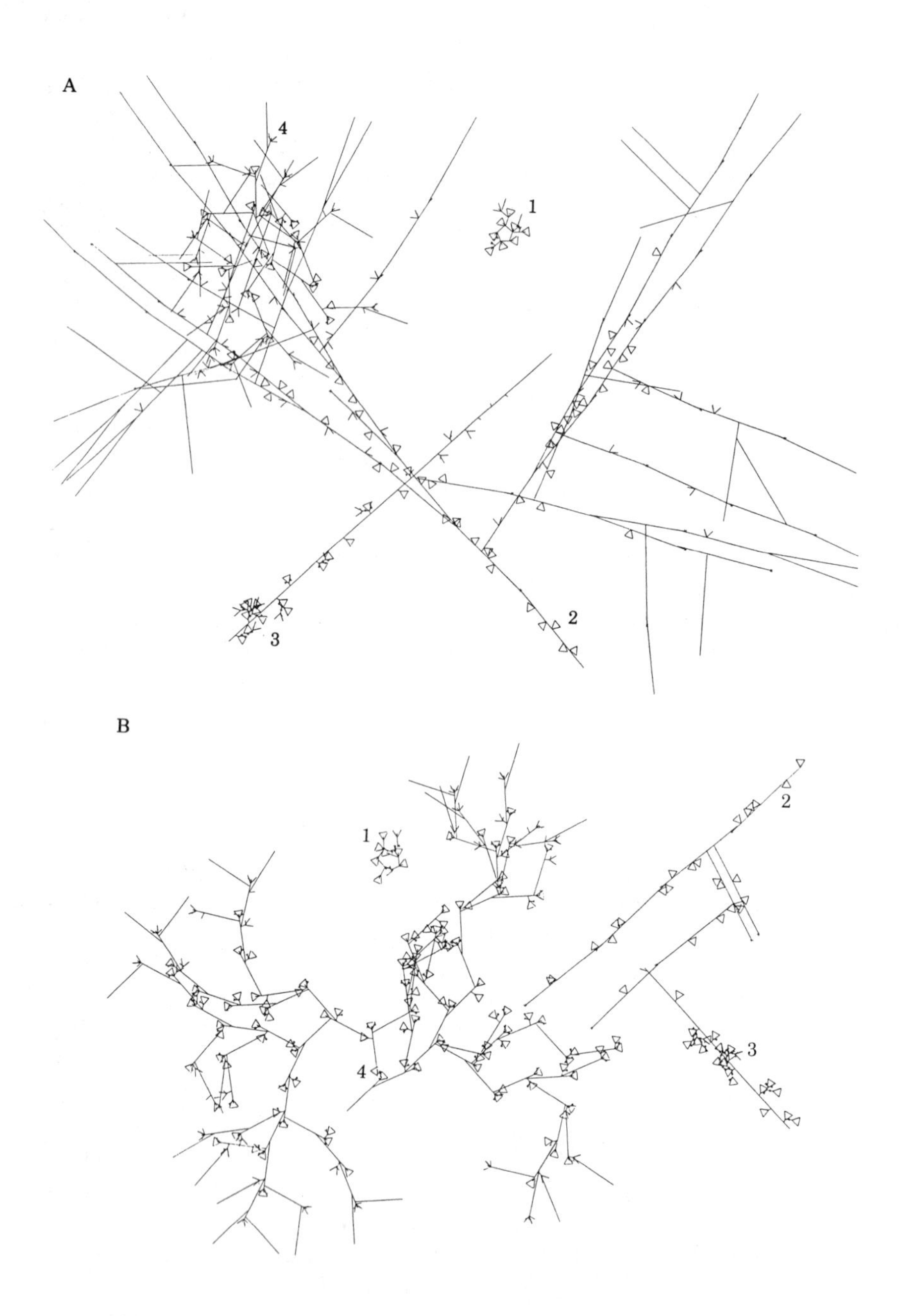
A
4
1
2
3
B
1
2
4
3

◀ **FIGURE 4.** A. Simulated clonal growth of four separate bamboo branching patterns developing without any interaction. Lines represent underground rhizomes; triangles represent orthotropic aerial shoots. (1) *Bambusa* sp. (based on No. 1, Figure 3B). A pachymorph, phalanx system. (2) *Phyllostachys* sp. (based on No. 9, Figure 3B). A leptomorph, guerrilla system with a dormancy in the activation of aerial shoot meristems. (3) *Shibataea* sp. (based on Nos. 7 and 8, Figure 3B). A leptomorph, guerrilla system capable of developing limited lateral phalanx systems. (4) *Yushania* sp. (based on No. 6, Figure 3B). A coarse-grained phalanx, apparently the most successful in simulated competition with the former three types. B. The simulation exercise shown in A, replicated this time with competition for space occupancy. Pattern No. 4 has proved the most successful in this instance.

FIGURE 5. Ulam's packing game (Ulam, 1966) extended to the forty-fourth generation.

aid to the understanding of the origin of morphological organization. Indeed, they can be used to create hypothetical situations in which plant form, or behavior, is shown to represent an essential element of plant population dynamics.

SIMULATION EXERCISES

The hypotheses derived from the observation of clonal branching patterns can be summarized as follows:

1. Site location will either utilize a minimum amount of spacer to exploit a given area or utilize an economical amount of spacer to explore a larger area, or both.
2. The degree of linearity or compactness in a given organism has evolved over time and reflects the grain of its natural habitat, that is, the size of obstacles and the space between them.
3. Interference between sites in the same clone will be minimized.
4. The basic pattern will be able to respond to environmental interference (passive, the substrate; or active, other organisms) by deploying otherwise dormant meristems that will reiterate the initial pattern.

To illustrate these hypotheses relating to expected attributes of clonal organisms, a simple exercise is demonstrated and an extension to it proposed.

A simple exercise

A comparison between investment of resources into many short rhizomes with correspondingly many, large, rooted sites (the clump) or investment into few, long, rhizome spacers with few, small, rooted sites (the runner) provides a simple simulation exercise.

Figure 6A shows a stylized representation of a typical clumping plant based on a ginger (*Alpinia speciosa*; Bell, 1979). A triangle represents a shoot–root site occupied during one growing season and in budget terms costing an arbitrary 40 units to grow. Having existed for one season, it is deemed to have assimilated sufficient resources to provide 92 units for the next season, resources that are spent on two more triangles plus their locating rhizome segments representing 6 units each. (The relative cost of rhizome and aerial shoot–root in this exercise is based loosely on theoretical dry weight comparisons.)

This computer-generated plant will grow season by season, new shoots developing from the old, branching angles being based on those of *A. speciosa* shoots and rhizomes "rotting" after five seasons. Two triangles can occupy the same site if they are contemporary, but a triangle will fail to develop if its site is previously occupied. If a shoot

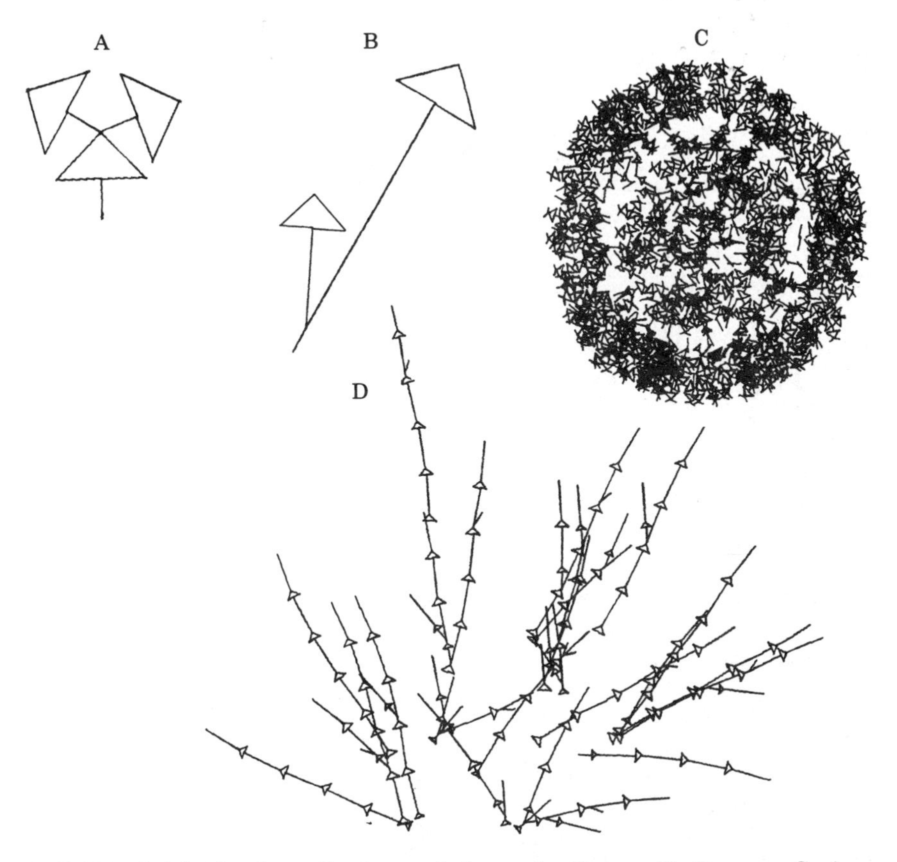

FIGURE 6. The basic stylized morphology. A. Clump. B. Runner. C. An 18-year-old clump. D. An 18-year-old runner. (A-D not to scale.)

fails to develop because its site is not available, the excess energy (40 units) activates a dormant bud and produces a smaller rhizome–triangle combination which, if successful, reiterates the basic model, the budgeting being correctly balanced. An 18-year-old clump developed according to these rules is seen in Figure 6C. The fairy ring formation is a pleasant vindication of the model; it is a familiar feature of real plant clones.

The alternative mode of growth, that of the runner, is exemplified by a second caricature plant (Figure 6B), based on a sedge (*Carex arenaria*; Noble et al., 1979). The shoot–root site for this plant has exactly the same cost and potential as the shoot–root site of the clumper (40 units producing 92 units), but its morphology is different. Each triangle bears one long rhizome segment, costing 43 units, with a new

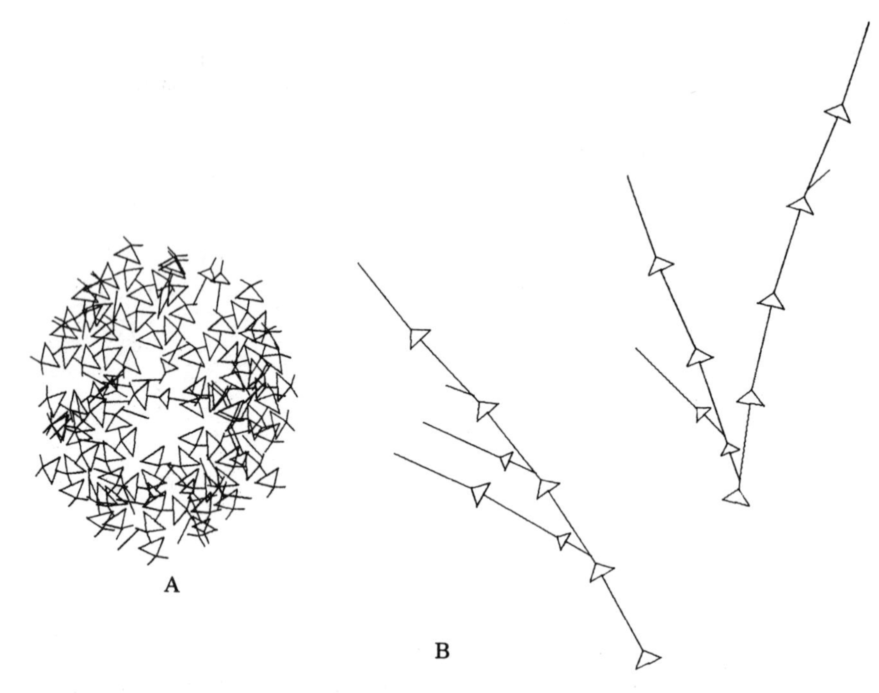

FIGURE 7. The resulting pattern of a single (control) plant after eight "seasons" of growth. A. A solitary clump plant. B. A solitary runner plant.

triangle at its distal end. Nine units are thus stored from each new shoot per season, to be summed (in the average case) every three seasons to produce one reiteration worth 27 units. Reiteration also occurs if a triangle is balked, in keeping with the behavior of the clump species (Figure 6D). Thus, a successfully established triangle of either species, clumper or runner, has the same productivity, but its potential is deployed in the first case mostly toward more sites of assimilation at the expense of mobility and in the second case toward exploration at the expense of costly nonproductive rhizome material.

A replicated (ten) series of simple experiments was conducted as follows, each experiment lasting for eight "seasons":

1. Solitary clump plant (Figure 7A).
2. Solitary runner plant (Figure 7B).
3. Ten clump plants in competition (Figure 8A).
4. Ten runner plants in competition (Figure 8B).
5. Ten clump plants in a static "stony" environment with stone sites unavailable for shoot–root establishment (Figure 8C).
6. Ten runner plants in a "stony" environment (Figure 8D).
7. Five clump plants with five runner plants (Figure 8E).
8. Five clump plants with five runner plants in a "stony" environment (Figure 8F).

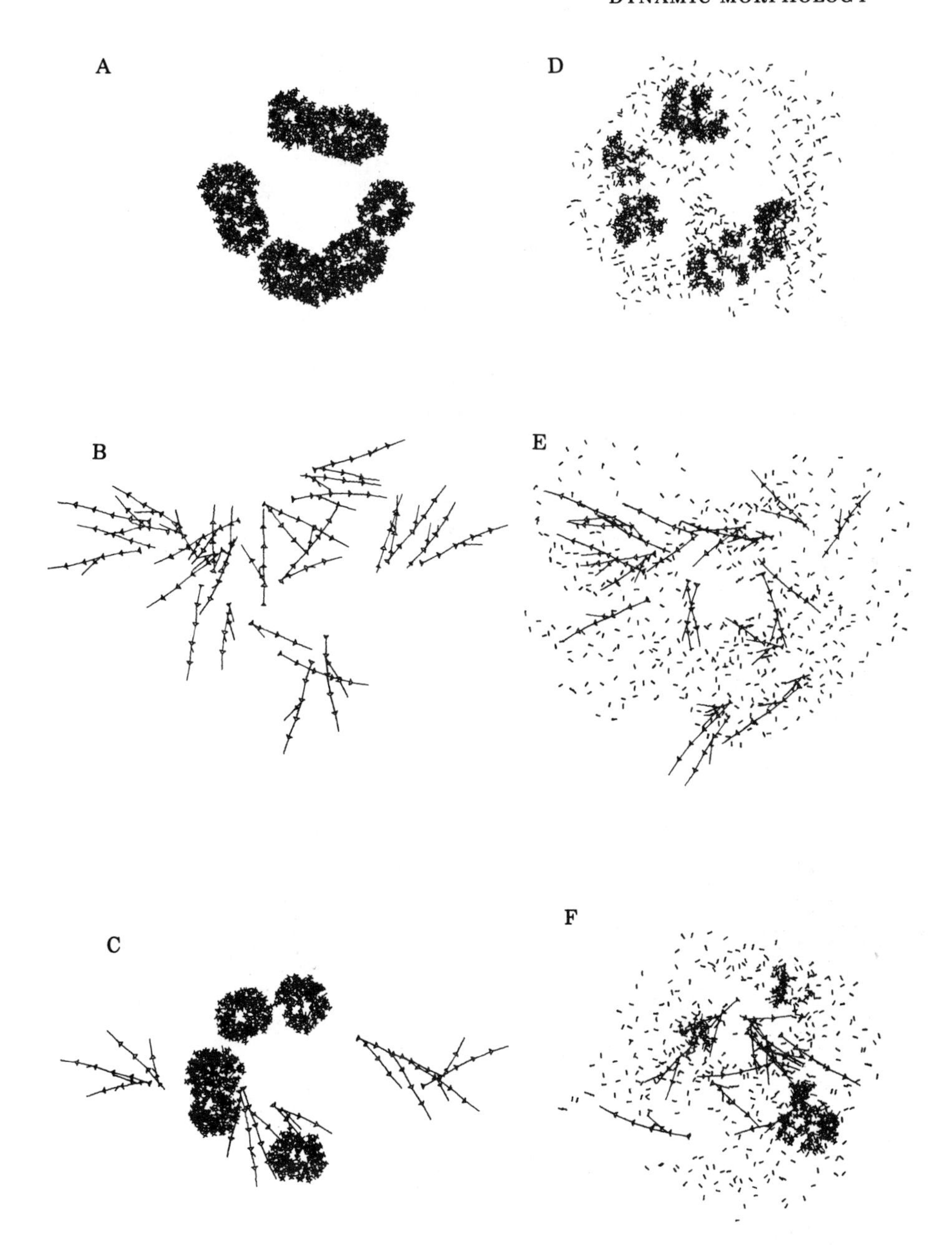

FIGURE 8. Representative replicates of a growth exercise after eight years' development. A. 10 clumps. B. 10 runners. C. 5 clumps and 5 runners. D. 10 clumps in a "stony" environment. E. 10 runners in a "stony" environment. F. 5 clumps and 5 runners in a "stony" environment. (See text for details.)

Performance of plants in the various replicates and treatments was available in the form of productivity based on the numbers of successful triangles and rhizome segments produced during the eight seasons, weighted for cost. The two controls (solitary plants grown without interference) gave the potential for clumper and runner under ideal conditions. Plant areas were calculated from graph plot hard copy.

Results are presented in a simple form in Table 1 and confirm what one might expect for real plants of these types, without the complication of numerous unknown variables. Clumpers are more productive than runners, are not affected by runners meandering around them, but are severely limited by static obstacles. Runners are moderately and similarly depleted by intraspecific competition, by interspecific competition, and by inanimate objects.

This type of theoretical morphological exercise, conducted in more detail, can evaluate rapidly the effect of small changes in "behavior" of plant populations and perhaps throw light on their ecological properties. It would be possible, for example, to test the hypothesis of Cooper and Kaplan (1982) that a coin-flip based organization might prove to be a successful strategy in particular situations, rather than an inflexible rigid plan.

TABLE 1. Performance of plants (productivity) in the various treatments of the growth exercise.

Treatment	Productivity[a]
Runners competing with runners	82
Runners competing with clumps	68
Runners competing with stones	64
Runners competing with both	66
Clumps competing with clumps	78
Clumps competing with runners	91
Clumps competing with stones	43
Clumps competing with both	44

[a]Productivity measured as a percentage of the appropriate control. [A runner has ¼ the productivity of a clump; a runner covers 4 times the area of a clump.]

An extended exercise: evolution of pattern

Paradoxically, the second exercise should represent the tantalizing precursor to the first and can be introduced by quoting from Holland's (1975) *Adaptation in Natural and Artificial Systems:*

> This book's main objective has been to make it plausible that simple mechanisms can generate complex adaptations; however, the book will have fulfilled its role if it has communicated enough of adaptation's inherent fascination to make the reader's effort worthwhile.

This second exercise is designed to monitor the evolution of "successful" branching patterns from an arbitrary and random set of initial rules. Different clonal patterns will be made to compete for one environment, which may or may not be homogeneous, and if heterogeneous, may or may not be static in its patchiness.

The computation will commence with the creation of a number of different "organisms," each having drawn its "rules of growth" at random from a "bank" of rules (or "genes"). These rules will govern such attributes as branching angles, branch lengths, location and type of daughter branches, degree of symmetry (Bell et al., 1979). Among these components will be all the rules identified from the bamboos and gingers, for example, but they will not be present in the presumed successful combination found in these extant systems. The simulation will proceed for a finite time span, at the end of which an organism will be deemed to be sexually mature. (This achievement alternatively can be related to size of individual.) Some clones will expand at the expense of others, and during the sexual phase that now follows the more successful patterns will shed more rules into a system of recombination from which will emerge a new generation of additional organisms with new pattern allocations. These processes will be continually repeated, fewer and fewer rules surviving in the population, more and more conservative patterns developing. In theory, the ginger–bamboo combinations might appear but not necessarily persist. Provision could be made for the inclusion of rule mutation. Pattern phenomena not included in the initial selection (such as previously unrecorded branching angles) would replace existing rules at the time of reproduction. The rate and novelty of such mutations could be varied. Patterns might emerge that are present in the fossil record or reflect gingers still to come. The prospect does hold a certain inherent fascination; the morphologist should not be confined to a nineteenth century image!

CHAPTER 3

DYNAMICS OF PLANT POPULATIONS FROM A SYNECOLOGICAL VIEWPOINT

Eddy van der Maarel

INTRODUCTION

This chapter will deal with interrelationships between synecology and population ecology. It will do so on a broad level under implicit, and at places explicit, reference to Harper's (1977) textbook on plant population biology. Synecology is the study of the relationships between plant communities and their environment. The term originated in European phytosociology (Braun-Blanquet, 1964, see also Westhoff and van der Maarel, 1978) to indicate one of a number of ecological subdisciplines, such as synmorphology (dealing with composition and structure), synphysiology (dealing with the functioning of plant communities, later to be known as ecosystem ecology), syndynamics (dealing with the changes in time), and syntaxonomy (dealing with classification and nomenclature). As Egler (1954) pointed out, the whole of phytosociology is in fact known as community ecology or synecology by most American ecologists.

This chapter will deal with synecology in the latter broad sense, but it is written by an ecologist with a European phytosociological background. While Anglo-American synecology has always been related to population or at least to autecology, European phytosociology remained a rather independent ecological science until recently. As Harper (1982) stated correctly, the phytosociologist has a geographical rather than a functional approach and is interested in species and area rather than in individuals and pattern.

We find this difference in attitude reflected in the development of phytosociology. If we must give a simplified picture, we conclude that the first decades were largely devoted to classification; in the 1940s and 1950s, environmental analysis and ecological characterization of communities became more apparent, whereas in the last 15 years the emphasis shifted more and more toward succession and fluctuation.

At the same time, numerical methods for a much more thorough analysis were developed, both within Anglo-American synecology and European phytosociology (cf. Whittaker, 1978a,b; van der Maarel et al., 1980). This twofold development toward ecological and dynamical analysis, respectively, is well reflected in the expansion of a leading textbook in the field (from Ellenberg, 1956, to Mueller-Dombois and Ellenberg, 1974).

The development of plant population ecology within phytosociology, which is largely the theme of this contribution, is directly related to these shifts in emphasis, as we will see later.

FROM SYNECOLOGY TO POPULATION ECOLOGY: LINES OF DEVELOPMENT

Gradient analysis

A first major line of development is through gradient analysis, an approach along Gleasonian traditions which was largely developed by Whittaker (1967, 1978c). The leading idea was that plant species are distributed individualistically (Gleason, 1926; McIntosh, 1976) along environmental gradients. This distribution pattern is usually presented as a series of bell-shaped, broadly overlapping species population curves with separated optima (e.g., Whittaker, 1960, 1967; also described by Harper, 1977). Such patterns are interpreted as the result of competitive exclusion or competitive niche replacement. As Werner and Platt (1976) showed, niche overlaps may become reduced under more mature conditions. Clearly, this is one example of an area of common interests to vegetation and population ecologists.

Whittaker combined Gleason's individualistic approach with his view of the plant community as a more-or-less discontinuous or at least discernible unit in the field amenable to classification (Whittaker, 1962). This is a rather unique viewpoint: the individualistic approach is usually linked with the continuum concept of vegetation (Curtis and McIntosh, 1951; McIntosh, 1967). It is significant that he presented a synthesis of both approaches to a European phytosociology symposium under the title "The population structure of the plant community." Various phytosociologists adopted similar views and promoted the acceptance of Whittaker's concepts, and finally Westhoff and van der Maarel (1978) incorporated it in their definition of the plant community, as follows: "A phytocoenose is defined as a part of a vegetation consisting of interacting populations growing in a uniform environment and showing a floristic structure that is relatively uniform and distinct from the surrounding vegetation."

Ordination

The multivariate approaches to vegetation analysis which became so popular in the 1970s include a series of ordination techniques (e.g., principal component analysis). Through ordination, vegetation samples—or plant species—are arranged in few-dimensional spaces in which the floristic similarities and dissimilarities are represented in a summarized way, keeping as much as possible of the multidimensional variation in the data set (see Whittaker, 1978b; Orlóci, 1978; and Gauch, 1982 for definitions and recent surveys of vegetation ordination).

The resulting axes of floristic variation are first judged as to their mathematical efficiency. In most cases, the first three (or four) axes explain 50–60% of the total variance; and in cases of areas with one overriding environmental gradient (e.g., a lake shore gradient), the first axis alone may account for up to 40–70% of the total variance. Next, the ecological effectiveness is inspected by correlating the axes, or diagonal trends, with measurements of environmental factors or with the distribution of species and species groups. Usually through isolines or isocoenes (see van der Maarel, 1969), clear species patterns arise, though they suggest nonlinear responses to underlying gradients (see later).

Quite commonly, a good correlation is found with general, so-called conditional environmental factors, such as altitude or a general moisture factor. But, much of the floristic variation remains unexplained. Correlation with so-called operational factors, such as resource gradients or stress gradients would probably be clearer, but measurements of such factors or gradients are not generally available (van der Maarel, 1976). Nevertheless, ordination approaches may lead to population studies in the same way that direct gradient analysis does.

Species performance along gradients

The interpretation problem mentioned is at least partly related to the problem of the nonlinearity of species performance along environmental gradients. Because of this nonlinear distribution, the ordination diagrams of most of the ordination methods show some distortion (see Austin and Noy-Meir, 1971; Austin, 1976, for a discussion; and Orlóci, 1978; Gauch, 1982, for reviews).

In the context of population ecology, the following twofold problem is most interesting: (1) Are we measuring the right factors and plotting them in the right way? (2) Are we expressing species performance in the most effective way by simply taking cover degree, as is usually done? Austin (1981, 1982) and Austin and Austin (1980) showed how in a mixture of grassland species experimentally grown under different levels of nutrition, wedge-type performance curves arise if the nutrient factor is plotted geometrically and the performance measure is a relative one. Van der Maarel (1976) foresaw wedge-type curves in relation to gradients of environmental stress.

Clearly such experimental and theoretical approaches of species performance are an important element in "searching for a model in vegetation analysis" (Austin, 1980), and obviously population ecology can contribute to vegetation science in this search.

Vegetation dynamics

Succession studies have played an important part in vegetation science from the beginning. However, it was mainly through the study of permanent plots that vegetation scientists became aware of population dynamics as a powerful help to studies on succession. The study of permanent plots has been carried out profusely in European seminatural and postcultural vegetation. At least four European symposia have been held in recent years, and their proceedings (Faliński, 1978; Beeftink, 1980; van der Maarel, 1980a; Poissonet et al., 1981) have produced a tremendous amount of factual evidence. Three general conclusions can be drawn, at least, which are relevant for our present framework.

1. The floristic composition of most vegetation types changes from year to year, often without a clear tendency toward a following step in a succession; and only seldom do they return to a floristic state which had been attained earlier.
2. If a drastic change in environmental conditions occurs (drought, fire, inundation, stopping of grazing), the vegetation shows an after-reaction to such a change, an after-reaction that may last for many years, while the cause as such is no longer evident from the floristic composition. This has much to do with what is known as the historical factor.

3. During a sequence of years of change, many species disappear; whereas some of them never return, other species appear and become established, and still others occur only as ephemerals. This makes the vegetation scientist aware of the (populational) processes of extinction and immigration.

The following examples may illustrate these general conclusions and at the same time demonstrate the use of multivariate analyses in vegetation dynamics.

Fluctuation and succession in a grassland. This example concerns the dynamics of an Australian grassland under sheep grazing (Austin et al., 1981). Permanent plots were followed from 1949 to 1968 in a *Danthonia semiannularis* grassland where different grazing intensities were maintained. Quantitative floristic data of 324 quadrats for six years were subjected to a numerical classification on the basis of a divisive cluster analysis. Table 1 presents a summary of community types. Communities A–D characterize vegetation types of early establishment and of years of low rainfall; E–G are typical of wet years, and H–I of late, dry years. Figure 1 shows the diagram of axes 1 and 2 of a principal components analysis of the data. The first component, accounting for 21% of the total variance, can be interpreted as a general temporal trend (see the arrows), which both grazed and ungrazed quadrats undergo. Whether this is real succession or "prolonged

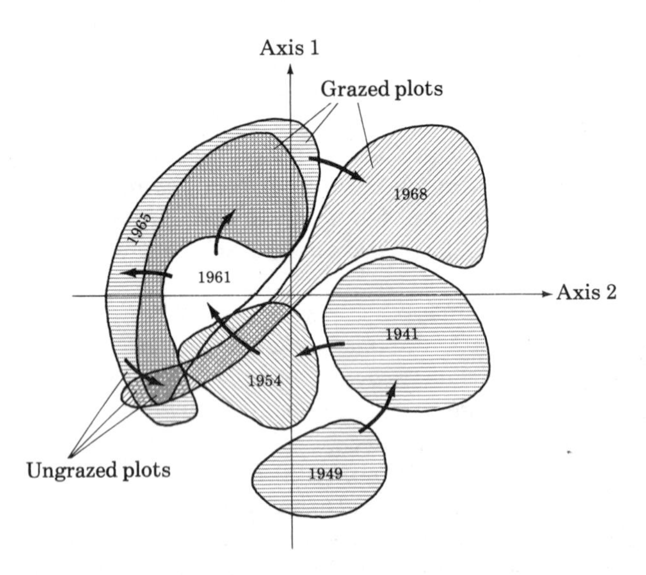

FIGURE 1. Diagrammatic distribution of grassland quadrats from six years on the first two axes of a principal components analysis. (From Austin et al., 1981.)

TABLE 1. Numerical classification of grassland quadrats over six years of successional analysis[a].

Community type	Species[b]	FREQUENCY OF OCCURRENCE (%)[c]					
		1949	1951	1954	1961	1965	1968
A	−*Plantago varia* −*Medicago polymorpha*	80.1	2.2	9.6*	1.5	6.6	—
E	+*P. varia* −*Hypochaeris radicata* −*Hedypnois cretica*	12.4*	72.7	4.7	0.6	1.8	8.3
B	+*P. varia* +*M. polymorpha* −*Avena fatua* −*Euphorbia drummondii*	23.7*	10.3	36.1	11.3	17.0	1.5
C	−*P. varia* +*M. polymorpha* −*A. fatua* +*E. drummondii*	8.8*	—	89.6	1.6	—	—
I	+*P. varia* +*H. radicata* +*Isoetopsis graminifolia*	—	—	6.0	59.1	24.2*	10.7
H	+*P. varia* +*H. radicata* −*I. graminifolia*	—	—	3.2	42.4*	49.8	4.6
D	−*P. varia* +*M. polymorpha* +*A. fatua*	—	0.7	7.6	20.8*	52.1	18.8
F	+*P. varia* −*H. radicata* +*H. cretica* −*I. graminifolia*	—	23.6	9.7	2.1	26.4*	32.2
G	+*P. varia* −*H. radicata* +*H. cretica* +*I. graminifolia*	—	7.9*	1.9	0.5	4.7	85.0
Rainfall for May-July (mm)		264	715	166	276	293	449

[a] From Austin et al., 1981
[b] +, Species present; −, species absent
[c] The values in the table are the frequencies of occurrence of each community type in a given year as a percentage of total occurrences. Year of primary occurrence underlined; year of secondary occurrence marked with an asterisk (*).

fluctuation" is unclear. The second component, accounting for 17% of the total variance clearly reflects the seasonality, that is, the variation in winter rainfall (cf. Table 1).

The same study also included some population dynamics data, which we will not discuss here.

Secondary succession in an abandoned orchard. The second example is concerned with a sudden (experimental) change: The understory of an apple orchard on a loamy soil in the Netherlands was treated in different ways after the trees were taken away in 1969 (van der Maarel, 1980b). Plots of 10 × 7 m were established with various degrees of mowing intensity and various forms of fertilization; in one quadrat sod cutting was applied and some quadrats were completely left intact. Details of the treatments are shown in Table 2. Floristic

TABLE 2. Summary of treatments applied to an abandoned orchard.[a]

Plot	Mowing	Years	Treatment	Years
1	S[b]	1970→		
2	—[c]			
3	S	1970→		
4	S	1970→	Marl[f]	1971–1975
5	S	1970→	Peat[g]	1971–1975
7	S			
8	S	1972→	Sod cut	1971
9	(S)[d]	1970–1972	Furrowed	1973
10	S	1970→	NPK[h]	1971–1973
11	S	1970→		
12	S	1970→	Ca, Mg[i]	1971–1975
13	(S)	1970–1972		
14	(S)	1970–1972	Urea[j]	1973–1975
15	S	1970→		
16	S	1970→(not 1973)	Urea	1973–1974
17	S	1970→		
18	S	1970→	Urea	1973–1974
19	(S)	1970–1972		
20	S	1970→		
PB	3x[e]	1970→		

[a] From van der Maarel (1980b).
[b] Mown in summer.
[c] Not mown since 1970.
[d] Mown in summers 1970–1972; not mown since 1973.
[e] PB, Plots between large plots: mown 3–4 times a year.
[f] 2 kg/m^2/yr.
[g] 1.5 liter/m^2/yr.
[h] ASF pellets, 0.15 kg/m^2/yr.
[i] Calcium and magnesium carbonate, 0.7 kg/m^2/yr.
[j] 0.15 kg/m^2/yr.

analysis of four subplots of all 20 plots plus 3 between-plot areas were subjected to agglomerative clustering and principal component ordination.

Figure 2 shows the dynamical relations of the plots; it can be seen how some plots are more or less floristically stable (e.g., 2 and 20) in accordance with the stable management they receive, whereas others move considerably through the ordination diagram in relation to the type of applied management; for example, plots 8, 10, and 12 (mown) notably develop in the direction of plot 20; plot 8 (top soil removed + mowing) moves to an Arrhenatherion-type grassland much more rapidly than plots which had been mown only. Positions and movements of plots can be interpreted with the help of species occurrences. Table 3 shows the behavior of the four leading species in six plots. For example, it can be seen that (1) in plot 2 (not mown) there is a clear persistence of *Urtica dioica*, whereas in plot 1 (mown) the same species steadily decreases through time; conversely, *Arrhenatherum elatius* increases; (2) *Heracleum sphondylium* appears to behave independently of the major treatments; (3) quite characteristically, if *A. elatius* and *U. dioica* are both present, they behave oppositely under the major treatments. The data on species richness (Table 3) show, among others, these trends: the species number in mown plots is around 20, compared to 15 in the unmown plots. Fertilization leads to a rapid decrease in species number (together with a rapid increase in the cover of *U. dioica*), but the reverse relation occurs again after stopping fertilization; also there appears to be a correlation between the occurence of high quantities of *H. sphondylium* and a relatively low number of species in mown plots.

This study shows the application of numerical methods in detecting vegetation changes and suggests that these, in conjunction with analyses of single species behavior, may be quite revealing in the study of dynamical aspects of vegetation.

Primary succession in a dune area. The third example refers to a development of vegetation over a much longer period of time in the (calcareous) dunes near Oostvoorne, the Netherlands. These dunes developed in a number of zones since Roman times. Up to 1910 the then-formed dunes were heavily grazed, in the end to the extent that most of the Medieval dunes had become almost barren, and the seaward dunes were losing their protective power against the sea. Then in a very short time the cattle were removed from the area and a rapid succession, which can be called quasi-primary, started in these dunes. In 1959 the vegetation of the whole area was analyzed and

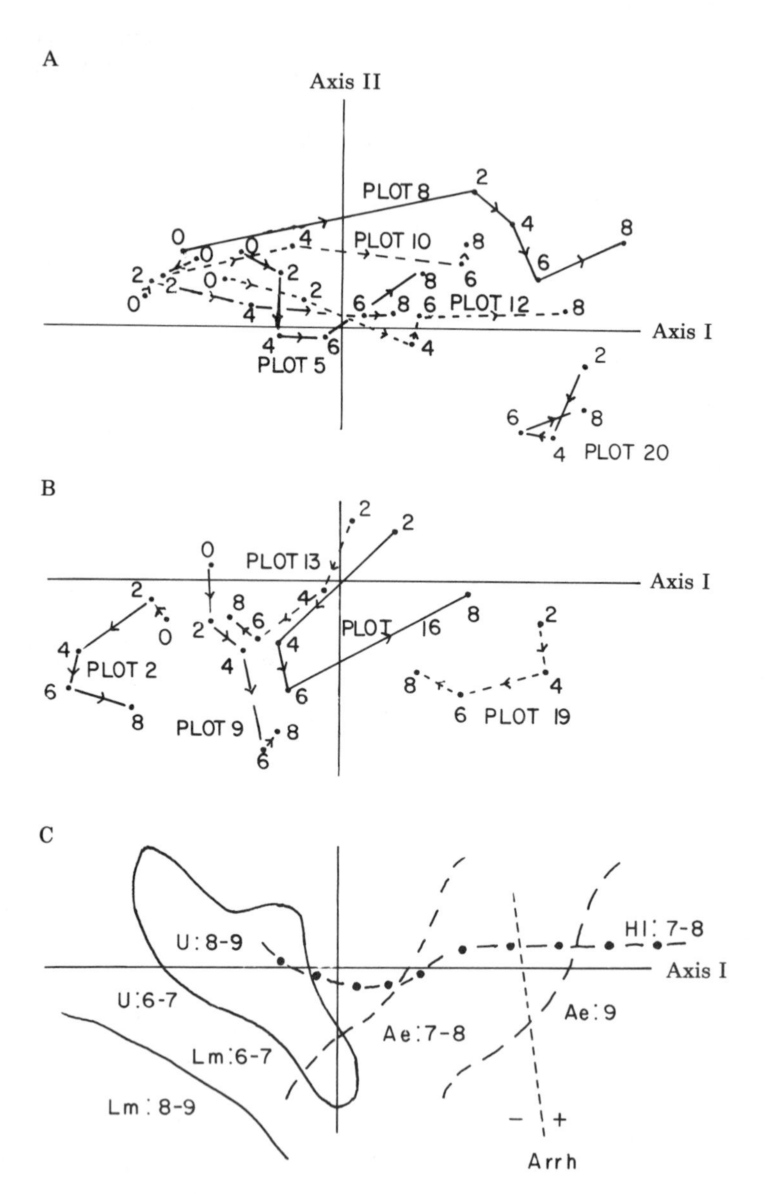

FIGURE 2. A principal component ordination of subplots of the experimental plots set up in an orchard after tree removal; axes I and II are shown. A. Positions of mown plots (5, 8, 10, 12, and 20). B. Positions of mown plots (2, 9, 13, 16, and 19). C. Isolines for some important species: *Lamium maculatum* (Lm), *Urtica dioica* (U), *Arrhenatherum elatius* (Ae), and *Holcus lanatus* (Hl), and line separating areas with (+) and without (−) representation of characteristic species (other than *A. elatius*) of the alliance Arrhenatherion (Arrh.). (From van der Maarel, 1980b.)

TABLE 3. Occurrence of four leading species and species richness in six representative experimental plots from 1970 to 1978[a].

	Plot 1									Plot 8								
	70	71	72	73	74	75	76	77	78	70	71	72	73	74	75	76	77	78
Arrhenatherum elatius	0	0	0	1	1	3	5	5	6	0	0	2	2	3	0	6	7	7
Urtica dioica	6	6	5	5	3	4	3	3	3	8	0	1	1	1	0	0	0	0
Lamium maculatum	6	7	6	3	2	4	3	3	3	4	0	0	0	0	0	0	0	0
Heracleum sphondylium	2	1	2	5	5	3	3	6	5	5	2	1	1	2	3	3	6	6
Number of species	24	18	20	18	19	26	19	23	23	24	30	25	24	23	28	20	20	21

	Plot 2									Plot 13								
	70	71	72	73	74	75	76	77	78	70	71	72	73	74	75	76	77	78
Arrhenatherum elatius	0	0	0	0	0	0	0	3	4			0	0	0	2	0	0	0
Urtica dioica	5	7	5	6	5	5	7	5	6			3	5	5	4	5	6	7
Lamium maculatum	5	7	6	9	9	9	8	8	8			2	2	4	4	6	6	7
Heracleum sphondylium	5	1	5	7	5	1	5	1	2			5	7	6	4	2	3	5
Number of species	16	11	16	14	14	15	15	15	15			20	19	19	19	15	19	17

	Plot 16									Plot 20								
	70	71	72	73	74	75	76	77	78	70	71	72	73	74	75	76	77	78
Arrhenatherum elatius			5	7	6	5	8	8	8			9	8	9	8	9	8	9
Urtica dioica			3	6	9	9	7	3	3			1	2	2	2	0	0	0
Lamium maculatum			0	0	0	0	5	0	0			0	0	0	0	0	0	0
Heracleum sphondylium			6	8	7	2	3	3	5			7	7	5	2	2	5	3
Number of species			18	17	10	15	16	18	22			13	16	13	24	12	18	20

[a] From van der Maarel (1982b). For details of the treatments applied to each plot, see Table 2.

mapped (scale 1:2500) (van der Maarel and Westhoff, 1964). In 1980 the vegetation mapping could be repeated in the same way (van der Maarel et al., in press). Moreover, air photographs from 1934 to 1980 could be interpreted. Finally in 1981 an age-structure analysis of the important woody species in the area was performed on the basis of 1600 wood cores (van der Maarel et al., in press). From these studies, it became apparent that six zones over a width of 1200 m can be distinguished with the following characteristics and dynamical trends:

1. Dunes originated after 1926: mainly a large system of primary dune

slacks and low dunes, which were still almost unvegetated in the 1940s. During the 1950s dune marsh developed, and since the 1960s *Betula* woodland (a mixture of *Betula verrucosa* with *B. pubescens*) began to develop.

2. Dunes originated after 1910: again largely a system of primary dune slacks with *Betula* woodland up to 8 m and, since the 1960s, also an *Alnus glutinosa* woodland. The original *Hippophae rhamnoides* scrub is being replaced by taller scrub with *Ligustrum vulgare* and *Crataegus monogyna*.
3. Former coastal dune zone forming the seaward dunes until 1910: relatively high dunes which largely remained open with patches of *Ammophila arenaria*, scrub of *Sambucus nigra* and *Hippophae rhamnoides*, with a clear succession into *Crataegus* scrub.
4. The Medieval dunes behind zone 3 where an extensive low scrub had developed in the 1940s. Here the overall development is toward tall *Crataegus monogyna–Rhamnus catharticus* scrub, with tall *Betula* woodland in damp places.
5. The older Medieval dunes in which tall scrub, mainly of *Crataegus*, and woodlands with *Betula* and *Populus tremula* existed already in the 1950s. Here the succession goes on toward tall scrub on places which formerly had low scrub, and a general increase in the extent of *Rhamnus catharticus* is noticed. The development of the already established tall woody vegetation is slow to almost stagnant. There is no further growth of the dominant species, the understory becomes poorer in species and the first signs of death of the dominants are being noted.
6. Roman-time zone with dune grassland surrounded by both tall and low *Crataegus* scrub, which is advancing at the expense of the grassland. This is related to the stopping of grazing (in the 1930s).

This very complicated successional pattern can be understood in some more detail by means of the age structure data of major woody species.

The age pyramids showed that no species population had individuals older than 70 years (with very few exceptions). This directly confirms the historical development as sketched above. The population structure of many species showed irregularities, notably, a shortage in age classes between 10 and 20 years. These could partly be explained by the combined effect of summer drought and high rabbit grazing in some critical years.

Figure 3 shows the age structure of the most important shrub, *Crataegus monogyna*, in the six zones indicated above. We see how the rate of population development gradually decreases from younger to older zones. It is also clear that there was an almost general standstill in the 1960s and a comeback in the most recent years. This picture is

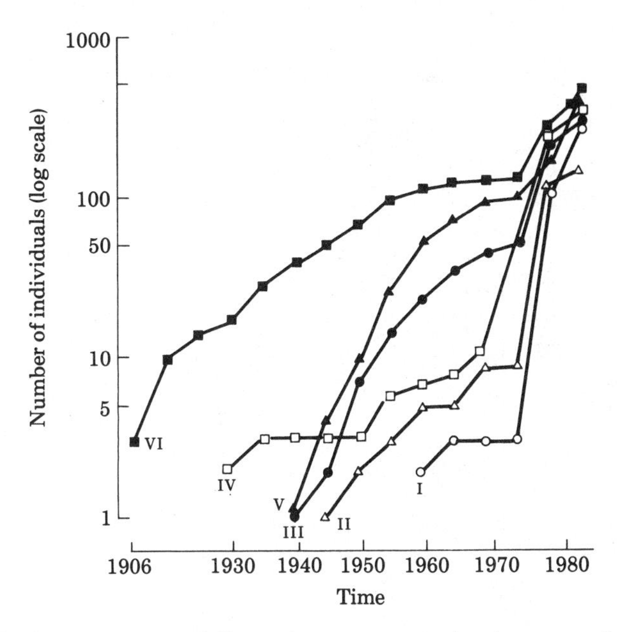

FIGURE 3. Age structure of *Crataegus monogyna* in six successional zones (I–VI, young to old) in the dunes near Oostvoorne. (From van der Maarel et al., in press.)

remarkably similar to the innovative graph Harper (1977) presented for *Juniperus osteosperma* (Figure 4).

Figure 5 presents the age structure of two species in the tall *C. monogyna* scrubs in zone 5. This species has a rather regular distribution, with a peak around 20–30 years. There is new recruitment: 80 seedlings, and 100 saplings less than 4 years old. *Rhamnus catharticus*, *ticus*, also with new recruitment, shows a "younger" profile and could well be considered a next successional species. (The problem of death indicated above involves only a part of the entire *Crataegus* scrub zone!)

This partial information may at least suggest the perspectives vegetation ecologists will get if they combine their succession studies with population dynamics!

Succession modeling

A different approach in vegetation dynamics is even more directly related to population dynamics. It is the modeling of succession through the prediction of establishment, growth, and death of the

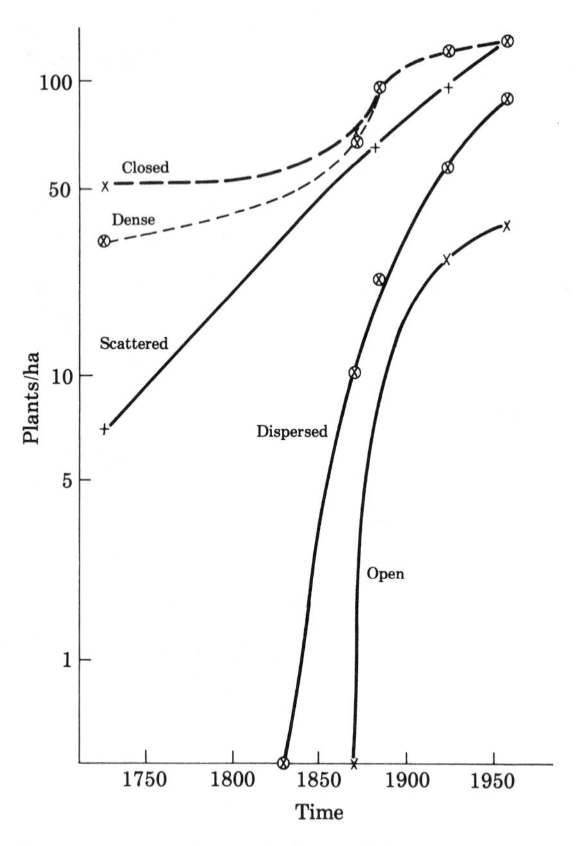

FIGURE 4. Age structure of populations of *Juniperus osteosperma* as indicated by the ages of survivors in 1956. (From Harper, 1977.)

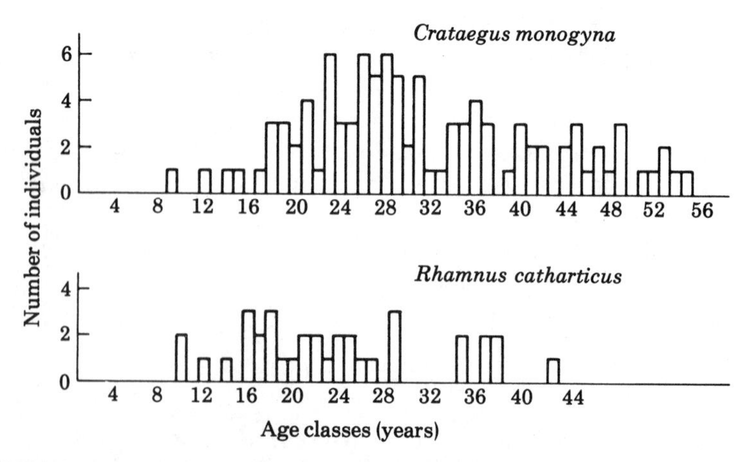

FIGURE 5. Number of individuals of different age classes of *Crataegus monogyna* and *Rhamnus catharticus*, in the *Crataegus* scrubs in zone V of the dunes near Oostvoorne. (From van der Maarel et al., in press.)

major species in plant communities, especially forests. We may mention the models of Botkin and Shugart (see Shugart and West, 1980; West et al., 1981) in which forest succession is predicted on the basis of the different development rhythms of the dominant species. Another relevant approach is that of Slatyer, Connell, and Noble (Connell and Slatyer, 1977; Noble and Slatyer, 1980) which especially models the regeneration of trees after disturbance (e.g., fire).

In this approach, a number of life-history attributes (termed *vital attributes*) pertaining to the potentially dominant species in the community are defined. The vital attributes define what they call *species types*. The interactions among various species (based on their species types) then yield in their models a replacement sequence which depicts the major shifts in composition and dominance which occur following an episode of disturbance. In their approach, a given successional pathway may be displayed "but as a result of logically determined interactions between species rather than as a form of community ontogeny" (Noble and Slayter, 1980:20).

In summary we may conclude that the study of vegetation dynamics is a very obvious way to population ecology (see, for example, Miles, 1979) with disturbance as a major driving force in dynamics (Grime, 1977, 1979a).

Diversity studies

A last aspect of vegetation study leading to population studies is the analysis of diversity. After Whittaker (1965) introduced the concepts of dominance and diversity in an integrated approach, many vegetation scientists took diversity studies up in their analyses. For our purposes two aspects are especially important. The first is the development of species diversity during the course of succession. This can be expressed in changes in the form of the so-called dominance–diversity curves (cf. Whittaker, 1969). Figure 6 shows an example from an old field succession, in this case abandoned vineyards of increasing age (Houssard et al., 1980). The early succession stage shows an almost geometrical series, whereas later stages approach the log-normal distribution. However, the entrance of dominants in later stages causes a temporary return to the geometrical series, which is supposed to express more open conditions with less competition between species on the same resources and a more straightforward "niche-pre-emption" (cf. Werner and Platt, 1976). Hubbell (1979) showed through simulation that a more mature balanced forest community with a log-normal type of dominance–diversity relation is driven back to a geometric type through repeated disturbance.

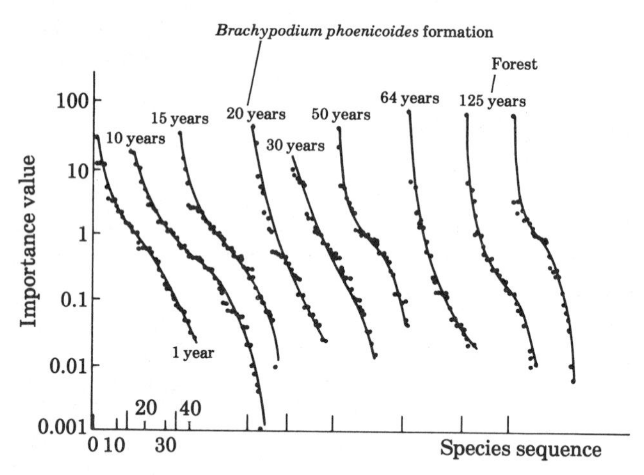

FIGURE 6. Dominance–diversity curves for successional communities in Le Causse de la Selle. The points represent the species, plotted as a relative cover importance against the species rank in the sequence of species from most to least dominant. For clarity, the figures have been arbitrarily spaced out. Positions of their origins on the abscissa are indicated by vertical lines along the border of the figure. (From Houssard et al., 1980.)

The second aspect is that from descriptive studies of diversity, one easily enters the population dynamics area of competition studies. A representative example of a field-oriented competition experiment is that of Beeftink (in press) with *Plantago maritima* and *Limonium vulgare* in a well-developed stand of the association named after the two species: Plantagini-Limonietum. The usual replacement series was laid out in the field. One of the outcomes was that the performance of *P. maritima* was better in the presence of *L. vulgare*. This may have to do with a lower salt concentration in the rhizosphere of *L. vulgare*, which takes up much salt (and exudes it again); this may favor *P. maritima*, which is less salt tolerant.

This is just one example of a possible positive interaction between species as a contribution to the maintenance of diversity. Other types of competitive interactions and their bearings on diversity are mentioned by Dirzo (Chapter 7).

Many new approaches are on their way; we cannot discuss them all. I only refer here to some important studies on niche differentiation in space and time (Grime, 1977, 1979a,b; Grubb, 1977; Braakhekke, 1980; Berendse, 1981).

THE SIGNIFICANCE OF POPULATION BIOLOGY FOR SYNECOLOGY

The significance of population biology has become clear, especially from such a standard work as that of Harper (1977). To open such a new field, or at least a discussion about it, requires cooperation and integration of the two disciplines. Table 4 provides a list of chapter titles in Harper's book and the corresponding synecological approaches. This shows how much I had to leave untouched in this chapter. Let us still mention five important fields of research in which the population

TABLE 4. Correspondence between population biology and synecology.

Population biology	Synecology
Seed rain	• Accessibility; Species–area relations • Species introduction
Dormancy	• Phytosociology of seed–rhizome banks • Colonization
Recruitment of seedling populations	• Microheterogeneity
Density, influence on yield, mortality Density, influence on form, reproduction	• Performance; importance value studies
Species mixtures: space	• Coexistence studies
Species mixtures: time	• "Timesharing" of resources
Limiting resources	• Synecology *s.s.*; studies on ecological indicators
Mechanisms of interaction	• Association analysis; pattern
Defoliation Seasonality; search; choice Role of grazing animal Predation of seeds and fruits Pathogens	• Analysis of structure—crown cover • Grassland ecology
Role of predation	• Seminatural vegetation studies
Natural dynamics:	
Annuals, biennials	• Ecology of ephemeral communities and of ruderals
Herbaceous plants	• Ecology of grasslands and of heathlands
Woody plants	• Succession
Reproduction; growth	• Life-form ecology
Reproduction; life cycle	• Phenology
Community structure; diversity	• Gradient analysis; dominance–diversity studies
Natural selection	• Paleoecology • Environmental dynamics

ecologist and the vegetation ecologist can profit from each other's work, or at least where the vegetation ecologist can very well use data and theories from the population ecologist:

1. Immigration–extinction.
2. Seed bank and rhizome bank.
3. Microassociation, micropatterns, and coexistence.
4. Dominance–diversity relations and structure of vegetation.
5. And last, but not least, population dynamics and succession.

Let us hope that vegetation scientists will realize the profits of cooperation with population ecologists, especially in the field of vegetation dynamics.

I should have ended with a synecological quotation from Darwin, but I did not really look for one because I expected not to find any. By the time of Darwin's death, the first textbook on community ecology had yet to be written. Instead I dared to change the two famous adages in a synecological way:

Striving for co-existence
Survival of the fitting

CHAPTER 4

THE ANALYSIS OF DEMOGRAPHIC VARIABILITY AT THE INDIVIDUAL LEVEL AND ITS POPULATION CONSEQUENCES

José Sarukhán, Miguel Martínez-Ramos, and Daniel Piñero

INTRODUCTION

Plants not only stay quietly in one place to be counted and measured by ecologists as they grow, reproduce, and die; they are also endowed with the ability to grow, reproduce, and die at rates that vary widely among individuals within the same population.

The display of this enormous variability has been a source of profound interest for man from the beginnings of civilization, when the earliest attempts at domesticating plants began. It has also been a source of puzzlement and study for biologists, the greatest of whom pointed out that:

> Selection acts only by the accumulation of slight or greater variation, caused by external conditions, or by the mere fact that in generation the child is not absolutely similar to its parents. Man, by this power of accumulating varia-

tions, adapts living beings to his wants . . . [and that] in nature we have some *slight* variation occasionally in all parts; and I think it can be shown that changed conditions of existence is the main cause of the child not exactly resembling its parents.

Letter from C. Darwin, Esq., to
Prof. Asa Gray, Boston, U.S.,
Down, September 5th, 1857.

Later, plant biologists delved deeper into this variability (e.g., Salisbury, 1942; Harper et al., 1970; Harper, 1977 and references therein; Bradshaw, 1965).

The seminal work by Kira et al. (1953) and Yoda et al. (1963) [reinterpreted and expanded later, principally by White and Harper (1970), Kays and Harper (1974), and White (1981)] has shown that variability in vegetative behavior has environmental constraints that enable the prediction of the average weight or size of individuals in even-aged, crowded populations. Size or weight distributions around the different mean values show a clear skewness. The work of Koyama and Kira (1956) with *Erigeron*; Obeid et al. (1967) with *Linum usitatissimum*; Ogden (1970) with several annual weed populations; Ford (1975) with *Tagetes patula*; Hiroi and Monsi (1966) with *Helianthus annuus*; Mohler et al. (1978) with *Abies balsamea* and other temperate arboreal species, and the many examples reviewed by White (1980) show that very marked size hierarchies are established among individuals of even-aged monocultures.

However, the occurrence of size hierarchies is not constricted to planted, even-aged monocultures. It has also been ubiquitously recorded for natural populations (mostly uneven-aged) of short- and long-lived species in both mono- and plurispecific communities (Leak, 1964; Day, 1972; references in Harper and White, 1974; Werner, 1975; Crisp and Lange, 1976; Hett and Loucks, 1976; Cook, 1980; Franco and Sarukhán, 1981; Kohyama, 1981; Knowles and Grant, 1983).

The generally spotty nature of studies on age or size structure in plant populations and the frequent inability to explain individual variance in vegetative and reproductive performance, made the demographic approach a very welcome contribution to the study of plant populations in the late 1960s and early 1970s (Tamm, 1956; Sagar, 1959; Harper, 1967; Sarukhán and Harper, 1973). This new approach triggered a cascade of actuarial studies with plant populations that had widely different habits and that grew in a multitude of environments (e.g., Baskin and Baskin, 1974; Sharitz and McCormick, 1975; Jefferies et al., 1981; Klemow and Raynal, 1981; Symonides, 1977; West et al., 1979; Van Valen, 1975; Hartshorn, 1975; Sarukhán, 1980; Piñero and Sarukhán, 1982; Bullock, 1980; Yadav and Tripathi, 1981; review by Silvertown, 1982b).

Although age was usually sought as a natural population vector, it was soon realized that age could be a poor predictor of vegetative, and especially reproductive, performance of individual plants; size or "stage growth" as an estimation of the vegetative status of an individual was found often to be better correlated with its demographic behavior (see Harper and White, 1974; Werner, 1975; Werner and Caswell, 1977; Kawano, 1975; Barkham, 1980; Bullock, 1982). However, when both relatively accurate age estimates and growth stages are used, the understanding of population dynamics and the responses of individuals to environmental factors becomes much greater. That vegetative status could vary rather independently from chronological age and could be a good indicator of a population's structure and dynamic stage was realized early by Soviet plant ecologists, who have produced an abundant literature on the demography of many herbaceous species (Rabotnov, 1960, 1969, 1978a; Uranov, 1960, 1975; Smirnova, 1967, 1968; Zhukova, 1961; references in the review by Gatsuk et al., 1980).

Demographic studies soon revealed certain general patterns of population behavior among plant species (namely, in the types of mortality patterns shown by species with different life-histories) and suggested the probable role played by physical or biotic factors of the environment. The average patterns of population behavior described different life-history traits.

Different life tables and population models have been derived for many of the species studied to date (e.g., Sarukhán and Gadgil, 1975; Hartshorn, 1975; Van Valen, 1975; Leverich and Levin, 1979; Callaghan, 1976; Bullock, 1980). All forms of representation of population flux correspond to average conditions of all individuals observed at different age or size classes, during the period of observation, from one or several localities. The vision conveyed by most actuarial studies of plants is that of the ideal plant for an age or size class, in an ideal year or period, and in an ideal site. The individual variance around the mean behavior either in the onthogenic, spatial, or temporal dimensions is virtually always absent from these studies.

It is the objective of this chapter to show the importance of individual variability in demographic parameters in explaining the components of individual fitness: survivorship, growth, and reproduction. We shall also attempt to explore the demographic consequences of individual variability and discuss some of the results of demographically interpretable studies dealing with the genetic or environmental explanation of such variability.

THE COMPONENTS OF FITNESS

Most modern population biology studies have concentrated on the question of defining, estimating, and explaining a measure of evolutionary advantage of individual organisms resulting from the basic demographic attributes of survivorship and reproduction. This evolutionary advantage is determined by the action of the whole environment and its fluctuations in time and space on the phenotypes in question (Figure 1), in relation to other phenotypes in the population. Within this context, it is vital to know (1) the effect of the environment on the survivorship, growth, and reproduction at different stages of the life cycle of each individual in a population; (2) the correlative changes between growth, reproduction, and survival resulting from compromises in the utilization of limited resources; and (3) the degree to which such individual responses to the environment are genetically determined and, therefore, potentially inheritable by the progeny left by each individual. Details of these three points are only very partially known, often based on data which only look at one of the corners of this triangle, details fundamental to the understanding of the evolutionary consequences of population dynamics.

Differential survivorship

The first indication of the presence of individual size-dependent mortality in plant populations came from studies of single-species, even-aged plantations growing under high-density stress. It is well known now that under these conditions a thinning process develops in

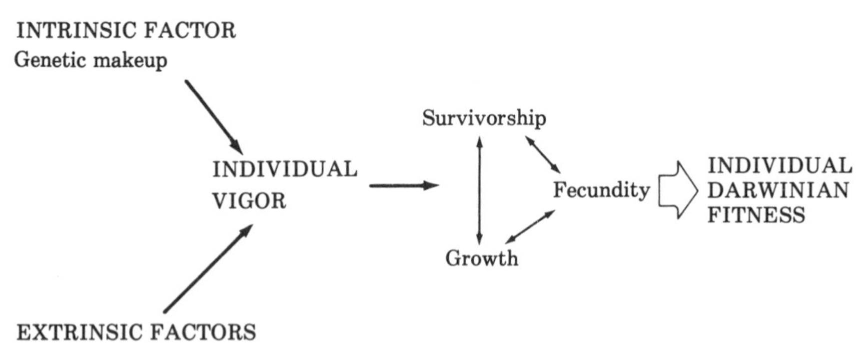

FIGURE 1. The factors intrinsic and extrinsic to an individual plant that determine its vigor and, consequently, its Darwinian fitness.

which, as plants increase their mean weight, mortality occurs, particularly among individuals in the lower end of the weight distribution. This density-dependent process in which mortality is differential and size-dependent has come to be known formally as the "self-thinning rule" or the "−3/2 power law" (Yoda et al., 1963; White and Harper, 1970). This rule, which is one of the few general rules in plant population biology, has been observed to apply in a large number of species from a wide spectrum of taxa, life forms, and environments (White, 1980).

In naturally occurring populations, size plays an important role in differential survivorship from very early in the life cycle of plants. A general hypothesis proposing that maternal expenditure on future progeny is adjusted to the predictability and availability of resources has been put forward by Lloyd (1980c). According to this hypothesis, plants control their maternal investment on the fruits (and eventually seeds) they will bear by determining sequentially the number of flowers produced, the development of the ovaries, the maturation of developed ovaries (or fruits), and finally the number of ovules that develop in each ovary as seeds. The on–off switching of energetic investment at each stage is determined by the level and predictability of resources available at that stage.

Genetic differences of the zygotes may lead to the production of size hierarchies in seeds through a differential allocation of maternal resources (Westoby and Rice, 1982). However, the seed vigor may be the result of factors extrinsic to the zygote's genotype, as is the case for *Lupinus texensis* (Schaal, 1980b), in which seed vigor is determined by the position of the ovule in the pod: those proximal to the peduncle of the fruit are the ones which will develop the larger seeds. Other examples of position-dependent seed size are presented by Janzen (1969) and Aker (1982).

The size (vigor) of seeds may have in some cases a definite influence in determining the performance of adult individuals (e.g., Salisbury, 1942; Harper and Obeid, 1967). Schaal (Chapter 9) observed a wide variation in individual weight within and between families of seeds produced by different mothers. The variation had a strong maternal effect, less than 10% of it being due to genetic causes. Larger seeds had significantly greater germination rates, germinated more, and produced more vigorous seedlings with a higher survival than smaller seeds.

Differential survivorship of seeds can also be influenced maternally through the amount of energy invested in the structures which protect the seeds in the fruit. Hare (1980) found that in *Xanthium strumarium*

(Compositae) predation of its seeds by insects decreased significantly with the size and thickness of the burr. Additionally, seed size has been shown to have a genetic component among different populations. Apparently in this case predators induce a selection toward increasing fruit size (and seed size?), attacking smaller burrs, which also have thinner walls and are therefore more susceptible to egg-laying. Seed size was found to be a better predictor of the susceptibility of being predated than either morphological or chemical traits (see Hare and Futuyma, 1978; and also Bridwell, 1918). However, Janzen (1969) has found the contrary for several legume species whose seeds are attacked by bruchids. In these species, small seed sizes are selected for because proper development and emergence of oviposited bruchids is not possible in small seeds.

Seed size is not, however, always correlated with differential germination (e.g., Cideciyan and Malloch, 1982, for *Rumex* spp.) or with seedling vigor (Solbrig, 1981, for *Viola* spp.). In other instances, shape rather than size is a source of differential survivorship. Dimorphic seeds are frequent in the family Compositae. Venable (Chapter 8) describes the occurrence of significant differences in the survival of seedlings originated from ray or disk achenes for two composites.

Seed size is clearly subjected to conflicting selective forces: from the compromise confronted by the mother on leaving an optimal number of progeny with maximum resources per seed (Westoby and Rice, 1982) to those which influence dispersal mechanisms, predator defense, germination ability, and vigor in a given temporal and spatial environment. Janzen (1969) has proposed in this context certain tradeoffs between seed size and number in the presence of predator pressure. Along the same line, a reduction in seed size could be the result of selective forces acting on traits favoring ample dispersal, as seems to be the case of pioneer tree species that colonize forest gaps in the tropics (Howe and Estabrook, 1977; Vázquez-Yanes, 1981; Brokaw, 1982). Howe and Richter (1982) studied the seeds of *Virola surinamensis* (a tropical forest canopy tree), and their results support the hypothesis that variation of seed size within and between crops of different parents could be caused by alternating pressures to (1) increase the probability of colonizing favorable sites with small seeds dispersed by endozoochory or (2), in the absence of dispersal agents, to increase the probability of survival of seedlings in conditions of environmental stress by bearing large seeds which produce robust seedlings.

The effect of environmental factors (especially biotic) on seed traits defined by maternal influence or otherwise, can be studied at the level of the plant–animal interface (e.g., Janzen, 1976; Dirzo, Chapter 7). In these studies, differential herbivory is simulated by removing different portions of the maternal capital in the seeds. In general, these studies show that the greater the amount of maternal resources removed the greater the chances of future mortality at the seed and seedling stages. Also, the removal of seed capital produces a marked size hierarchy (both in height and number of leaves) among emerging seedlings. Patterns similar to those found in experimental studies have been observed under natural conditions for seedlings of *Nectandra ambigens*, an emergent species in neotropical rain forests, attacked by larvae of a fly (*Pteticus cyanifrons*) and beetles (*Pagiocerus frontalis* and a curculionid). Mortality is a function of how much seed capital is lost to the predator, the smaller seeds losing a greater proportion of it than larger seeds. Percentage germination and seedling vigor (height, number of leaves, and biomass) also show hierarchies (B. Córdova and J. Sarukhán, unpublished data).

Selection should act toward increasing maternal investment in seeds in environments where low levels of resource availability (i.e., light) affect seedling establishment. This has been found to be the case for shade-tolerant species in Malaysian forests (Ng, 1978); these species have large seeds that germinate rapidly and produce vigorous seedlings with high survivorship. In general, under suppressive forest-floor conditions, larger seedlings do have greater chances of survival and a greater ability to recover from accidental physical damage or defoliation by herbivores (see Dirzo, Chapter 7). We have found evidence of this in seedling cohorts of *Astrocaryum mexicanum*, a dominant understory palm of tropical rain forest in southeast Mexico (Piñero et al., 1977; Sarukhán, 1978). As shown in Figure 2, two-year-old individuals possessing three or more leaves had greater probabilities of surviving to their fifth year of life than individuals with fewer leaves. The same pattern was found for the category of infants (age classes between 1 to 8 years) and juveniles (age classes between 9 to 15 years) (cf. Figure 4); a larger standing leaf area means a greater survivorship chance for an individual, under light-limited conditions.

Intergenotypic differences have been shown for germination and seedling emergence (Eagles and Hardacre, 1979; Nelson, 1980) and seedling vigor (Voight and Brown, 1969; Fakorede and Ojo, 1981).

Differential mortality occurring among young individuals of temperate species and attributable to plant size or biomass has also been documented profusely (Werner, 1975; Cook, 1979b, 1980; Solbrig et al., 1980; Solbrig, 1981; Gross, 1981; Parker, 1982; Bazzaz, Chapter 16) (Figure 3).

The time required for germination and establishment seems to be crucial in determining the obtention of resources (Bazzaz et al., 1982), especially in environments open to colonization. Cook (1980) found that early recruitment in *Viola sororia* populations resulted in greater

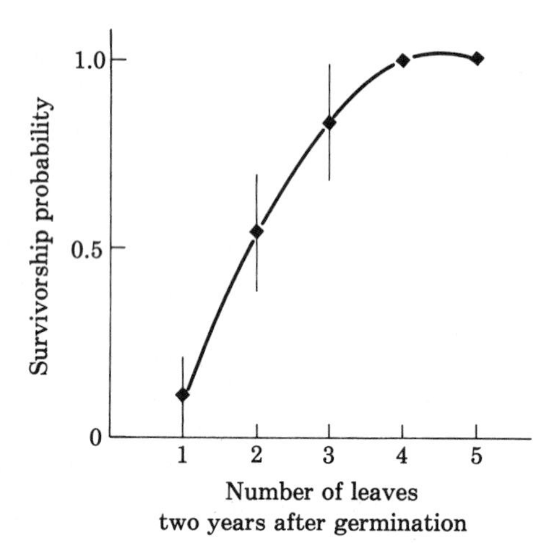

FIGURE 2. Survival probabilities to the fifth year for seedlings of the same cohort of *Astrocaryum mexicanum* as a function of the number of leaves that presented two years after germination. Data are means ± SD. The curve was eye-fitted.

vigor (i.e., individual weight) and consequently greater probabilities of survival than late recruitments of the same cohort (but see Venable, this volume). Experimental studies in *V. sororia* (Solbrig, 1981) showed that seed size is not correlated with either speed of germination or seedling growth rate. Field and laboratory studies indicated that the former factors as well as the probability of attaining a large size (and therefore greater expectations of survival and reproduction) depend to a greater extent on environmental factors rather than on genetic factors. Germination in favored microsites (nutrient rich; low density and predator-stress) may determine greater survival and reproductive success in phenotypes expressing larger individual size.

A similar situation was found by Fowler and Antonovics (1981a) and Antonovics and Primack (1982) in their studies on populations of *Salvia lyrata* and *Plantago lanceolata*; in these studies phytometers representing different groups of half-sibs were used. In *P. lanceolata*, however, there was evidence that some genotypes do better in some sites than in others, suggesting the existence of polymorphisms maintained by microspatial heterogeneity, much in the same way as those found earlier by Turkington and Harper (1979b) for *Trifolium repens*. At this scale, intraspecific density appeared as a factor defining the differences in individual performance and survival (Fowler and Antonovics, 1981a). Mortality was concentrated in individuals with lower leaf areas as a result of high growth rates in the summer.

It becomes clear that the "sifting" effect of the soil micro-

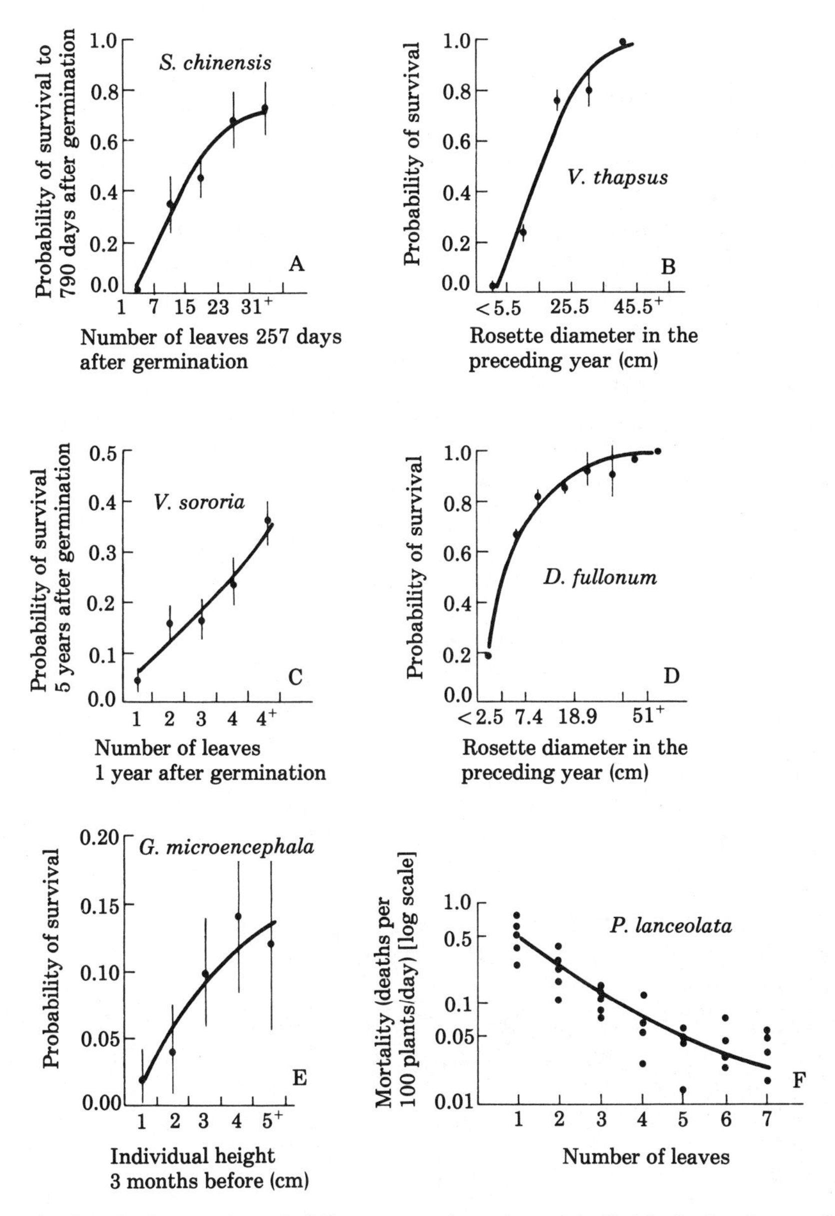

FIGURE 3. Survival probabilities as a function of individual size for seedlings. A. *Simmondsia chinensis*, a shrub of the Sonoran Desert (F. Molina and A. Castellanos, unpublished data). B. *Verbascum thapsus* (from Gross, 1981). C. *Viola sororia*, a herbaceous perennial (from Solbrig, 1981). D. *Dipsacus fullonum*, a biennial (from Werner, 1975). E. *Gutierrezia microcephala*, a desert shrub (from Parker, 1982). F. *Plantago lanceolata*, a herbaceous perennial (from Antonovics and Primack, 1982). Values for A through E are means ± SD. For F, points indicate different study sites. All curves were eye-fitted.

environment (Harper, 1977; Antonovics and Levin, 1980) acting on a genetic pool in the seed bank has a "leveling" effect. This can erase differences that could have been generated by highly different genotypes germinating in a totally homogeneous substratum. Hence, superperformers at the seedling level may well be inferior genotypes in a favorable microsite.

Limitations by physical factors (e.g., water availability) have been identified as an important mortality factor at early stages of life (cf., for example, the detailed studies of Steenbergh and Lowe, 1977, in saguaros). Parker (1982) followed the fate of cohorts of *Gutierrezia microcephala* (Compositae) in an arid grassland. Plants of less than 3 cm in shoot height did not survive to their first year whereas larger plants were strongly affected by herbivory: large investment in vegetative growth freed plants from death risks by drought but exposed them to serious damage by herbivores. In this circumstance, if herbivory acts as a constant mortality factor, size-dependent defoliation may act as a selective force for moderate shoot growth rates and for an increase of resource allocation to the root system (Parker, 1982).

Clear evidences of genotype-dependent differential mortality have been reported in *Trifolium repens* (Dirzo and Harper, 1982b) and in cultivars of *Phlox drummondii* (Bazzaz et al., 1982). A polymorphism involving cyanogenic and acyanogenic morphs within a population of *T. repens* seems to be at least partly maintained as a result of (1) increased mortality of the cyanogenic morphs in areas of low herbivore pressure where acyanogenic morphs do better and (2) greater mortality of acyanogenic morphs in areas of high herbivory stress (Dirzo and Harper, 1982a). Dirzo (Chapter 7), on the basis of the study of Cody (1966), argues that cyanogenic morphs may have less competitive ability as a cost of assigning a disproportionately higher amount of their resources to chemical defenses. Bazzaz et al. (1982) found in comparisons between cultivars and wild forms of *Phlox drummondii* that there were genotype-specific survivorship curves for both forms and that curves were determined by environmental conditions such as nutrient status and degree of competition.

Differential survivorship in adult plants is even more scantily documented than in seeds and seedlings. General mortality patterns for mature populations are known for numerous species, but the variance for each age or size class is mostly unknown. Data on age-specific mortality rates and their variability for populations of *Astrocaryum mexicanum* are shown in Table 1. The variance within each age class decreases with age, a finding suggesting, in addition to the existence of higher mortality risks, a much lower environmental predictability for seedlings and infant palms than for immature and mature stages (Piñero and Sarukhán, 1982).

Part of the interindividual variability in survival in infants and

TABLE 1. Survivorship probabilities for different individual stages in *Astrocaryum mexicanum.*[a]

Stages	Survivorship probabilities $\overline{X}$	SD
Seedlings and infants (1 to 8 years)	0.48	0.15
Juveniles (9 to 19 years)	0.86	0.09
Immatures (20 to 39 years)	0.95	0.04
Matures (≥ 40 years)	0.95	0.03

[a] Figures are data obtained from six 600-m^2 plots, during six years (1975–1981).

juveniles for a given age category is related to the size of the palm and more directly to its leaf area. For infant plants with eight leaves, the average survival probability is close to 1, whereas for those with three or four leaves, it is between 0.5 and 0.7 (Figure 4A). Equally, for juvenile palms (Figure 4B), survivorship during a four-year period is significantly greater ($D_{max} = 0.27$; $P < 0.01$; Kolmogorov-Smirnov text) for plants with four to six or more leaves, five being the most common number, and is poorer for individuals with three leaves or less; these constitute 45% of the deaths observed in this category.

The causes of mortality in adult individuals of *Astrocaryum mexicanum* are largely unknown to us, although one-third of the mature individuals (30 in total) which died after seven years were killed by direct hits by a falling branch or tree. Also, recently dead palms showed significantly ($P < 0.01$) fewer leaves ($\bar{x} = 10.20 \pm 3.30$) than living palms ($\bar{x} = 12.50 \pm 3.27$); this suggests that a depletion of resources could enhance the probability that an individual will die. Thus, for a given age-class, there is no doubt that individual vigor (in this case, leaf number) is crucial in determining mortality risks.

Differential growth and reproduction

Interactions between growth and reproduction are very complex and understood only at a rather superficial level. The mutual effects of growth and reproduction on each other have represented a fertile medium on which much of man's agronomic technology has been based. The empirical knowledge of the result of the interactions is ample, although it is only recently that systematic, experimental studies have started to disentangle growth and reproduction interactions and the effects of environmental constraints on them. For this

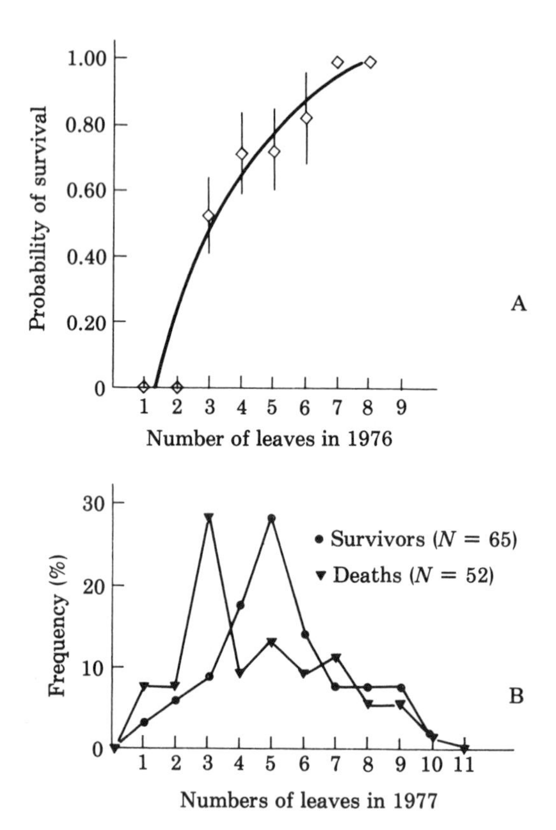

FIGURE 4. The relationship between number of leaves and survival in *Astrocaryum mexicanum*. A. Infants (1–8 years old) between 1976–1981. Means and SD are expressed. B. Juveniles (9–15 years old) between 1977 and 1980. Curves were eye-fitted.

reason, in this section we shall discuss variability in growth and reproduction simultaneously.

That individuals in even-aged populations under density stress develop size hierarchies implicitly indicates the existence of differential growth rates among them. Differential growth has been observed also in naturally occurring plant populations, but its demographic implications have been seldom documented. As a result of different growth rates, reproductive rates are necessarily affected and differentiated among individuals. Little information exists on individual reproductive variance in naturally occurring populations; moreover, the interaction of growth on reproduction and of this on further growth and survival have only fractionally been investigated in a demographic context.

Evidence for the genetic basis of differential growth rates and reproduction is slight. Burdon and Harper (1980) found that growth rate in *Trifolium repens* appeared to be yet another trait for which genetically based variation exists in naturally occurring populations; they point out that average growth rate estimates for a population obscure individual variability. Law (1979) and Law et al. (1977) found genetically based variation in the effects of reproduction on further growth, reproduction, and survival in *Poa annua*. Also, Primack and Antonovics (1981) found a genetic basis for variation in components of seed yield on *Plantago lanceolata*.

However, other causes have also been documented as determinants of individual variability in growth and reproduction. Gottlieb (1977) argues that for highly plastic annual plants such as *Stephanomeria exigua* ssp. *coronaria*, extreme variation in size and growth rates are not due to genetic differences and that these differences do not constitute the basis of evolutionary changes. Gibbs and Harrison (1976) have shown that viral diseases have an important impact on plant yield and growth form; often viral diseases are not self-evident, especially in highly variable natural populations, and their effects can be ignored or mistaken as caused by other factors.

Aker (1982), working with the desert plant *Yucca whiplei*, found a significant correlation between basal area of individual rosettes and the number of mature fruits produced by them, although other components of reproductive effort (ovules per capsule or seed weight) were not at all or only inconclusively correlated to plant size. Bentley and Whittaker (1979) and Bentley et al. (1980) reported that seed number per plant and seed weight are significantly affected by the effect of grazing on *Rumex crispus* and *R. obtusifolius* by a chrysomelid beetle. Milton et al. (1982) found that larger trees (based on diameter at breast height and crown diameter) of *Ficus yoponesis* and *F. insipida* produce larger fruit crops and at shorter intervals than smaller-sized trees and that these differences are sustained under differing environmental conditions.

Studies on the individual variability in both vegetative growth and reproduction in a demographic context have been carried out by Piñero and Sarukhán (1982) in populations of *Astrocaryum mexicanum*. Because of the strictly monopodic mode of growth of *A. mexicanum*, gains in size are achieved by virtue of new leaf production at the apical meristem, which in turn adds height to the trunk of the palm. Gains in height vary with the age of the individuals (Figure 5), an average of 1.5 to 6.5 cm per year from seedling to mature plants. This is equivalent to an average of 0.5 to 2.3 leaves per year, respectively. There are negligible differences in the mean number of leaves per individual produced once plants reach the mature stages.

A tenfold variation in the net height gains is shown by individuals

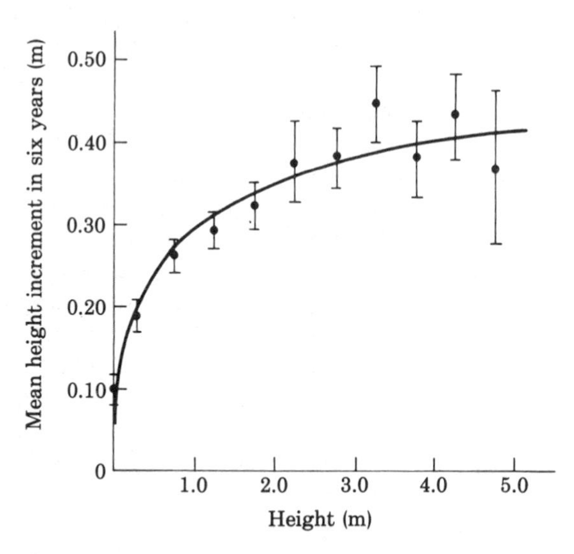

FIGURE 5. The general pattern of height gains for mature individuals of *Astrocaryum mexicanum* as a function of plant height. Data are means ± SE for each 0.5-m height class. Curve adjusted to a logarithmic model ($G = 0.30 + 0.08 \ln H$).

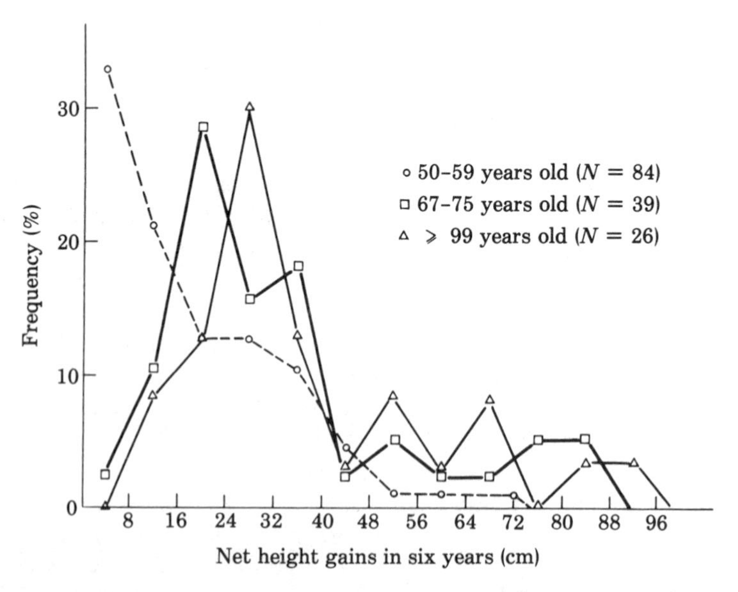

FIGURE 6. Height-growth differences within three groups of different ages of *Astrocaryum mexicanum*. Data are relative frequencies of individuals in different increment classes.

within three different age categories (matures of 1 to 1.5 m height; matures of 2 to 2.5 m; and matures taller than 4 m) during a six-year period of observations (Figure 6). Although about two-thirds of all individuals have a fairly similar (modal) net gain, some 30% in each category show gains two or three times greater than the modal gain. This difference, if maintained through long periods, would result in a continuously increasing advantage of certain individuals of older (taller) age categories over younger ones through the positioning of their crowns higher along the light gradient. The ever-changing nature of the forest canopies accounts for the great spatial dynamism that individuals experience many times during their life times.

Variability in reproduction in *A. mexicanum* arises from two sources: (1) the probability that a mature palm will reproduce; and (2) the number of fruits produced. Both vary amply with the age of the mature palms (Figure 7), with a clear tendency to increase as plants become older (or taller).

However, within each age category of mature plants (with an age interval equivalent to only 6–16% of the age class) there are markedly

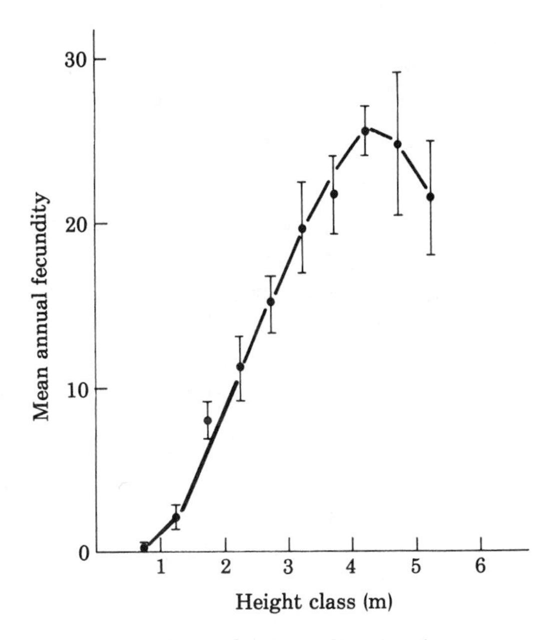

FIGURE 7. The pattern of yearly fecundity in *Astrocaryum mexicanum*. Data are means ± SE for height classes of 50 cm, based on seven years of observations. Curve was eye-fitted.

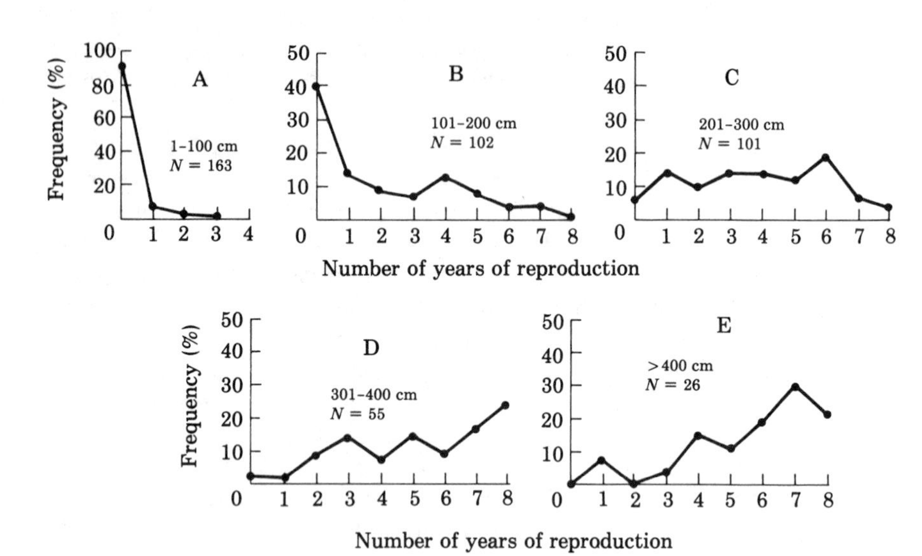

FIGURE 8. Relative frequencies of the number of reproductive years in five height classes (A–E) of *Astrocaryum mexicanum.*

different individual behaviors. Figure 8 shows the frequencies of number of years of reproduction for all individuals in different age categories for a period of eight consecutive years. Only five broader categories are illustrated for ease of presentation, although the original categories of the flux model given by Sarukhán (1978) follow similar patterns. It is evident that certain groups of individuals show higher frequencies of reproduction than others for each age class. A decreasing percentage of nonreproducing individuals occurs as the age category is greater until the last one (401 cm), in which all individuals reproduce at least once in eight years.

Individual fecundities also vary from 10-fold to almost 25-fold for palms of the same age category (Figure 9). Variability is much higher in older than in younger age classes. If we consider the overall average fecundity per individual from all sites and years of observation (31 fruits per palm) as a standard, then 83% of the individuals between 50 and 59 years of age will bear less than the average; this figure is 59% for individuals of 67 to 75 years and only 43% for those between 99 and 106 years old. What these figures may reflect is the fact that at the lower levels of the forest canopy, where younger individuals have their leaf crowns, a majority of the palms are growing in environmental spots that allow only below-average fecundities, whereas more than one-half of the individuals between 99 and 106 years of age experience environments allowing above-average fecundities.

The rate at which an individual palm grows seems to affect in a dif-

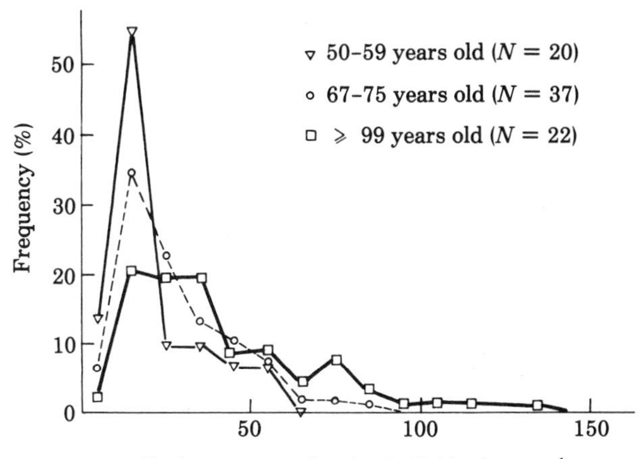

FIGURE 9. Relative frequencies of the annual production of fruits by reproductive individuals in three age classes of *Astrocaryum mexicanum.*

ferential way its fecundity. Figure 10 shows the correlation between the individual height and the number of infrutescences produced in seven years for three groups of palms: slow growers (3–20 cm height increase in six years), moderate growers (21–40 cm), and fast growers (more than 40 cm). In addition, recall that taller (older) palms reproduce more actively than shorter (younger) palms and that number of fruits per palm is more variable in older than younger individuals. The significant differences ($P < 0.05$) between these groups occur only for individuals between 1 and 2 m in height (the first two height categories in Figure 10). At least for younger, suppressed individuals, there may be a trade-off in vegetative versus reproductive investment. When the annual leaf production per individual is compared to their probability of reproducing (Figure 11), it appears that years of high leaf production are followed, in general, by years in which a relative decrease in reproduction occurred, whereas years when high probabilities of reproduction were attained were preceded by years of low leaf production.

A measure of plant vigor (or age, if age is closely correlated to vigor) is clearly a better predictor of an individual's vegetative and reproductive performance. Obviously, with this measure of vigor, a specific relation of the individual with its physical and biotic environment is normally implicated.

In an inductive way, the vegetative, and especially the reproduc-

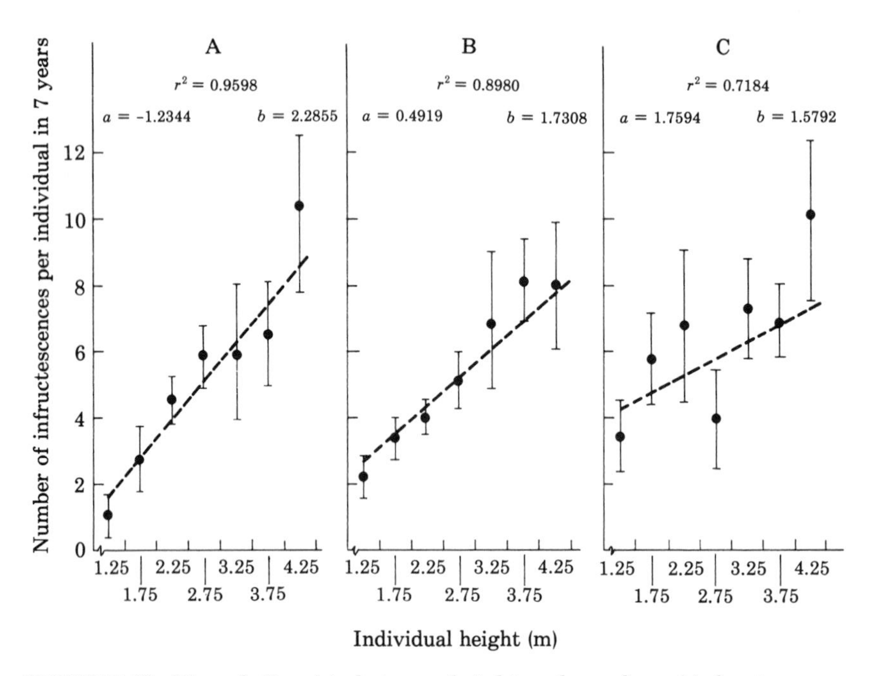

FIGURE 10. The relationship between height and number of infructescences produced per individual in seven years in *Astrocaryum mexicanum*. A. Individuals with low growth rate (< 20 cm of net height gain in six years). B. Medium-growth-rate individuals (21 to 40 cm). C. High-growth-rate individuals (> 40 cm). Data are means ± SE for eight classes of 50 cm each. The parameters of linear regressions are given as intercept (a) and slope (b).

tive behavior, may suggest aspects of the structure of the environment that may be relevant in determining such behaviors. The repetitivity with which an individual reproduces from one year to the next may be a good correlate of favorable environmental conditions. An analysis of reproductive frequencies for individuals of *A. mexicanum* of different size classes or ages and of different community stages suggests some of the ways in which favorable environments are distributed both within different community stages (Figure 12A) and in a vertical gradient for plants of different age (or size) (Figure 12B). The distribution of individuals among the different reproductive frequencies in the stable site is uniform ($\chi^2 = 7.3$, $P > 0.2$) while in the nine-year-old forest gap there is a J-shaped distribution of frequencies, with the most repetitively reproducing trees being more abundant. Analyzing how reproductive palms of different ages (heights) perceive the environment, it becomes clear (Figure 12B) that younger individuals, placed lower down in the vertical gradient, show a distribution of frequencies skewed toward the low extreme, whereas the taller palms

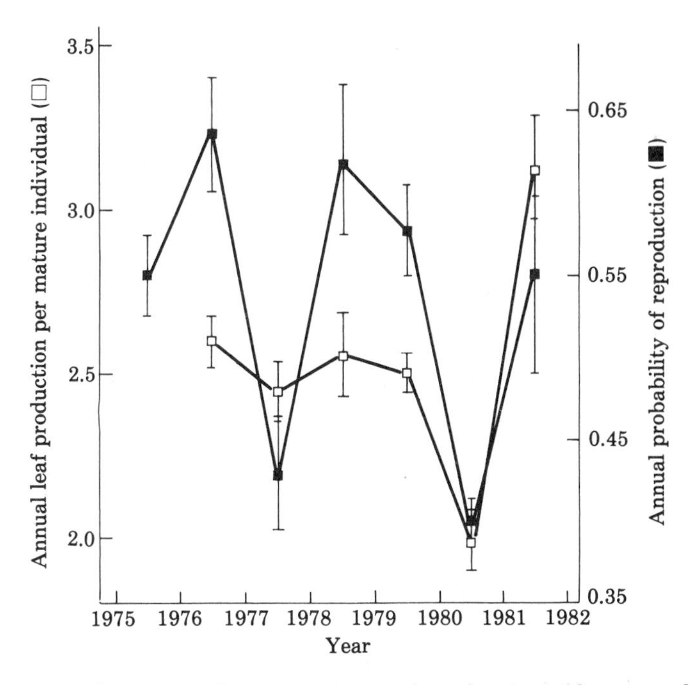

FIGURE 11. The temporal pattern of annual production of leaves and the annual probability of reproduction in mature plants of *Astrocaryum mexicanum*. Data are means ± SE.

show also a J-shaped distribution of frequencies, similar to that of the young-gap situation (Piñero and Sarukhán, 1982).

The former real distributions of the frequency with which reproductive individuals of *A. mexicanum* bear fruit suggest different model situations like those depicted in Figure 13. If environmental patches intermediate from the viewpoint of reproduction were more frequent, a normal distribution of reproductive frequency would be expected. This is not the case either for all individuals in stable or gap sites or for palms of different sizes and hence positions in the vertical gradient of the forest. A lower or higher proportion of favorable patches is therefore likely to be occurring. A high frequency of favorable patches would generate reproductive frequency distributions like that of Model 3, which would correspond to that observed for the taller palms. If the favorable patches are infrequent, a skewed distribution like that of Model 1 would be expected, with the distribution corresponding to that of the younger group of reproductive individuals (1–2 m tall).

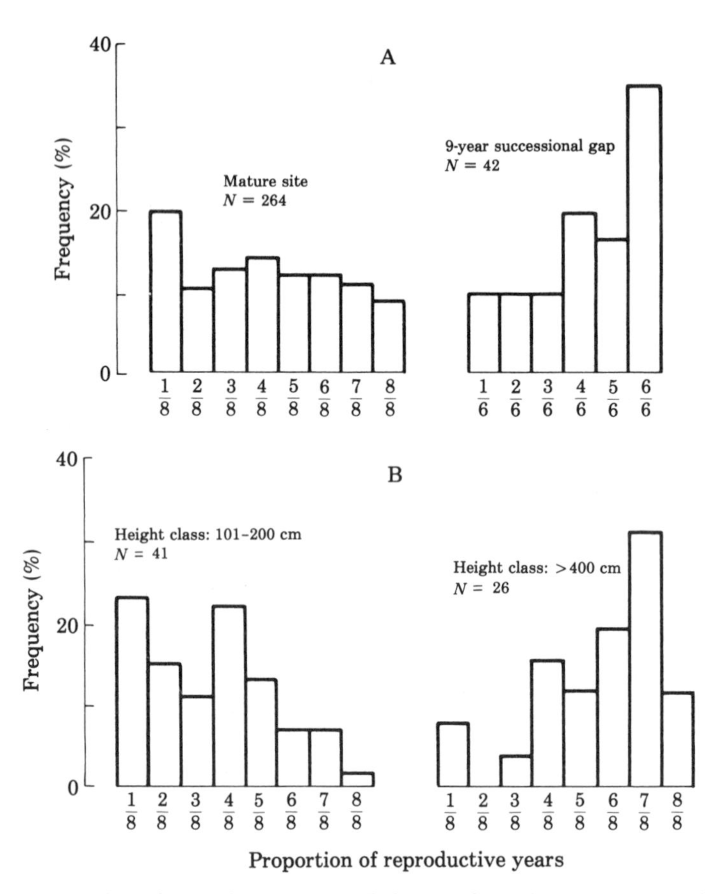

FIGURE 12. The relative frequencies of the number of years of reproduction of individuals of *Astrocaryum mexicanum* growing in (A) mature sites and a forest-gap, and in (B) two different height levels of the mature sites.

Finally, if the proportion of favorable, intermediate, and unfavorable patches is the same, then a distribution like that of Model 4 would be attained. This distribution occurs for all reproductive palms in the mature sites (Figure 12A), but it results from the combination of Models 1 and 3 for younger and older palms, respectively. It is difficult to think of one or a combination of several environmental factors that would have such even spatial distribution in a highly complex community like the tropical rain forest.

Of those physical environmental factors more likely to affect probabilities of reproduction of *A. mexicanum* from one year to the next, soil and light appear the most plausible. Because most soil characteristics, including nutrient levels, are not likely to change drastically from one year to the next, but particularly because of the

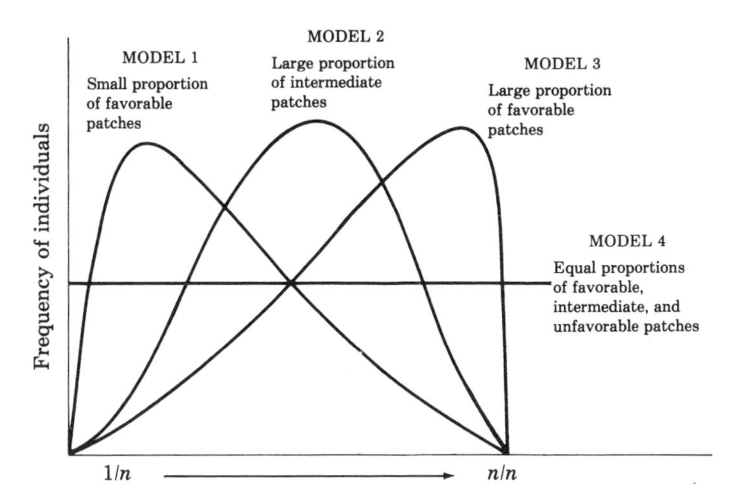

FIGURE 13. A hypothetical model of the reproductive behavior of *Astrocaryum mexicanum* in relation to the proportion of favorable environmental patches. The reproductive behavior is taken as the proportion of n reproductive events for a given plant. The assumptions of this model are that (1) the reproductive individuals are regularly distributed in the area, (2) the age structure of reproductive palms is uniform, and (3) the probability of reproduction is almost constant for the ages considered. For *Astrocaryum mexicanum* we have evidence to support all these assumptions.

different fruiting frequencies of young versus old palms within one piece of forest, light appears as a more likely factor influencing the probabilities of a palm reproducing in a given year. It also would adjust better to the assumptions of patchiness made for the models in Figure 13. Forest gaps with large openings would represent those large, aggregated patches of "favorable" environment for reproduction. This would be similar for those older palms which place their crowns in a higher place in the light-exhaustion gradient within the forest. (Imagine light as a stream of resource being sifted successively—in quantity and quality—by intervening tree canopies as it penetrates the forest from the top, until a few, separated trickles reach the forest floor.)

Individual variability in growth and reproduction in a demographic context has also been studied by Kohyama (1981) in *Abies veichtii*, Solbrig (1981) in *Viola* spp.; Bullock (1982) in *Compsoneura sprucei* and other neotropical trees, and Peters (1983) in *Brosimum alicastrum*.

Studying *Abies* forests, Kohyama (1981) found considerable individual variability in age at first reproduction and in reproductive patterns. His data show that age at first reproduction may be related to tree size or age (both well correlated). Also Solbrig (1981) has shown that individual fecundity is strongly correlated to plant size which, in turn, is substantially affected by competition and the physical environment. Not only size, but also sex in dioecious plants may affect individual growth rates and reproduction. Bullock (1982), studying populations of *Compsoneura sprucei*, a neotropical tree, found that individual size was better correlated with fecundity in males than in females.

In another tropical gynodioecious tree, *Brosimum alicastrum*, Peters (1983) found also that, in addition to size and age, sex influenced individual growth rate; taller trees flower earlier and produce more fruits than shorter individuals. However, no differences were found in growth rates as a result of sex differences in *Aralia nudicaulis*, a temperate herbaceous perennial (Bawa et al., 1982), whereas flowering frequency was affected by sex differences.

Lloyd and Webb (1977) proposed that these sex-induced differences reflect a cost which is associated with reproduction, so that a high energetic investment by females leads to a higher mortality risk and/or to a lower growth rate relative to males. Following these lines of reasoning, Bawa et al. (1982) think that the cost in reproduction may also be reflected in a more prolific flowering by males.

In studies of the population biology and population genetics of plants, it has been customary to try and establish the extent of the genetic determination of components of fitness, particularly those related to reproductive behavior (e.g., Primack and Antonovics, 1982). The converse (i.e., the effects of reproductive behavior on the genetic structure of the population) has been a lesser used approach. Bullock (1982) found for four dioecious tropical arboreal species a high dominance of the reproductive output for very few individuals, that is, only one tree accounted for more than 20% of all flowers produced. Because of the length of the study, it was not known for how long reproductively dominant trees would remain so. Analyzing data for *A. mexicanum*, we found that the same 58 individuals account for over 43% of the ca. 33,000 fruits produced during seven years of observation; the remaining 56% of the fruits were contributed by constantly different mixtures of some 130 other individuals every year. We have no data at the moment to establish any genetic basis for the continued reproductive behavior of these individuals. Also, we have reported elsewhere (Piñero and Sarukhán, 1982) that overreproductive palms are associated with overreproductive neighbors, that is, there seems to be a spatial factor that would clump highly reproductive individuals together. But it is clear that if such "reproductive dominance" by the same individuals lasts for longer than the seven-year period we have observed, there might be an important effect on the genetic structure of the population. The highly dynamic nature of the forest canopy suggests that such dominance by the same few individuals may not be very long-lasting and that other individuals may take over as frequent reproducers.

GENERAL DISCUSSION

It is evident in the literature reviewed in this chapter and in other contributions to this book (e.g., Venable, Bradshaw, Schaal, Levin) that the ability to attribute individual variation in plant populations to genetic or environmental factors is still very limited. The discernible patterns, if any, indicate a greater environmental than genetic influence on the individual variability of demographic parameters. We believe that this difficulty arises partly from the highly dynamic nature of the environment (both physical and biotic) acting on a population on a life-time span, and partly from the dynamic response of plants (plasticity), which themselves are changing in size (height, crown size, root volume, etc.) and therefore modifying their response to that part of the environmental spectrum that they face at any given moment of their lives (see Jefferies, Chapter 17). The poorly known interrelationships between vegetative growth and reproduction and their demographic consequences in naturally occurring populations add to the difficulties of defining the genetic determination of individual variance.

The "genetic dominance" by a few individuals found in several plant populations provides an interesting way to explore the genetic determination of individual traits relevant to survival and reproduction, although here, too, environmental factors may obscure the situation by means of individual plasticity and the dynamic nature of such factors.

More rigorous studies of what constitutes the relevant environment for individuals of a population are needed to understand several key factors underlying individual variation in growth and reproduction. This is particularly complicated when a single arboreal individual in a forest may "sample" in its lifetime, not only differences in competitive interactions with its neighbors and the temporal fluctuations of the physical environment, but also different parts of it with the passage from seed to seedling, sapling, treelet, etc. Very probably not all "environmental stages" are equally relevant demographically speaking, but certainly it is necessary to determine how each of them shapes the final contribution of the individual to the dynamics of its population.

Much of the plant physiological literature available is concerned with phenomena observed either at the suborganismic level or in conditions with little direct relevance to that of a plant amidst its neighbors in the field. The exciting developments of new field methodologies enabled by the technological revolution in electronics makes it realistic, for the first time, to call for observational and experimental work with whole plants (see Mooney and Chiariello, Chapter 15) living under natural field conditions.

The real influence of the environment surrounding an individual will only be adequately understood when the demographic consequences of different physiological traits on survival, growth, and reproduction are known. This could only be done on a whole-plant level, under natural field conditions, and within a populational context.

On the other hand, a plethora of population genetics studies has mostly been concerned with the genetic characterization of populations. However, the very demographic expression of such changes in fitness (i.e., survival, growth, and reproductive rates) are seldom incorporated in those studies.

The study of an individual's variability in time and relative to that of its neighbors appears, to our view, as the source for understanding population-level dynamics. It is our contention that the studies of individual variability in demographic parameters within a demographic context provides not only a very sound basis for understanding patterns of population behavior, but also is the crossroads where plant population genetics, plant physiology, and plant demography may interact in the most fertile way.

CHAPTER 5

LOCAL-SCALE DIFFERENTIATION AS A RESULT OF COMPETITIVE INTERACTIONS

Roy Turkington and Lonnie W. Aarssen

INTRODUCTION

Population ecology is to a large extent a study of natural selection. Genetic variation within a population necessarily dictates that some individuals will leave more descendants than others. Consequently, all levels of biological organization are affected, and the relative frequencies of their components are in a state of continuous flux. Experimental ecology investigates the resultant patterns and the mechanisms and processes by which these patterns are generated. The scope of this chapter is defined by its title: the pattern we seek to describe is local-scale population differentiation and one of the mechanisms by which it is generated—competition. Topics of related interest are treated elsewhere in this volume (see Antonovics, Bradshaw, Jain), and also by Hamrick (1982).

The description of intraspecific adaptive differentiation in plants is not new, and various aspects of the literature have been reviewed by McMillan (1960), Bennett (1964), Heslop-Harrison (1964), Langlet (1971), and Hamrick (1982). Langlet (1971) cites 111 references to various aspects of intraspecific differentiation which are pre-1900, six of which are from the 1700s. For example, in 1756 the Count de Buffon wrote (in French), "and if each type is further examined in different regions, varieties will be found responding in terms of size and form; to varying degrees all take on a colour from the region" (in Langlet, 1971).

In many cases where patterns of intraspecific differentiation are described, the mechanisms generating the differences are either ignored or only referred to in a speculative manner and often the differences occur over large distances. For example, the classic studies by the Carnegie group (e.g., Clausen et al., 1940, 1948; Clausen and Hiesey 1958a) on *Achillea millefolium* and *Potentilla glandulosa* described differences along a transect from coastal California to approximately 3350 meters in the Sierra Nevada. Numerous examples of differentiation on a smaller scale are reviewed by Heslop-Harrison (1964), but it was not until the late 1950s and early 1960s that attention became more focused on differentiation over distances measured in meters or centimeters (largely as a result of the work of Bradshaw and his coworkers (e.g., Bradshaw, 1959, 1960, 1972; Snaydon, 1963; Jain and Bradshaw, 1966) and emphasis shifted to intrapopulation differentiation. Obviously, in a chapter considering differentiation as a result of competition, we are necessarily restricted to differentiation that has occurred among individuals living in close proximity, but not necessarily among individuals of only one species.

Many species of plants occur over broad ecological and geographical ranges. In such species we might expect a complex of variant ecotypes, each uniquely adapted to a local microenvironment. In contrast, wide distributions may be maintained by species consisting of individuals having wide tolerance ranges and capable of surviving in diverse environments (see Baker, 1965). When competition is invoked as a mechanism by which local differentiation arises, implicitly we are referring to the specialist ecotypes that have evolved in response to microenvironmental heterogeneity so common in nature, rather than to the "general purpose genotype" (Baker, 1974).

Competition and its effects have occupied the minds of ecologists ever since Darwin (see reviews by Harper, 1961, 1964; Donald, 1963; R. S. Miller, 1967; Risser, 1969; McIntosh, 1970; Ayala, 1970, 1972; Arthur, 1982). Darwin (1859) made the statement "we have reason to believe that species in a state of nature are limited in their ranges by the competition of other organic beings." Nevertheless, in spite of Darwin's emphasis, studies of intraspecific or intrapopulation differentiation in plants have bee dominated by what Harper (1977) has called "Wallacian" forces, or abiotic factors of the environment. These include soil moisture (Hamrick and Allard, 1972; Linhart and Baker, 1973), serpentine soils (Kruckeberg, 1951), the degree of wind exposure

(Aston and Bradshaw, 1966), heavy metal contaminated soils (Jain and Bradshaw, 1966; Antonivics et al., 1971; Wu and Antonovics, 1976), fertilizers (Snaydon, 1970; Snaydon and Davies, 1982), date of snowmelt (Waser et al, 1982), proportion of sandstone in parent material (Morrison and Myerscough, 1982), and temperature and irradiance (see Teramura and Strain, 1979). Other studies have shown population differentiation in response to grazing (Kemp, 1937), topography (Schaal, 1975; Hamrick, 1979), disturbance (Solbrig and Simpson, 1974), stage of succession in old fields (Hancock and Wilson, 1976), and trampling (Warwick, 1980). The importance of adaptations to physical factors is unquestionable. It is surprising, however, that plant ecologists still tend to emphasize physical environmental factors as selective forces since Darwin's philosophy, which has been central to the development of evolutionary theory, emphasized the role of interactions between species. Indeed, Darwin (1859) said that there is a "deeply-seated error of considering the physical conditions of a country as the most important for its inhabitants; whereas it cannot, I think, be disputed that the nature of the other inhabitants, with which each has to compete, is generally a far more important element of success." These are what Harper (1977) has called "Darwinian" forces of selection. Burdon (1980) has argued that the tremendously variable nature of the local biotic environment may act to produce and maintain a high level of variation in plant populations [see parallel arguments by Glesener and Tilman (1978) for terrestrial animals]. Even though plant-plant interactions might be expected to produce the necessary conditions which would result in local-scale population differentiation, there have been remarkably few attempts to search for and document such effects.

The scope of this chapter is therefore to consider the role of plant-plant interactions in producing specialized ecotypes among individuals living in close proximity. In natural communities it would be impossible, indeed meaningless, to divorce the interaction of biotic and abiotic forces of selection, but in this chapter emphasis will be focused on studies where biotic forces are paramount.

PATTERNS OF LOCAL POPULATION DIFFERENTIATION

In 1972 Bradshaw wrote about the evolutionary consequences of being a plant and stated that "we can deduce that it is selection and particularly the pattern of selection which determines the patterns of differentiation in plant populations. If the environment varies then so will the plant population." Bradshaw then proceeds to document the literature on plant population differentiation in response to changing environmental (mostly edaphic) conditions (see also Snaydon, 1970). Nature seems to have imposed few restraints in equipping plant species to differentially respond to the local heterogeneities of the environment, and it appears that polymorphisms are the rule rather than the exception. In an attempt to categorize these patterns, most of which may be loosely called polymorphisms, there will, no doubt, be a certain degree of arbitrariness.

Morphology

Schaal and Levin (1978) demonstrated character divergence within subpopulations of *Liatris cylindracea* growing only ca. 10 m apart. B. Trenbath (unpublished data) sampled seeds of *Avena fatua* from areas which had been under different crop rotations for 50 years. Individual seeds of *A. fatua* from continuous wheat plots average 70% heavier than those from a wheat–pea rotation, but about 25% fewer are produced; the proportion of all seeds that are viable is nearly three times greater in the continuous wheat population. McNeilly (1981) grew seed-derived populations of *Poa annua* (from three open and three closed habitats) with *Lolium perenne* and demonstrated differentiation in plant size over short distances. Watson (1969) collected plants and seeds of *Potentilla erecta* from an *Agrostis-Festuca* grassland and from an immediately adjacent *Molinia caerulea* grassland. The different environments selected different growth forms from this population of *P. erecta* within a distance of 2.7 m. Other examples of neighbor-correlated differences within a single population are provided for *Trifolium repens* (Turkington and Harper, 1979a) and *Medicago sativa* (Turkington, 1979).

Physiology

This is perhaps the most difficult area to investigate because adaptive characters in plants are frequently physiological and require special techniques. As early as 1887, Cieslar (in Langlet, 1971:677) had established the existence of physiological varieties within species of trees, chiefly Norway spruce, larch, and oak, from different altitudes in the Tyrol. These differences, however, occurred over large distances.

Shaver et al. (1979) documented quantitative differences in growth and phosphate-uptake kinetics within a population of *Carex aquatilis* from a variety of ice-wedge polygon microhabitats separated from one another by only a few decimeters. Similar patterns have been observed by Teeri (1972) in *Saxifraga oppositifolia* between adjacent beach ridge and meadow sites in a polar desert. Individuals of *Plantago lanceolata* from adjacent sunflecked and open sites separated by less than 8 m show substantial differences in the photosynthetic response to temperature and irradiance (Teramura and Strain, 1979). Linhart and Baker (1973) demonstrated intrapopulation differentiation of physiological response to flooding in a population of *Veronica peregrina*. Davies and Snaydon (1973a,b,c) detected physiological differences among populations of *Anthoxanthum odoratum* in response to calcium, aluminum, and phosphate.

In a stimulating article, Angevine and Chabot (1979) draw attention to the "biotic stresses" faced by seeds and seedlings, and they argue that a variety of "germination syndromes" have evolved in response to these biotic stresses. In this context, Westoby (1981) argues that diversified seed germination behavior should depend on dispersal and on the spatial scale of competition.

Finally, Burdon and Harper (1980) showed a continuous range of relative growth rates (RGR) within a single population of *T. repens*. Burdon and Harper argue that biotic forces of selection which dominate local evolution may explain these within-population differences in RGR.

Demography

Cook (1979b) synthesized the present status of theory and research on patterns of juvenile mortality and recruitment in plants. He focuses attention on the need to study the demographic behavior of different genotypes within populations of a species undergoing divergent selection. Unfortunately, such studies are rare. McNeilly and Antonovics (1968) demonstrated displacement with respect to flowering time in populations of both *Agrostis tenuis* and *Anthoxanthum odoratum*. Law (1975) and McNeilly (1981) have shown that populations of *Poa annua* from open habitats suffer greater mortality as a consequence of density-dependent stress than populations from pasture and lawn habitats. Law et al. (1977) have shown that populations of *P. annua* from pasture and old lawn situations have evolved a longer life cycle and may behave as perennials.

Fowler and Antonovics (1981b) transplanted individuals of *Plantago lanceolata* and *Salvia lyrata* into plots of otherwise unmanipulated vegetation which were only meters apart. Large differences in the means of several demographic parameters, such as survivorship, mortality rates, and leaf and inflorescence numbers, among the plots were detected and these differences were related to spatial variations in density. Grant and Antonovics (1978) and Keddy (1980) have also shown a correlation between demographic parameters and the density of neighbors.

Population structure

Solbrig and Simpson (1974) investigated three populations of *Taraxacum officinale* which were dominated by four biotypes, two of them accounting for 70% of all plants. The relative proportions of the biotypes varied between populations, and they had different life-history strategies. Solbrig and Simpson attributed the observed polymorphism to differences in disturbance between the three sites and concluded that intraspecific competition must have been an important aspect of the dynamics of the population.

Badger disturbances on tall-grass prairies constitute a limiting resource for a guild of fugitive plants. Platt and Weis (1977) showed that the composition of the fugitive species guild varies locally and that the types of competitive interactions occurring among individuals influences guild composition.

Temporal polymorphism—Cytotypes and biotypes

Lewis (1976) conducted a cytoecological study of a single population of *Claytonia virginica* which has diploid, triploid, and tetraploid races. Sequential population sampling showed that the relative frequency of ploidy levels varied over a six-week period, and Lewis argued "that each ploidy level has a unique phenological niche that lowers competition among plants of differing ploidy levels and thereby permits cytotypes to coexist."

On a longer time scale, Hancock and Wilson (1976) monitored biotype frequencies of *Erigeron annus* from adjacent one-, two-, and four-year-old old fields; they documented age-related differences which depend on the successional environment. It appears that plant succession involves intraspecies succession as well as a change in the dominant species, and temporal variation may be as important as spatial heterogeneity.

Competitive ability

One would intuitively expect that since so many ecologists covertly believe that competitive interactions between organisms shape the world, prodigious efforts would have been spent in the search for changes in competitive ability as a result of competition. Certain aspects of this topic have been much debated by zoologists (Pimentel 1961, 1968; Pimentel et al., 1963, 1965; Leon, 1974; Gill, 1974, Lawlor and Maynard Smith, 1976). Botanists have been strangely silent and rather than investigating changes in competitive ability per se have instead been more concerned with winners and losers in competition, and r- and K-selection in relation to competitive ability. Perhaps this apparent lack of effort is a legacy passed down over this past 50 years from the classical theory of community dynamics set forth by Lotka and Volterra. This theory and most of its modern outgrowths have failed to take into account the genetic heterogeneity of the interacting populations (Leon, 1974).

Lovett Doust (1981a) collected individuals of *Ranunculus repens* from a woodland and an adjacent grassland site where both populations were assumed to have originated from the same population about 12 years before. The woodland clones were apparently more tolerant of intraclonal competition than grassland clones. Lovett Doust argued that the lower tolerance of intraclonal interference in grassland clones may be the result of selection pressures which have favored the ability to compete with other species, notably grasses, rather than the ability to tolerate intraclonal or intraspecific competition.

McNeilly (1981) showed that *Poa annua* exhibits a spectrum of behavior with *L. perenne* that is consistent with the view that ecotypic differentiation has occurred with respect to competitive ability. Similar evidence for genetically based differences in competitive ability have been presented by Law et al. (1977) and Warwick and Briggs (1978b) for *P. annua*, Hill and Shimamoto (1973) for *L. perenne*, and Bogaert (1974) for *Festuca pratensis*.

From within a 1-ha pasture in North Wales, Turkington and Harper (1979b) sampled *T. repens* from within clumps dominated by four different grasses. The study clearly showed genetic variation among clones in their performance with different species of grass in the sward. Turkington (1979) traced changes in fine-scale specialization in response to neighbors in *T. repens* and *M. sativa*.

The *Drosophila* of plant biology: *Trifolium repens*

Harper (1977:708) documents the various characters for which *T. repens* is polymorphic (see also Burdon, 1983; Turkington and Burdon, 1983). Its biology makes *T. repens* an eminently appropriate species for the types of study described in this chapter. Burdon (1980) investigated the intraspecific diversity of a population of *T. repens* from a 1-ha permanent pasture in North Wales. Fifty plants were assessed for a wide range of characters which had been shown previously to be of importance in determining the survival of *T. repens* or of other plant species. Correlation analyses between individual characters showed that the majority of characters assessed appear to be independent of one another. More remarkable, the vast majority of the 50 plants assessed differ from one another with respect to at least one character, and usually several characters; the average clone differs significantly in 3.3 characters. Unfortunately, Burdon was not able to establish beyond doubt the genetic origin or maintenance of this overwhelming variation. He discounted selection and evolution on a microscale as attributable to physical factors because previous studies (Turkington and Harper, 1979a) had shown the physical environment of the field to be remarkably uniform. Instead, Burdon argued that the local component of environmental variation is determined almost exclusively by biotic factors such as the identity, age, and size of neighbors and the degree to which they compete. "In a permanent pasture containing a large number of different plant species, the local environment clearly may change radically not only from place to place in the field but also from time to time. The tremendous temporal and spatial variation in the *nature and intensity of competitive interactions* [emphasis added] which results from this patchiness represents a complex array of local directional selective forces which, in an otherwise uniform environment, are likely to dominate the immediate fitness of individuals and *produce evolution on a micro-scale* [emphasis added]. . . . The tremendously variable nature of the local biotic environment may thus act to produce and maintain a high level of variation in the plant population" (Burdon, 1980).

Meanwhile, L. W. Aarssen and R. Turkington (unpublished data) were conducting essentially the same type of investigation, but were measuring 15 characters of 100 clones, of each of *T. repens*, *L. perenne*, and *Holcus lanatus*, from each of four adjacent pastures (2, 21, 40, and 65 years old). In general, their studies confirmed the overwhelming amount of variation for *T. repens* in the oldest pasture but could not concur with Burdon's (1980) arguments concerning the origin and maintenance of this variation. Rather, their results showed that 11 characters measured in the four populations of *T. repens* showed a significant ($P < 0.05$) and progressive decline in variation with increasing age of origin of the pasture. Also, preliminary data suggest that *L. perenne* is following a trend similar to that of clover.

EVOLUTION OF COMPETITIVE RELATIONSHIPS

We will not dwell on aspects of a topic that has been reviewed by Antonovics (1978). Nevertheless, the question must be asked: In what ways do plant–plant interactions influence intrapopulation differentiation with regard to competitive ability? That is, a study of evolutionary changes in plant–plant competitive relationships is needed.

There are a number of different evolutionary solutions to the problem of a struggle for limited resources. A large number of the experimental studies have been done with animals, particularly *Drosophila* and various other flies (Pimentel et al., 1965; Seaton and Antonovics, 1967; Pimentel, 1968) and unfortunately the results are not always applicable to plant populations. Not surprisingly, most studies have been done with laboratory populations; in natural populations, spatial and temporal heterogeneity at even a local scale imposes a complexity of interaction that is difficult to disentangle.

Selection resulting in niche differentiation

Obvious responses to the struggle for limited resources are migration or death. The most noteworthy feature of studies on changes in plant–plant interactions as a result of competition is the variability of the results: Many fit the expectation of improved performance through divergence. Just as frequently, competitive performance declines or remains unchanged (see Antonovics, 1978). While these responses are interesting, they will not be considered further. The conclusion of most theoretical studies (e.g., MacArthur and Levins, 1964, 1967; Lawlor and Maynard Smith, 1976) is that populations will tend to diverge in their ecological requirements so that they increasingly use different resources and so reduce or avoid competition (see Antonovics, 1976b). In plant studies this has been termed selection for ecological combining ability (Harper, 1964).

The central feature of this mechanism is that competitive interaction is reduced or virtually eliminated as a consequence of selection for niche differentiation. If competition is avoided, more efficient exploitation of the environment may result and two plant species will have a higher total yield in combination after selection than before selection.

In recent ecological literature, this theme has become nearly axiomatic (see Cody, 1974a; Grubb, 1977; Diamond, 1978; van den Bergh and Braakhekke, 1978) even though most evidence is circumstantial with little empirical support (Wiens, 1977; Connell, 1980).

Clear evidence in support of this hypothesis in plant populations is provided by the work of Allard and Adams (1969a) with barley. They developed a highly heterogeneous barley population over 18 generations by crossing, in all possible combinations, 31 different varieties. Eight families were selected, each from a single random plant taken from this heterogeneous population. Individuals sampled from each of the eight families were grown in pure and in mixed stands. Of the various combinations of genotypes, 40% showed significant increases in yield when grown in mixtures versus when grown in pure stands. This is a predictable outcome since the genotypes used had a history of as many as 18 generations of selection in a highly heterogeneous population of genotypes. A parallel outcome was demonstrated by Joy and Laitinen (1980) with *Trifolium pratense* and *Phleum pratense*, and Remison and Snaydon (in Snaydon, 1978b) in populations of *A. odoratum, Dactylis glomerata, H. lanatus*, and *L. perenne*. Martin and Harding (1981) showed that the total reproductive rates of sympatric mixtures of *Erodium obtusiplicatum* and *E. cicutarium* were higher than those of allopatric mixtures.

There have apparently been no previous attempts to trace the *changing* competitive relationships between two species in one particular community. Aarssen (1983a) addressed this question using a series of three contiguous but different-aged pastures (2, 21, and 40 years since sowing), each of which was initially sown with the same species composition. In an area of each field where both *H. lanatus* and *T. repens* were abundant, a pair of individuals were collected as immediately adjacent neighbors. After multiple clonings, the species were grown together in a de Wit (1960) replacement series. *Holcus lanatus* and *T. repens* showed a trend of decreasing regression slope in the ratio diagram as well as a shift in the position of the regression (Figure 1A). An unstable behavior (slope greater than 1.0) is illustrated for the two-year-old genet pair and this shifts to a stable behavior (slope less than 1.0) in the pairs from the two older pastures. This is interpreted as niche differentiation.

Aarssen (1983a) collected all possible two-species pairs of natural neighbors (of *D. glomerata, H. lanatus, L. perenne, Poa compressa*, and *T. repens*) from each of the same three pastures. After cloning, 25 cuttings from each clone were planted in mixture with 25 cuttings of its natural neighbor. Successive harvests were carried out for one year and yearly cumulative total dry weights measured. To reflect the reciprocity of competitive effects between neighbors, a component yield ratio (CYR), Y/Y', was calculated where Y and Y' are the total yearly cumulative dry weights of the lower- and higher-yielding components of a mixture, respectively. CYR values are strictly relative; values closer to 1.0 indicate that the two components are more balanced in their yields in mixtures than are combinations having a value closer to zero. Changes in the nature of reciprocal competitive effects for a given neighboring species pair are encompassed by two parameters: (1) changes in total yield of the combination, and (2) changes in the relative contribution to total yield from the two components as reflected by the CYR. In the *D. glomerata* and *H. lanatus* combination, the total yield of the mixture increased with increasing pasture age and the CYR also increased (Figure 2A). These results support the hypothesis of niche differentiation in response to competition.

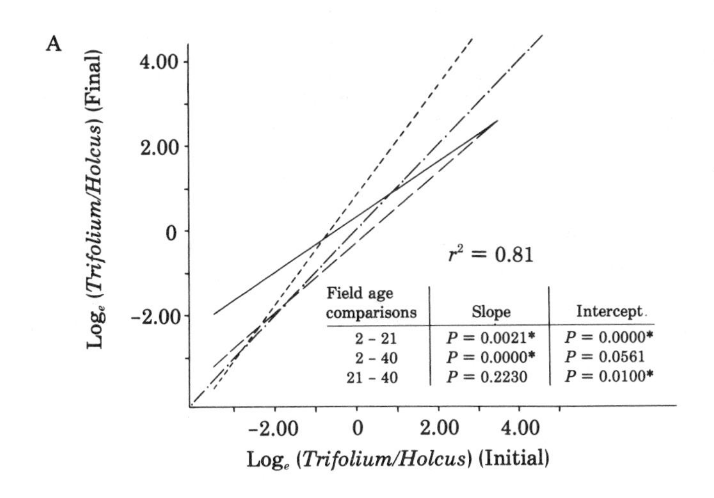

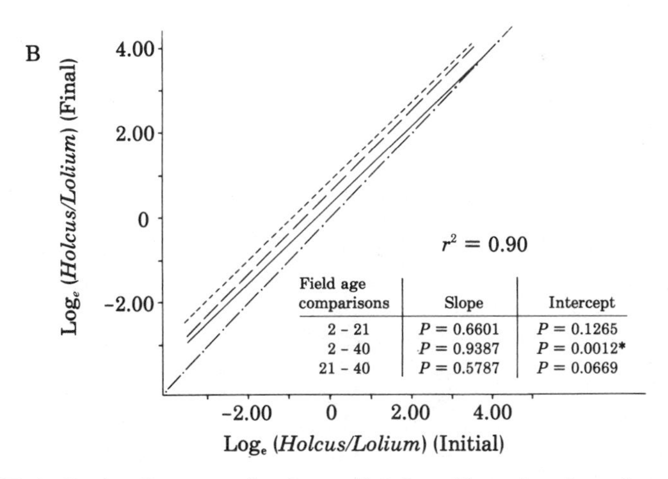

FIGURE 1. Ratio diagrams for immediately adjacent pairs of genets. A. *Holcus lanatus* and *Trifolium repens*. B. *Holcus lanatus* and *Lolium perenne*. *F* tests for the homogeneity of variances after log-transformation were not rejected. Probability levels are from an analysis of variance for significant differences in slopes and intercepts of the regressions (* significant at $P < 0.05$) for all possible pairwise comparisons of field age. A coefficient of determination is given for each multiple linear regression. Pasture age: ---, 2 years; ———, 21 years; ———, 40 years.

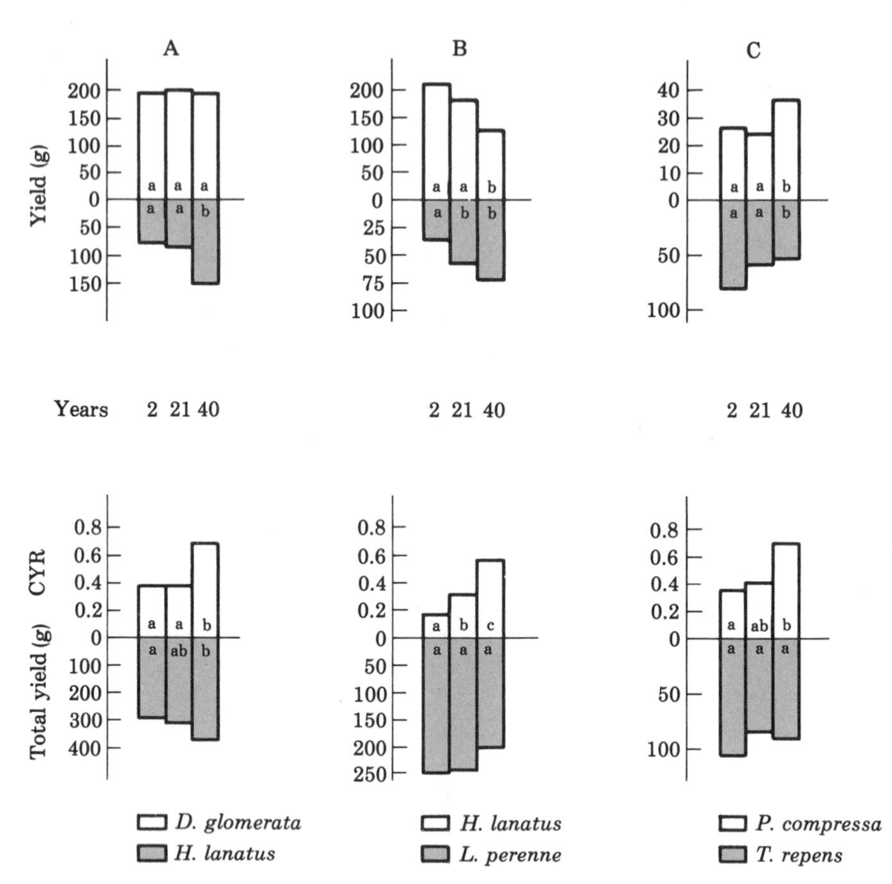

FIGURE 2. Comparisons of individual component yields, total yields, and component yield ratios (CYR) for each genet pair type. A. *Dactylis glomerata* and *Holcus lanatus*. B. *Holcus lanatus* and *Lolium perenne*. C. *Poa compressa* and *Trifolium repens*. Plants were collected from different aged pastures (2, 21, and 40 years) and grown in competition in a mixture diallel. In each case, histograms represent a mean of three replicates. In comparing the three field ages, 2, 21, 40 (with respect to component yield, total yield, or CYR), those which do not share a common code letter (a, b, or c) are significantly different at the $P < 0.05$ level based on Scheffe's multiple comparison test. F tests for the homogeneity of variances were not rejected and all analyses were performed on untransformed data.

Selection resulting in balanced competitive abilities

Aarssen (1983a,b) considers an alternative consequence of selection in response to competition. In this hypothesis, instead of natural selection resulting in niche differentiation and hence avoidance of competition, it reduces the difference in relative competitive abilities. Any genotype that is a superior competitor acts as a selective force to increase the competitive ability of its own inferior competitors. Superiority should therefore alternate between (and among) members of the two populations and local neighborhoods would be constantly engaged in a fine tuning process that alters the way members respond to each other. This selection mechanism may be identified when the CYR increases but the total mixture yield does not change after selection. This is in contrast to an interpretation of niche differentiation which requires that the total mixture yield increases as a consequence of selection.

An extensive search of the literature yielded no references to the possibility of such a mechanism operating. In our own work, however, we have from three studies preliminary evidence which lends support to the argument. First, a de Wit replacement design was set up as described in the previous section using the grasses *H. lanatus* and *L. perenne* from each of the three pastures. These two species demonstrate (Figure 1B) largely a frequency-independent competitive relationship in all pairs. This indicates that in all three pastures, *H. lanatus* and *L. perenne* have widely overlapping niche requirements and this degree of overlap does not change, as there is no significant difference in the three regression slopes. *Holcus lanatus* has an advantage in mixtures at all relative frequencies, but this advantage diminishes with increasing pasture age. This indicates that the relative difference in competitive abilities between the two species is decreasing with pasture age.

Second, in the diallel analysis described earlier, two of the species pairs (*H. lanatus*-*L. perenne*, and *P. compressa*-*T. repens*) revealed results consistent with the hypothesis. In both pairs the CYR increased with pasture age, but the total yield in mixtures was not significantly different between pastures (Figure 2B).

Third, genets from four different pairs of physical neighbors of *L. perenne* and *T. repens* were collected from widely separate locations in the 40-year-old pasture from areas where both species were abundant. All material was cloned and the clovers and grasses were grown together in pots in all of the 16 possible interspecific mixtures for one year. The mixtures were clipped periodically. Cumulative dry weights were calculated after one year and a CYR was calculated for all mixtures. Each clover genet had its highest yield when grown with its natural grass neighbor, but each grass genet had its lowest yield when grown with its natural clover neighbor. There are higher CYR values among natural neighbors than among mixtures not involving natural neighbors (Figure 3). However, consistent with the hypothesis, com-

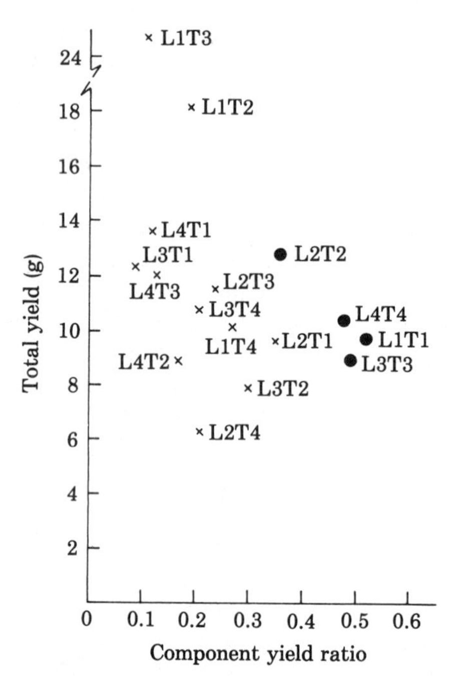

FIGURE 3. Relationship between total yield and component yield ratio for different genet type combinations of *Lolium perenne* and *Trifolium repens*. The values for the naturally neighboring genet pairs collected from the four sites in the experimental field are circled. All values are means of three replicates.

bined yield was not highest among the natural neighbors. These results also lend strong support to the notion of biotic specialization at the genotype level; no longer do we walk across a field of grasses and dicots, but across a pasture of genotypes, each one with its own Christian name!

The complexity of changing competitive ability

The results, presented and referenced, show that there are many and varied solutions to the problem of competition among plants. The changes demonstrated in component yield ratios and total yields of mixtures of pairs of natural neighbors from different-aged pastures suggest that competition is an important force of natural selection. Moreover, much of the data did not corroborate the commonly assumed notion that selection pressures from competition usually result in the evolution of ecological combining ability (niche differentiation) in the component species of a community. Of the species combinations studied, a more common finding was selection for more balanced competitive ability. Selection for niche differentiation may, however, be preceded by selection for more balanced competitive abilities (A. D. Bradshaw, personal communication). Intuitively, one might expect that an inferior competitor may be selected to maximize its fitness until such a point that the two competitors may be quite similar in relative competitive ability. At this stage, the expected process of niche differentiation may begin. Both alternatives underscore the necessity of investigating changing patterns rather than extrapolating from a single pattern because selection for balanced competitive abilities can only be detected when changing patterns are investigated. When these results are superimposed upon evidence for competitive exclusion, niche convergence, and abiotic effects, we are presented with a formidably complex situation.

OVERVIEW

Examples of population differentiation over short distances in both animal and plant populations are common in the literature (see Snaydon, 1970, 1978; Bradshaw, 1972; Antonovics, 1971, 1976a,b, 1978). For example, populations of *A. odoratum* differ over distances of less than 1 m (Snaydon and Davies, 1976); *C. aquatalis* populations over a few decimeters (Shaver et al., 1979); and Aarssen (1983a) has shown differentiation in *T. repens* and *L. perenne* populations in situ.

The rate of evolutionary change of populations has also received extensive study. Some of the most rapid changes recorded occurred in ten years in *M. sativa* (Turkington, 1979), eight years in a hybrid population of *Helianthus* (Stebbins and Daly, 1961), and six years in *A. odoratum* (Snaydon and Davies, 1982). Bradshaw (1972) and Snaydon (1978b) refer to a number of studies in which considerable changes in the genetic structure of populations occurred within the lifetime of a single generation.

Most environments vary spatially and temporally and plant populations respond to this heterogeneity by phenotypic plasticity (Bradshaw, 1965) or by population differentiation. One question is frequently asked: Under what circumstances will selection favor specialization to one microenvironment, and under what circumstances will it favor the generalist that can occupy a range of environments? There are apparently no consistent characteristics of either the environment or species which favor genetic divergence. A large number of the examples reviewed so far are for perennial species and it is apparent that perennials are favored in local-scale differentiation. Yet, some of the most detailed examples of population differentiation are for the annuals *A. fatua* and *A. barbata* (Jain, 1969; Allard et al., 1972; Hamrick and Holden, 1979), *V. peregrina* (Linhart, 1974), and *Oenothera laciniata* (Ellstrand and Levin, 1982). Similarly, most of the examples described concerned populations in relatively stable environments, yet *V. peregrina* shows population differentiation in vernal pools where, particularly at the periphery, conditions fluctuate erratically (Linhart, 1974).

The situation is complicated when one considers that Sulzbach (1980) demonstrated that rapid changes in *Drosophila* may be highly dependent on the particular population used. In addition, Pianka (1978) distinguishes the evolutionary outcome of intraspecific and interspecific competition. The presumed products of intraspecific competition are mostly demographic in nature, such as rectangular survivorship curves, delayed reproduction, and decreased reproductive output. But the most far-reaching evolutionary effect of interspecific competition is diversification. Pianka's conclusions corroborated the findings of Linhart (1974) but do not support the findings of McNaughton and Wolf (1970).

Levins (1968) addressed the above question in a theoretical way. He concluded that when adjacent microenvironments are relatively similar with respect to the ability of individuals to phenotypically respond the generalist is favored. When microenvironments are sufficiently different, selection favors specialization to either one microenvironment or the other. This notion is integrally linked with the concept of environmental grain (Levins and MacArthur, 1966). Most plants, by virtue of their relative immobility, tend toward a coarse-grained use of microenvironments, that is, they spend disproportionate amounts of time in only one or a few of the numerous microenvironments available. If these adjacent microenvironments are sufficiently different, selection may act disruptively; and if movement between the microenvironments is slight, selection may result in divergence of the population into different morphs. Population differentiation is more likely to occur when environmental factors are stable in time but variable in space. Nevertheless, if a population is subject to environmental changes and the duration of a single environmental condition is more than or equal to the population's generation time, genetic polymorphism may occur (Bradshaw, 1965).

These various relationships between life span and stability of environment must act in concert, yet some kinds of compromise are involved. The time scale of change in relation to longevity is of paramount significance. For example, the patterns described in the last section from the 40-year-old pasture should not be extrapolated to, say, forests. While similar patterns may develop over extensive time periods, they would certainly not occur within 40 years. A distinction must also be drawn between stability and predictability. In a stable system the environment may be changing very slowly in a directional manner. Likewise, a long-lived perennial, in say a pasture, will be exposed to various environmental fluctuations during its lifetime. Here, although these changes are cyclic rather than directional, they are nevertheless predictable. At the periphery of Linhart's (1974) vernal pools, the environmental conditions are said to "fluctuate erratically," but at least these erratic conditions are predictably different from the "relatively predictable" conditions of the center of the pool some 5 m away; the reported population differentiation occurred between these two environments, and not within them. In contrast, an unpredictable environment continually imposes upon its inhabitants new evolutionary challenges which are neither cyclic nor directional. When this occurs, the response to selection is such that each of the transient environments has no effect in producing directional changes, especially divergence, within the populations; this is what Heslop-Harrison (1964) has called evolutionary "hunting." So presumably, one of the reasons we might not expect differentiation among populations of annuals is that annuals tend to exploit unpredictable and transient habitats. If an annual species should persist in adjacent, but predictable, environments that are quite different, then a priori we should expect population differentiation (see, for example, Law et al., 1977; Warwick and Briggs, 1978b).

The role of breeding systems in population differentiation has been frequently discussed (Baker, 1953; Heslop-Harrison, 1964; Bradshaw, 1972; Warwick and Briggs, 1979). The general conclusion is that predominantly outbreeding species maintain higher levels of intrapopulation variation than predominantly inbreeding species, while inbreeders exhibit a greater degree of population differentiation than outbreeders. This relationship is, however, not always consistent, and there is wide variation in results. There have been examples of population differentiation reported in an effectively asexual annual, *O. laciniata* (Ellstrand and Levin, 1982); clonal perennials, *T. repens* (Turkington and Harper, 1979b), *R. repens* (Lovett Doust, 1981a), *C. aquatilis* (Shaver et al., 1979); almost completely self-incompatible perennials, *A. odoratum* (Snaydon, 1978b), *S. lyrata* (Fowler and Antonovics, 1981b); obligate outcrossers, *P. lanceolata* (Fowler and Antonovics, 1981b), *L. cylindracea* (Schaal and Levin, 1978); and predominantly self-pollinated annuals, barley (Allard and Adams, 1969a), *A. fatua* (Imam and Allard, 1965). Again, it may not be so critical to discern whether differentiation takes place any more readily with one breeding system or another, but rather to discern whether there are differences in the breeding system between the morphs in the population. For example, Antonovics (1968a) demonstrated that populations of *Agrostis tenuis* and *Anthoxanthum odoratum* tolerant to heavy metals had greater self-fertility than adjacent nontolerant populations.

Ford (1981), working on the apomict *T. officinale*, states that *Taraxacum* agamospecies have had clonal life spans of up to 12,000 years with little opportunity for recombinants to occur during that period. One would intuitively predict that this would ultimately lead to extinction, and yet dandelions have achieved considerable ecological success. Ford argues that local differentiation in his populations cannot be considered to represent specialization to particular current microenvironments. Rather, they represent a range of preadapted genotypes each with high specialization, but they are maintained in proportions relevant to the microenvironment that they occupy. A "change in proportion of the various grasses occupying the meadow might be expected to change the composition of the dandelion population by altering the frequency of the component agamospecies" (Ford, 1981).

Whatever the explanation, it is clear that differentiation is widespread among plant populations. Indeed, we may well be advised to ask the question: When and under what circumstances do populations not show differentiation? Palmer (1972) found no correlation between date of commencement of flowering of *Trifolium arvense* in seven sites and environmental differences. Lammerink (1968) and Dunbier (1972) studied populations of *Medicago lupulina* in various habitats, and on sunny and shady sides of a valley, and found no difference in flowering time. McNaughton et al. (1974) detected no evidence for the evolution of heavy-metal tolerance in populations of *Typha latifolia* which were rooted in soils having a 385-fold difference in zinc concentration. Warwick and Briggs (1979, 1980b) could provide no evidence of distinct "races" of *A. millefolium* in lawns or grazed areas and only limited evidence for differentiation in different grassland types. They concluded that phenotypic plasticity is important. Guries and Ledig (1982) concluded that populations of Pitch Pine (*Pinus rigida*) are only weakly differentiated. Nevertheless, it must be considered that different species may differentiate in a variety of ways—even in ways which are different from the ones we have measured.

Finally, Murray et al. (1982) studied microhabitat differences among genetically similar species of land snails belonging to the genus *Partula*. As many as four species may be found within a single 10-m square, coexisting without apparent hybridization. These authors conclude that "despite their close genetic relationships the species nevertheless differ in their distribution within the available habitat, indicating that pervasive genetic divergence is not a prerequisite for ecological differentiation." Similar studies have not been reported for plants, but such studies could be rewarding.

If the foregoing experimental results and theoretical considerations appear contradictory, it only illustrates the complexity of the evolutionary significance of competitive interactions in plant populations.

CONCLUSION

The major focus of this chapter has been on one biotic component of the environment—competition brought about by plant–plant interactions. This biotic component has profound influence in determining the fitness of a genotype, especially at high densities. Further, interactions between individuals are important in structuring communities and in maintaining genetic diversity in populations. Our discussion may be summarized in three conclusions.

A new approach

The crucial question left unanswered by previous studies of biotic specialization concerns the precise selection mechanisms by which species adjust to their environment of neighbors, resulting in more compatible behavior in their presence. Previous interpretations of biotic specialization concern "ecological combining ability" implying an evolved niche displacement. This confers some measure of escape from competition, with the result that species yield higher in mixture. This interpretation seems to apply, for example, to the data of Remison and Snaydon (in Snaydon, 1978b) but is inappropriately applied to the pasture population of *T. repens* described by Turkington and Harper (1979b). There was no attempt in the latter study to test for reciprocal specialization in the grasses, so it is not certain whether this is evidence for coadaptation or simply one-way specialization by the clover.

The best reflection of the ways in which neighbors respond reciprocally to each other must come from investigations involving individuals which actually interact in nature. Insight into the selection mechanisms producing biotic specialization is only possible through studying how the reciprocal responses between natural neighbors in a community change through time.

A new perspective

Traditional thinking treats the species as more of a taxonomically distinctive unit than as a wide-ranging collection of ecologically different individuals, and there is much current interest in changing this view (Antonovics 1976b; Raven, 1976; Harper, 1982). The crucial contest is not "species pitted against species," but rather "genotype against genotype."

Biotic specialization at the genotype level, such as that presented, calls for an organism-centered view of community structure and evolution (Aarssen and Turkington, 1983). The traditional view defines a community as a collection of populations occupying a given area, usually thought to affect the distribution and abundance of one another. A more refined approach regards the community as a montage of evolving neighborhoods which converges on the organism as the pivotal unit of interaction (e.g., MacMahon et al., 1978, 1981). This view is especially suited to sessile organisms such as plants because any given individual will interact with others in only a very local neighborhood. The results presented suggest that, given sufficient genetic variation, microevolutionary forces may be so precise that the properties which determine reciprocal adaptation in local neighborhoods may be neighbor-specific even at the scale of different genotypes of the same species.

Coexistence—a new view

The differences between species have become a universal touchstone for studies of species coexistence. Investigations repeatedly and almost automatically entail a search for, or an interpretation involving, some "important" difference in niche between coexisting competitors as though it were somehow remarkable to have found that one species has not competitively excluded the other. Indeed, Slobodkin (1961:122) stated that "if two species persist in a particular region it can be taken as axiomatic that some ecological distinction must exist between them." More recently, Pontin (1982:75) stated that "stable coexistence when sharing resources depends on there being some difference between the species." However, the evidence presented earlier indicates that coexistence may have alternative explanations (Aarssen, 1983b): (1) selection for niche differentiation, and (2) reciprocal selection for balanced competitive abilities. These findings should broaden our perspective in studies of coexistence, and it should no longer be necessary to insist that some difference in niche requirements supplies the only explanation. In the mechanism based on balanced competitive abilities proposed above, the presence of, or the potential to continuously generate, relevant genetic variants with changing relative competitive ability in each species, permits a stable coexistence in a preserved interaction that is maintained by persistent reciprocal selection (coevolution). This process of competitors "tracking each other" (involving mutual shifts in gene frequencies) parallels that in theories for the evolution of coexistence in predator–prey (e.g., Pimentel, 1961; Rosenzweig, 1973); plant–herbivore or plant–pollinator (Ehrlich and Raven, 1964; Gilbert and Raven, 1975); host–parasite (Pimentel et al., 1963); and model–mimic (Sheppard, 1975) systems but does not impose the assumption of frequency-dependent genetic alterations of intraspecific versus interspecific competitive ability as in the genetic feedback model of Pimentel (Pimentel et al., 1965; Pimentel, 1968).

An understanding of species coexistence requires a greater emphasis on the variability within species with respect to both fundamental niche requirements and relative competitive abilities, concurrent with an awareness of the differences between them. Efforts are needed to reveal how and under what circumstances natural selection affects each of these two components and when it is likely to affect one more than the other.

Two final points may be made at this stage. First, selection for more balanced competitive abilities provides reconcilation of the apparent paradox between "convergent adaptation" to a common environment and "divergent adaptation" to other members of the community (Antonovics, 1978). Second, coexistence through niche divergence alone leaves unanswered the question of coexistence of individuals of the same species, or more particularly genetically identical neighbors produced by vegetative means. Selection for more balanced competitive abilities, however, has broader application, as it can explain the coexistence of *individuals*, irrespective of taxonomy.

The data in the studies mentioned in this chapter show very clearly that the interrelationships of species mixtures are transient and change as a result of natural selection. These changes are coevolutionary and involve adjustments by both of the species; and although this coevolution brings about apparent stability, it does not necessarily result in an increased yield of the mixture. These studies have also attempted to interpret evolutionary events by experimentally investigating the process of evolution, rather than extrapolating from its products. But while the studies may have been partially successful in bridging some gaps, we still know virtually nothing about the physiological bases of such processes or the demographic principles which operate under such conditions (see Raven, 1976, 1979; Mooney and Gulmon, 1979; Cook, 1979b). A detailed investigation of the individuals in carefully selected populations in carefully selected natural or seminatural sites, involving demographic, genetic, and physiological studies, provides the best prospect of progress in plant population biology.

CHAPTER 6

SOME EVOLUTIONARY ASPECTS OF PLANT–PLANT INTERACTIONS

Subodh Jain

INTRODUCTION

Interactions among plants living as neighbors occur in a large and complex variety of forms, and accordingly, their evolutionary consequences are also highly diverse. In order to circumscribe the scope of this chapter, we should recognize the following concepts: A plant population may comprise several ecological neighborhoods described in terms of certain environmental descriptors relative to adaptive responses. Neighbors may show density-dependent (redefined in terms of detailed spatial patterns of distance) mortality, reproduction, and other characteristics which can be described in terms of intraspecies and interspecies competition. If different genotypes are competing as neighbors for some common resource, the outcomes would either be recorded as changes in genotypic frequencies (and such changes would provide estimates of relative fitnesses) or some measure of niche divergence presumed to lead to minimal interference between neighbors. Genotypes in a neighborhood interact as parents of the next generation through outbreeding, sexual selection, and so forth, and most likely result in a substructure of gamodemes (genetic neighborhoods) with genetic relatedness (Ennos and Clegg, 1981; Ritland, in press). Here, too, plant–plant interactions might involve heritable fitness variation and show evolutionary changes in the patterns of mating system, dispersal of progeny through pollen and seed flow, and an overall increase or decrease in such interactions. Moreover, it must be clearly recognized that both density-dependency and frequency-dependency aspects of natural selection would provide an important clue to the presence and magnitude of plant–plant interactions.

Thus, we have a framework of dealing with highly structured plant populations with smaller spatial units of ecological and genetic interactions. Hull (1980) pointed out that "evolution through natural selection requires an interplay between replication and interaction" of any given organizational unit, namely, organism, linked gene system, population, group, or deme. Many biologists have mistakenly followed Williams (1975) literally in rejecting group selection by making it an alternative to Darwinian individual selection. In population genetics theory, the notion of interdeme selection has been widely accepted in relation to the role of local founder events, shifting balance theory of evolution due to Wright (1970), population heritability (Slatkin, 1981b), and selection for behavioral traits, among others. Michod (1982) recently reviewed the theory of kin selection, as discussed in standard population genetic theory, which deals with the interactions among related individuals. Here, frequency-dependent selection is explicitly defined in terms of pairwise intergenotypic interactions as a fitness component for each individual genotypic class. In fact, it is important to realize that an experiment designed to detect such interactions must estimate fitnesses and show them to be somehow functions of relative genotypic composition of neighborhoods. Since neighborhood effects are likely to become more intense with greater crowding and increased genetic variance in fitness interactions, such experiments would benefit from varying densities and frequencies. Note that this is how de Wit (1960) developed his replacement series method for detecting competitive interactions. However, as noted by Spitters (1979) in his elegant essay, theory of such competitive interactions can meaningfully deal with only simple genetic situations (i.e., those involving only a few genotypes in pairs or three-way groupings), whereas with greater complexity and more diffuse interactive systems, the weighted marginal fitnesses of individual genotypes might increasingly drop the interaction terms. An empirical scientist would then have to verify in any given situation whether interactions are critical in dealing with certain spatial units of plant stand, certain levels of genetic variation, and with the overall evolutionary objectives of his research. Quite similar problems arise in a population geneticist's mind while dealing with the theory of simple polymorphisms vis-à-vis multilocus systems. Here, we shall briefly explore the consequences of frequency-dependent selection in plant populations with the use of some examples.

NATURAL SELECTION IN GYNODIOECIOUS POPULATIONS

Plant populations of a species might frequently vary in the proportion of females (effectively male-steriles) and hermaphrodite individuals. For nearly half a century geneticists have written on this subject; many examples have been studied for the mode of inheritance (i.e., genic, gene-cytoplasmic, number of loci and their interaction), and the primary evolutionary problem of the conditions (i.e., fitness advantages of females) required for the maintenance of gynodioecy is still a subject of numerous recent publications. Ross and Weir (1975), Ross (1982), Charlesworth and Charlesworth (1979), and others have pointed out the relative roles of mixed selfing and random mating of hermaphrodites, heterozygote advantage, resource allocation to seed output, and the exact mode of inheritance. In general, these models of gynodioecy as well as most others dealing with the evolution of breeding systems involve plant–plant interactions first through matings between neighbors and second through sib-competition arising within progenies of different mother plants dispersed within rather limited distances.

Specifically, in barley populations developed from composite crosses propagated by the breeders, male sterility was introduced in order to enhance recombination rates in this predominantly selfing species. Jain and Suneson (1966) showed that seed-set on the male-steriles was frequency-dependent such that an increase in the proportion of male-steriles over the range of 1% to 10 or 12% in the population significantly decreased the average seed set and the associated outcrossing potential in such populations. They further showed that although increased recombination rates yielded higher levels of newly arising quantitative genetic variation, selection did not efficiently utilize this variation as noted apparently by the lack of any proportional increase in gain under selection. Thus, models of gynodioecy might find the seed-set factor to be crucial in the maintenance of polymorphism at the locus for gynodioecy, but it need not always imply a greater scope for the sib-competition mode of selection to favor an outbreeding mechanism.

In another series of studies on gynodioecy in *Limnanthes douglasii* (R. Kesseli and S. K. Jain, unpublished data), seed-set on females was not affected by the frequency of females or even by their distinctly patchy distribution in space unless it reached rather high levels, namely, 20 or 30% females within local patches. The availability of pollen-donors or pollinator-efficiency did not appear to be critical factors associated with the seed output. In general, females showed only a slight average gain in seed output based on a larger number of flowers per plant and not on a higher seed number per flower. Several other comparisons should be noted: (1) Whereas male-steriles are completely outbred, the hermaphrodites had nearly 10 to 25% selfing, as measured by using genetic markers. (2) The hermaphrodites in populations without male-steriles had only 2 to 5% selfing which raises some interesting queries about the pathways in the origin of gynodioecy (Charlesworth and Charlesworth, 1979; Ross, 1982). (3) Average estimates for heterozygosity of allozyme loci were 3–5% and 8–12% in the progenies of hermaphrodites and females, respectively. (4) Even though this seems like a small difference, an inbreeding depression study clearly showed evidence for significantly larger effects of inbreeding in the gynodioecious populations. (5) Several life-history components (e.g., seed dormancy, seedling survivorship) might also be important in favoring male-steriles. Gene flow in *L. douglasii* appears to be highly leptokurtic and restricted (R. Kesseli, unpublished data), but it would be interesting to verify if progenies of male-steriles would tend to have greater dispersability in relation to both higher pollinator efficiency and a wider gene pool of pollen donors. Patchiness in the distribution of different gene combinations and use of founder colonies will allow further tests on the role of sib-competition and dispersal. Overall, polymorphism for sex allocation and correlated floral traits offers highly valuable tools for such ecological genetic research. Here, selective forces as well as resource allocation arguments are rather easier to identify and to assess quantitatively. (See Figure 1 for a summary of key features that would be required to describe the system.)

There is an "information explosion" in the areas of evolutionary ecology of sexual versus asexual reproduction and outbreeding versus inbreeding systems. An excellent theoretical treatment was provided by Maynard Smith (1978), who examined the belief "that sex and recombination are favored in a variable and unpredictable environment" and showed it to be too simplistic or demanding of very special situations. He specifically developed a model of sib-competition which also required frequency-dependent selection in favor of the rarer genotypes for maintaining polymorphisms in a large population. The outcomes clearly depended on the assumed characteristics of environmental heterogeneity and of dispersion. In the models of finite, small populations allowing random events to generate multilocus associations (hitchhiking generating linkage disequilibria), there is short-term gain in population fitness through recombination. These findings and numerous others, reviewed by Maynard Smith (1978a) and others,

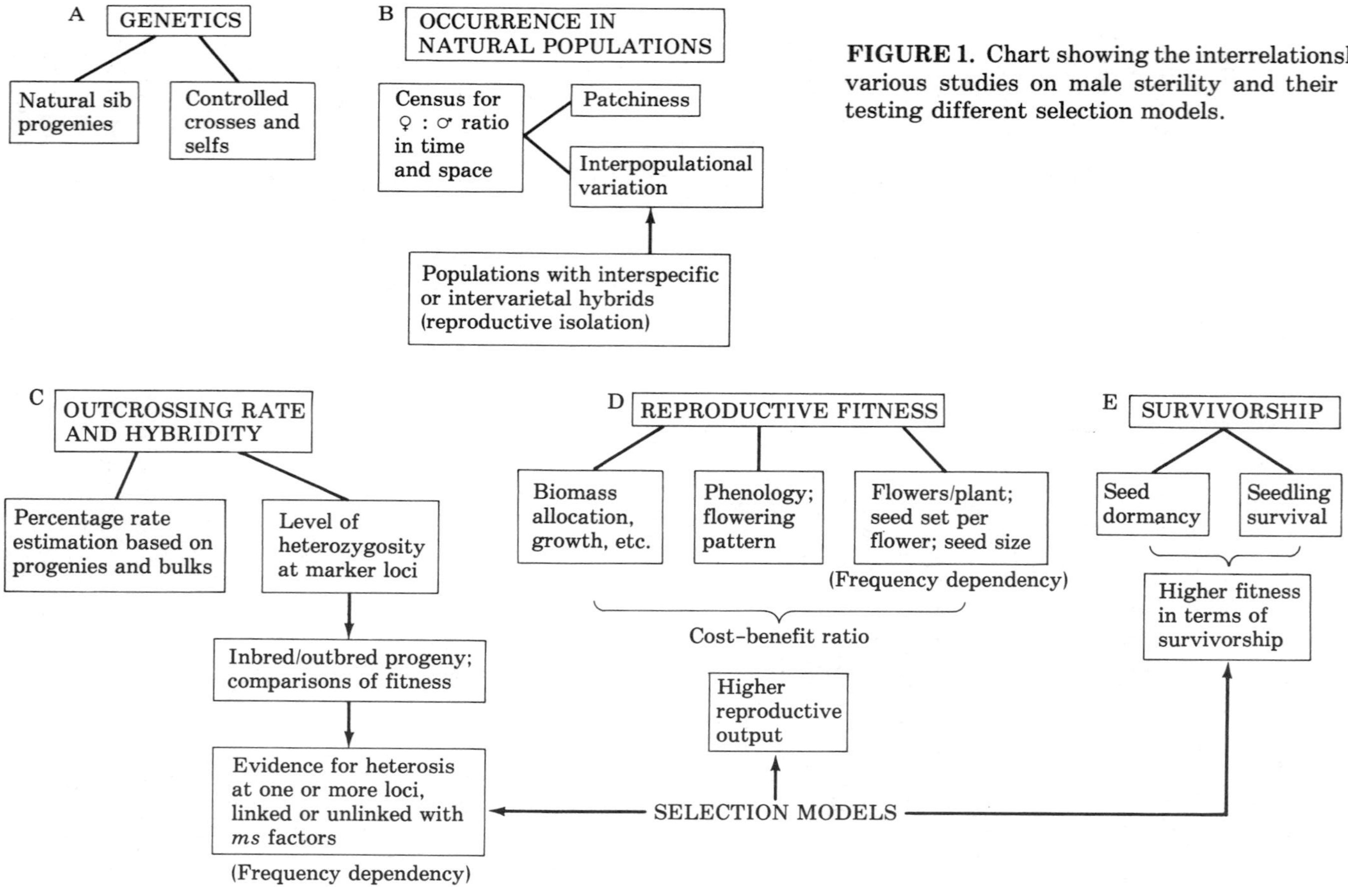

A GENETICS

Natural sib progenies

Controlled crosses and selfs

B OCCURRENCE IN NATURAL POPULATIONS

Census for ♀ : ♂ ratio in time and space

Patchiness

Interpopulational variation

Populations with interspecific or intervarietal hybrids (reproductive isolation)

C OUTCROSSING RATE AND HYBRIDITY

Percentage rate estimation based on progenies and bulks

Level of heterozygosity at marker loci

Inbred/outbred progeny; comparisons of fitness

Evidence for heterosis at one or more loci, linked or unlinked with *ms* factors

(Frequency dependency)

D REPRODUCTIVE FITNESS

Biomass allocation, growth, etc.

Phenology; flowering pattern

Flowers/plant; seed set per flower; seed size

(Frequency dependency)

Cost–benefit ratio

Higher reproductive output

E SURVIVORSHIP

Seed dormancy

Seedling survival

Higher fitness in terms of survivorship

SELECTION MODELS

FIGURE 1. Chart showing the interrelationships of various studies on male sterility and their use in testing different selection models.

clearly show in our present context that (1) many population genetic and ecological variables may be interwoven in such analyses, and (2) experimental data are far behind these theoretical developments. Recently, the evolution of dioecy has been discussed such as to define the so-called genetic (outbreeding advantage) and ecological (sexual selection, resource allocation) hypotheses (e.g., Givnish, 1980; Thomson and Barrett, 1981). I hope it is easily seen that any evolutionary hypothesis has to provide for genetic variation and the mechanism of natural selection and that geneticists are not likely to avoid ecological factors in proposing overdominance or frequency-dependent selection. This dichotomy is superficial and appears to delay the long-awaited joint studies. However, several recent studies provide interesting attempts to gather both genetic and ecological information. Waller (1980) discussed the heritability of the dimorphic chasmogamy-cleistogamy system in *Impatiens*; several environmental factors influence the relative proportion of cleistogamous progeny produced by an individual; and the evolutionary fate of outbred versus inbred progeny might depend on dispersion and environmental pattern, just as predicted by a sib-competition model. Both density and genotypic frequency variables should also be examined, as argued earlier. Antonovics and Ellstrand (in press) tested the frequency-dependent advantage of sex in a study of *Anthoxanthum odoratum* colonies. Various aspects of patchiness, optimal outcrossing rates, plant–plant interactions through pollinators, and various schemes of sexual selection are reviewed by Willson (Chapter 13), Waser and Price (in press), Turner et al. (1982), Levin (1978a), and Lloyd (1980c), among others. In vegetatively propagated species or habitually apomictic species, population structure might be extremely dominated by highly patchy clonal dispersion patterns (Jackson et al., in press; Kershaw, 1964), which are now beginning to be investigated using genetic variation assays. In general, the whole area of localized plant–plant interactions in terms of genetic variation and fitness parameters has now become exciting.

SELECTION IN MIXTURES AND COMPOSITE CROSS POPULATIONS

Plant breeders are often interested in the potential role of intergenotypic competition in altering the response to selection aimed toward the production of pure lines or inbred lines. A large number of studies have attempted to measure competitive interactions by growing mixtures of parental lines in various field plot designs. Spitters

(1979) concluded from his extensive review that delaying breeder's selection for yield until a later generation of a cross is not handicapped by intergenotypic interactions. However, Allard and Adams (1969b) found in wheat and barley that natural selection favored facilitation (i.e., mutual advantage) among competing genotypes such that positive interactions even enhance the retention of genetic diversity. Earlier, Mather (1969) had proposed models of selection through competition and cooperation so much so that Griffing (1977) developed a breeding method designed on the basis of group selection. That mixtures or other kinds of gene pools might evolve toward favorable interactions (coevolution, coadapted genotypes) is of special interest in evolutionary ecology (see Turkington and Aarssen, Chapter 5), although the mechanisms are not too well explored. Several multiline varieties of oats have shown their yield advantage under evolving host–pathogen interactions in terms of more durable disease resistance or lower disease incidence in a given season (Frey et al., 1975); here, a mixture can even use the physical barrier of "spore trapping" through the presence of resistant and susceptible individuals to a pathogen race in adjacent neighborhoods. Bremermann (1980) proposed a model of the evolution of sex based on host–pathogen interactions. Other models have invoked frequency-dependent selection and certain features of stabilizing selection. However, again, there are too few reports of such studies in natural populations to assess their ecological consequences.

Several long-term analyses of genetic changes in composite cross populations of barley and bulk-hybrid populations of lima bean have shown heterozygote advantage at a few marker loci to be involved in maintaining genetic variation under high level of selfing. On closer examination, heterozygote advantage seemed to be greater in populations with the lowest levels of heterozygosity. In order to verify this in terms of neighborhood competition, Singh (1972) developed barley populations with three levels of heterozygosity at locus *R/r* (rough versus smooth awn texture); the three genotypes (*RR, Rr, rr*) were planted in known spatial patterns and frequencies; then, individual plant data enabled estimation of relative fitnesses in terms of survivorship and seed yield. These estimates (Table 1) show that heterozygote advantage is due to frequency-dependent seed yield differential. Jana et al. (1973) reported cyclic advantage for the heterozygotes at the same locus. In both experiments, clearly, neighboring genotypes interact to favor heterozygotes surrounded by the homozygotes. Ecological models for such an outcome would be based on some concept of niche divergence or similarity. A formal simulation model was developed by Schutz and Usanis (1969) to find conditions for stable polymorphisms; Jain and Jain (1969) further showed theoretically that with frequency-dependent selection on homozygotes, even underdominance

TABLE 1. Evidence for frequency-dependent heterozygote advantage at locus *R*/*r* in barley.[a]

Input frequency of *R*/*r*	MEAN SEED NUMBER PER PLANT			AVERAGE RELATIVE FITNESS		
	RR	*Rr*	*rr*	*RR*	*Rr*	*rr*[b]
0.05	168.1	180.8	161.3	1.04	1.12	1.0
0.15	170.6	180.0	163.5	1.04	1.10	1.0
0.25	156.2	170.9	163.0	0.95	1.05	1.0

[a] From Singh (1972).
[b] W_{rr} = 1.0 for computing relative fitnesses.

(heterozygote disadvantage) would be admissible under certain stable equilibria. In other words, a rarity advantage model is of special interest to both population geneticists and ecologists dealing with natural selection. Although Roughgarden (1979) emphasized only density-dependent selection in relation to r- and K-selection primarily, Hedrick (1983) has elegantly reviewed many aspects of frequency-dependent selection as well. For example, he discussed a model to show in terms of a measure of the rates of loss of variation in finite populations that frequency-dependent selection is more potent than overdominance in maintaining genetic polymorphisms.

In another series of barley populations, Jain et al. (1981) also discovered selection against heterozygotes at the *Bb* locus for pericarp color. Polymorphism at this locus is invariably lost at Davis; an experiment was designed to use male sterility and manipulation of genotypic frequencies in synthesized populations such that relative proportions of the less fit genotype (*BB* or *Bb*) are artificially increased each generation. In all cases the frequency of allele *B* showed a steady decline; and although relative fitness of *BB* and outcrossing rate were slightly frequency-dependent, heterozygotes *Bb* were at 30% or more fitness disadvantage averaged over all neighborhood arrangements (Tables 2 and 3). Tillering and relative seed yield were the main fitness components. Here, heterozygotes seemed to have suffered because of late maturity and poor seed set conditions without any neighborhood effects from the homozygotes.

MICRODIFFERENTIATION AND THE ROLE OF SUBDIVISIONS

The patterns of genetic variation in two coexisting wild oats, *Avena fatua* and *A. barbata* have provided some interesting comparisons

TABLE 2. Analysis of variance of fitness components for the experiment on selection against heterozygotes at the locus *B/b* (for pericarp color) in barley (see text for details).[a]

Source	Degrees of freedom	MEAN SQUARES[b]			
		No. of plants per row	Plant height	No. of tillers per plant	No. of seeds per plant
Blocks	3	50.3	57.62*	21.78	20.6
Genotypes	2	2239.5**	226.64**	51.68	1480.5**
Interaction	6	31.3	4.40	9.74	36.2

[a] From Jain et al. (1981).
[b] Significance levels: *, $P < 0.05$; **, $P < 0.01$.

TABLE 3. Estimates of means (± SE) for traits associated with fitness of heterozygotes at locus *B/b* in barley.

Component of fitness	GENOTYPE		
	BB	*Bb*	*bb*
Percentage viability	0.88	0.24	1.00
Plant height	84.4 ± 1.38	71.2 ± 1.71	84.4 ± 1.18
Days to flowering (index 0 to 8)	6.5 ± 1.4	6.0 ± 1.7	4.9 ± 1.2
No. of tillers per plant	22.7 ± 1.09	14.5 ± 1.45	21.3 ± 1.08
Percentage late tiller/plant	0.02	0.03	0.0
No. of seeds/plant	129 ± 10	88 ± 6	146 ± 11

[a] From Jain et al. (1981).

(Jain, 1975). In *A. fatua* both pure and mixed stands show almost ubiquitous morphological and allozyme polymorphisms at each of the scales of sampling (namely, 0.5 m, 10 m, 100 m, or between sites several kilometers apart). In contrast, *A. barbata* showed regional pattern of presence versus absence of polymorphisms as well as highly patchy distribution of morphs even at the smallest scale (Rai and Jain, 1982). Both species are predominantly selfers and live under a parallel diversity of environments. However, several greenhouse experiments suggest that the interactive role of heterozygote advantage and microsite differentiation might have a more sensitive threshold in *A. barbata* than in *A. fatua* such that the loss of polymorphism within a given patch is easily triggered by small changes in frequency, density, or some habitat-related factors; even predators' apostatic choices might be involved. Given such a mosaic of small monomorphic patches, the aggregate properties of a deme or population unit would further depend on plant-plant interactions in the context of interpatch selection. Endler (1977) provided a thorough discussion of the patchiness parameters and their potential role in geographical patterns of variation. Population ecologists would find here a wealth of genetic variation examples in order to utilize genetic polymorphisms in their analyses of evolutionary models. Patch dynamics has assumed a central place in the studies of many plant communities; one finds that plant–plant interactions between species are somewhat traded here for more intense intraspecific contacts (S. K. Jain and K. Rice, unpublished data); this provides several divergent views of the genetic structures in relation to species' place in the community. Hedrick et al. (1976) noted that patchiness "implies discontinuities on many scales in time and space," discontinuities that exert powerful influences on the "distributions of species, their interactions, and their adaptations."

GENETIC BASIS OF INTERSPECIES COMPETITION

Only two sorts of examples will be reviewed. The evolution of self-fertility along a step cline such as to avoid interbreeding and production of less fit hybrids between two adjacent local races is an example of the evolution to avoid intergenotypic interactions. Antonovics and Levin (1980) extensively reviewed many examples of the origin of reproductive barriers and new taxa through similar competitive interactions; certain forms of character displacement fit under this category. The evolution of inbreeding in the rarer of two coexisting species was also illustrated by Levin's (1978b) own work in *Phlox* spp. Another kind of evolutionary change under interspecies competition was described as genetic feedback by Pimentel (1964), who showed that a species pair might coevolve such that the role of interspecies versus intraspecies competition alternates as species densities (here, equated to their relative proportions) oscillate; different genotypes seem to be favored under these two kinds of competitive interactions. A classic demonstration of the role of specific genetic variants in the outcome of interspecies competition was provided by the work of Lerner and Dempster (1962); they showed that the apparently indeterminate outcome of competition in *Tribolium* was caused by uncontrolled genetic variation present in the experimental stocks. Several interesting examples of this serendipitous result in plants have now been reported and are reviewed by Turkington and Aarssen (Chapter 5). Wu and Jain (1979) tested the competitive interactions between two series of populations of *Bromus mollis* and *B. rubens* (populations that showed frequency- and density-dependent interactions) so as to predict their stable coexistence (Figure 2). However, in natural stands we often found *B. rubens* to have patchy distribution with a high turnover rate of patches. In contrast, two species of *Avena* showed the role of genetic polymorphism in stable coexistence to be more or less in agreement with some of the observations in nature (Yazdi-Samadi et al., 1978). More recently, Martin and Harding (1981) also found a signifi-

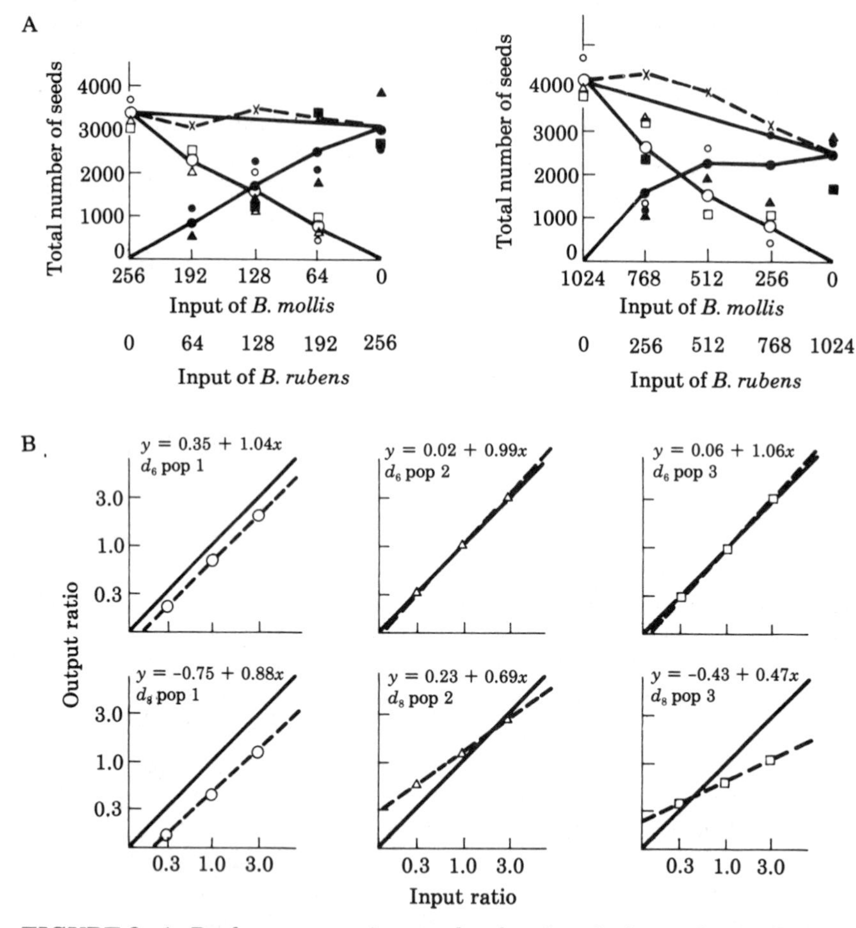

FIGURE 2. A. Replacement series graphs showing the interrelationships between *Bromus mollis* (open symbols) and *B. rubens* (solid symbols) with respect to the total number of seeds produced per pot. Solid line, expected; broken line, observed. (●, ■, ▲) refer to populations 1, 2, and 3 of *B mollis*, and (○, □, △) correspondingly refer to three populations of *B. rubens*. B. Ratio diagrams based on data in A for d_6 = 128 and d_8 = 256 as two of the highest plant densities. (After Wu and Jain, 1979.)

cant role of genetic variation in the outcome of interspecies competition in *Erodium*. All of these observations simply remind us to recognize the complex nature of interactions at different taxonomic, geographical, and genetic organizational levels. Harper (1977) devoted a full chapter of his book to the mechanisms of interactions between species, starting with an incomplete list, which named 13 different items. With his pioneering laboratory studies, he and his students provided evidence for many of them. Although he noted the dearth of field studies, the ecological and evolutionary properties of a community led him to conclude that "diversity exists as the somatic and phenotypic variation within a genet, differences of age and between genotypes of a species as well as diversity at the species level." A theoretical approach to the integration between hierarchical levels of this diversity, as attempted by Pimentel (1964) and Harper (1977), is a challenging problem. In a keynote address Harper (1978) further developed the arguments for such an integrated approach in applied ecology as well.

CONCLUDING REMARKS

It has probably become self-evident that the title of this chapter covers almost the whole of ecological genetics and that, not surprisingly, discussion is too sketchy or inadequate. However, a few main points should be reiterated. Any population with fitness variance might have a heritable component that is due to intergenotypic interactions and that will be expressed in relation to the neighborhood concept. Whether population subdivisions are described in ecological terms or through genetic descent, they would often fulfill Hull's (1980) criteria of being units of replication (over generations) and of integration among interacting neighbors. Evolution by natural selection might be operationally described at the level of such groups. Plant populations, in fact, offer numerous opportunities for fruitful research in this area. Spatial dispersion, modes of dispersal curves, choice of mates, sex ratio variation, and many ecological sources of patchiness together offer the ecological theater in which genetic parts of the evolutionary play must be fully recognized. A series of population genetic models, such as those of Wright (1969), Levins (1970), Wade and McCaulay (1980), Michod (1982), and Maynard Smith (1978b) among many others, provide some theoretical leads that warrant further developments. The evolutionary consequences are often considered in terms of species coexistence, maintenance of genetic polymorphisms, patterns of geographical variation, and evolution of mating systems. The demographic consequences were not dealt with here, but mechanisms of density regulation, evolution of life-histories, and the role of subdivided populations in avoiding the risks of extinction come to mind.

Dobzhansky's famous dictum, "Nothing in biology makes sense except in the light of evolution," would have certainly pleased Charles Darwin; however, I shall venture to guess that if he had the tools of Mendelian genetics to analyze variation (for which he said that "laws of variation are only dimly understood") he would have gone out to measure the forces of natural selection using genetic polymorphism at least once or twice in his own characteristic style. Also, I would presume that he would have written about the genetic basis of plant–plant interactions with far more naturalistic flair than is found in the few examples I have cited. However, as noted by Emerson (1960), "he [Darwin] did not adequately apply natural selection to whole group or population units in contrast to his theory of natural selection of individuals." Happily, this leave us with some reason to feel refreshingly creative about our discussions today.

CHAPTER 7

HERBIVORY:
A Phytocentric Overview

Rodolfo Dirzo

INTRODUCTION

Ecology is the science whose major concern is the understanding of the interrelationships between organisms (both plants and animals) and their environment (both physical and biotic). Though it might appear rather naive to initiate this chapter by saying what ecology is, I find it appropriate, if only to emphasize the fact that plant ecologists quite frequently tend to ignore the potential importance of some elements of the biotic environment of plants; this ignorance is particularly evident in the case of some elements such as plant predators or herbivores.

Neglecting the potentially important role herbivory may have on the population biology of plants is wholly unjustified in the light of the emphasis early ecologists like Charles Darwin placed on the study of animals as fundamental components of the population biology of plants. It is not difficult to find remarks pointing in this direction in the work of Darwin, for example (from *Origin*):

> if we wish in imagination to give the plant the power of increasing in number, we should have to give it some advantage over its competitors, or over the animals which prey on it.

This quotation contains the statement that plant predators have the potential to affect the dynamics of the plant population; or (also from *Origin*):

> on a piece of ground three feet long and two wide, dug and cleared, and where there could be no choking from other plants, I marked all the seedlings of our native weeds as they came up, and out of 357 no less than 295 were destroyed, chiefly by slugs and insects

which states with real, experimental data, the demographic effect of herbivores on plant populations.

It is surprising, therefore, that despite this intellectual legacy of more than 100 years, apart from the study of pollination, of which ecological and evolutionary literature provide a plethora of magnificent examples, the study of the plant–animal interface when the animal acts as a predator has been virtually ignored. This has been so, also despite the fact that the agronomic literature (e.g., on biological control) contains numerous examples of the potentially catastrophic role of herbivores on plant populations.

While predation in the animal–animal interaction has been a source of ecological study for a very long time, and in fact, the dynamics of the system were initially modeled as early as the 1920s by Lotka and Volterra, a comparable attempt to produce models for the plant–herbivore system has not been made until very recently by a few workers (e.g., Noy-Meir, 1975; Caughley and Lawton, 1981). However, such attempts are, as yet, no more than a superficial glimpse into the nature of the system, and what is urgently needed now is (1) to detect the basic biology of the plant–herbivore interface and (2) to get empirical data of phytocentric relevance.

In this chapter I will attempt to highlight some of the components of the interaction, particularly as seen from the viewpoint of the plant. To do so, I present and discuss some work of my own and of other authors, work that is analyzed at (1) an ecological and (2) an evolutionary level—in that order.

THE LEVELS OF INTERPRETATION OF THE INTERACTION

To start with, I wish to make a distinction between the two possible levels of interpretation of any plant–herbivore interaction. These two levels are defined by what Baker (1938) called proximal and ultimate factors determining the present behavior of organisms. For example, we can investigate, either experimentally or in the field, the relative effects of grazing on a range of food plants by a given herbivore. Let us assume that the differential effects of grazing may cause differences in seed production. These effects will represent short-term behavioral responses of both the plants and the herbivore. The proximal interpretation of this interaction is made in terms of present characteristics of the plant (e.g., its ability to allocate quickly resources to and from different plant parts) and present characteristics of the herbivore (e.g., its mobility and olfactory response). However, the differences in seed production investigated this way are not necessarily the result of evolutionary interactions between the ancestors of the plants and the herbivore that have left their descendants behaving as they do at present. Thus, it may be dangerous to assume that the urticating hairs of some tropical vines, which at present inhibit fruit-eating by monkeys (in South East Mexico), are the evolved consequence of ancestral experience of fruit predation by monkeys. It may well be that the urticating hairs evolved as a response to a very different herbivore or, indeed, to another quite different selective pressure!

Nonetheless, if a herbivore does feed on the fruits of a plant, it can have a profound effect on its population dynamics, even though this effect is not the result of evolutionary interaction. Therefore, it is quite valid to envisage the role of herbivory, first, on a *proximal* (ecological) level and, second, on an *ultimate* (evolutionary) level.

PROXIMAL (ECOLOGICAL) INTERACTION

Figure 1 is a schematic representation of the elements involved in the role of herbivory on the dynamics of plant populations. Three components can be distinguished (see Lubchenco and Gaines, 1981): the first is the likelihood that an individual plant will be contacted by herbivores. If the plant is encountered, the second element considers the likelihood that the animal will consume at least part of it. The third element considers the consequences of herbivory to the individual plant and, because we necessarily have to view this in the context of other plants in the population, this becomes the expected reduction of the plant's fitness as a result of herbivory.

In this chapter I will focus specifically on the third component of the diagram and its extension to the population and community level. For components 1 and 2, it will suffice to say that these given probabilities depend on a number of attributes of both the plant and the herbivore, as shown in Figure 1. These two components of the interaction have been reviewed by other authors (e.g., Feeny, 1976; Rhoades and Cates, 1976; Gilbert, 1977; Lubchenco and Gaines, 1981).

Given that (1) the herbivore detects a plant and (2) it consumes at least part of it, the consequence on plant fitness will be a function of at least the following variables:

1. The plant's phenostage (that is, whether it is a seed, a seedling, or a mature plant in a given phenological state).
2. How the plant is organized in terms of the items of consumption by the herbivore (e.g., leaves, buds). This we could call, generically, the *modular structure* of the plant (Harper, 1977; White, Chapter 1).
3. The quality of tissue damaged.
4. The quantity of tissue damaged.
5. In a more general sense, how these factors determine, for an in-

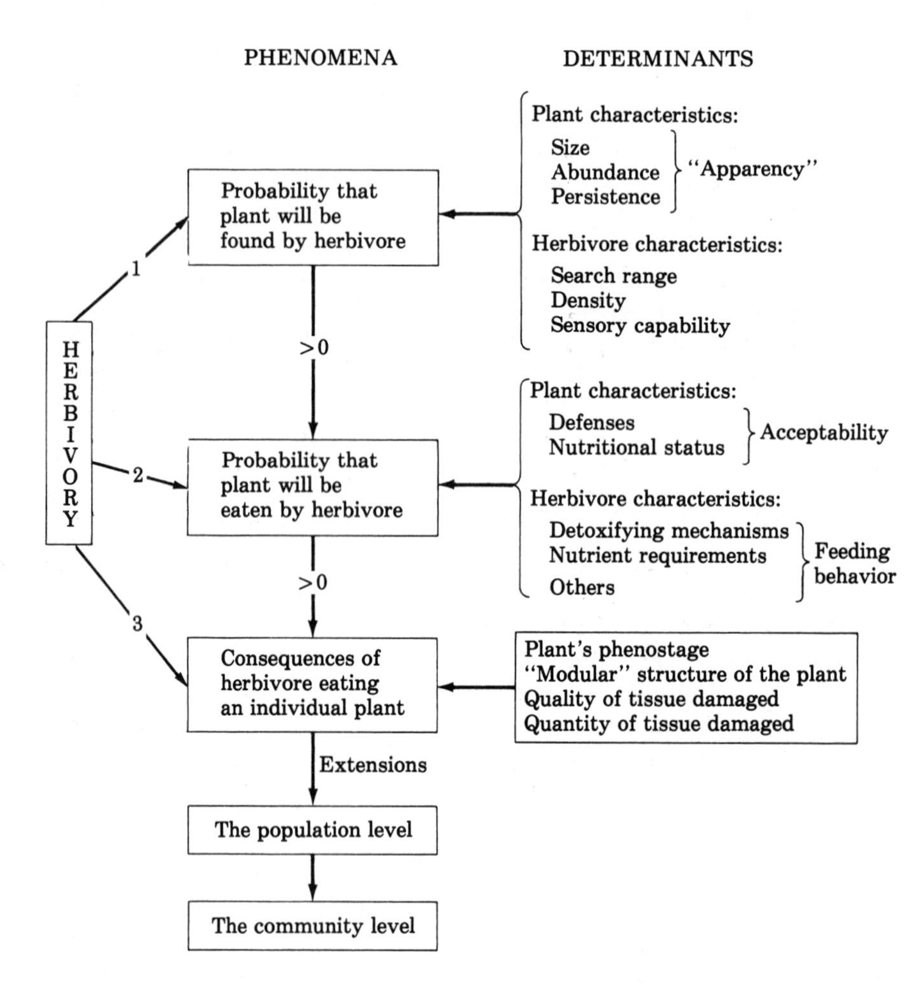

FIGURE 1. Schematic representation of the elements involved in considering the role of herbivory on the dynamics of plant populations (see text for details).

dividual plant, the cost of damage by herbivores in relation to undamaged plants.

Let us discuss each of these variables separately.

The plant's phenostage

The effects of herbivores on their individual host plants can vary depending on the age and/or physiological status, that is, the phenostage of the plant that is damaged. Some animals do kill the plant on which they feed but they usually do so only when the plant is in a juvenile

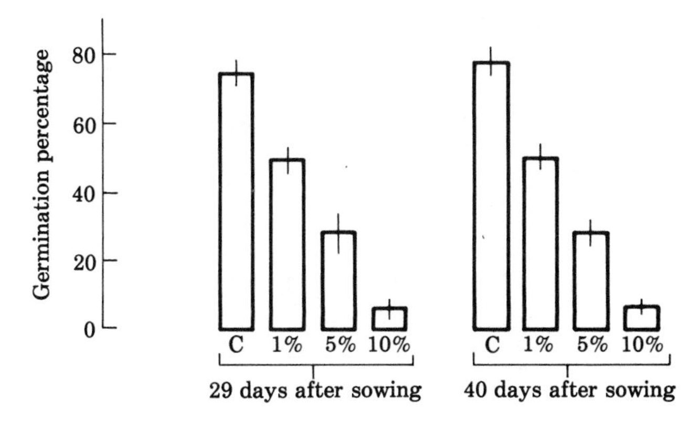

FIGURE 2. The germination of seeds of *Omphalea oleifera* subjected to various levels of artificial seed predation. Germination percentages are given for two recording dates. Values are means ± SE.

stage (e.g., ovules just fertilized, seeds, or seedlings). In this case (death), all of the plant's fitness is lost, and herbivory may then, in a sense, be equivalent to predation in the animal–animal interaction.

By far, the most extensively documented situation of this kind of effect on the plant is postdispersal seed predation. An outstanding example of this situation is reported in the work of Janzen (e.g., Janzen, 1971, 1980) for tropical systems for which seed losses due to predation can be extremely variable (ranging from 0 to 100%); the main predators are seed beetles in the Bruchidae, Curculionidae, Cerambycidae, and Scolitidae, as well as several mammals (e.g., Heteromydae and Sciuridae).

However, even at this very vulnerable phenostage of the plant, not all of the attack on individual seeds will necessarily kill the "plant" (seed); a seed may survive the attack (provided the embryo is not eaten), depending on the proportional seed biomass lost to the animal. This point is nicely illustrated by data gathered for the seeds of *Ompahlea oleifera*, a tropical tree growing in Southeast Mexico. Following Janzen's (1976) methodology, different intensities of seed predation were mimicked (R. Dirzo and A. Vargas-Mena, unpublished data) by drilling holes of different sizes into the seeds of *O. oleifera*; this accounted for a proportional seed removal of 1, 5, and 10% of fresh weight. We germinated these seeds and obtained germination percentages (Figure 2). The ability to germinate is increasingly reduced with the intensity of damage. Additionally, the experiments suggest that under natural conditions (other things being equal) there would be an

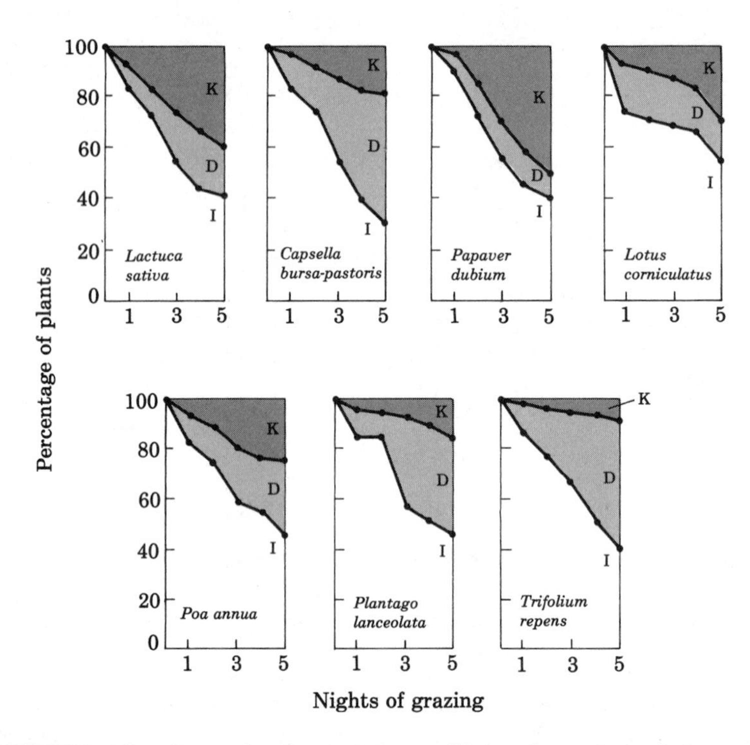

FIGURE 3. The alternative fates of the seedlings of a variety of plant species during a period of grazing by the slug *Agriolimax caruanae*. I, Intact; D, damaged; K, killed. (After Dirzo and Harper, unpublished data.)

increasing advantage for those phenotypes that experience the lowest levels of damage.

For seedlings (the other presumably vulnerable phenostage of the plant's life cycle), it is more difficult to locate relevant literature for natural systems. Perhaps not surprisingly, one of the few available reports is that of Darwin (1859): he calculated that 82% of seedling death was due solely to herbivory by slugs and insects grazing on plots of newly emerged seedlings of weeds.

Some illustrative data on the impact of herbivory on the seedling phenostage emerge from work with terrestrial mollusks and their host plants (R. Dirzo and J. L. Harper, unpublished data). Seedling populations of seven species were grown experimentally as monocultures; slugs were allowed to graze on them during several nights and recordings were made of the fate of seedlings in each population. Three fates were distinguished (Figure 3): intact (I), damaged (D), and grazed and killed (K); it is clear that not all of the slug–plant contacts were lethal, but the extent to which contacts were lethal to the plant varied enormously between species.

These data hint at the range of lethal herbivory that can occur in seedling populations, at least for this sort of system. It is quite interesting that these differences in the levels of mortality due to herbivory depend very much on subtle things such as relative plant size [cf. *Capsella bursa-pastoris* (big seedlings) and *Papaver dubium* (tiny seedlings)] or quick acquisition of chemical defense (e.g., cyanogenesis in *Trifolium repens*).

On the whole, mature plants are much more resistant to damage by herbivores, even under conditions of complete or nearly complete defoliation. But this resistance is not by any means universal; furthermore, there are other ways in which plant fitness may be considerably reduced without actually being killed, for example, when intense defoliation reduces the plant's reproductive capacity (Rockwood, 1973).

In summary, then, the harm done to an individual plant will rarely be simply measured by the amount of tissue removed. A unit weight removed may mean, if the plant is a small seedling, inevitable death (e.g., *P. dubium* in the previous example), whereas the same unit weight removed from an adult plant, or even from a large seedling (e.g., *C. bursa-pastoris* in the previous example), may be an almost insignificant fraction of a senile and almost useless leaf. Thus, it follows that in many instances herbivory should be studied at a finer plant-level.

The modular structure of plants

For many plant–herbivore systems, particularly when the herbivores are small and short-lived relative to the plant, the most appropriate way of treating the system is by considering the plant as a metapopulation of parts (White, 1979; Caughley and Lawton, 1981). Thus, the relevant search image and food unit for a herbivore feeding on an entity with modular construction may be the plant module (e.g., the leaf) and not the whole genet. Similarly, for considerations of the probability of a plant being found by the herbivore, it might be much more useful to work out the density and spatial arrangement of modules (e.g., leaves) in a population than to know the actual number of plants.

Surprisingly, this approach has not been attempted by plant ecologists working on natural systems, although some information can be obtained from agronomic studies (e.g., Gutierrez et al., 1979) or from studies with a more zoocentric emphasis (M. H. Williamson, cited in Harper, 1977:22).

If one considers the plant as an integrated population of parts with many of the attributes of a conventional population (age structure, birth and death rates, survivorship patterns; see Bazzaz and Harper,

TABLE 1. The dynamics of populations of leaves of *Psychotria chiapensis* under different levels of artificial defoliation.

		Control	5% Defol.	25% Defol.	75% Defol.
(a)	Number of leaves at t_0	74	76	60	76
(b)	Number of leaves at t_4	216	187	118	151
(c)	Net change	142	111	58	75
(d)	Rate of increase	2.92	2.47	1.97	1.99
(e)	Σ borne leaves	194	281	215	196
(f)	Σ lost leaves	52	170	157	121
(g)	Leaves surviving from t_0 to t_4	40	21	9	24
(h)	% survival	54	27.7	15	31.6
(i)	Expected time for complete turnover (months)	21.8	13.9	11.8	14.6
(j)	Total number of leaves produced	268	357	275	272
(k)	% mortality	19.5	47.7	57.1	44.5
(l)	% natality	72.3	78.7	78.1	72

1977; Harper, 1977; Bazzaz, Chapter 16), one can attempt to investigate the ways in which herbivory may alter the dynamics of the population of parts. An attempt at such an approach comes from work by B. Zagorin and R. Dirzo (unpublished data) with seedlings of tropical trees grown under a series of experimental conditions. We artificially defoliated the plants at three levels of intensity and then analyzed the response of the plants in terms of modular units (leaves). Table 1 shows a population flow chart, similar to the ones developed for whole plants [cf. Sagar's flowchart (in Harper, 1967) and Sarukhán and Harper (1973)] and shows the population dynamics of leaves under different degrees of herbivory. Some interesting features of the table are (1) the net change, which indicates the great dynamism of the leaf population and which tends to be lower in the defoliated treatments; (2) the population rate of increase which also appears to be negatively affected by herbivory. Unfortunately, these results are only descriptive and not amenable to statistical analysis because all replicates were pooled to increase sample sizes. However, and perhaps more importantly in the context of this volume, this table shows that it is quite feasible to describe the effects of herbivory on the dynamics of plant modules.

Another set of data at the modular level from the same study is the survivorship of different cohorts of leaves of *Psychotria chiapensis* (Figure 4). The survivorship of four different cohorts of leaves in the four defoliation treatments is shown. It is quite evident that the risk of death increases and that life expectancy decreases with the increase in

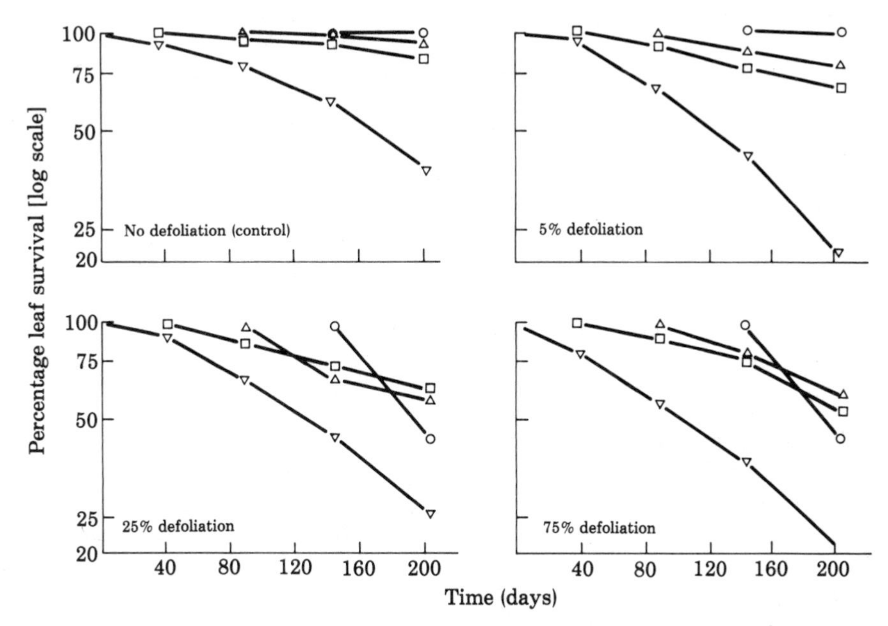

FIGURE 4. The survivorship of four cohorts of leaves of *Psychotria chiapensis* under four regimes of artificial defoliation. ▽, First cohort; □, second cohort; △, third cohort; ○, fourth cohort.

intensity of defoliation. Undoubtedly, studies on herbivory now require the joint expertise of ecophysiologists and module-demographers (see Mooney and Chiariello, Chapter 15; and White, Chapter 1).

In the field I have also been following the fate of individually marked leaves in terms of levels of herbivory per leaf and survival (R. Dirzo, unpublished data). For the five species I have been studying in detail, I have determined herbivory scores of 1–25% leaf area eaten (Figure 5). Comparisons of the fate of damaged and undamaged leaves from one recording date to a subsequent one (Table 2) show that the chances of a leaf surviving are significantly decreased if, on a previous date, it had been recorded as damaged by herbivores. Although this pattern is not always significant, for most recording dates this tendency holds for all species under study. The data suggest that these apparently trivial levels of herbivory can be of great relevance for seedlings that, like these, remain "sitting" on the forest floor for several years under suppressed conditions and usually with only two or three leaves. Consideration of the modular structure of plants has a natural linkage with the next topic to be discussed.

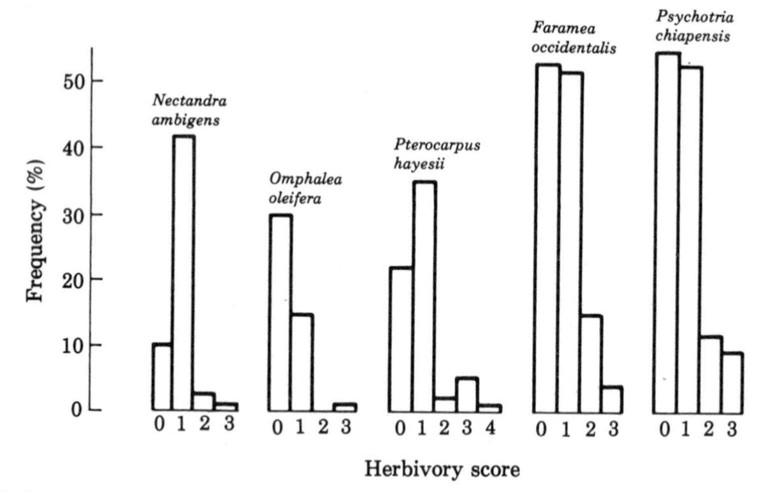

FIGURE 5. Relative frequencies of herbivory scores on the leaves of seedlings of a variety of species. Herbivory scores (in percentage of leaf area eaten): 1, 1–25; 2, 25–50; 3, 50–75; 4, 75–100.

TABLE 2. Contingency analysis (A) for the fate (survival or death) of damaged and undamaged leaves of *Pterocarpus hayesii* between two subsequent recording dates; (B) a summary of results of the same analysis for the other species under study.

A. Contingency analysis for *P. hayesii.*[a]

Leaves	Damaged	Undamaged	Σ
Surviving	12 (18.5)[b]	21 (14.5)	33
Dead	16 (9.5)	1 (7.5)	17
Σ	28	22	40

B. Summary for all species.

Species	χ^2	P
Nectandra ambigens	6.62	<0.025
Omphalea oleifera	2.12	n.s.[c]
Pterocarpus hayesii	10.20	<0.01
Faramea occidentalis	3.73	<0.05
Psychotria chiapensis	1.32	n.s.

[a] $\chi^2 = 10.20$; $P < 0.01$.
[b] Expected numbers in parenthesis.
[c] n.s., Nonsignificant.

Quality of tissue damaged: herbivory to different plant parts

Elaboration of this topic would have to come from the argument that different plant parts, or given parts of different age, have a different value for the plant (Harper, 1977; McKey, 1979); that is to say, on a per unit weight basis, different parts of the plant do not contribute equally to its fitness. Most of the available information relevant to this topic concerns the relative contribution to fitness of leaves of different age. If leaves of different ages do not equally contribute to the fitness of the plant, one should expect the effects of herbivory on the individual plant to be strongly dependent on the age of the leaves that are eaten.

McKey (1979) indicates that relevant studies in this context can be divided into three categories:

1. Studies that measure photosynthetic rate, carbon export, or any other measure of production by leaves of different ages.
2. Studies that compare the effects of simulated herbivory on leaves of different ages.
3. Studies that compare the effects of herbivores with preference for leaves of different ages.

Net production. Studies of net photosynthetic rate in relation to leaf age have usually led to the conclusion that maximum photosynthetic rate and maximum export of carbohydrate from a leaf do occur soon after the leaf attains maximum size, after which, photosynthetic efficiency and carbohydrate export decline. Studies by crop physiologists (Figure 6) do support in general terms such generalization. Likewise, 1-year-old needles of *Pinus resinosa* supply more photosynthate to expanding shoots than do 2- to 3-year-old needles (Dickman and Kozlowski, 1968).

Moreover, Kozlowski (1973) has also suggested that young leaves are more active in the production of hormones. Thus, the greater production of photosynthate and hormones should mean that young-age-class leaves make a greater contribution to the plant's fitness than do older ones and, therefore, the physiological consequences of herbivory will depend on the type of foliage that is damaged.

Herbivory simulation. Accordingly, artificial defoliation studies have shown that the removal of older leaves has a much less damaging effect on the plant than does the removal of young foliage. In *Astrocaryum mexicanum*, selective defoliation on the old, in-

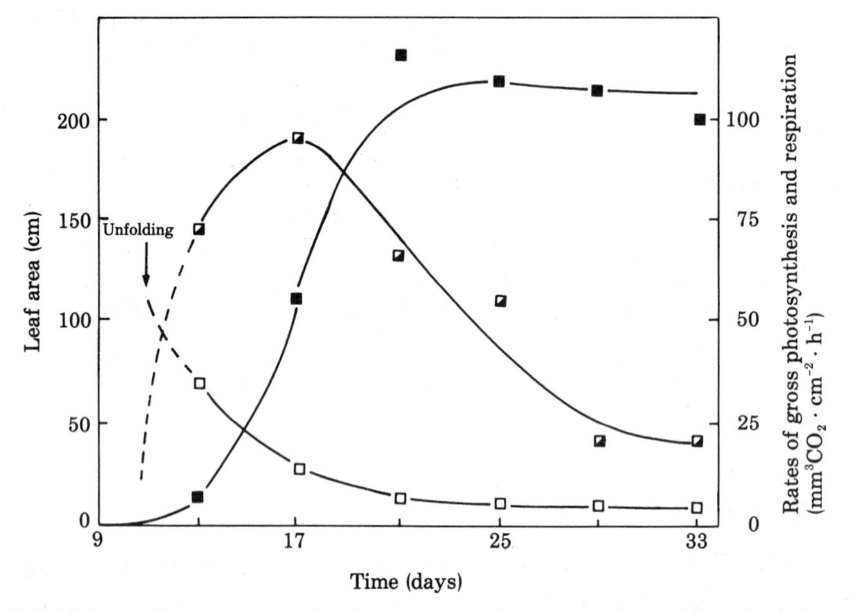

FIGURE 6. The changes in leaf area (■) and in rates of photosynthesis (◪) and respiration (□) with respect to age of the second leaf of cucumber. (After Hopkinson, 1964.)

termediate, or young third of the palm's leaf population (Mendoza, 1981) indicates that removal of the oldest foliage has a smaller detrimental effect in terms of the net foliar change of the plants (Table 3) than defoliation involving the leaves of young (and intermediate) age class. Similar results have been obtained with other plant species (see McKey, 1979 and references therein).

Herbivore preference. In contrast, it is much more difficult to locate in the ecological literature studies that evaluate the effects on the plant of herbivores preferentially feeding on foliage of a given age. On the whole, it appears that most herbivores investigated prefer young tissue, and this finding includes animals as different as Koala bears (R. Dirzo, personal observation), numerous groups of insects (e.g., winter moth caterpillars; Feeny, 1970), and spider monkeys (R. Dirzo, personal observation). Unfortunately, in most of the reported studies, no emphasis is given to the differential effect caused by herbivores preferentially taking young or old tissue.

As a corollary to this section, it could be hypothesized that within an organism, defenses against herbivory should be allocated in direct proportion to the tissue or plant part that confers the greatest fitness to the individual plant. McKey (1974, 1979) has explicitly suggested

TABLE 3. Effects of artificial defoliation on the net foliar change as a function of the age class of the leaves in the palm *Astrocaryum mexicanum*.[a]

Defoliated age class[b]	Net foliar change[c]
Young	0.90 ± 0.21
Intermediate	1.06 ± 0.16
Old	1.63 ± 0.19*

[a] After Mendoza (1981).
[b] Defoliation level = 66% of leaves removed.
[c] Net foliar change = leaves produced − leaves abscised · ind^{-1} · $year^{-1}$. Data are means ± SE; *, significantly different at a significance level of $P < 0.05$.

that since damage to young leaves and buds should result in a greater loss in plant fitness, tannins, resins, alkaloids, and other defensive compounds should be present in greater concentrations in these valuable parts. More studies are needed to evaluate the validity of this generalization. If anything, it is clear that the effect of damage inflicted by a herbivore is strongly dependent on *what part* of the plant is eaten.

Quantity of tissue damaged

It should be evident now that only under very special circumstances should herbivory be evaluated solely on the basis of the quantity of tissue eaten. An ideal study of the impact of herbivory on the plant should (1) measure the amount of tissue eaten (2) taking into account the plant's phenostage; (3) describing how, through modification of its modular structure, the whole plant is affected; and (4) describing how much this effect is accounted for by the quality of removed tissue.

Such an ideal study is not in the literature nor indeed is it easy to do. This is an area for collaboration between physiologists and plant and animal ecologists. Having seen the basic phenomenology involved when considering the individual plant, we can attempt to explore how this translates at the populational level.

Herbivory on the individual and its effect on the plant population

Direct death of the prey in the animal–plant interaction is rather uncommon (as we have seen); more frequently, only parts are eaten and the plant is left to regenerate the animal's meal. Although regeneration can occur, a consequence of this is that the position of the plant in the hierarchical organization of the population may be damaged (Harper, 1977; Bazzaz, this volume). It follows that the role of the grazing animal will generally have to be seen in the context of these postherbivory effects. This situation is summarized in Figure 7. In a population in which grazing is nonselective, the effects will be simply to reduce the photosynthetic area and to bring the plants to an earlier stage of growth; by contrast, in a situation in which herbivory is selective, the effect is not only to reduce growth of the damaged plant. In

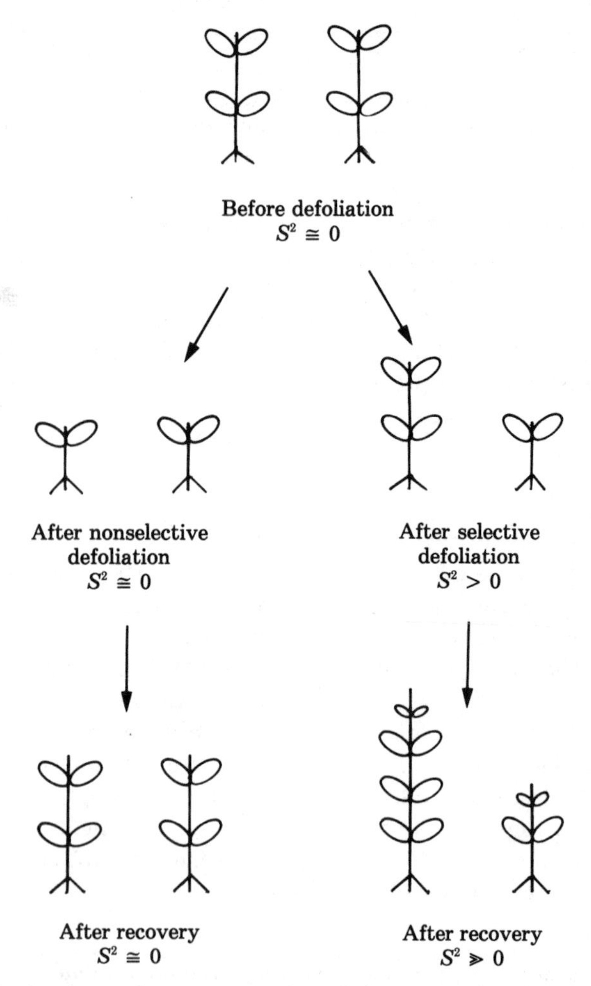

FIGURE 7. A schematic representation of the outcomes of a plant–plant interaction after an episode of selective and nonselective herbivory. Below each situation is shown the variance (S^2) for plant size (e.g., biomass) to indicate how herbivory may generate a hierarchy of plant sizes in the population. (After Harper, 1977.)

addition, plants that escape grazing are left to interact with less aggressive neighbors. These postherbivory effects are likely to have consequences on either growth, survival, or reproductive capacity.

Illustrative examples of the combined effects of herbivory plus plant competition at the intraspecific level come from some experimental studies. In the experiment with mollusks referred to previously (Figure 3), after a period of grazing on seedling monocultures of seven species, slugs were removed from the populations and all the plants that survived, both grazed and ungrazed, were marked and analyzed for dry weight production after a postgrazing (recovery) period. Figure 8 shows the resulting frequency distribution of dry weight for each species. The lower bars in the histograms show the proportion of grazed plants falling into each of the size classes, after the recovery period. With the exception of the populations of *Lotus corniculatus*, and to a lesser extent of *Plantago lanceolata*, the grazed individuals tended to be overrepresented in the smaller or smallest size categories

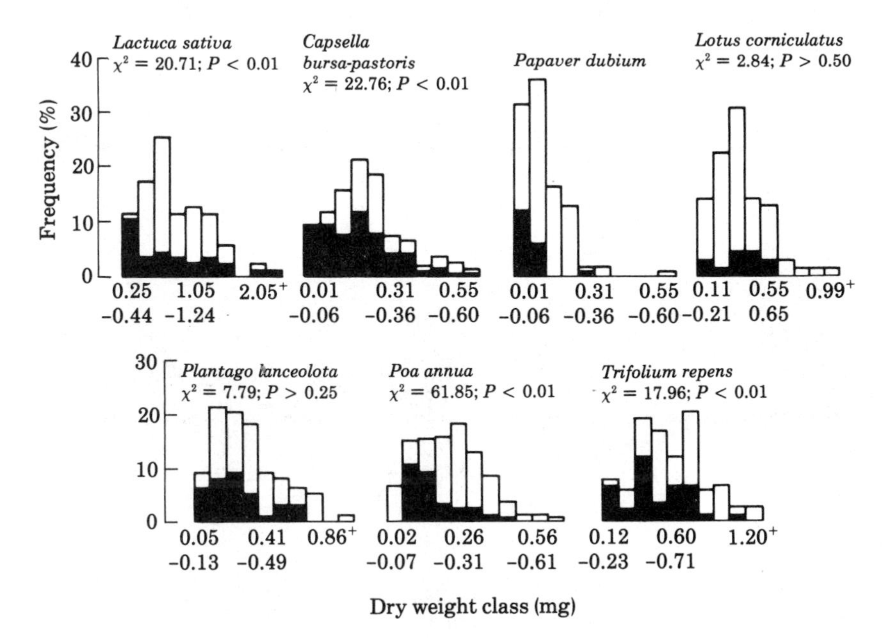

FIGURE 8. The frequency distribution of dry weights of seedlings of seven plant species after a recovery (postgrazing) period. The χ^2 values correspond to tests of the distribution of grazed (bottom bars) and ungrazed (upper bars) seedlings in each weight class (see text for details). (After Dirzo and Harper, unpublished data.)

of the population; with these two exceptions, χ^2 comparisons of the number of individuals in each size class were significant, indicating a greater probability for the grazed plants to belong to the subordinate size class in the population hierarchy (R. Dirzo and J. L. Harper, unpublished data).

The experiments of B. Zagorin and R. Dirzo (unpublished data) referred to previously (see section on The Modular Structure of Plants) are quite illustrative of the way herbivory may interact with other environmental stresses typical of certain systems. In the tropical rain forest, tree-falls are a common event and they produce light gaps in the forest (see Hartshorn, 1978) in which seedlings suddenly get released from the stress of shade. We hypothesized that herbivory could have quite a different meaning depending on whether the plant was growing in the shade or in a gap. We then designed an experiment in which we included two light conditions (shade and sunshine), competition (i.e., two contrasting densities), and three levels of artificial herbivory. Figure 9 shows plant survival in all the applied treatments for *O. oleifera*. In the absence of light stress, plants at low density experienced no mortality at all; plants at high density experienced some minor mortality correlated to the intensity of defoliation. In the shade, defoliation had no effect at all provided the plants were at low density;

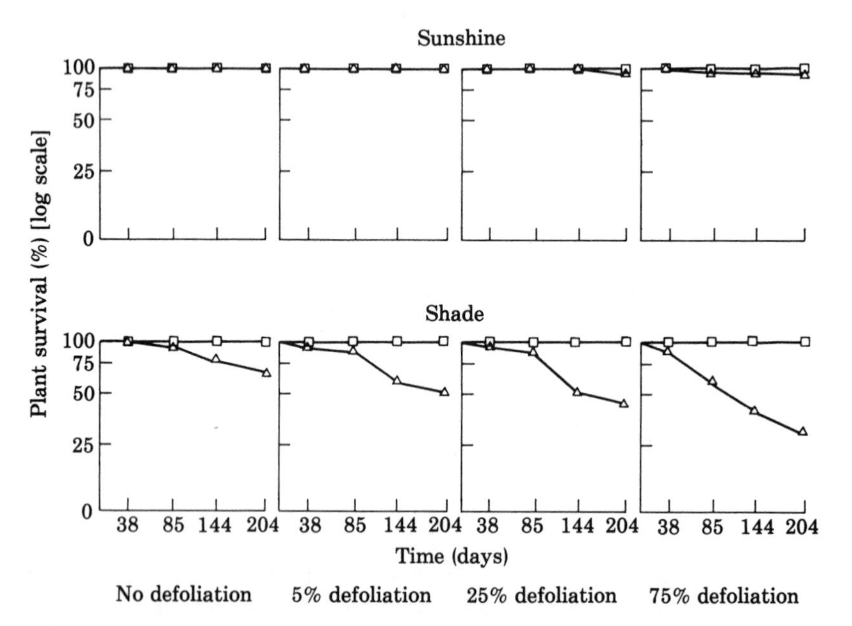

FIGURE 9. The survivorship of seedlings of *Omphalea oleifera* grown under two light conditions (sunshine and shade); two plant densities, low (□) and high (△); and four defoliation treatments (0, 5, 25, 75% leaf area removed).

however, under conditions of high density, defoliation increasingly reduced plant survival. This experiment illustrates clearly the combined effect of herbivory plus plant competition in the presence of another relevant stress from the physical environment.

In the experiments on artificial seed predation mentioned earlier (see Figure 2), the seedlings that emerged from the seed that experienced 0 (= control), 1, and 5% seed weight removal were grown together (i.e., in competition) under experimental conditions. Some preliminary results on the growth of these seedlings are shown in Figure 10 for *O. oleifera*. When subjected to competition, seedlings that emerged from control and 1% damaged seeds showed no difference in their mean height whereas those that emerged from 5% damaged seeds had a markedly reduced growth.

It is not difficult to envisage the relevance of these differences in plant performance under the conditions of crowding and light stress that these seedlings usually experience in the understory of the forest. The experiment also illustrates the subtle role herbivory or seed preda-

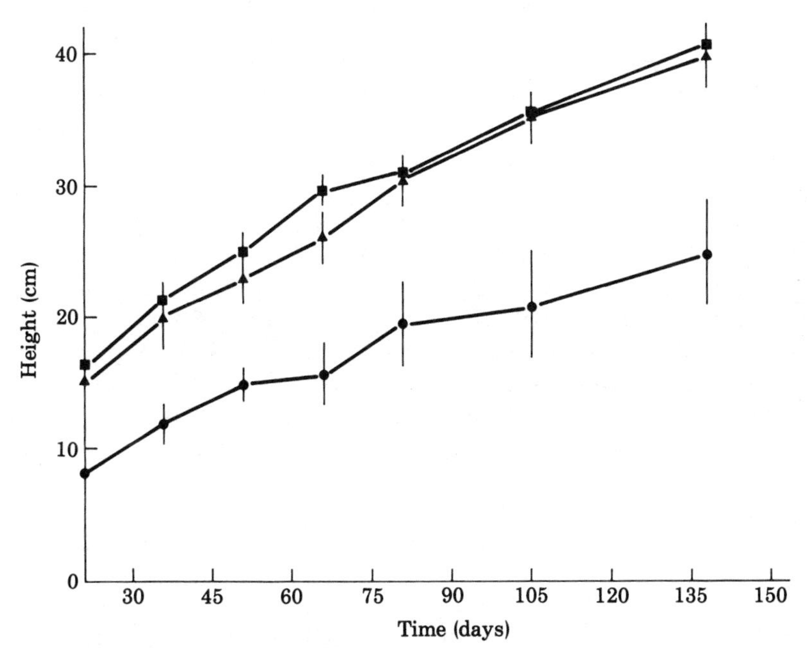

FIGURE 10. The growth of seedlings of *Omphalea oleifera* that emerged from seeds subjected to three levels of artificial seed predation: ■, no seed weight removed (control); ▲, 1% seed weight removed; ●, 5% seed weight removed.

tion may play in conjunction with plant competition on the population dynamics of plants.

This situation of herbivory coupled with plant–plant interactions could be extended to that in which the biological matrix of a plant is determined by a host of neighbors that are not only conspecific but are of other species as well. Considerable attention has been devoted to this experimental topic probably because of (1) its relevance for numerous applied problems in grassland management (introduction to and withdrawal of herbivores from particular communities, etc.) and (2) the argument that grazing may play a fundamental role in shaping and maintaining the structure of plant communities.

Relevant information began to be produced early in this century when Tansley and Adamson (1925) used fences to exclude rabbits from chalk grasslands and followed the changes in the vegetation after rabbit removal. More recently, reviews have been written by Harper (1977), Whittaker (1979), Caughley and Lawton (1981), Lubchenco and Gaines (1981), and Ceballos (1982). These reviews place emphasis on the details of plant population dynamics, as affected by herbivory in combination with interspecific competition. This type of study is well represented by the excellent work of Sibma et al. (1964), a study involving mixtures of oats and barley grown in replacement series (Figure 11). The plants were grown in containers of soil infested or non-infested with the root nematode *Heterodera avenae*. The barley was resistant to the nematode and the oats were susceptible. In the absence of *H. avenae*, oats were very aggressive toward barley in the mixture (relative crowding coefficient, $K_{ob} = 6.0$) but in the presence of the nematode, the situation changed completely: the interaction became almost equally balanced ($K_{ob} = 1.3$).

Ceballos (1982) carried out some work with the slug *Agriolimax caruanae* and two of its host plants: *Capsella bursa-pastoris*, which is highly acceptable to this slug, and *Poa annua*, which is wholly unacceptable (Dirzo, 1980a). He grew populations of both species in monocultures and 50% mixtures with 0, 2, and 4 slugs per experimental plot; he then measured the performance of the populations. Figure 12 shows that the relative replacement rate (RRR) of *P. annua* toward *C. bursa-pastoris* was very high and increased with the density of slugs; the opposite was true for the RRR of *C. bursa-pastoris* toward *P. annua.*

An interesting contrast of the work of Ceballos with respect to that of Sibma et al. (1964) (cf. Figure 11), Cates (1975), Dirzo (1980b), and others is that in his study, the weaker competitor was also the more acceptable diet to the herbivore. In this situation, differential herbivory may lead to complete suppression of the weaker competitor, whereas in the other studies preferential herbivory on the stronger competitor tends to produce balanced mixtures.

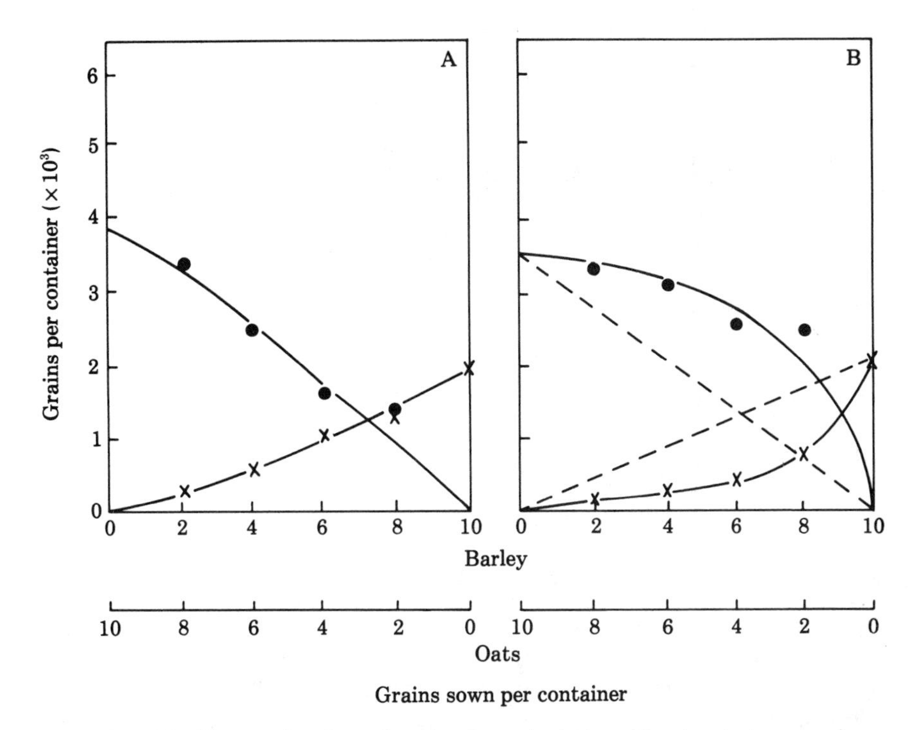

FIGURE 11. The production of grains by oats (●) and barley (×) grown in a replacement series experiment in the presence (A) and absence (B) of the nematode *Heterodera avenae*. (After Sibma et al., 1964.)

These contrasting results can be extended to the situation of multispecies mixtures. Experiments in which herbivores or predators are removed from the community (e.g., Tansley and Adamson, 1925; Paine, 1966; Harper, 1969; Lubchenco, 1978) have shown that under certain circumstances herbivores generate and maintain diversity whereas under other circumstances they reduce it. It now appears that the outcome of the situation depends on the relationship between competitive capability of the food plants and herbivore preferences (see Lubchenco, 1978). When the competitive dominant is also the preferred food, herbivory prevents competitive exclusion and diversity is high; when the preferred food *is not* the competitive dominant, diversity is decreased.

The situation in which preference is correlated with superior competitive ability is particularly interesting, for this could be interpreted as a trade-off between competitive ability and antiherbivore resource

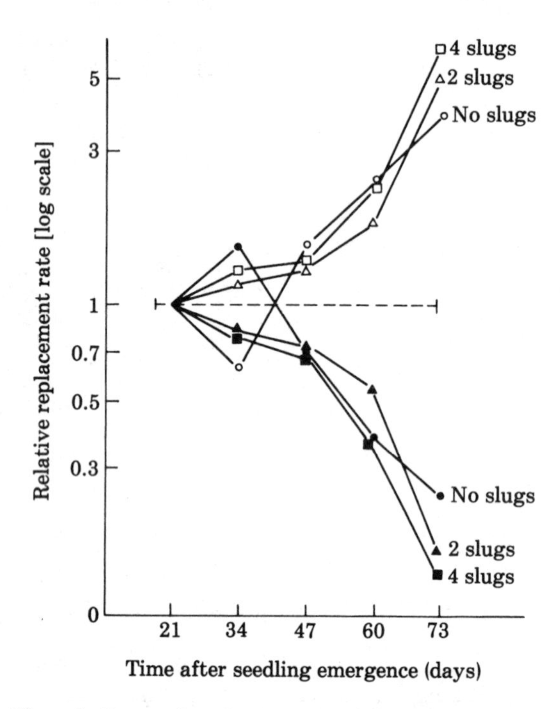

FIGURE 12. The relative replacement rates of *Capsella bursa-pastoris* toward *Poa annua* (solid symbols) and of *P. annua* toward *C. bursa-pastoris* (open symbols) at different times after seedling emergence and under three densities of the slug *Agriolimax caruanae*. (After Ceballos, 1982.)

allocation (defense). Plant polymorphisms for some presumed antiherbivore trait provide a good system with which to explore such trade-offs. In *Trifolium repens* there are cyanogenic forms that are well protected against grazing by slugs and snails (Dirzo and Harper, 1982a); acyanogenic forms, in contrast, are readily eaten but show some signs of being potentially better competitors (Dirzo, 1980b; Dirzo and Harper, 1982b). Competition experiments coupled with artificial herbivory on the two morphs of *T. repens* (Figure 13) can be intrepreted in support of the trade-off hypothesis. If such trade-offs are widespread in nature, they may represent an evolutionary process underlying species balance via herbivory. This evolutionary interpretation leads to the final section of this chapter.

ULTIMATE (EVOLUTIONARY) INTERACTIONS

Evolutionary interpretations for the plant–herbivore interaction have been suggested by various authors in several contexts. For example, Stanley (1973) claims that grazing was the ecological factor that

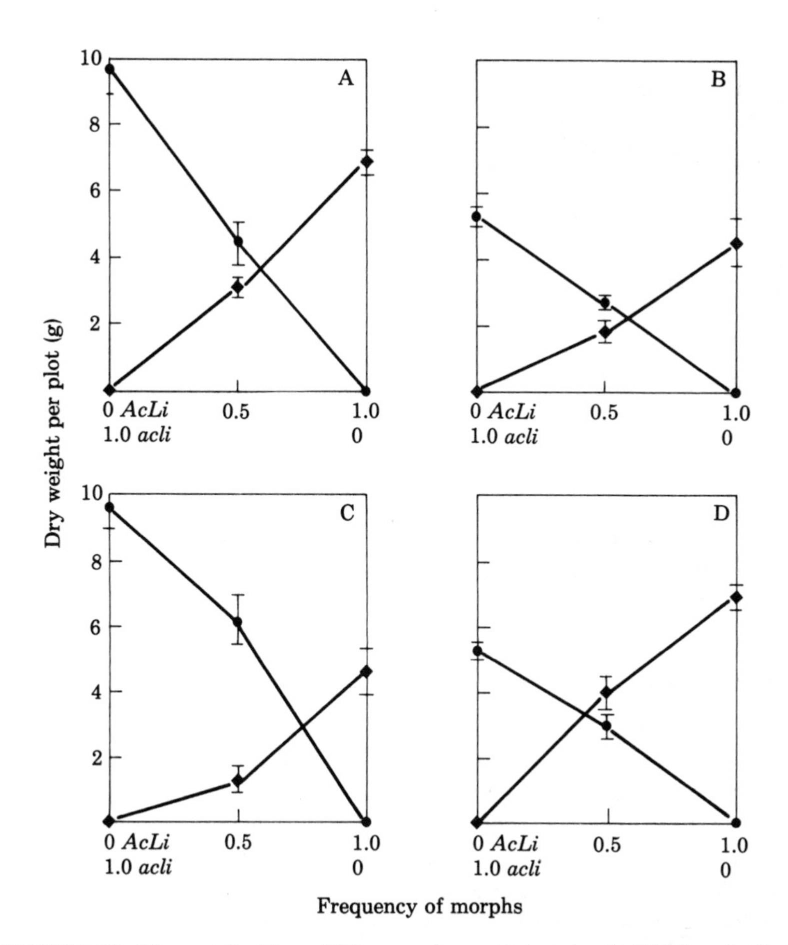

FIGURE 13. The production of biomass by cyanogenic, *AcLi* (diamond), and acyanogenic, *acli* (circle), morphs of *Trifolium repens* grown as monocultures and 50:50 mixtures under four treatments of artificial defoliation. A. Both morphs undefoliated. B. Both morphs defoliated. C. Only *AcLi* defoliated. D. Only *acli* defoliated.

destroyed the competitive monopoly responsible, in his view, for the poor diversity of organisms of the late Pre-Cambrian. He argues that at that time, most organisms were limited by interspecific competition for resources until grazing finally came on the scene and, by reducing competition among the producers, allowed the existence of many more types of plants, which in turn released a jet of evolution of both consumers and producers.

Other evolutionary interpretations are more modest and can be grouped into three main categories:

1. The correlation between plant life-history and strategy of defense by the plant (Feeny, 1976; Rhoades and Cates, 1976; Rhoades, 1979; McKey, 1979).
2. Coevolution between herbivores and their food plants (Ehrlich and Raven, 1964; Gilbert and Raven, 1975; Jones, 1973; Harborne, 1978; Dirzo and Harper, 1982a), as well as other evolutionary interpretations, correlated with coevolutionary arguments [e.g., the number of herbivores associated with given plant taxa (Southwood, 1962; Janzen, 1968)].
3. The interaction between herbivores and intraspecific variants within a given plant population.

Categories 1 and 2 have been extensively reviewed by other workers (see references therein) and are out of the scope of this chapter. In contrast, category 3, which is the most directly relevant in the context of the present volume, has been very poorly explored. In the following sections I will present some recent findings with intraspecific variants that might be of some evolutionary significance.

Heteromorphic life-history in algae

A number of algae typical of the intertidal zones of rocky shores alternate between two major forms in their life cycle: an upright filament or blade and a nonupright, fleshy crust. The functional and evolutionary significance of such heteromorphic life cycles has been interpreted recently as an adaptation to fluctuations (spatial or temporal) in grazing pressure (Lubchenco and Cubit, 1980; Slocum, 1980). Observations and experimental removal of herbivores along rocky shores of New England and Oregon (Lubchenco and Cubit, 1980) and Washington (Slocum, 1980) suggest that the upright morphs are adapted for high rates of growth and reproduction when grazing pressure is low, whereas the crustose morphs are well adapted for surviving in areas and through times of high grazing pressure. Presumably the combination of both sets of advantageous attributes overcompensates for the disadvantages inherent to each of the morphs (e.g., the blades are very susceptible to herbivory and the crusts are very susceptible to shading and overgrowth); and the strategy (heteromorphology) is selected for if the appropriate set of tactics (blade or crust) is used in the situations in which they are likely to be successful (see Stearns, 1976; Harper, 1977). Thus, for this kind of plant–herbivore system, "possessing heteromorphic phases and some control over their expression, may increase fitness, both in each phase and in the complete cycle of the individual plant" (Slocum, 1980:108). Most of these findings and interpretations are summarized by Lubchenco and Cubit (1980) in a list of the predicted life-history patterns in relation to the mode of grazing pressure (Table 4).

Genetic polymorphism for a defensive trait

This type of intraspecific genetic variation (especially when only one or few loci are involved) is particularly useful for the exploration of a number of aspects of some evolutionary significance. The common background of similarity between the morphs enables a realistic evaluation of the adaptive value of the presumed defensive trait (see Jones, 1973; Dirzo and Harper, 1982a). Some of the possible consequences of resource allocation to defense were mentioned earlier for *Trifolium repens*, the white clover. For this species, a detailed study was made of the interaction between the clover morphs and terrestrial mollusks (the main clover herbivores) in the field (Dirzo and Harper, 1982a). An analysis of the distribution of mollusks (slugs and snails) in the field, showed that there are some areas of intense activity of these herbivores, whereas in other areas slugs and snails are absent, or very rare. At the same time, an analysis was made of the distribution of cyanogenic and acyanogenic morphs throughout the same field (Figure 14). An interesting correlation emerged from these two surveys (Table 5): in areas of intense activity of mollusks there was a tendency for

TABLE 4. Predicted life-history patterns of some algae typical of intertidal rocky shores under different modes of selective pressure due to grazing.[a]

Mode of grazing pressure	Predicted strategy and tactics
Constant and light	Isomorphology: advantage of competitively superior morphs (uprights)
Constant and heavy	Isomorphology: advantage of grazing-resistant morphs (crusts)
Variable and predictable	Heteromorphology: seasonal alternation of production and predominance of each morph
Variable and unpredictable	Heteromorphology: continuous production (though not survival) of both morphs

[a] After Lubchenco and Cubit (1980).

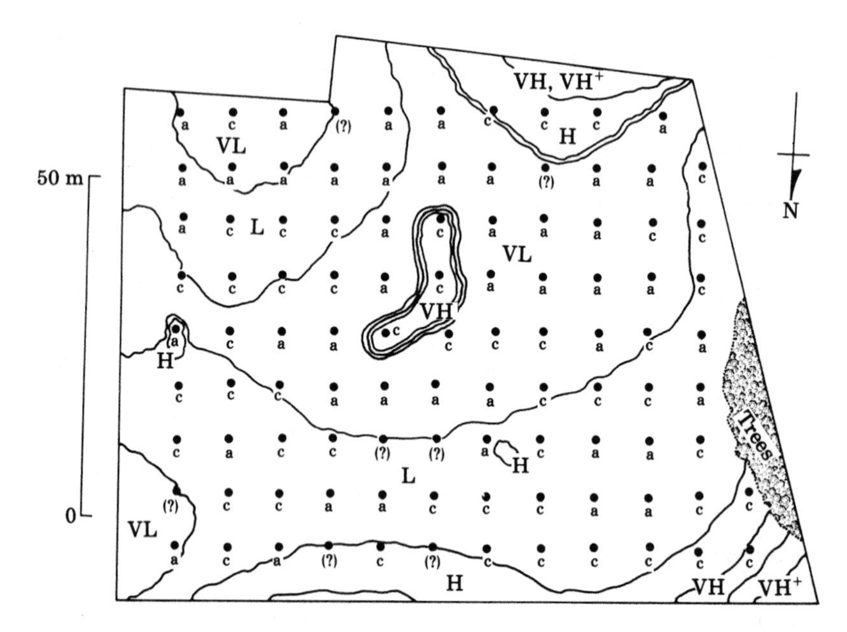

FIGURE 14. Contour map of the density of active mollusks at Henfâes Field, Aber, North Wales. The values shown by the contour lines are the mean number of mollusks observed at each sampling site (shown by).The mnemonics for mollusk density are VL, very low; L, low; H, high; VH and VH+, very high. The letters below each site indicate the morph of sampled clover plants: cyanogenic (c) and acyanogenic (a) (of three different phenotypes); (?) indicates clover samples lying on the boundary between two adjacent areas. For details, see original paper. (After Dirzo and Harper, 1982a.)

TABLE 5. The number of samples of the two phenotypes of *Trifolium repens* and their distribution in areas of different density of active mollusks (shown in Figure 14).[a]

Mollusk density	PLANT PHENOTYPE[b]		
	Cyanogenic	Acyanogenic[c]	Total
Very high + high	10 (5.8)	1 (5.2)	11
Low	29 (23.7)	16 (21.3)	45
Very low	10 (19.5)	27 (17.5)	37
Total	49	44	93

[a] After Dirzo and Harper (1982a).
[b] Expected values in parentheses. $\chi^2 = 18.7$; $P < 0.001$.
[c] The acyanogenic samples include three different morphs incapable of producing the cyanogenic reaction.

overrepresentation of cyanogenic morphs whereas acyanogenic morphs were underrepresented; the reverse was found in areas where mollusks were rare or absent. Thus, it appears that the spatial heterogeneity created by the uneven distribution of herbivores may create suitable or unsuitable patches for the establishment and survival of the clover morphs.

This type of event may ultimately be, at least partly, responsible for the maintenance of genetic variants within a given plant population.

The critical test for this type of system would be to expose the morphs of the plant in question to the host of environmental variables—besides herbivory. This test should be done under natural conditions, and the fate of individual plants should be followed carefully. Dirzo and Harper (1982b) attempted to conduct such a test, again with white clover. Differences were found for plant survival, leaf and stolon production, and susceptibility to damage by pathogens, herbivores, and physical stress (e.g., winter temperatures). More importantly, this study reveals that a realistic evaluation of the role herbivory may play in the population biology of plants should be made in the context of other features (both physical and biotic) of the plant's environment.

To conclude, I would like to say to the plant population ecologist that the incorporation of animals into his plant systems offers a fascinating area for exploration; I think that the increasing willingness to consider herbivores as an important component of the population biology of plants will be an extremely rewarding field in the science of plant ecology and is, besides, an appropriate means of commemorating the work of Charles Darwin.

CHAPTER 8

USING INTRASPECIFIC VARIATION TO STUDY THE ECOLOGICAL SIGNIFICANCE AND EVOLUTION OF PLANT LIFE-HISTORIES

D. Lawrence Venable

INTRODUCTION

Life-histories represent an interesting bridge between the fields of ecology and evolution because they are directly interpretable ecologically as adaptations for survival and reproduction. Demographic parameters are also interpretable as components of fitness so that the strength of natural selection on life-history variation can be measured. Thus, life-histories represent an important and convenient focus for the study of the evolution of ecological phenomena, but their utility for investigations depends on the feasibility of collecting accurate demographic information in natural settings. For this reason, plants, as ubiquitous sessile organisms easily marked and relocated, should play an important role in evolutionary ecological studies of life-histories.

In this chapter I shall first briefly discuss the strengths and weaknesses of different ways of investigating life-histories. Second, I shall discuss what we know and would like to know about the nature of intraspecific variation, and finally I shall discuss the demographic interaction of this variation with environmental factors important as selective agents.

WAYS OF STUDYING LIFE-HISTORIES

A variety of methods can be used to investigate the ecological significance and evolution of plant life-histories. These include mathematical modeling, comparative surveys, detailed comparison of a few species, and study of intraspecific variation. Each of these methods has certain advantages and disadvantages with respect to the others. Mathematical models may be used to formulate testable hypotheses about the ecological consequences and evolution of life-histories (Levins, 1966; Cody, 1974b; Maynard Smith, 1978b). They are powerful tools for discovering the logical consequences of assumptions about the biology of a system. Furthermore, they enable us to go beyond our data to explore hypothetical situations via simulation. Optimization models, in particular, depend on correctly identifying a measure of fitness that is actually maximized by evolution and also on an appropriate selection of constraints. Mathematical models are sometimes vulnerable to the criticism of adaptive storytelling. This criticism, leveled by Gould and Lewontin (1979), refers to the fact that it is often easy to develop a plausible adaptive story in keeping with one set of hypothetical assumptions but to forget that there may be other equally plausible explanations.

The comparative survey can be useful to uncover general patterns in nature such as those between seed size and habitat (Salisbury, 1942; Baker, 1972b) or between dispersal structures and habit (Venable and Levin, 1983). Yet it suffers from a basic weakness in that it is correlational rather than experimental in approach (Huff, 1954, Chapter 8). Furthermore data on the potentially important confounding factors are frequently difficult to obtain when a large number of species is surveyed.

With comparisons of a few related species, one can attempt to evaluate hypothetical causal mechanisms that could only be inferred from correlational studies (e.g., Wilbur, 1976, 1977; Schaffer and Schaffer, 1977, 1979). One can obtain data to test for potential environmental confounding factors and often control for them experimentally. Of course, there may be confounding differences due to biological differences between species other than those involving life-history. Even with closely related species, we investigate the population dynamics of the products of evolution. We cannot necessarily infer the dynamics of the raw material in its natural selective milieu (see Hickman, 1979).

Intraspecific comparisons present some unique opportunities for the study of the ecology and evolution of life-histories. The relatedness of individuals of the same species and especially of the same population provides some control for confounding differences due to traits irrelevant to the life-history question at hand. With intraspecific life-history variation, demographic differences can be directly interpreted as components of fitness. Selective agents can be manipulated and their effects on the fitness of different life-history variants can be directly measured. Intraspecific life-history variation can often be partitioned into genetic and environmental components to give some idea of evolutionary potential or history (Falconer, 1981; Mather and Jinks, 1977). For example, high within-population genetic variation for a life-history trait, such as annuality–perenniality, might imply potential for evolution by individual selection. A high genetic variation between populations suggests evolutionary divergence, although one cannot discriminate between natural selection and drift as the diversifying agents. The important implication here is that intraspecific life-history variation provides the opportunity for monitoring the dynamics of life-history change within a natural ecological context.

Use of intraspecific variation to study life-history ecology and evolution does have disadvantages. There is always the possibility that results from a single species do not represent general ecological or evolutionary dynamics. Another potential problem is the lack of interesting life-history variation at the intraspecific level. Yet the history of plant genecology has been a discovery of more and more intraspecific variation including life-history traits (see Table 1). Another problem arises from the suggestion that intraspecific variation in life-histories may be largely nonadaptive with respect to the selective factors hypothesized in life-history theory (Stearns, 1980). This is because of physiological, developmental, and genetic constraints. Stearns cites the example of the Procellariform birds which are morphologically constrained to lay a single egg at a time. Plants are typically not so constrained. Modular construction with plasticity in the number of modules usually provides developmentally, physiologically, and allometrically simple ways to vary life-history traits. Even if a fruit is constrained to produce a single seed, the number of fruits per plant is flexible. Genetic constraints such as negative genetic covariance are potential factors retarding response to selection on life-history characters but have not often been successfully studied for plant life-history traits (but see Primack and Antonovics, 1981, 1982). Most kinds of balancing selection that can explain high levels of isozyme polymorphism of nonneutral alleles at a single locus could ex-

TABLE 1. Some examples of intraspecific variation in life-history traits. Intraclass correlation coefficient (t) and heritability (h^2) are given when available.

Character	Distribution of variation	Genetic versus environmental control	Citation, organism, and technique
Germination	Interpopulational	t = 0.51, 0.59, 0.43, 0.21, 0.23	Schmidt (1982) *Phlox drummondii* Reciprocal transplant
Survivorship		t = 0, 0, 0, 0.19, 0.30, 0, 0, 0, 0, 0, 0.35, 0.24, 0.21, 0.30, 0	
Fecundity		t = 0.56, 0.52, 0.33, 0, 0, 0, 0, 0, 0, 0.48, 0.43, 0.11, 0, 0, 0	
Finite rate of increase		t = 0.74, 0.33, 0.14, 0, 0, 0, 0, 0, 0, 0.58, 0.28, 0, 0, 0, 0	
Seedling survivorship	Interpopulational (annual, perennial, and intermediate forms)	t = 0.30	Oka (1976) *Oryza perennis* Transplants to natural sites
Adult survivorship		t = 0.37	
Seed bank (no. of viable buried seed)		t = 0.21, 0.43	
Height	Intrapopulational (central peripheral)	N.S.[a]	Linhart (1974) *Veronica peregrina* Greenhouse
Seeds/plant		Genetic component	
Seed weight		Genetic component	
Days to flower		N.S.	
Branches/plant		Genetic component	
Percentage biomass vegetative		Genetic component	
Percentage biomass in seed		Genetic component	
Total biomass		Genetic component	
Germination rate		Genetic component	
Total germination		N.S.	
Competitive ability	Intrapopulational	Genetic component	Solbrig and Simpson (1974) *Taraxacum officinale* Greenhouse
Seed set		Genetic component	

TABLE 1. (continued)

Character	Distribution of variation	Genetic versus environmental control	Citation, organism, and technique
Survival Fecundity	Intrapopulational (central and marginal)	Genetic component Genetic component	Grant and Antonovics (1978) *Anthoxanthum odoratum* Reciprocal transplants
Survival Growth Fecundity	Interpopulational	$t = 0$ $t = 0$ $t = 0$	Antonovics and Primack (1982) *Plantago lanceolata* Reciprocal transplants
Inflorescences/ plant Capsules/ inflorescence Seed/capsule Seed weight Yield	Intrapopulational	$h^2 = 0.49$ $h^2 = 0.56$ $h^2 = 0$ $h^2 = 0$ $h^2 = 0.44$	Primack and Antonovics (1981) *Plantago lanceolata* Growth chamber
Prereproductive time	Intrapopulational Interpopulational	$t = 0.48$ $t = 0.21$	Law et al. (1977) *Poa annua* Garden transplant
Plant diameter	Intrapopulational Interpopulational	$t = 0.25$ $t = 0.20$	
Age-specific reproduction	Intrapopulational Interpopulational	$t = 0.15\text{-}0.52$ $t = 0.05\text{-}0.26$	
Flowering time	Interpopulational Intrapopulational	$t = 0.61$ $t = 0.72$	Jones (1971) *Arabidopsis thaliana* Garden transplant
Rosette width	Interpopulational Intrapopulational	$t = 0.79$ $t = 0.26$	
Plant height	Interpopulational Intrapopulational	$t = 0.59$ $t = 0.17$	
Days of flowering Tiller no. Height Seed number	Intrapopulational	$t = 0.75, 0.69, 0.49, 0.33, 0.44, 0.40$ $t = 0.34, 0.15, 0.27$ $t = 0.22, 0.58, 0.46, 0.61, 0.50$ $t = 0, 0, 0.28$	Imam and Allard (1965) *Avena fatua* Garden transplants

TABLE 1. (continued)

Character	Distribution of variation	Genetic versus environmental control	Citation, organism, and technique
Days to flowering	Intrapopulational		Kannenberg and Allard (1967) Garden transplant
	Lolium multiflorum	$h^2 = 1.89$	
	Avena fatua	$h^2 = 0.55$	
	Festuca microstachys	$h^2 = 0.38$	
Height at maturity	*L. multiflorum*	$h^2 = 0.67$	
	A. fatua	$h^2 = 0.48$	
	F. microstachys	$h^2 = 0.37$	
Tiller no.	*A. fatua*	$h^2 = 0.26$	
	F. microstachys	$h^2 = 0.40$	
Leaf number Leaf birth Leaf death	Parapatric (adjacent woodland/grass)	Genetic/ environmental interaction	Lovett Doust (1981c) *Ranunculus repens* Reciprocal transplant
Tiller no. Height No. leaves No. spikes Spikelets/spike Seeds/clone No. rhizomes Rhizome length Rhizome weight Root weight Tiller weight	Between subpopulations	Genetic component for most	Silander and Antonovics (1979) *Spartina patens* Common environment
Seeds/plant Seeds/spikelet Spikelets/ inflorescence Spikelets/plant Inflorescences/plant Yield/plant Seed weight	Interpopulational	N.S. in reciprocal transplants, genetic component for some in a growth chamber	Clark (1980) *Catapodium rigidum; C. marinum* Reciprocal transplants and growth chamber
Phenology of runner production Branch crown production No. flowers No. fruit Achene weight Fruit weight	Interpopulational	Genetic component in all	Hancock and Bringhurst (1978) *Fragaria vesca* Greenhouse

TABLE 1. (continued)

Character	Distribution of variation	Genetic versus environmental control	Citation, organism, and technique
Total dry weight Shoot dry weight Root: shoot ratio Stolon length Rate of stolon extension No. flower heads, florets/head	Intrapopulational	t = 0.11–0.81	Burdon (1980) *Trifolium repens* Greenhouse
Emergence Seedling weight Relative growth rate	Interpopulational	$h^2 = 0.54$ $h^2 = 0.45$ $h^2 = 0.05$	Fakorede and Oko (1981) *Zea mays* Garden
Germination time	Intrapopulational	$h^2 = 0$	Arthur et al. (1973) *Papaver dubium* Garden
Seed dormancy	Interpopulational	$t = 0.76$	Morley (1958) *Trifolium subterraneum*
Seed dormancy	Intrapopulational	Response to selection to different degrees	Hilu and de Wet (1980) *Eleusine indica; E. coracana*
Seed dormancy	Intrapopulational	$h^2 = 0.13$	Witcombe and Whittington (1972) *Sinapis arvensis*
Dormancy	Intrapopulational (4 populations)	$h^2 = 0.50$	Jana and Naylor (1980) *Avena fatua*
Seed and tuber dormancy	Interpopulational	Significant genetic variation and positive genetic correlation	Simmonds (1964) *Solanum tuberosum*

[a] N.S., Nonsignificant.

plain the maintenance of positive genetic correlations as well as heritable variation for traits contributing significantly to fitness. These factors include frequency-dependent selection, spatial and temporal variation in selection, and differential selection between the sexes or life stages. The question of genetic constraints on adaptive life-history variation in plants is an interesting one deserving to be more fully investigated.

EXPERIMENTAL INVESTIGATIONS ON THE NATURE OF VARIATIONS IN LIFE-HISTORY

A survey of the literature on plant life-histories (Table 1) indicates that intraspecific variation exists for a wide variety of traits including germination, age-specific survivorship, fecundity, finite rates of increase, seed bank dynamics, seed yield and yield components, and module dynamics (leaf number, rates of leaf birth and death) (see Table 1 for examples). Differentiation in these traits has been documented frequently at the intrapopulational and interpopulational levels. Techniques used to investigate genetic versus environmental components of life-history variation vary from transplants to common natural, garden, or growth chamber environments to natural reciprocal transplant experiments.

Although original citations often treat the genetic component of variation as a "yes-or-no" question, I have tried to extract quantitative information from analysis of variance tables in the form of heritabilities (h^2) or intraclass correlation coefficients (t). Heritabilities should reflect the additive genetic component of phenotypic variance. The h^2 values reported in Table 1 were determined by the original investigators and vary somewhat in meaning as a result of differences in experimental design and rigor. They should, therefore, be interpreted cautiously. The intraclass correlation is an estimate of the degree of genetic control from the proportion of the phenotypic variance attributable to the significant differences between groups (families or populations). This is the same between-group variation that the original investigators took as evidence of some genetic control. Since the groups are routinely randomized across the study environments, significant between-group differences should be due to genetic causes. The t's are related to heritabilities by a constant whose value depends on the relatedness of group members and the breeding structure of the population (Falconer, 1981, Chapter 10). These studies demonstrate ample manipulable variation for use in demographic experiments focused on life-history selection. Indeed, a number of the species in Table 1, such as *Trifolium repens* (Burdon, 1980), seem ideally suited for selection experiments.

In addition to demonstrating the existence of manipulable variation, we would like information on the distribution and control of life-history variation to tell us something about evolutionary potential.

Are the patterns of selection documented in nature likely to result in rapid evolutionary adjustments? How canalized are the traits? To understand evolutionary potential, we need simultaneous information for a variety of life-history traits in a single organism. We need to know about the distribution of variation among individuals and populations and the control of the expression of variation by genetic, developmental, and environmental factors. In an investigation of the hierarchical levels of variation, Law et al. (1977) demonstrated variation in *Poa annua* in survival, size, and reproduction within and among habitat types, populations, and families grown in an experimental garden. Selection intensities and genetic covariance are most easily interpreted at the between-individual, within-population level. Between-group selection and between-group genetic variance-covariance are relevant to the study of higher-order selection phenomena, such as group selection.

Information on the control of life-history variation should preferably come from heritability experiments in the field, reciprocal transplant experiments, and the controlled manipulation of important environmental factors. The evolutionarily relevant heritability value is the one calculated in the field with normal levels of environmental variation. The importance of this consideration can be seen by comparing the controlled environment and field investigations of Primack and Antonovics (1981). Coefficients of variation were 17 to 170% larger for field-collected plants than growth chamber-grown plants for three out of four yield components in *Plantago lanceolata*. The fourth yield component, weight per seed, is a relatively invariant life-history character (Harper et al., 1970). A reciprocal transplant study found significant differences in seedling survival of *P. lanceolata* "among small adjacent quadrats placed in an area chosen for its apparent uniformity" (Antonovics and Primack, 1982). Assuming that these large variances documented in the field are due to environmental effects, the additive genetic component of the phenotypic variation is likely to be considerably smaller in natural environments than in controlled environments. This means that potential rates of evolution calculated from greenhouse experiments are likely to be great overestimates.

Reciprocal transplant experiments are desirable for the study of the control of life-history variation because the expression of genetic variation may vary from one environment to another. These genotype-environment interactions are evident in the results of Schmidt (1982) (Table 1). Schmidt collected detailed demographic information on seed banks, germination, survival, and fecundity for reciprocal transplant populations of *Phlox drummondii*. The variance in life-history traits due to the site of origin is significantly different in the different transplant sites (the multiple t values in Table 1 are values for different transplant sites). In some sites there is no significant variation attributable to site of origin, whereas in others more than half of the life-history variation is attributable to site of origin.

Further insights into the environmental component of variation can be gained by manipulating important environmental variables in order to study plastic variation in life-histories. If plastic variability increases fitness, it provides an immediate nongenetic avenue of adaptation. Plasticity increases the environmental component of variance, thereby reducing heritable variation, yet plasticity itself may be under partial genetic control (see Jain, 1978).

If the kinds of information outlined above can be simultaneously collected on several life-history traits, it is possible to calculate genetic and environmental components of covariance at intrapopulational and interpopulational levels. With this information, life-history trade-offs can be understood in terms of components due to different control factors. For example, in a favorable environment big plants might produce many big seeds so that seed size and number would possibly be correlated. Yet some aspect of the environment, perhaps shading, might induce developmental shifts such that plants of similar biomass might produce fewer large seeds in the shade and more small seeds in the sun. A genetic correlation due to linkage and pleiotropy might cause simultaneous selection for more and bigger seeds to operate very slowly and selection for more seeds alone to produce smaller seeds. Law (1979) demonstrated trade-offs between first- and second-year reproduction and reproduction and growth in *Poa annua*. Since families from a variety of populations were grown in a common garden, a genetic component to the trade-off is implied. Primack and Antonovics (1982) studied the phenotypic and genetic variance–covariance structure for reproductive effort, growth rate, age at anthesis, number of vegetative offshoots, vegetative biomass, and total plant biomass for *Plantago lanceolata*. They demonstrated that differences in reproductive effort in the field were largely environmentally determined. A cost of seed reproduction in terms of negative correlations with vegetative growth and vegetative propagation was only evident in controlled conditions, a finding implying a context-dependent correlation structure. Within- and between-population correlations were often in contrasting directions, and environmental and genetic correlations were not always of the same sign, a finding indicating a structure of conflicting correlations at different hierarchical and control levels. Such information deepens our insight into the evolutionary significance of life-history trade-offs and enables us to consider quantitative life-history variation in its proper multivariate context.

In summary, the evidence indicates that a considerable amount of intraspecific variation in life-histories exists and that a component of this variation and covariation is under genetic control. The expression of the genetic component varies with the environment, and the environmental component of variance is often larger in natural sites than in controlled environments. Furthermore, the genetic and environmental variance–covariance structures may be different from one another at the between- and within-populational levels.

The distribution of variation among individuals and populations, the control of variation by genetic and environmental factors, and the continuity or discreteness of intraspecific variation in life-histories suggest a variety of interesting ecological and evolutionary questions. Can the different patterns of life-history variation be considered to be adaptive strategies? When is plastic control superior to genetic determination? What environmental regimes select for varied life-histories or for more continuity or discreteness in life-history variation? What are the implications of not tracking the environment genetically? These questions require different information from that obtained in most traditional genecological investigations. These have usually been aimed at establishing that variation exists, that it is correlated with habitat, and that it is under genetic control (Heslop-Harrison, 1964). To better understand life-history evolution and the meaning of variation patterns, we should describe the kinds of variation present, their causes, and their distribution among hierarchical levels of organization. This may tell us as much about the limits to evolutionary tracking of the environment as about the fine tuning of adaptation (Antonovics, 1976a).

THE ENVIRONMENTAL INTERFACE

Unlike isozyme variation, life-history variation is directly interpretable ecologically. Much theoretical interest revolves around questions of how different environmental patterns affect the evolution of life-histories. We want to know how density-dependent versus density-independent regulation affects selection for different life-histories (e.g., Pianka, 1970). What are the consequences of environmental variability in space and time (e.g., Schaffer, 1974)? What are the consequences of the predictability or uncertainty of variability (e.g., Venable and Lawlor, 1980)? How do disturbance, competition, and stress differentially impact life-history variation and select for different life-history traits (e.g., Grime, 1979a)?

For the purposes of understanding how different environmental regimes result in the evolution of life-history patterns, it is useful to separate conceptually questions about selection at the environment–phenotype interface from questions about the phenotype–genotype interface and between generation transmission rules. The genetic variance–covariance structure of life-history traits is complicated and still poorly understood. When we conceptually isolate this underlying genetic structure, treating it temporarily as a black box, we can focus our attention on the rich implications of demographic data for understanding how selection operates in nature. Techniques exist for developing analogs to Wright's adaptive topography in phenotype space from demographic data. Selection for different traits can be partitioned into direct and joint effects using multiple regression techniques (Arnold and Wade, in press; Lande and Arnold, in press; Lande, 1982b; Pearson, 1903). Viewed in this way the interaction of different life-history phenotypes with the environment gives us valuable information about the "opportunity for selection" as opposed to the "response to selection," which depends additionally on the relationship between phenotypic and genotypic variation.

Several studies of plant populations have attempted to assess the role of density-dependent and density-independent regulation as selective agents for life-histories. Solbrig and Simpson (1974, 1977) tested the prediction that density-dependent regulation selects for competitive, low seed-set life-histories whereas density-independent regulation selects for poorly competitive, high seed-set life-histories, a prediction in keeping with ideas on r- and K-selection. An isozyme analysis detected four principal apomictic biotypes of *Taraxacum officinale* in a field site in Michigan. Under a variety of light, moisture, pH, and competitive conditions in the greenhouse, the "r-selected" biotype consistently produced more seed, but the "K-selected" biotype consistently won in competition. Density-independent regulation was applied to some garden plots by artificial defoliation or removal of entire plants, whereas density-dependence was simulated by leaving plots undisturbed. After four years, the survival differential in the undisturbed plots was 0.18:1 in favor of the "K" biotype and the biomass differential was 0.10:1. In the disturbed plots the survival differential was reversed, being 1:0.10, and the biomass differential was 1:0.06 in favor of the "r" biotypes. This study demonstrated the efficacy of density-dependent and density-independent regulation in selecting for life-history differences using an artificial selection regime operating on a genetically simple system of apomictic biotypes.

Warwick and Briggs (1978a,b) studied lawn and flowerbed populations of *P. annua* that exhibit differences in growth form and days to anthesis. The differences between the lawn and flowerbed environments were interpreted in terms of r- and K-selection since flowerbed populations were postulated to experience more regulation by density-independent weeding, whereas lawn populations experience more competition. After establishing a genetic component to differences in growth form and days to anthesis, three selection experiments were conducted. First, plants grown from seed were compared to plants grown from cuttings of adult plants. Seed-grown plants were significantly more variable than clones of adults, and the life-histories of adult clones were skewed toward either the lawn or the flowerbed form, depending on their population of origin. These differences were interpreted as indicative of selection and gene flow. Second, clipping experiments were performed to simulate the lawn environment, and the effects on vegetative, floral, root, and total biomass were measured. Selection coefficients against the erect garden form were between 0.13 and 0.68. Third, in a simulated flowerbed environment, selection against prostrate lawn types was 0.77 using the percentage of dry weight allocated to floral parts or 0.44 using total dry weight per plant allocated to floral structures. As the authors noted, the effects of competition were ignored, and reciprocal transplants would be needed to more fully appreciate selection differentials in the lawn and flowerbed environments. Although we may question the relevance of selection for erect versus prostrate growth form and time to anthesis in lawn and flowerbed environments to r- and K-selection in more natural settings, the authors attempted to document life-history selection by subjecting intraspecific life-history variants to different selective agents and measuring their responses.

Another set of investigations used a different approach to explore the effects of density on the evolution of life-history traits in *P. annua* (Law et al., 1977). Variances for survival, size, and age-specific reproduction were partitioned within and between habitat types, populations, and families. The authors were able to infer genetic differences at all levels. The significant life-history differences between "opportunist" and "pasture" populations are in the directions predicted for r- and K-life-history patterns. Yet these data are basically correlational and post factum. Critical experiments would measure selection or the outcome of competition in these environments. To this end Law (1975) sowed seed from a wide range of natural populations together under high- and low-density conditions. Some small response to selection was reported after 21 months for a few reproductive characters such as age at first inflorescence, but strong patterns were not evident at the end of the experiment.

To test the role of density dependence in life-history selection, we might want to use simple techniques such as those described in Keddy (1981) to quantify density-dependent versus density-independent regulation in different populations or subpopulations. One could then sow the life-history variants in a reciprocal design and measure fitness differences demographically. Analyses along the lines of Arnold and

Wade (in press) would provide information on the intensity of selection that could be related to quantitative genetic models of life-history evolution.

A reciprocal transplant experiment between forest and edge populations of *Impatiens pallida* (Schemske, in press) detected selection for length of juvenile life stage. Although no significant difference in germination was documented, survival to reproduction and time to reproduction were different and the 12-day time to reproduction difference was genetically based. Inside the forest, all plants were attacked by a Chrysomelid beetle in the genus *Rhabdopterus*. The native forest population flowers earlier and accomplishes considerable seed set before complete defoliation by the beetle, whereas the late-flowering edge plants suffer serious reduction of reproductive output via defoliation. Interestingly, virtually no juvenile mortality was documented in the forest habitat (Deevey type I survivorship) whereas a constant mortality (Deevey type II) was documented up to the time of frost death in the edge population. This system illustrates how increasing adult mortality selects for an earlier onset of reproduction as predicted in the life-history models of Murphy (1968). Furthermore, this experiment fully uses the advantages afforded by intraspecific variation for elucidating the selection–phenotype interface.

Sohn and Policansky (1977) studied some of the selective consequences of seed versus vegetative reproduction in *Podophyllum peltatum*. From demographic data they were able to construct transition matrices to compare the probability of extinction of individuals that always produce fruit to that of individuals that always fail to produce fruit. According to their data, the eventual extinction of constantly fruiting individuals is certain whereas the extinction probability for sexual plants that fail to produce fruit is 0.55. Although not dealing with the difficult problem of comparing fitness due to seed versus vegetative reproduction, this analysis does quantify one aspect of the cost of seed reproduction.

At this point I would like to further illustrate the use of intraspecific variation to investigate life-history phenomena by focusing on a set of experiments aimed at elucidating dormancy and establishment strategies. Early events such as germination and establishment often determine fitness. Thus, they may drive life-history evolution for annual plants especially, but also for perennials. A successful germination behavior must anticipate the beginning of favorable periods for growth and reproduction. Early germination risks facing inadequate conditions for survival; a killing frost may yet occur or perhaps a dry warm spell will precede reliable rains. On the other hand, late germination risks facing suppression by successful early germinators. A subordinate position in a size hierarchy, even if not resulting in death, will result in reduced reproduction. The net result of these conflicting demands is a life-history trade-off, with later germination favoring the survival component of fitness and earlier germination favoring the reproductive component. The optimal germination time occurs when the product of these two opposing fitness components (probability of survival × expected reproduction given survival) is maximized (Figure 1A). Late germination is a conservative behavior which sacrifices some potential for reproduction and population growth under good conditions for greater survival or persistence under bad conditions. Early germination is a high-risk behavior which may yield low fitness and population crashes under bad conditions but yields a high reproductive payoff and rapid population growth under favorable conditions. Selection should favor a conservative germination behavior in harsh environments since the survival probabilities would be shifted downward for seeds germinating on early dates (Figure 1B). In contrast, selection would be expected to favor a high-risk germination behavior as intraspecific or interspecific competition becomes a more important factor. This is because competition reduces the size and reproductive potential of the subordinate members of a size hierarchy, many of which are subordinate because they germinated late (Figure 1C).

Arthur et al. (1973) document this trade-off between survival and reproductive capacity as a function of germination time in *Papaver dubium*. Some seeds of this annual plant germinate in the autumn and some the following spring. The relative success of autumn and spring germination varies widely from place to place and year to year, depending largely on the harshness of the winter. In a harsh winter all autumn-germinating seeds may die, a situation resulting in absolute fitness superiority of the spring germinators. Autumn germinators are much larger than spring germinators and realize a larger fitness advantage when a mild winter permits their survival.

I have used species with somatic seed polymorphism to explore the evolution of germination strategies. Somatic seed polymorphism permits a high degree of control in selection experiments since the variation is not only intrapopulational but is contained within the progeny of single individuals. The genetic background is more or less constant, so any demographic differences are attributable to seed biology. Furthermore, the variation (seed types) is manipulable and easily measured.

A set of experiments was performed with *Heterotheca latifolia* Buckl. (Compositae), a winter annual that germinates in the autumn in central Texas and flowers, fruits, and dies in the following summer and autumn. *H. latifolia* produces dimorphic seeds which have a number of morphological, anatomical, and physiological differences af-

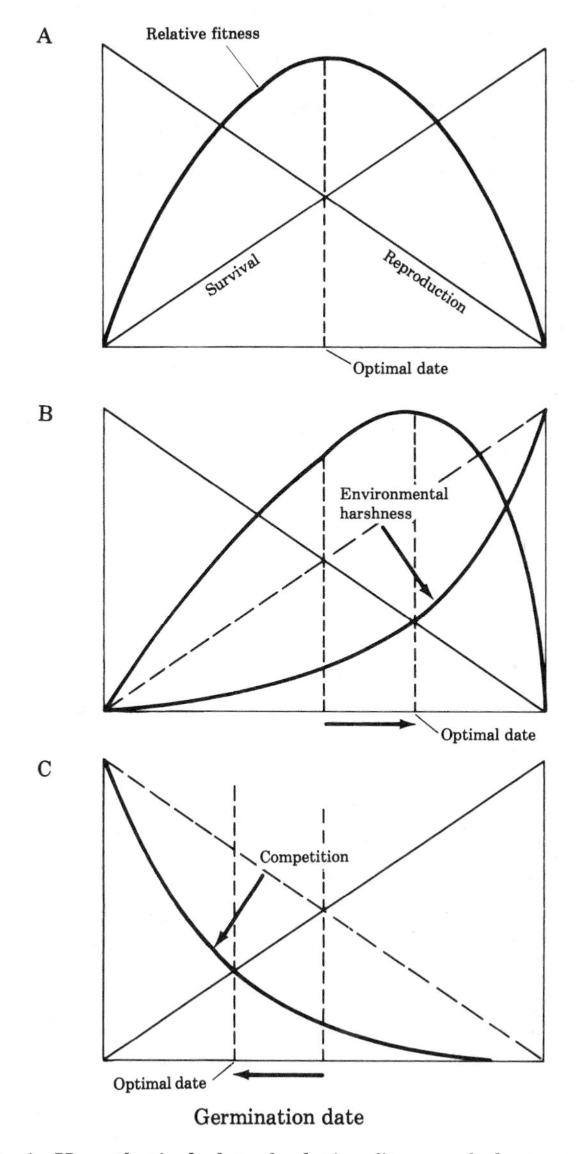

FIGURE 1. A. Hypothetical plot of relative fitness of plants germinating at different times preceding the onset of favorable conditions for growth and reproduction. B. Harsh environments should lower the survival probability at times before the reliable start of the growing season, causing maximum fitness to be associated with later germination. C. Increasing intra- or interspecific competition reduces the reproductive potential of later germinators, causing fitness to be maximized at early germination dates.

fecting their germination biology. Disk achenes have a high-risk germination strategy in that they germinate faster and have higher germination percentages than ray achenes (Venable and Levin, in press a). Ray achenes have a conservative germination strategy, germinating later under more restricted conditions and maintaining a dormant seed bank as a "hedge" against environmental uncertainty. They lose dormancy with age and are inhibited from germinating by darkness and burial. When achenes of known type are sown in randomized rows in natural habitats, it is possible to measure the relative fitness of these two germination strategies under realistic natural conditions (Venable and Levin, in press b). Fitness can be broken down into demographic components to see how the germination strategies actually give rise to fitness differences. In one such field experiment in a favorable environment, it was found that a germinating disk seedling had twice the expected fecundity of a germinating ray seedling. The reason for this difference was partly due to significant survival differences (Figure 2) and partly to reproductive differences of survivors (Figure 3). Survival differences were restricted to the seedling establishment phase. Late rains in the fall of 1977 with no ensuing hot dry periods resulted in suc-

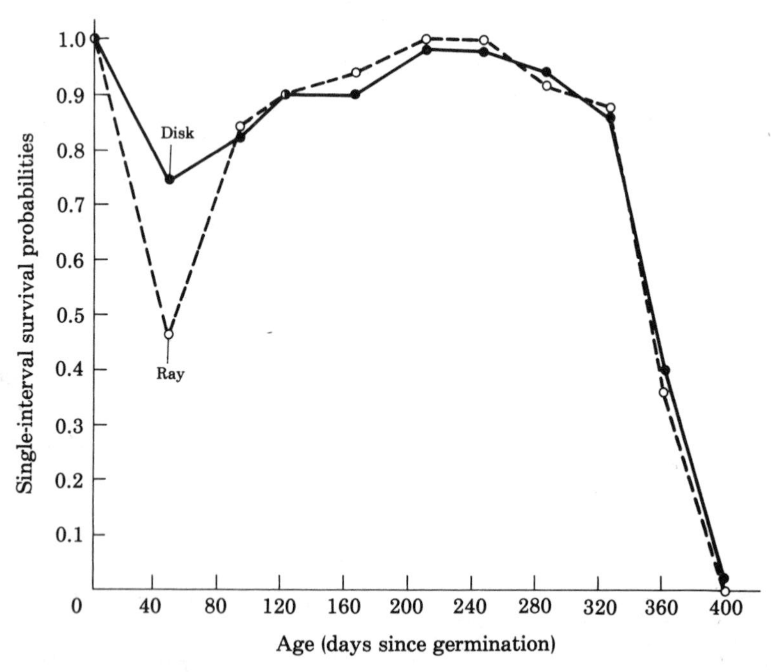

FIGURE 2. Single-interval survival probabilities for *Heterotheca latifolia* calculated as the probability of surviving from the previous census to the current one.

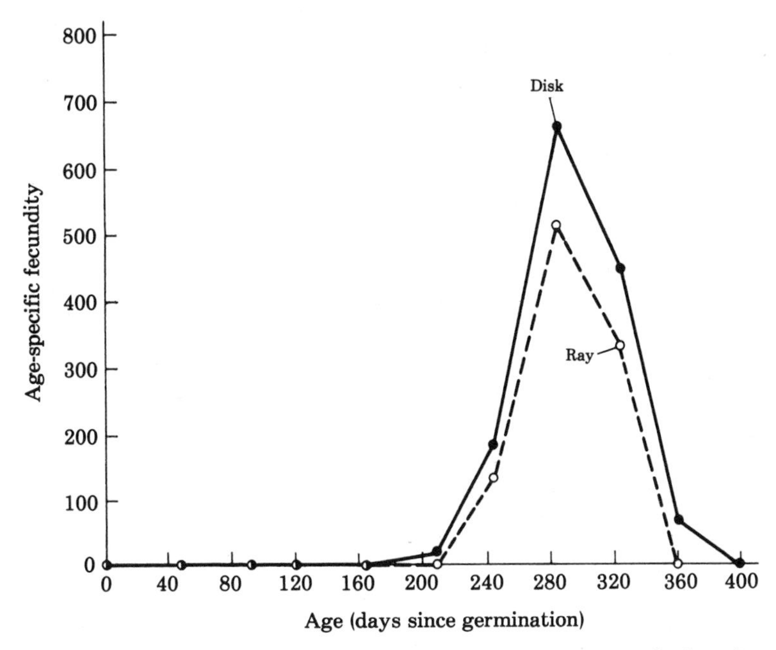

FIGURE 3. Age-specific fecundity for *Heterotheca latifolia* calculated as the average number of achenes produced during a census interval per individual alive at the beginning of the interval.

cessful survival of the high-risk disk achene strategy. The reproductive differences are due to larger plant size for the high-risk disk achenes.

To determine how the demographic consequences of these germination strategies vary when harshness and competition differ, four watering regimes were applied to mixed sowings of seed types in a greenhouse experiment. Drought treatments affect competition indirectly by causing differences in germination, survival, and, thus, density. As the environment becomes less harsh and competition more important, selection differentials change from favoring the conservative ray achene germination strategy to favoring the high-risk disk strategy (Figure 4).

The second system I have used to investigate germination strategies is *Heterosperma pinnatum* Cav. (Compositae), an annual which germinates in the spring and early summer in the central highlands of Mexico and flowers, fruits, and dies in the following autumn. Like *H. latifolia*, it exhibits somatic seed polymorphism, but the morphs are

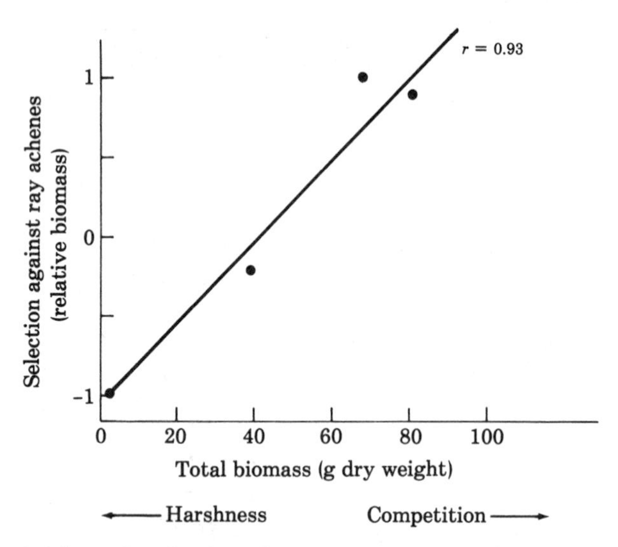

FIGURE 4. Plot of selection against ray (+) or disk (−) achenes of *Heterotheca latifolia.* Biomass is taken as a measure of fitness and selection against ray (conservative) achenes measured as $1 - W_{ray}$, whereas selection against disk (high-risk) achenes is $-(1 - W_{disk})$. Selection coefficients are plotted against total biomass for different drought treatments of mixed sowings in the greenhouse.

not constrained to ray versus disk florets. Instead, central versus peripheral florets make different achene types with some intermediates. The polymorphism is associated with germination differences, with the central achenes losing dormancy earlier in the season and germinating more readily and over a broader range of temperatures than the peripheral achenes. Since the achene type is not determined by floret type, the ratio of achene types is not constrained by floral biology. The proportion of central high-risk achenes varies from a mean of 3 to 58% in different populations, and the between- and within-population phenotypic variances pooled over 32 populations are nearly equal. Greenhouse experiments have indicated that genetic variation exists in the ratio of high-risk to total achenes per head (A. Búrquez and D. L. Venable, unpublished data). In order to measure the relative fitness of the conservative and high-risk germination strategies, reciprocal transplant experiments were carried out across an aridity gradient of natural *H. pinnatum* sites over a 100-km geographic range (E. Morales and D. L. Venable, unpublished data). The conservative, peripheral achenes were selected against in sites with high seedling establishment, and the central, high-risk achenes were selected against in sites where mortality was high (Figure 5). The ecological causes of these selection differentials can be seen by break-

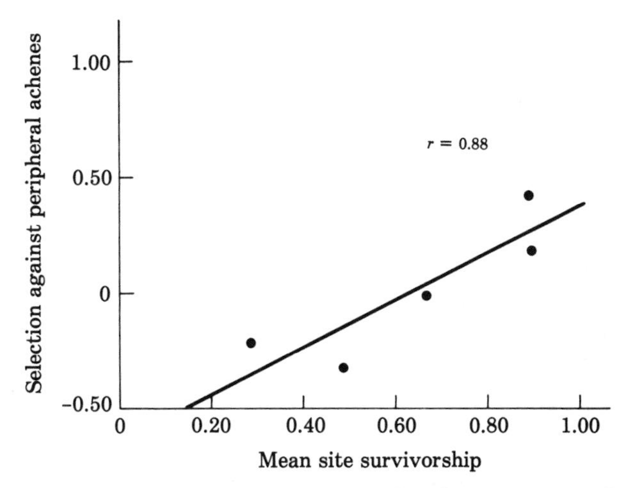

FIGURE 5. Plots of selection against peripheral (+) or central (−) achene types of *Heterosperma pinnatum*. Fitness is defined as the product of the probability of germination × survival probability × plant size. Selection against peripheral (conservative) achenes is measured as $1 - W_{peripheral}$ whereas selection against central (high-risk) achenes is $-(1 - W_{central})$. Selection coefficients are plotted against the expected survival of a germinating achene (disregarding type).

ing fitness down into demographic components in the different populations. For example, in the Tula, Hidalgo site the fitness of the high-risk achene type was 0.67 relative to the conservative peripheral type (Figure 6). Germination occurred in two flushes in the spring, with roughly twice as many central achenes as peripheral achenes germinating in the first flush and twice as many peripheral achenes germinating in the second flush (the overall number of achenes germinating was roughly equal for the two flushes). An extended drought between the flushes resulted in ca. 10% survival of early germinators, whereas survivorship of late germinators was ca. 80%. With such high mortality of early-germinating achenes, the conservative peripheral type was favored.

As this investigation illustrates, it is possible to measure the ecological events that affect the success of life-history variants in natural settings and interpret the selective impact. We hope to be able to use this understanding of the environmental–life-history interface to predict the evolution of seed ratios in *H. pinnatum*. Such predictions can then be compared with actual differentiation patterns observed in the field to test our understanding of life-history evolution.

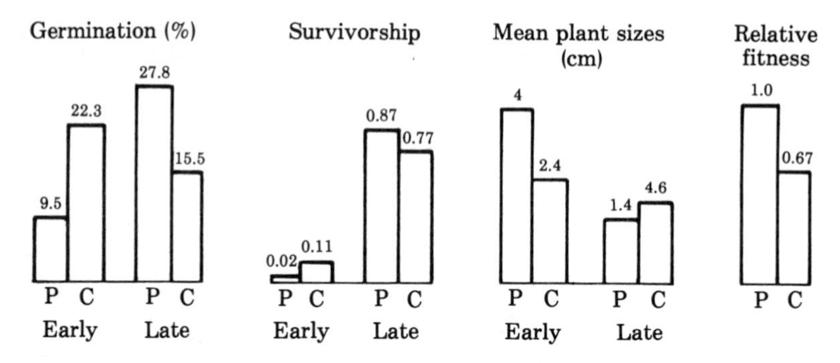

FIGURE 6. Percentage germination, survivorship, plant size, and relative fitness calculated from the product of the three for peripheral (P) and central (C) achene types of *Heterosperma pinnatum* at the Tula, Hidalgo site in 1982, for early and late germination flushes.

CONCLUSIONS

1. Intraspecific variation permits us to monitor the dynamics of plant life-history evolution in natural ecological contexts. Demographic differences can be directly interpreted as components of fitness, and thus ecological causes of selection on phenotypic variation can be investigated. Furthermore, it is often possible to gain some understanding of the underlying genetic variance–covariance structure for life-histories and make inferences regarding potential or past responses to the natural selection dynamics we can measure.
2. Considerable intraspecific variation for a wide variety of life-history traits has been documented at the within-individual and within-population levels as well as between populations. Some additive genetic component to the variation at these different levels has been implicated in most genetic investigations. Ideally, variation should be studied simultaneously for a variety of life-history traits at several levels of hierarchical organization. This would, among other things, permit an understanding of life-history trade-offs in terms of genetic and environmental correlations. Although such comprehensive investigations have seldom been undertaken, they are necessary to put the investigation of the genetics of life-history variation in its proper multivariate context. Pioneering attempts have suggested that considerable complexity exists. Genotype–environment interactions, a considerably greater environmental variation component in nature than under controlled conditions, and complex genetic and environmental variance–covariance structure, have all been implicated.

3. The dynamics of natural selection are relatively easy to investigate with plant populations which tolerate considerable manipulation in natural sites without loss of realism. Questions about the selective interaction of the environment with phenotypic variation can be separated from the questions about the phenotype–genotype interface and the rules of heredity. This conceptual separation is completely consistent with quantitative genetic models of evolution. Ideas about how life-histories evolve in response to different environments can be tested by confronting life-history variation with the hypothetical selective agents in demographic experiments. Although holding much promise for a better understanding of life-history evolution, this approach has been taken by relatively few investigations, and many questions about the evolution of plant life-histories remain to be asked.

CHAPTER 9

LIFE-HISTORY VARIATION, NATURAL SELECTION, AND MATERNAL EFFECTS IN PLANT POPULATIONS

Barbara A. Schaal

INTRODUCTION

Much attention in modern plant population ecology has been focused on the variation and evolution of life-history features. Many life-history details underlie the basic patterns of reproduction and survivorship within a population. Moreover, a thorough knowledge of such demographic characters is central to an understanding of the numerical dynamics of plants and to the prediction of changes in population numbers in relation to the environment, competitors, predators, or other factors. Widespread recognition of the central nature of demography for plant ecology has stimulated a burgeoning literature which details the life-history features of almost every kind of plant species (for reviews see Solbrig, 1980; Silvertown, 1982b).

The life-history approach is also central in evolutionary biology. The demographic features that are studied in population ecology are equivalent to the components of fitness used in population genetics. The breakdown of fitness components into their age-specific parts provides potentially great insight into the mechanisms of genetic change within populations. It is axiomatic that for genetic change and evolution to occur, there must be variation for given traits, and this variation must have a genetic basis. Demographic studies of plants reveal variation at many levels, both within and among species. A significant part of theoretical ecology has been devoted to explaining and predicting major patterns of life-history adaptation (e.g., Stearns, 1977; Istock, 1982; Lande, 1982a). The magnitude of life-history variation within a species is necessarily much less than detected among the totality of plant species. Yet, within-species differences in the pattern of survivorship or in the mode of reproduction occur, presumably reflecting adaptation to differing environments (Law et al., 1977; Keddy, 1981; Jefferies et al., 1981). Variation within populations for life-history features has received relatively less formal attention. Such variation is often casually observed or anecdotally reported. All plant populations show demographic variation; some plants survive and reproduce extensively, other plants succumb to the forces of mortality early, even before germination.

Determining the basis of demographic variability is central to an understanding of the evolutionary genetics of plant populations. The interface of genetics and evolution of life-histories is receiving increased attention (e.g., Dingle and Hegmann, 1982). Is the variation that we see among plants within populations environmentally induced? In this case, differences in life-history features are a result of phenotypic plasticity. Whether or not a plant is successful in terms of survivorship and reproduction is thus the result of chance, the chance dispersal of a seed to a favorable site. This type of environmentally induced variability is important for the numerical aspects of populations and for determining a general adaptive strategy, but it cannot be the substrate for future evolution of life-history features. If, however, phenotypic differences in survivorship and reproduction have an underlying genetic basis, then future adaptive evolution within a population is possible. These alternatives can be examined in another way. Have evolutionary processes occurred in the past so that populations are "fixed" for the genes which determine life-history features, or does genetic variability for these characters still remain within plant populations?

Almost certainly the underlying basis for the observed demographic variation in plant populations is a combination of both environmentally induced phenotypic plasticity and genotypic variation.

Not all evolutionary adaptation has occurred in the past, as a purely environmental explanation presupposes. Nor is all variation due to differences in genotype since many plants are noted for their enormous ranges of phenotypes resulting from environmentally induced plastic responses. The task for the evolutionary biologist is to sort out the relative importance of the genetic versus the environmental component of variation. Life-history characters are complex; both survivorship and reproductive functions occur over the entire life span of a species, and interactions between the two functions, such as trade-offs, may occur. We might then expect the interaction of environmental and genetic variation to be correspondingly complex. How important is genotypic variation in relation to environmental variation? Is the relationship between genetic adaptation and phenotypic plasticity constant, or does their relative importance vary during the life cycle?

It is in this framework that the following analysis of life-history variation is cast. The work addresses several issues relating to life-history variation within plant populations. First, life-history variation within populations will be documented. This variation will be considered in terms of both genetic and environmentally controlled maternal effects and their subsequent influences on offspring fitness. Then, the age-specific influence of genotype on fitness will be measured, by a comparison of inbred and outbred plants. Finally, the interaction among maternal effects, genotype, and environmental variation will be considered in concert to gain insight into the age-specific aspects of natural selection in plant populations.

The work described below is predominantly on the legume *Lupinus texensis,* the Texas bluebonnet. *L. texensis* is a winter annual, endemic to the calcareous soils of central Texas. The species is wide-spread across its range and forms large, showy populations of many thousand plants. Up to 30 inflorescences of 7 to 60 flowers are borne on terminal racemes. Fruits bear a mean of 5.02 seeds and range from 0 to 7 seeds per fruit. Seed output per plant varies from 0 to over 200 seeds per individual. Seeds show large variation in size; viable seeds range from 10 to over 60 mg, with mean seed weights per population ranging from 13 to 46 mg, depending on population location. Flowering is indeterminate, and growth continues during favorable conditions, although plants eventually become senescent. The species is protandrous and is predominantly outcrossed; however, manual self-crosses will seed in the greenhouse. *L. texensis* is highly variable for many characters. Populations show high levels of variation and differentiation for allozymes and for ribosomal DNA sequences. Morphological characters, such as leaf shape, texture, and pubescence vary within and between populations. Both flower color and fragrance are variable, and there are striking polymorphisms in both seed color and patterning.

INTRAPOPULATION VARIATION IN SURVIVORSHIP AND FECUNDITY

Plants in their native habitats show great variation in the reproductive and vegetative functions. Commonly most plants produce just a few seeds and a few plants produce many seeds (Salisbury, 1942; Harper et al., 1970; Leverich and Levin, 1979; Schaal, 1980b). Attempts to sort out reproductive variation in the field into genetic or environmental components have met with variable results. For example, *Stephanomeria exigua* ssp. *coronaria* shows no allozyme differences between plants with high and low levels of reproduction, a finding suggesting environmentally induced variation (Gottlieb, 1977). In *Liatris cylindracea*, however, allozyme differences are associated with the reproductive components of fitness (Schaal and Levin, 1976). Manipulative studies have been somewhat more successful in partitioning genetic and environmental variation. Local differentiation for demographic features has been demonstrated on a microgeographic scale, in studies using phytometers (Fowler and Antonovics, 1981b); and genetic variability for life-history characters has been demonstrated within populations of *Poa annua* in a common garden experiment (Law et al., 1977). The present study examines variation of life-history features in a greenhouse population of *L. texensis*. Greenhouse studies have been chosen since our major interest is in genetic variation. Each experimental method has inherent advantages and disadvantages. Greenhouse studies reduce the environmental component of variation (although by no means eliminate it; hence the necessity of random placement of plants on benches), thus making genetic variation easier to detect. The major disadvantage of such studies is that caution is necessary for any inferences to field situations.

An analysis of demographic variation among half-sibling groups in a greenhouse suggests ample life-history variation within native *L. texensis* populations. Thirty seeds were collected from ten plants growing in a natural field site in Texas. Progeny from a single plant are at least half-sibs and most likely range up to full-sibs. Plants were grown from seed and followed throughout their life spans in a greenhouse. Age-specific mortality and reproduction were measured weekly, and life table values calculated. Half-sibling groups from the original ten plants showed many differences, not only in life-history features, but also in morphology, such as leaf shape and overall habit, which ranged from tall erect to low spreading plants. Variation of demographic characters among and within half-sibling groups were compared.

The range of age-specific survivorship curves is shown in Figure 1. Three of the half-sibling groups, representing the two extremes and an intermediate survivorship curve, are shown. Not surprisingly, the survivorship curve of *L. texensis* in the greenhouse is Type I, where most of the plants survive to reproduction. One expects greenhouse plants in general to have a Type I curve, since greenhouse conditions presumably are close to optimum with an absence of competitors, pests, and water or nutrient stress. In the case of *L. texensis*, the survivorship curve in field populations is also Type I after germination. The survivorship curves of the half-sibling groups differ slightly but significantly. One major area of difference is in the shape of the curve immediately after germination. Some groups experience significant mortality, others none. Plants were analyzed for a cause of death; plant pathologists could find no apparent cause of mortality and suggested that the plants were of "inferior" genotype. After this initial time of mortality, survivorship curves show little difference during most of the life span; nearly every plant survives up to the twentieth week of life. Differences in the curves during this time are due to differences in establishment during the first few weeks after germination. Survivorship curves were again variable during the time of flowering and the beginning of population decline. Some half-sibling groups declined earlier than did others. This decline occurred at the height of summer heat and could perhaps represent differential response to heat or some other factors. In general, the survivorship curves of the ten half-sibling groups are very similar but do exhibit statistically significant variation ($P < 0.05$) during two relatively short periods in the life cycle. We can conclude that the half-sibling groups show differential survivorship responses, even in the greenhouse environment. Thus, the ten field plants show variation in the viability components of fitness.

Reproductive differences among the half-sibling groups were large (Figure 2). These differences were due in part to the different morphologies which tend to be associated with specific groups. The low spreading plant form showed less reproduction and more vegetative growth than did the other plant forms. Those half-sibling groups which contained a predominance of the low, vegetative forms, such as groups A-4 or A-6, were less reproductive than the other groups. These differences in reproduction are shown in Figure 2, which plots age-specific values of reproduction. Reproduction is measured as m_x where m is the average number of ovules produced per individual during the age class x. Age-specific reproduction is weighted by the age-specific survivorship value l_x in order to show the relative, age-specific contribution to total reproduction (Table 1). Ovules are used in this study as a measure of reproduction. Not all ovules in *L. texensis* can become seeds; ovule values are used to determine relative reproductive rates.

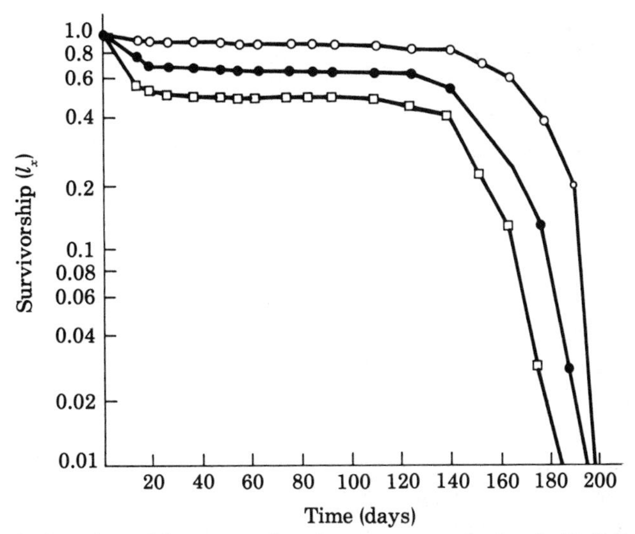

FIGURE 1. Survivorship curves for three representative half-sibling groups of *Lupinus texensis.*

To use seed production in this case would add environmental variation. Since pollen must be carried from anther to stigma and since the efficiency of pollination varies with the type of cross and the pollinator, measures of seed production include variation due to many nongenetic factors.

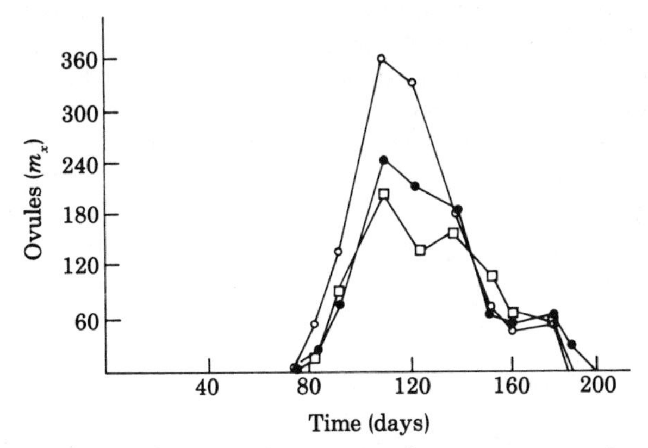

FIGURE 2. Age-specific reproduction in three representative half-sibling groups of *Lupinus texensis.* Reproduction is measured as number of ovules.

TABLE 1. Estimates of the net reproductive rate for half-sibling groups of *L. texensis*.

Half-sibling group	R_0
A-1	709.9
A-2	495.0
A-3	377.4
A-4	352.9
A-5	443.4
A-6	341.7
A-7	607.2
A-8	561.8
A-9	368.5
A-10	385.6
Mean ± SE	464.3 ± 41.87

The time of initiation, the duration, and the time of cessation of flowering for all ten groups were the same (Figure 2). In addition, all half-sibling groups reached maximum reproduction at the same time. Reproduction begins between days 75 and 80 after germination and peaks about two weeks later. After peaking, reproduction gradually declines; the rate of decline is variable among groups (Figure 2). Differences among the groups are manifest in the intensity of reproduction; the most reproductive half-sibling group had greater reproduction during every age class. Reproductive differences can be summarized by determining R_0, the net reproductive rate, which is the total average number of offspring produced per individual. R_0 is determined by $R_0 = \Sigma l_x m_x$. Table 1 lists R_0 values for the ten groups. R_0 ranges from 341 to 709 ovules. Thus, potential reproduction is twice as much in some half-sibling groups as in others. The mean R_0 value of 465 ovules converts to an R_0 for seed production of 1.86 based on a field study of ovule to seed ratios (Schaal, 1980b). The results of the reproductive analysis suggest that variation, even in the uniform greenhouse environment, occurs among half-sibling groups for reproduction. It should be remembered that the ten half-sibling groups in this analysis represent a sample of ten individuals from a single natural population. The occurrence of different morphologies, with corresponding differences in reproductive and vegetative expenditures, suggests that genetically determined biotypes may occur within the population.

MATERNAL EFFECTS

The preceding study suggests that genetic variation for life-history characters exists in native *L. texensis* populations. The results of this study are not definitive, however. Consider the variation in survivorship curves among the half-sibling groups. Differences in survivorship are established at the very beginning of the life span, immediately after germination. These differences in survivorship are then retained throughout the life cycle of the plants. Such variation in early survivorship may, of course, be due to genotypic differences among the seedlings. However, these differences in survivorship could also be solely a function of the size of the seed the plants came from; differences in survivorship among half-sibling groups may be a consequence of seed collections from plants with different mean seed sizes. Several studies, including one of *L. texensis*, have shown that the size of a seed may strongly influence the overall fitness of a seedling, including the probability of survivorship (Schaal, 1980b). Seed size may be a function of the genotype of the zygote, but it may also be due to a maternal effect.

The maternal plant has many influences on its offspring. It contributes one-half of the genome of the progeny. In most plant species, it contributes all of the chloroplast genome. And, the maternal plant can be responsible for many seed characters, such as color. Maternal effects, as used here, are a result of the nutritive or nurturing role the seed-producing parent has on the developing seed. This relationship of maternal plant to developing seed is analogous to the placental relationship that mammalian mothers have to their offspring. How well a plant fulfills this maternal or nutritive role may strongly influence offspring fitness, as it does in animal species. The basis of maternal effects can include both genetic and environmental components. Seed size may be related to maternal genotype and it may be a response to environmental favorability. Maternal effects certainly are involved with offspring fitness during the time the embryo exists as an immature, developing seed. Potentially, maternal effects can influence offspring fitness far beyond the time of parturition, that is, the time when mature seeds are released from the parent plant.

The following section will document maternal effects via seed size and will demonstrate the influence of seed size on subsequent offspring fitness. First, how does seed size influence offspring fitness? The relationship between seed size in *L. texensis* and seed germination, seedling survivorship, and seedling biomass is shown in Table 2. Small seeds are less likely to germinate than are larger seeds (42 and 100% germination, respectively). Seedlings from small seeds have lower survivorship than do seedlings from large seeds (85 and 100% surviving, respectively). And, the biomass of seedlings from small seeds is less than that of seedlings from large seeds (8.0 and 35.5 mg, respectively). In addition, seedlings from large seeds grow most rapidly. In Table 2

TABLE 2. Seedling survivorship and biomass as a function of seed weight.[a]

Seed size class (mg)	Number of seeds	Percentage germination	Percentage surviving	Mean seedling biomass (mg)
< 17	31	41.9	84.6	8.09 ± 1.1
18–22	81	90.1	94.5	12.94 ± 0.33
23–27	120	94.2	99.1	16.52 ± 0.28
28–32	120	95.8	98.3	19.40 ± 0.32
33–37	107	95.3	100.0	21.22 ± 0.36
38–42	98	96.9	95.8	26.22 ± 0.42
43–47	50	98.0	100.0	31.37 ± 0.54
> 48	52	100.0	100.0	35.48 ± 0.54
Mean		92.6	98.2	21.57 ± 0.31

[a]From Schaal, 1980b.

there is a threefold variation in seed weight, yet a fourfold variation in seedling biomass, due to the more rapid rate of biomass accumulation in large seedlings. This relationship between seed size and fitness components is a direct result of the greater nutrient resources of large seeds (Harper, 1977). Similar results have been obtained for other plant species, and in general, there exists a clear relationship between seed size and some components of fitness, although the strength and specific nature of the relationship can vary from species to species (Cideciyan and Malloch, 1982).

We are interested in what part of seed-size variation and its subsequent effect on seedling fitness is due to a maternal effect. Seed size can be determined or influenced by the genotype of the zygote. Seed size can also be affected by the maternal plant, either by the genotype of the seed-producing parent or by an environmentally induced plastic response of the maternal plant. The heritability of seed weight provides some information on apportioning the variation of seed weight into genetic versus environmental components. Heritability indicates what portion of the total variance in seed weight is due to the additive genetic component of variation. Heritability of seed weight in the population of *L. texensis* being studied here is low, $h^2 = 0.09$ (Schaal, 1980b), indicating that less than 10% of the total variation is due to the additive genetic component. This implies that most of the seed-weight variation is due to environmental variation, which in turn suggests a strong maternal effect, specifically one associated with the mother's phenotypic plasticity. The environmental influences on seed weight can be seen even in the seed weights within a given fruit. The seed which is closest to the peduncle is always smaller than the other seeds in the legume (Schaal, 1980b), presumably because of nutrient limitations. Such position effects on seed size are observed in many other species (Harper, 1977). In general, the environmental component of seed-weight variation can be quite high. In *Helianthus*, a sixfold difference in seed weight is observed as a plastic response to plant density (Clements et al., 1929). In other species, however, heritability of seed weight may be quite high; this is particularly evident in cultivated species. Heritability of seed weight in *Glycine soja* is 0.93 (Harper, 1977:670), a finding indicating a much stronger genetic component of variance in seed size than for *L. texensis*. Finally, it should be noted that many plant species show remarkably little variation in their seed weights (Harper, 1977). In such species, maternal effects via seed size will contribute little to the variance in offspring fitness. In general, in plants where seed weight is variable, seed weight is greatly influenced by both the environment and the genotype of the maternal plant, an observation suggesting a strong role for maternal effects on seed size and subsequent seedling fitness. Discussion of the effects of the zygote genotype on fitness will be dealt with later.

The previous studies discussed above demonstrate that seed size strongly influences fitness and suggest that seed size is in large part a maternal characteristic. We are now in the position to ask: How strongly and for how long are the maternal influences exhibited in the next generation? This was addressed in a demographic analysis of maternal effects throughout a greenhouse generation of *L. texensis*. Fifty-four plants were grown from field-collected seeds, acquired from the same natural population as in the previous studies. At reproductive maturity each plant was crossed repeatedly. Half of the crosses for each plant were self-pollinations; the other crosses used an outcross, mixed pollen load. Nineteen plant families were chosen for further analysis, a plant family being the progeny of one initial maternal plant and including both selfed and outcrossed progeny. The number of seeds from each family was adjusted so that equal numbers of selfed and outcrossed seed were represented. Seed weights per family varied, and mean seed weights ranged from 22.9 to 56.1 mg. A total of 550 progeny individuals (275 from self-crosses and 275 from outcrosses) were analyzed further. These individuals were germinated and the resulting plants followed throughout their life spans in the greenhouse. Age-specific survivorship and fecundity were measured weekly, and life table values were calculated as in the previous studies. Demographic features were related to the mean seed size of the initial families. The relationship between demography and breeding system is considered later.

In many species, the first observed demographic relationship to seed size is an expected negative correlation between seed size and seed number. This relationship is usually explained as a result of fixed amounts of energy being allocated to reproduction. However, no significant correlation between seed size and number is detected in *L. texensis* (Table 3). Both the percentage of seed set and the number of seeds per legume show no significant correlation with seed size. This lack of relationship is most likely a reflection of indeterminate growth and flowering and suggests that reproduction in *L. texensis* may not always be resource limited. Seed size appears to have little feedback on the parent plant.

Although seed weight shows no relation to seed number in *L. texensis*, seed weight is significantly correlated with components of fitness after seed parturition. The percentage of seed germination and the time of germination are related to seed size (Table 3). Larger seeds germinate in higher frequency and earlier than do small seeds. Plant size is positively correlated to seed weight through the first four weeks after germination (Table 4). By Week 6, no further relation between seed and plant size is detected. Survivorship values (Table 5) are also related to seed size during the first four weeks after germination; after Week 4, again no significant correlation is detected; the value of the correlation coefficient between seed size and survivorship gradually decreases throughout the life cycle (Table 5; Figure 3). There is no correlation between seed size and the total life span ($r = 0.10$, $P > 0.05$).

The total reproduction for the entire life span of each plant family as measured by R_0 (the net reproductive rate) is not significantly correlated with mean seed weight ($r = 0.36$, $P > 0.05$), though the correlation coefficient seems relatively large; this finding suggests that some component of total reproduction might be associated with seed weight. To examine this possibility, overall reproduction was divided into its age-specific components (m_x), and the age-specific reproductive contributions were tested for relation to seed size. In this case, a significant relationship between seed weight and age-specific reproduction

TABLE 3. Correlation coefficients between offspring seed size and several measurements of seed and seedling production.

Seed size versus:	r	P
Percentage seed set	−0.25	n.s.[b]
Seeds/fruit	−0.36	n.s.
Percentage germination	0.49	<0.05
Time of germination	0.61[a]	<0.05

[a] Spearman's correlation coefficient.
[b] n.s., Not significant.

TABLE 4. Correlation coefficients between seed weight and plant size, as measured by the number of leaves or the number of stems.

Plant size	r^a
No. of leaves at 2 weeks	0.76*
No. of leaves at 4 weeks	0.59*
No. of leaves at 6 weeks	0.19
No. of stems at 8 weeks	0.02

[a] Significance level: *, $P < 0.05$.

was detected for age classes 7 through 10 (Table 6), the ages of most intense reproduction (cf. Figure 2).

The results of this study indicate that maternal effects as measured by seed size affect both the time and probability of progeny germination. Maternal effects also influence seedling size and survivorship until the fourth week after germination. No maternal effects are detected in survivorship beyond the fourth week. The occurrence of a significant association between seed size and reproduction later in life, during the time of maximum reproduction, is most likely an effect

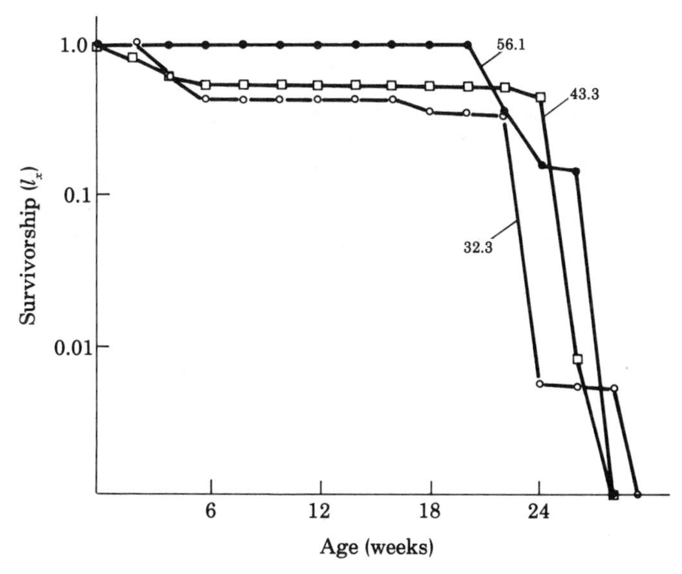

FIGURE 3. Survivorship curves of three plant families of differing seed weights (mg) in *Lupinus texensis*.

TABLE 5. The correlation coefficient between seed size and survivorship (l_x) at different times.

Time (weeks)	r^a
2	0.56*
4	0.45*
6	0.20
10	0.18
14	0.18
18	0.18
22	0.17
24	−0.06
26	−0.17
28	−0.37
Whole life span	−0.10

[a]Significance level: *, $P < 0.05$.

TABLE 6. Correlation between seed size and age-specific reproduction.

Age class	r^a
4	0.30
5	0.16
6	0.34
7	0.56*
8	0.56*
9	0.62*
10	0.49*
11	0.34
12	0.23
13	0.10
14	0.21

[a]Significance level: *, $P < 0.05$.

secondary to the size hierarchy established early after germination. Plants which are large early in their life tend to be the most reproductive individuals later in many plant species (see Leverich and Levin, 1979). The same phenomenon is probably occurring in *L. texensis.* Seedlings which are produced from large seed are themselves large, therefore they eventually reproduce more. Thus, maternal effects have a strong influence on offspring fitness through the fourth week after germination, and the influence may extend beyond this period. The time following parturition of a new seed, and up to one month after its germination, includes important parts of the life cycle of the plant. This period involves events such as seed dispersal, dormancy, seed germination, and seedling establishment. These are times in the life cycle of a plant which may be the most critical for success (Harper, 1977). It is noteworthy that the fitness of an individual during these selectively critical times may be, in very great part, a function not of that individual but rather of the maternal plant.

PLANT GENOTYPE AND FITNESS

So far we have demonstrated a potentially large influence for both genotypic and environmental maternal effects on the fitness of plant offspring. Next we ask: How and when does the genotype of the progeny, that of the individual, affect its own fitness? In the previous study of maternal effects, each family group consisted of 50% progeny from self-crosses and 50% progeny from outcrosses. The selfed and outcrossed groups are genetically distinct, even though they have the same maternal parent. The inbred plants are far more homozygous than the outbred plants. In a species, such as *L. texensis*, where outcrossing is the predominant mode of reproduction in nature, inbreeding depression is expected in the progeny of self-crosses, since presumably inbreeding does not occur frequently enough to "purge" deleterious alleles from populations. Thus, differences in genotype among the inbred and outbred groups should be manifest as directly measurable differences in the components of fitness. The portions of the life cycle which are most sensitive to genotypic differences should then be detectable by a comparison of inbred and outbred progeny. The inbred and outbred groups should show mean differences in fitness components when the individual genotype is predominant in affecting fitness and when other factors, such as maternal effects or environmental variation are secondary. Of course, the fact that genotypic effects are not observed during part of the life cycle in such a comparison does not mean that genotype is unimportant or that selection on genotype does not occur; it only means that other factors are causing more fitness variation so that genotypic effects are not readily detectable. Finally, a comparison of inbred and outbred progeny will determine both time and the intensity of inbreeding depression in *L. texensis* and will provide an estimate of the fitness advantage of an out-crossing breeding system.

First we can ask: What effect does the breeding system have on seed set, seed weight, and germination? If inbreeding depression is expressed early in the life cycle, we expect a reduction in inbred seed relative to outbred seed in all these three aspects of fitness. In *L. texensis* there is little evidence for early inbreeding depression (Table 7). The percentage of seed set of selfed and outbred crosses does not differ significantly. The number of seeds per fruit produced from the two types of crosses is virtually identical. There are no significant differences between the mean number of viable or aborted seeds per fruit in the inbred versus outbred progeny, although there is a tendency of outcrossed fruit to have fewer aborted and more viable seed than inbred progeny. Mean seed weight between the two seed groups does not differ significantly, although, again, outcrossed seeds tend to be larger. Finally, there are no differences in the percentages of germination or in the times of germination between inbred and outbred seeds (Table 8). Thus, in early life, through the time of germination, we find no significant differences in fitness among inbred and outbred plants and no strong genotypic basis to which we can ascribe the variance in fitness at this stage of the life cycle. These results for *L. texensis* are not unique. No difference between rates of fruit maturation in selfed and outcrossed seeds was found in a study of *Cassia fasciculata*, another legume (Lee and Bazzaz, 1982).

TABLE 7. The production of seeds and seedlings by selfed and outcrossed progeny.[a]

	Selfed	Outcrossed
Percentage seed set	12.5 (n = 2708)	11.46 (n = 1571)
Seeds/fruit	4.57 (n = 214)	4.56 (n = 104)
Viable seeds/fruit	2.86	3.08
Aborted seeds/fruit	1.71	1.48
Seed weight	38.65 (n = 206)	39.58 (n = 223)
Percentage germination	91.55 (n = 206)	91.03 (n = 223)

[a]For all comparisons, $P > 0.05$.

An analysis of the components of fitness after germination begins to detect differences between the inbred and outbred progeny (Table 9). Plant size begins to show an influence of breeding system by four weeks after germination, but no influence of genotype is seen prior to that age. Breeding system shows a strong relationship to plant size, the outcrossed progeny being larger beyond four weeks (Table 9). Inbreeding depression (and thus genotype) also affects the total life span of *L. texensis*. Mean life span is 140 ± 77 days for selfed progeny, but it is higher (165 ± 63 days) for progeny of outcrosses. These differences in mortality are shown in Figure 4, which shows the age-specific survivorship curves for the two progeny groups. Thus, the viability component of fitness is markedly influenced by genotype, but only beyond the early part of the life cycle of *L. texensis*.

Reproduction is also strongly affected by the breeding system, as shown by the average number of offspring per individual: the net

TABLE 8. Proportion of seeds that germinated at different dates in selfed and outcrossed progenies.[a]

Date	Selfed	Outcrossed
5/17	0.912	0.951
5/18	0.068	0.093
5/19	0.005	0.005
5/20	0.000	0.005
5/21	0.015	0.000

[a]$\chi^2 = 5.62$; $P > 0.05$. χ^2 was calculated with the absolute numbers for the first two dates only.

TABLE 9. Plant size (as measured by the number of leaves or stems) in selfed and outcrossed progeny.

Plant size	Selfed	Outcrossed	t^a
No. of leaves at 2 weeks	2.62 (n = 167)	2.67 (n = 192)	0.317
No. of leaves at 4 weeks	8.94 (n = 154)	10.76 (n = 176)	4.12*
No. of leaves at 6 weeks	20.08 (n = 144)	23.44 (n = 172)	2.14*
No. of stems at 8 weeks	5.22 (n = 140)	5.56 (n = 166)	1.99*

[a] Significance level: *, $P < 0.05$.

reproductive rate (R_0). R_0, here calculated as the number of flowers per individual, is 88.4 for the progeny of self-crosses whereas R_0 for the progeny of outcrosses is much higher: 108.7 flowers. This difference in reproduction between the two groups is clearly illustrated in Figure 5, which plots the age-specific flower production for progeny from both

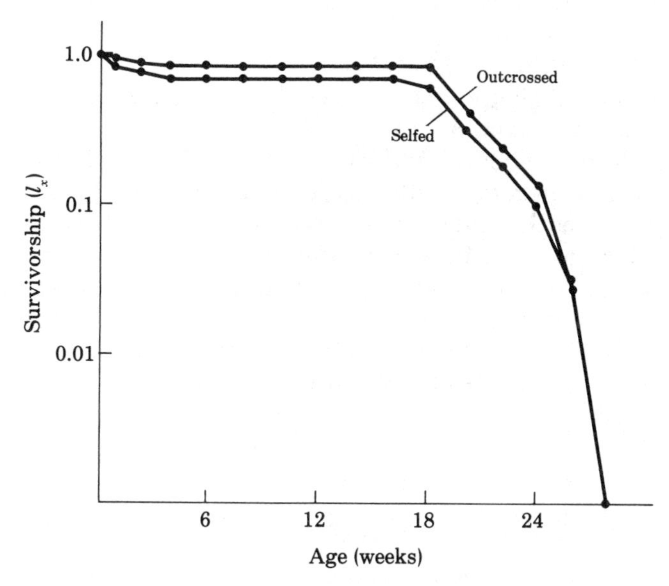

FIGURE 4. Survivorship curves for progeny of self-crosses and progeny of outcrosses in *Lupinus texensis*.

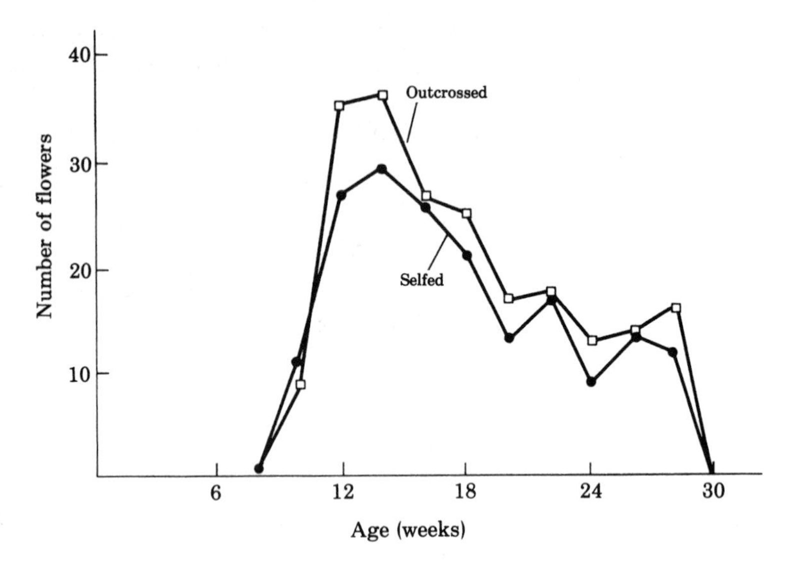

FIGURE 5. Age-specific contribution to reproduction for progeny of self-crosses and progeny of outcrosses in *Lupinus texensis*. Reproduction is measured as number of flowers.

self-crosses and outcrosses. Selfed and outcrossed progeny have the same time of initiation and the same duration of reproduction, although the intensity of reproduction is greater in the outcrossed group during nearly every age class.

A strong inbreeding depression clearly exists, and hence a genotypic effect on fitness occurs in *L. texensis*. Inbreeding depression is not significantly manifested until the fourth week after seed germination. By the time genotypic effects are manifested, a significant and critical part of the life cycle of *L. texensis* is complete. Clearly, the genotype must also be important during this early stage. Plants homozygous for lethals or alleles with major deleterious effects are no doubt selected against. In general, the abortion rate in plants, including *L. texensis*, is quite high, even under optimal, outcrossed conditions. Perhaps the suboptimal individuals among the developing progeny of self-crosses are being selectively aborted, and these additional abortions resulting from inbreeding are being absorbed into this already large "pool" of abortions. Such a process would act as a sieve and would leave the remaining seed just as viable and fit, regardless of cross. This assumes, of course, that many abortions otherwise are fully fit. Possibly only genes which affect seed characteristics or embryo features would be acted on in such an early system, since the effect of inbreeding and genotype on characteristics of mature plants are in fact seen later in life. An alternative explanation is that maternal effects or environmental variation predominate in affecting fitness during the early stages of life; genotypic variation is "swamped" out by the greater variability due to these other factors. Perhaps it is only when other influences on fitness are diminished that the genotypic differences among individuals are manifested. This is certainly possible in *L. texensis*. The time in the life cycle of *L. texensis* when maternal effects are no longer detected is exactly when genotypic effects become influential.

CONCLUSIONS

There is considerable life-history variation in *Lupinus texensis*, and this is observed both in viability and in fertility. As expected, variation can be apportioned into both genetic and environmental components. The relationships among environmentally based variability, genetically based variability, and their effects in influencing life-history features are clearly complex. The fitness of individuals in a given generation of plants is a function of the environment the plants themselves are exposed to, and their fitness is also a function of the environment their seed parents experienced in a previous growing season. With long seed dormancy, and with seed dispersal over far distances, it is possible to have very different environmental regimes exert influences on the fitness of one individual plant.

Genotypic effects on fitness are likewise complex. There is a genotypic component associated with how well a maternal plant can nurture the seed—the genotypic maternal effect. There is also a genotypic effect on the fitness of the plant due to that individual's own genotype. However, there are also potentially "grandfather effects" in plant species where plants may exist as separate sexes, either structurally (as in dioecy), or functionally (as in hermaphroditic plants which contribute pollen but do not set seed). In males of these species, although maternal effect genes are present, such genes which involve nurturing the developing embryo do not express their influence until the following generation, when daughters are called on to nurture their seeds. Their expression is delayed one generation. These genes, which affect seed development and which are from the male parent, are expressed in the fitness of the F_2. Such grandfather effects are clearly demonstrable in animal species (Reznick, 1981). If the sexes are not separate and if a plant is functionally hermaphroditic, then the genes influencing maternal effects are expressed in the F_1 generation.

Natural selection in plant populations must be exceedingly complex. The phenotype upon which selection acts may be underlaid by many different genotypes. We cannot always assume that selection acts on a single individual's genotype, for every character (although many traits may, of course, be this simple). Early in the life cycle, a significant portion of fitness is a function of maternal effects, which are, in turn, a function of the maternal genotype (in part). If the sexes are separate, male-borne genes with a potential maternal effect are expressed as grandfather effects. After early life, beyond seedling establishment, an individual's own genotype appears to be more important in determining fitness, and hence it is the individual's genotype rather than the maternal genotype that is affected by selection. Other factors not considered here add further complications. Selection also occurs on the haploid genome, both the male and female gametophyte. Selection at these stages of the life cycle must influence and interact with selection at other stages and genotypes.

Finally, how significant is genotypic variation in the face of environmentally induced variation? Which is more important in determining variation of fitness in the field? Phenotypic plasticity is extremely important in determining maternal effects, but its importance appears to diminish as progeny age. In the greenhouse, this aspect of phenotypic variation changes in influence and importance during the life cycle. As maternal influences on progeny individuals decline, the influence of their own genotype becomes correspondingly more important. Under field conditions a similar conversion in relative influences may also occur. Plants are faced with environmental heterogeneity throughout their life spans. We do not know the relative importance in field populations of environmentally induced variation versus genetically based variation in determining plant fitness. Is environmental variation so great that the genotypic differences we detect in the greenhouse are totally obscured? Is chance, therefore, of overriding importance in determining fitness of plants in the field? Certainly adaptive evolution must be occurring some of the time, so that genotypic differences must be important. Does selection occur periodically, or is it more or less a continuous process? How do natural selection, genotypic variability, and environmentally induced variability interact? These questions must be more fully addressed before a complete understanding of plant evolutionary processes can emerge.

SECTION II
THE INTERFACE BETWEEN ECOLOGY AND GENETICS

Consequences of the Genetic Structure of Populations
Consequences of Plant Breeding Systems

INTRODUCTION TO SECTION II

With the wisdom of hindsight, it would seem obvious that an integrated development of population ecology and population genetics should be one of the most appropriate ways of gaining understanding in evolutionary processes. Ecological phenomena determine genetic changes but are themselves modified by the changing composition of populations. However, on the whole, the two sciences (ecology and genetics) have had quite different origins and largely independent developments: A robust theoretical framework, achieved as early as the 1930s, preceded experimental and observational data in population genetics; in contrast, in population ecology a great deal of empirical data came first and a strong theoretical development did not arise until the 1960s (in particular, under the influence of R. A. MacArthur). In addition, population ecology has been dominated by zoologists to a large extent, while botanists have only recently begun to explore this field. This contrast has perhaps not been quite as marked in the case of population genetics, though it is evident that the major developments in animal population genetics have been highly theoretical, and based on laboratory studies—mainly with *Drosophila*. In the case of plant population genetics, the theoretical approach has not been anywhere as strong; however, empirical field information has received, in comparison to animal population genetics, more attention.

These historical vagaries have led presently to a rather complex state of development in plant population biology which, nevertheless, poses one of the most interesting challenges in the study of the plant sciences: the integration of population genetics and population ecology. The task is not an easy one, but it is one that warrants prompt attention. An attempt at achieving this goal is presented in the five chapters that constitute this section.

The first part of this section deals with an analysis of some of the consequences of the genetic structure of plant populations. Chapter 11 by Antonovics dedicates a great deal of space to a detailed and comprehensive historical overview of the study of genetic variations in populations. This overview not only relates historical facts, it is also a very rich conceptual description of the field, and a useful preamble to the population ecologist interested in the interface of ecology and genetics. There is an illustration of how maintenance of genetic variation within natural populations can be addressed, with major emphasis on the relevance of within-population variance to the ecology of the individual plant.

Cloning techniques with the grass *Anthoxanthum odoratum* are described in which genetically uniform and genetically variable "progenies" (tillers) can be produced and transplanted as phytometers. The experiments deal with (1) the measurement of individual fitness; (2) the comparison of fitness for genetically variable vs. homogeneous progeny; and (3) the evaluation of the techniques used to map individual phenotypes into fitness. The experimental design and the techniques discussed here illustrate the tremendous potential of plants for the study of natural selection in the wild.

Besides its valuable historical overview, this chapter offers a powerful approach to understanding the importance of genetic variation within populations; to evaluating the advantages of such variation for individual plants; and to obtaining direct (and precise) measurements of individual fitness in natural settings.

Bradshaw (Chapter 10) deals with the ecological significance of genetic variation between populations, providing examples that emphasize the well-established fact that genetic variation is immense and involves a wide variety of plant characteristics. He suggests that in fact the proper question might be, not what characters show evolutionary differentiation, but whether there are characters in any species which do not show differentiation among populations. As long as we are able to show that the observed differences enable each population to perform better in its environment than can other populations, we can safely suggest that these differences are of evolutionary significance. Appropriately, Bradshaw discusses the ways the population ecologist can attempt to assess this. He then discusses the paradigm of whether the ecological range of an individual species is attributable to extensive population differentiation or to a wide ecological amplitude; his conclusion is that it is necessary to apply approaches very different to those usually adopted (such as comparative study of species of wide ecological distribution). His studies, and those of his colleagues, with metal tolerant species clearly suggest that only certain species contain enough genetic variability to evolve tolerance, and that usually only those species evolve tolerance and occur in metal-contaminated soils. He suggests that, at least for this sort of environmental setting, the ability to show evolutionary differentiation determines ecological amplitude directly. The concept of genostasis is introduced. Genostasis takes place when dispersal continuously transports a species into habitats to which it is not adapted; since the distributions of most species are stable, it is presumed that natural selection must have taken most species to their selective limits where further adaptation is limited by lack of *appropriate* variability.

A major highlight of these two chapters on genetic variation is the detailed description of methodological approaches to this field of study. Undoubtedly, the population ecologist will find the more genetic discussion of Antonovics, together with the more ecological approach of Bradshaw, extremely stimulating, offering a challenging, wide-open field to explore.

A host of ecological phenomena can affect, to varying degrees, the genetic structure of plant populations. One such phenomenon that intuitively would be expected to have a profound effect on the sociology, demography and genetic structure of plant populations is immigration: the arrival of disseminales (pollen, seeds) into an existing population. Yet, surprisingly, this area has received very little attention. Levin (Chapter 12) attempts to fill this gap. He presents a detailed review of the state-of-the-art knowledge of immigration and discusses some of the factors affecting the outcome of immigration on genetic variation within and among populations. (It therefore constitutes an appropriate extension of the previous two chapters.) His review discusses what we know (or do not know) about (1) immigration rates; (2) patterns of immigration (whether it is random or not, whether it is via pollen or seed, and the consequences of different modes of gene transport); and (3) the fitness of the immigrants. A striking point of this review is that for most cases the actual levels of immigration are practically unknown and poorly understood—the main reason being that immigration rates are, in fact, extremely difficult to determine. Since a proper quantification of immigration within and among populations is a prerequisite for understanding the dynamics and rates of population divergence, this remains an area badly in need of sensitive and efficient methodologies. The analyses of the implications of immigration for population systems of this chapter (mainly based on theoretical models and computer simulations) show that a great number of factors (mutation, genetic kin structure of the migrants, migrant population size, etc.) have the potential to affect the outcome of a given level of immigration. Levin suggests that it is the study of the interplay of these factors that is likely to provide the best information about immigration as an evolutionary force.

The last part of this section, "Consequences of Plant Breeding Systems," includes two chapters on a slightly different subject to that dealt with by the previous chapters. They have to do with some aspects of reproductive patterns in plants, still with an emphasis on population ecology and its intermingling with population genetics. Willson's contribution (Chapter 13) deals with a rather general topic, mating patterns in plants, while Lloyd (Chapter 14) addresses a more specific theme: gender allocation in outcrossing cosexual plants.

Willson develops her arguments in two main areas: (1) a detailed discussion of the biological bases of nonrandom mating and (2) a survey of the consequences of some types of nonrandom mating in plants. The first part of her discussion shows that a great number of factors or phenomena can prevent random mating. These include the absence of cross breeding; inbreeding with relatives (due to the clumped distribution of relatives); rare events of mating (e.g., long-distance movements of disseminules, differential reproductive female success as a function of pollen-source distance); directional dispersal (due, for example, to wind or water currents); assortative and disassortative mating (matings between like or unlike phenotypes); and differential mating ability. Quite clearly, the results of these multiple sources of nonrandomness are widely different and complex; Willson discusses the consequences of nonrandom mating in plants in terms of: (1) population structure (perhaps the most documented consequence of mating pattern); (2) evolutionary potential, referred to here as long-term survival of the population on an evolutionary time scale; and (3) life-histories. She discusses this third aspect in relation to several other aspects of sex in plants: evolution of sex expressions, of progeny sex ratios in dioecious species, and resource allocation patterns in cosexual species. Thus, Willson's discussion offers an appropriate preamble to Lloyd's contribution. In particular, she discusses that several factors, such as inbreeding and gamete dispersal differentials, may cause deviations from the expected 50:50 allocation to male and female functions in cosexual plants—an expectation based on Fisher's sex ratio theorem.

Since the vast majority of seed plants are cosexual (having a single morph that combines both male and female functions), the problem of gender allocation in natural plant populations is one of particular relevance in the population ecology–genetics interface. Lloyd explores this problem, attempting to extract some patterns of allocation from the meager information available. This review indicates equal allocation, or deviations of varying extent toward greater maternal expenditure; also, that the allocation pattern is related to at least one ecological factor: the pollinating agent. He offers a theoretical assessment of the impact of a number of factors on gender allocation patterns in a homogeneous environment. Of the several factors explored, local seed and pollen competition do not appear to cause great deviations from equal allocation; moreover, it appears that paternal allocation may indeed facilitate the maternal function. Recurrent unilateral maternal costs—e.g., those incurred in the production of fruits and the maintenance of ancillary structures (such as pedicels during seed maturation)—result in deviations toward a higher maternal investment (as observed in several species), but this factor is likely to be of major importance only in cosexual species with multiseeded fleshy fruits. For animal-pollinated species, an upper limit on paternal fitness seems to be the most promising factor to explain the observed deviations toward maternal expenditure. On the contrary, it is expected that wind-pollinated species should have higher paternal allocations, and the few available data suggest that this is so.

The theoretical analyses presented in this section suggest the sort of empirical data that are necessary to begin to understand the patterns of gender allocation. This will undoubtedly assist researchers in explaining a number of aspects of floral biology and the ecological and genetic consequences for plant populations.

CHAPTER 10

ECOLOGICAL SIGNIFICANCE OF GENETIC VARIATION BETWEEN POPULATIONS

A. D. Bradshaw

INTRODUCTION

For Charles Darwin the variability existing within species was the crucial starting point for his arguments for evolution by natural selection. He was clearly overwhelmed by the extent of the evidence. "To treat this subject [variation in nature] properly, a long catalogue of dry facts ought to be given, but these I shall reserve for a future work" (Darwin, 1859). It was a pity that this future work never appeared because it is unlikely that the catalog would have, in any way, been dry. But it was plant physiology and morphology which was to gain from his extraordinarily logical mind, when he turned to botany for relaxation and recreation (Heslop-Harrison, 1964). So perhaps it is not improper for us to look at all this variation with the benefit of an extra hundred years and try to see what it means to us as ecologists. Darwin made it amply clear what it meant to theories of evolution.

There are two levels at which this variability can occur, within populations and between populations. Although Janis Antonovics

(Chapter 11) and I are examining these two levels of variation in separate chapters, it must be realized that they are just two parts of a continuous phenomenon in which variability becomes manifest and is then sifted by the processes of natural selection. The variability that occurs between populations has very clearly, in all but a few cases, been strongly influenced by natural selection—in other words, by the forces of the environment. It should therefore be of particular significance to ecologists because it represents the judgment of environmental processes on the variability of life, accumulated over many generations.

THE RANGE OF EVOLUTIONARY DIFFERENTIATION WITHIN SPECIES

With the benefit of the work of 100 years we now know that the extent of the variability which is to be found between populations within plant species is immense. The crucial step forward was taken when investigators such as Kerner (1891), and especially Turesson (1922, 1925), appreciated the value of the common garden technique to remove the direct effects of differing environments and reveal underlying genetically determined differences, an approach so simple that it seems extraordinary it had not been used by Darwin.

The differences found between populations can be on a very broad scale of distance (over hundreds of kilometers) or on a very local scale (over just a few meters); these distances make no difference to the magnitude of the differences which can occur between the populations. This is because the pattern of differentiation depends on the outcome of the interplay between natural selection (tending to enhance differences) and migration (tending to reduce differences); in plants, because of their essentially sedentary nature, migration can be easily overruled by selection (Bradshaw, 1972). As a result, patterns of differentiation of populations tend to follow patterns of environment very closely.

From an environmental point of view, it is clear that we can find differentiation within species in relation to, effectively, every environmental factor, on both broad and local geographical scales (Table 1). In most cases, a common garden technique reveals obvious morphological differences. However, for adaptation to soil factors, it has usually been necessary to examine the material under experimental culture conditions; morphological differences may not occur between populations from different soil types [e.g., in *Festuca ovina* in relation to calcareous and acidic soils (Snaydon and Bradshaw, 1961)] or may only partially run parallel with physiological differences [as in *Trifolium repens* (Snaydon and Bradshaw, 1962)]. This is only to be expected since morphological and physiological characters in these situations are almost inevitably controlled by different gene systems (Broker, 1963; Clausen and Hiesey, 1958a).

TABLE 1. Some examples of differentiation within species to distinct environmental factors, on both broad and local geographical scales.

Factor	Broad scale	Local scale
Climate	*Achillea millefolium*	*Agrostis tenuis*
	Prunella vulgaris	*Agrostis stolonifera*
	Pinus sylvestris	*Ranunculus repens*
	Lolium perenne	*Geranium sanguineum*
Soil	*Streptanthus glandulosus*	*Anthoxanthum odoratum*
	Lolium perenne	*Agrostis tenuis*
	Trifolium repens	*Plantago lanceolata*
	Capsella bursa-pastoris	*Festuca ovina*
Biotic	*Camelina sativa*	*Agrostis tenuis*
	Solidago virgaurea	*Trifolium repens*
	Plantago maritima	*Poa annua*
	Trifolium repens	*Potentilla erecta*

The characters that can be affected vary from the trivial to those of the whole growth of the organism, whether measured in terms of final yield [e.g., *Agrostis tenuis* (Bradshaw, 1959)] or growth rate [e.g., *T. repens, Lolium perenne*, and other species (Elias and Chadwick, 1974)]. They can be the life cycle of the organism, as in *Poa annua* from open and closed habitats (Law et al., 1977). Phenotypic plasticity can be affected, as in the remarkable differences in response of leaf growth to temperature in European populations of *L. perenne* (Cooper, 1964) and to submergence in aquatic populations of *Ranunculus aquatilis* (Cook and Johnson, 1968). Life cycles can themselves differ between populations in their plasticity, as in *Capsella bursa-pastoris* from stable and unstable environments (Sørensen, 1954). The examples are endless, and the proper question might be not what characters show evolutionary differentiation, but whether there are any characters in any species which do not show evolutionary differentiation among different populations. We will return to this point later when we have established the ecological significance of the differentiation.

THE ADAPTIVE SIGNIFICANCE OF DIFFERENTIATION

Since we presume these differences are the product of natural selection, we imagine that they are adaptive. But this could be a dangerous assumption because it is based on a circular argument within which there is no direct or even indirect evidence of adaptation, except that if the material has been growing in a particular set of environments it must be in some way adapted to them or else it would not have survived. The crucial question is whether the differences we find in different populations enables each to survive better in its own original environment than can other populations. This would be direct evidence of adaptation and would answer the critics of our concepts of adaptation (Gould and Lewontin, 1979; Harper, 1982), who suggest that much of what we interpret as adaptive may be present for historical rather than truly adaptive reasons. A convenient property of populations of a single species is that they often differ only in a very few characteristics. It is therefore easier to attribute differences in fitness to particular character differences and interpret adaptation with precision.

Since we are dealing with living material and want to assess fitness, we can best do this by the realistic ecological procedure of establishing the material we wish to examine in its natural environment and following its performance over a period of time. Since we wish to compare material from different environments, we can make the experiment complete, and critical, by a reciprocal transplant design. This technique was first used by Kerner (1891) and Bonnier (1895, 1920), but they applied it only to related species. We are indebted to Clausen et al. (1940) for showing us how powerful it can be for the assessment of populations. There is now a range of different types of reciprocal transplant experiments for us to consider, experiments which have different powers of resolution of adaptation.

Spaced-plant experiments

The simplest form of transplant involves taking the material into prepared gardens in two or more of the environments from which the material came and setting it out as spaced plants with all influence of other plants, either of the material being compared or of the native flora, removed. The technique has been used very widely by foresters as a simple means of assessment (Wright, 1962). It was adopted for evolutionary studies by Clausen et al. (1940, 1948) as an extension of Turesson's common garden approach, for *Potentilla glandulosa*, *Achillea millefolium*, and other species. It has the advantage of simplicity and allows, particularly, the effects of climate to be studied. But climatic effects inevitably are confounded with the effects of any parallel variation in soil, and the technique removes any compounding biotic influences of other plants or animals. However it revealed that startling differences in climatic adaptation exist in different populations of *P. glandulosa* and *A. millefolium* such that the extremes of the range of populations studied, an altitudinal transect of 3000 m in California, cannot survive in each other's habitats. Intermediate populations and an intermediate transplant station showed that all gradations of adaptation can be found.

Similar, but smaller, differences in adaptation over a much narrower range of environments (800 m) were found by reciprocal transplants in *A. tenuis* (Bradshaw, 1960). In this case the adaptation seems to be in relation not to generalized aspects of climate but to particular factors such as cold wind exposure in the upland transplant station and salt spray in the maritime station (Table 2). General adaptation to different soil types was tested for in further reciprocal transplant experiments but was not found. However, a spectacular adaptation was found to elevated levels of lead and zinc; populations from normal soils could not survive properly on soil containing lead and zinc, whereas populations from metal-contaminated habitats could grow excellently. We now know by further transplant studies that there is complete failure when populations from normal soils are transplanted into metal-contaminated sites (Smith and Bradshaw, 1979).

Artificial competition experiments

A variation of the reciprocal transplant experiment is to transplant material into a common environment but to let each population experience the reciprocal environment of others. The importance of

TABLE 2. Damage to different local populations of *Agrostis tenuis* in exposed coastal and upland environments.[a,b]

	PERCENTAGE OF LEAVES BROWNED			
	Coastal site		Upland site	
Population	Dec. 49	Sept. 50	Apr. 50	Apr. 52
Lowland				
Dale cliff (very exposed)	24.2	29.7		
Dale pasture (protected)	33.8	62.0		
Morfa coast (exposed)	29.7	44.0	20.5	34.8
Pen Dinas top (exposed)	26.7	55.2		
Pen Dinas back (protected)	25.1	65.2	25.3	37.3
Upland				
Disgwylfa N (exposed)			16.3	15.5
Disgwylfa S (exposed)			16.2	14.8
Plynlymon side (exposed)			12.5	16.5
Plynlymon top (very exposed)	58.0	73.0	13.2	16.5

[a] From Bradshaw (1960).
[b] Population differences significant ($P < 0.01$) on each occasion.

allowing competition of other material in the assessment of populations is very clear from work on the fitness of metal-tolerant plant populations. These usually survive well on normal soils in the absence of competition, and their fitness, measured by vegetative growth, is little different from normal nontolerant populations. But in the presence of competition provided by another species (*L. perenne*), a competition which would occur on normal soils, the fitness of the tolerant populations, relative to normal populations, declined considerably and progressively (Hickey and McNeilly, 1975). This work, among other things, demonstrates the cumulative nature of the effects of selection in perennial species.

Barley cultivars hardly count as natural populations, yet they do, of course, have different origins and related adaptations (Finlay and Wilkinson, 1963). The remarkable series of competition experiments in ten different environments involving eleven different cultivars sown and resown together over several generations in a single mixture, carried out by Harlan and Martini (1938), shows what remarkable differences in fitness can be revealed when competition is allowed to take effect (Table 3). These cultivars, which could all perform adequately in the absence of competition, could be eliminated in as little as four generations when in mixture, the cultivar remaining depending on the environment of the particular mixture. Similar experiments by Suneson and Weibe (1942) showed the same type of results. Perhaps the most interesting experiments are those on rice; in these experiments performance of cultivars in mixtures bore no relation to their performance in pure stands (Jennings and de Jesus, 1968).

TABLE 3. Final population composition of a mixture of barley cultivars grown in ten United States locations for 4 to 12 years.[a]

Cultivar	Va.	N.Y.	Minn.	N. Dak.	Neb.	Mont.	Idaho	Wash.	Ore.	Calif.
Coast and Trebi	89.2	11.4	16.4	31.2	44.8	17.4	42.0	30.0	1.2	72.4
Gatami	2.6	1.8	3.0	4.0	1.4	11.6	2.0	0.2	0	0.2
Smooth Awn	1.2	10.4	2.8	4.6	2.4	5.0	0	1.0	0.2	0
Lion	2.2	0.6	5.4	2.8	2.6	7.4	0.4	0.6	0	1.6
Meloy	0.8	0	0	0	1.4	0.8	1.6	1.2	0	5.4
White Smyrna	0.8	0	0.8	3.4	38.8	48.2	31.4	55.2	97.8	13.0
Hannchen	0.8	6.8	61.0	30.4	2.6	3.8	18.0	6.0	0.8	6.8
Svanhals	2.2	0.4	10.0	16.0	5.2	1.6	3.6	4.6	0	0.4
Deficiens	0	0	0	0.2	0.6	0	0.4	1.0	0	0.2
Manchuria	0.2	68.6	0.4	7.4	0.2	4.2	0.6	0.2	0	0

[a] From Harlan and Martini (1938).

It is surprising, therefore, that only recently have any competition experiments been carried out between natural populations. In populations of *T. repens* adapted to different soil conditions, the effect of competition is to enhance considerably the differences expressed in pure culture (Snaydon, 1962). In comparisons of metal-tolerant and normal populations of *A. tenuis*, competition between them reveals major differences not shown in pure stands (Cook et al., 1972) (Figure 1).

These experiments reveal simple differences in relative fitness between different populations. But natural environments provide competition not just between different genotypes of the same species, but between different genotypes of different species. In this case, although the presence of competition of another species may merely enhance differences between genotypes of the same species, as shown by Hickey and McNeilly (1975), it may reveal more subtle relationships, because we have the possibility of adaptation to these other species. Thus, in *T. repens*, competition experiments have shown localized adaptation to the presence of particular grass species (Turkington and Harper, 1979b) which is not revealed when the grass species are not present. But here we are dealing with differences which to many people would be considered to be within a population.

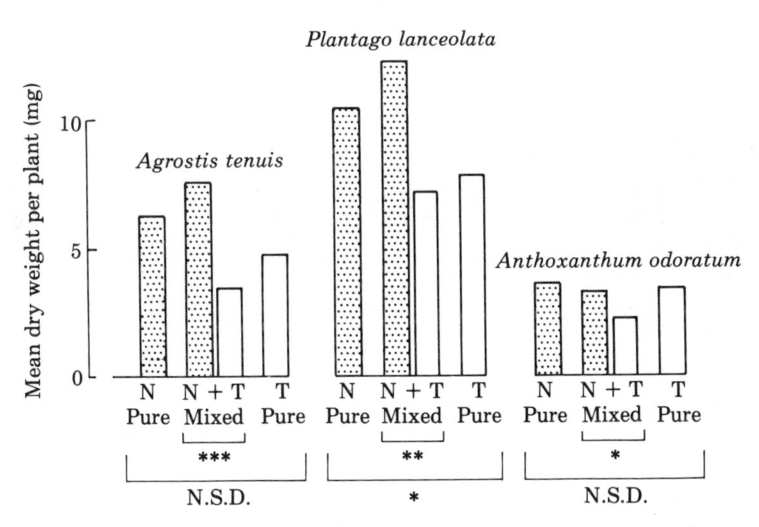

FIGURE 1. The growth of metal-tolerant (T) and normal (N) populations of *Agrostis tenuis, Plantago lanceolata*, and *Anthoxantum odoratum* in pure stands and in mixtures. *, $P < 0.05$; **, $P < 0.01$; ***, $P < 0.001$; N.S.D., not significantly different. (After Cook et al., 1972.)

Transplants into natural conditions

The fitness of material can really only be assessed properly in natural conditions in which the total effects of the environment can be allowed to operate. This was the approach of Bonnier, but his management of his experimental material does not allow valid conclusions to be drawn (Clausen et al., 1940). Nevertheless, with proper management and careful recording, populations can be followed to excellent effect. Thus, populations of clover which are adapted to different soils and which show equivalence when planted in competition on acid soil in pots in a greenhouse show startling differences when planted into an acid upland grassland (Snaydon, 1962).

This approach has been pursued to great effect by Davies and Snaydon (1976) in studies on the different populations of *Anthoxanthum odoratum* occurring in the Park Grass fertilizer plots. The performance of the populations was compared in reciprocal transplants (after clonal multiplication) into their original "native" and the alternative "alien" plots. In each plot the relative fitness of alien/native populations was determined. For all populations in all plots, the mean value was 0.67; but in many instances there were values < 0.5 (Table 4). The average half-life of populations in alien plots was 8 months; that of populations in native plots was 24 months. It must be remembered that the distances separating these populations in their natural habitats averages only 30 m.

Very similar results have been obtained for populations of *Ranunculus repens* that were reciprocally transplanted over a short distance between woodland and grassland environments (Lovett Doust, 1981c).

TABLE 4. Relative fitness of alien populations of *Anthoxanthum odoratum* for different characters, when alien and native populations are planted into the existing vegetation on different plots of the Park Grass Experiment, Rothamstead.[a]

	SURVIVAL (%)			TILLER NUMBER			DRY MATTER	
Plot	May 69	Oct. 69	May 70	May 69	Oct. 69	May 70	May 70	Mean
9L	0.33	0.26	0.23	0.42	0.45	0.43	0.40	0.36
9U	0.92	0.93	0.91	0.83	0.83	0.75	0.80	0.35
1L	0.71	0.58	0.60	0.58	0.77	0.77	0.65	0.67
1U	0.88	0.90	0.88	0.71	0.74	0.75	0.80	0.81
8L	0.69	0.50	0.46	0.76	0.80	0.69	0.61	0.64
8U	0.85	0.78	0.78	0.71	0.67	0.58	0.64	0.72

[a] From Davies and Snaydon (1976).

Several characters indicative of relative fitness of aliens were measured. The character integrating growth most completely (total length of stolon) gave relative fitness values of the alien of 0.2 in both environments, a result quite remarkable for populations which may have had separate existences for only 12 years.

The ultimate aim of this technique is to be able to establish the populations in reciprocal transplants and allow them to go through complete reproductive cycles. Where this has been done with *Phlox drummondii* (Schmidt in Levin, Chapter 12), the fitnesses of alien populations originating from not very different habitats was about 0.6.

Cumulative effects

The reciprocal transplant technique is a powerful method by which the fitness of populations can be compared. Obviously, a great deal more can be done by using the methods of population biology. But already it is clear that different populations of a species have very different adaptations to different environments and that it is commonplace for a population in an alien environment to have a fitness of only half that of the native population. Where the source environments are very distinct, the differences in fitness can be very much greater, as in the case of metal-tolerant populations in which relative fitnesses of < 0.1 are usual (Jain and Bradshaw, 1966).

What we must remember, however, is that these differences have almost all been determined over a single growing season. Whether we are dealing with annual or perennial material—it does not matter which—the effects of differential fitness are cumulative. Therefore, over a five-year period, a relative fitness of an alien population of, for instance, 0.6 will lead to a final relative fitness of 0.078. An indication of this cumulative effect in vegetative growth over only a few months is given by the data for metal-tolerant and normal populations of grasses of Hickey and McNeilly (1975). But it is equally apparent for the survival values of *A. odoratum* in Park Grass (Davies and Snaydon, 1976). The adaptive significance of population differentiation is inescapable.

EVOLUTIONARY DIFFERENTIATION AND ECOLOGICAL AMPLITUDE

All these comparisons examine only alien or native growth and survival. From the point of view of species, what we should be interested in is the ecological range of individual species—their occurrence and overall survival in a range of habitats. This means their ability to:

(1) invade different habitats;
(2) develop and reproduce in those habitats;
(3) persist in those habitats by surviving any environmental changes.

To what degree is the presence of an individual species in a range of habitats—a presence resulting from these three processes— enhanced or permitted because of evolutionary differentiation? We can only obtain clear indications from some of the previous evidence, in particular that for *Achillea* in California (Clausen et al., 1948) and *Agrostis* and *Festuca* on metal mine wastes in Britain (Smith and Bradshaw, 1979) (Figure 2). In these examples, even though they involved spaced plants, it is apparent that the species are composed of populations which will not survive in each other's habitats. If the species were composed solely of material of one of these populations, their ecological amplitude would be radically reduced. The species would only recover if evolutionary adaptation occurred—which of course has happened.

In these examples the material was established for a considerable period of time, satisfying condition 3 above. But the material was introduced artificially, so that the material did not have to show success in 1 and 2. The only work on the contribution of population differentiation to condition 2 is that on *P. drummondii* by Schmidt (Levin, Chapter 12). There is no critical work on the contribution of population differentiation to 1. We must surely conclude that if a species were represented by material of a single population and 1, 2, and 3 were operating, it would be restricted to a much narrower range of habitats than that suggested by *A. millefolium*. This restriction would be even more enhanced than we expect when we remember that the different populations would have to experience not only the total effects of the natural environment but also their accumulated effects over a long period.

It seems clear from these arguments that the ecological amplitude of a species must be affected by its ability to evolve populations adapted to different environments. This then raises the problem as to whether evolutionary differentiation is a component of the adaptive armory of all species or whether it is a particular attribute only of some. It is important to know whether the wide ecological amplitude of some species is the direct outcome of an above average capacity to show

FIGURE 2. Survival of populations in transplant experiments in natural environments. A. *Achillea millefolium* at high altitude in California. (After Clausen et al., 1948.) B. *Festuca rubra* on a zinc/lead mine in North Wales. (After Smith and Bradshaw, 1979.)

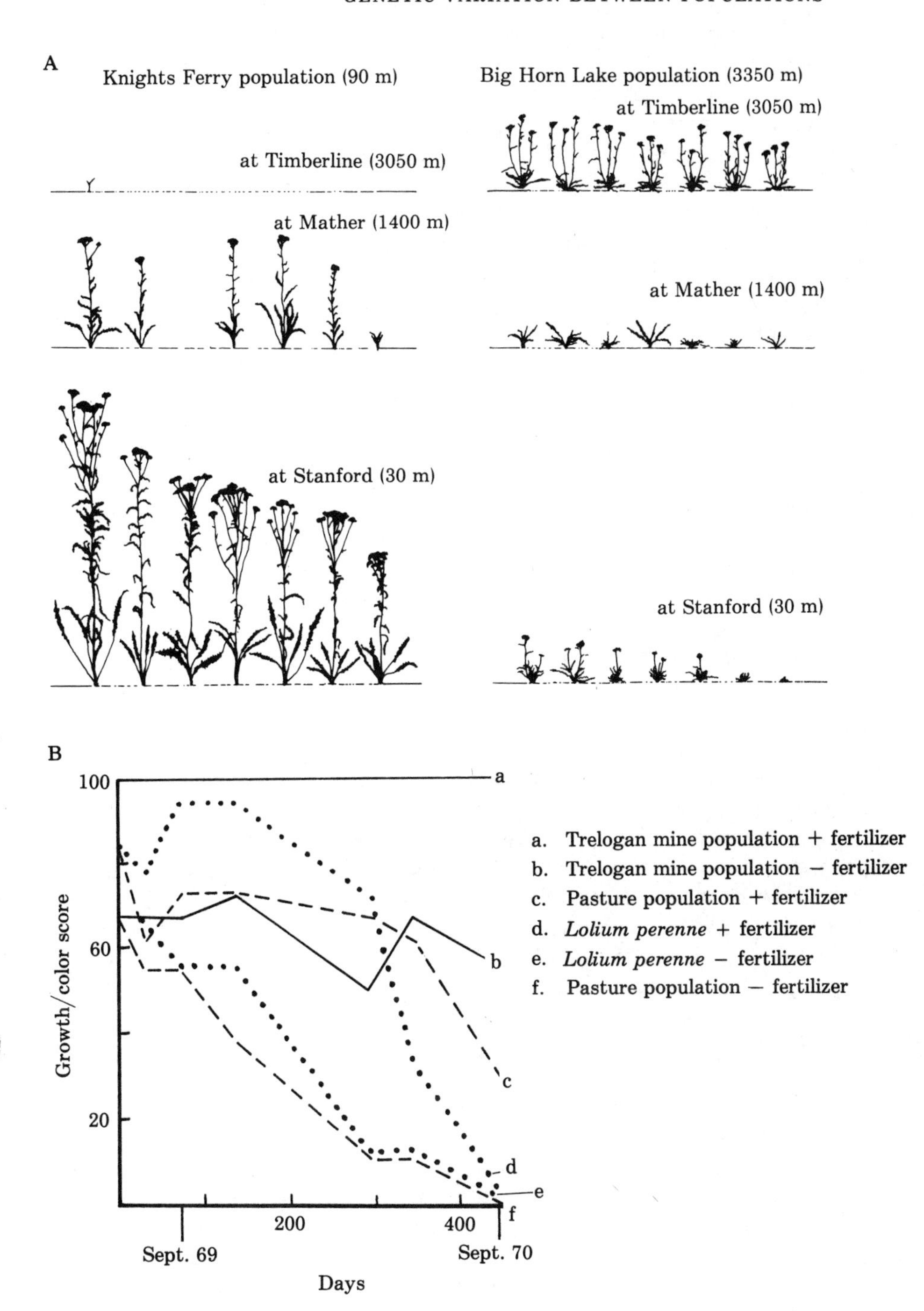
A
Knights Ferry population (90 m)
at Timberline (3050 m)
at Mather (1400 m)
at Stanford (30 m)
Big Horn Lake population (3350 m)
at Timberline (3050 m)
at Mather (1400 m)
at Stanford (30 m)
B
100
60
20
Growth/color score
200
400
Sept. 69
Sept. 70
Days
a
b
c
d
e
f
a. Trelogan mine population + fertilizer
b. Trelogan mine population − fertilizer
c. Pasture population + fertilizer
d. Lolium perenne + fertilizer
e. Lolium perenne − fertilizer
f. Pasture population − fertilizer

evolutionary differentiation or whether it is only somewhat assisted by it.

If we examine what is known about population differentiation in species of wide and narrow amplitudes, it is very clear that the best evidence for extensive population differentiation is in species of wide ecological amplitude such as *Prunella vulgaris* (Bocher, 1949), *P. glandulosa* (Clausen et al., 1940), *A. millefolium* (Clausen et al., 1948), *Pinus sylvestris* (Langlet, 1959), and *A. tenuis* (Bradshaw, 1959). However, we cannot from this conclude that these species owe their wide ecological amplitude to their extensive population differentiation. It is equally plausible that they owe their extensive population differentiation to their wide ecological amplitude. To solve this problem we must seek other evidence.

LIMITS TO EVOLUTIONARY DIFFERENTIATION

If evolutionary differentiation had no limits and could adapt a species to any environment, it follows that species would have universal distributions. But this is not so. From this it follows that evolutionary differentiation does have limits. To appreciate this we can examine evolutionary differentiation and colonization in one rather distinctive environment—that of metal-contaminated soils.

It is very clear from the earliest evidence of *Silene vulgaris* (Prat, 1934), that evolution of metal tolerance enables species to colonize metal-contaminated sites: without this tolerance the species cannot survive in these habitats. This is abundantly clear from later work (e.g., Smith and Bradshaw, 1979), as we have already seen.

It is also clear from analyses of the floras of metal-contaminated environments (see Antonovics et al., 1971 for a review) that only certain species are to be found in these environments, despite the fact that a wide range of species has had the opportunity to colonize them. In the case of a copper refinery site in Lancashire, only 4 species (now known to be 5) are to be found on the copper-contaminated soils and are tolerant, although originally at least 26 must have been present before copper contamination began (Bradshaw, 1975). Clearly only some species had the capacity to evolve tolerance and thereby had an ecological amplitude which allowed them to remain on the site as it became polluted.

It has been possible to demonstrate experimentally that the ability to evolve metal tolerance is limited in certain species and that this is because they do not possess the necessary genetic variation. Evolution of tolerance depends on the selection of appropriate variability occurring in normal populations (Walley et al., 1974). In species which do not evolve tolerance, this variability is not to be found (Khan in Bradshaw, 1971; Gartside and McNeilly, 1974). In the latter work there was an exception, *Dactylis glomerata*, which was found to possess variability for copper tolerance yet seemed not to occur in copper-contaminated habitats. However, copper-tolerant populations of this species have now been found.

Occurrence of metal-tolerant individuals in normal populations of species was originally tested by screening normal populations on metal-contaminated soils. A refinement has been to test the material in metal solutions, by supporting the seeds on polyethylene beads. Using this method, it has been clearly demonstrated (C. Ingram, personal communication) that only certain species contain variability in metal tolerance and that only among these species are those which do evolve tolerance and occur on metal-contaminated sites (Table 5). So for this environment at least, we have good evidence that the ability, or lack of it, to show evolutionary differentiation determines ecological amplitude.

TABLE 5. Percentage of copper-tolerant individuals found in normal populations of various grass species, in relation to the presence of the species on copper-polluted waste and whether the plants collected were tolerant of copper.[a]

Species	Percentage occurrence of tolerant individuals	Presence of species on mines		Tolerance of collected adult plants
		On waste	Margins	
Holcus lanatus	0.16	+	+	+
Agrostis tenuis	0.13	+	+	+
Festuca ovina	0.07	–	+	–
Dactylis glomerata	0.05	+	+	+
Deschampsia flexuosa	0.03	+	+	+
Anthoxanthum odoratum	0.02	–	+	–
Festuca rubra	0.01	+	+	+
Lolium perenne	0.005	–	+	–
Poa pratensis	0.0	–	+	–
Poa trivialis	0.0	–	+	–
Phleum pratense	0.0	–	+	–
Cynosurus cristatus	0.0	–	+	–
Alopecurus pratensis	0.0	–	+	–
Bromus sp.	0.0	–	+	–
Arrhenatherum elatius	0.0	–	+	–

[a] Unpublished data of C. Ingram.

This important principle needs to be tested in other situations before we can be certain of its generality. But there is no doubt that different species do not have the same evolutionary abilities. In order to breed resistance to particular diseases or physiological stresses, plant breeders often have to go outside the material they wish to improve to find the necessary genes (Allard, 1960). Evolution of resistance to herbicides is a recent phenomenon in plants. It appears that the ability to evolve resistance to particular herbicides is restricted to particular species (Bandeen et al., 1982; Gressel et al., 1982).

Limits to evolutionary differentiation will affect not only whether or not a species can exist in a particular type of habitat, but also what extreme of that habitat it can endure. This is again very clear in metal-contaminated habitats. Despite the evolution of metal tolerance there are areas in metal-contaminated habitats where metal contamination is too high to permit even the growth of metal-tolerant plants. There is a considerable body of evidence that shows that in selective situations a limit of progress under selection is always reached eventually because of a lack of further appropriate additive variation (see Falconer, 1981 for a review). Uncolonized areas in metal-contaminated habitats are therefore indications of limits to evolutionary differentiation within species which have evolved tolerance.

The next step in this argument relates to the stability of plant distributions. Dispersal is always transporting a species away from its normal habitats into habitats to which it is not adapted. As a result natural selection must constantly be acting in a directional manner, attempting to expand the ecological amplitude of the species. Since the distributions of most species are stable, we must presume that most species have been taken by natural selection to their selective limits. From this we can conclude that the determination of ecological amplitude by genetic flexibility is a universal phenomenon. It follows from this that most species, at least in their marginal populations, have reached a situation in which their further adaptation is limited by a lack of appropriate variability. These species are, therefore, in what can be termed a condition of *genostasis* (Bradshaw, 1983). We tend to think that nearly all populations contain considerable amounts of variability. But for the ecologically significant variation, they must be in a genostatic condition. It is this genostasis which gives species at least some of their particular ecological characteristics.

All this, of course, refers to existing situations. In a new situation, either where a species is dispersed into an environment which it has not previously experienced or where new environmental factors occur in a prexisting habitat of the species, the new selection pressures will be able to act on previously unselected variation, and rapid evolutionary adaptation can be expected. This could permit startling changes to the previous ecological amplitude of the species.

It is indeed true that the best examples of evolution are in new situations, usually due to man's activities, for instance, pollution, in both plants and animals (Bishop and Cook, 1981; Bradshaw and McNeilly, 1982). But evolutionary changes have also taken place when a species has been taken from one continent to another. Unfortunately the evolutionary changes which have occurred have rarely been properly documented. The best evidence of very substantial adaptive changes in ecological characters when a species or cultivar is taken from one environment to another is that of Sylven (1937). The magnitude and speed of the changes reported indicate the remarkable effect evolution can have on previous ecological amplitude. Other evidence is more circumstantial (Baker and Stebbins, 1965). But it seems likely that a lack of positive evidence is only because it has not been looked for.

In Chapter 1 of *Origin of Species*, Darwin makes much of the fact that altered conditions of life appear to be a cause of increased variability. We have often dismissed Darwin's knowledge of variablity and its causes. However, Darwin was an observer of unparalleled ability and shrewdness. It seems likely that Darwin, 100 years ago, did in fact observe that in the environments to which they have previously been accustomed species show genostasis, whereas in new environments they reveal variation which has not previously been subject to selection.

CONCLUSIONS

It is truly remarkable that Darwin was able, by deductions from his own limited observations, limited because they hardly included any experimental work, to weave the complex fabric of a theory which we still respect today and which remains the foundation of our evolutionary ideas. It is therefore important that we realize that in Chapter 4 of *Origin of Species*, where he is discussing Divergence of Character, Darwin says, "The more diversified the descendants from any one species become in structure, constitution, and habits, by so much will they be better enabled to seize on many and widely diversified places in the polity of nature, and so be enabled to increase in numbers." He does not explore this any further in relation to what can happen within a species, but it remains an admirable statement of the crucial part played by genetic variation in determining the ecological amplitude of species. Until now ecologists and physiologists have, perhaps unintentionally, tended to envisage that the range of habitats which a species occupies is related to its innate physiological tolerance, properties which can be measured in single genotypes. It is imperative that we now look for evolutionary causes.

As part of this consideration, it is then also important that we realize that, where in the last chapter of *Origin of Species* Darwin states that "I can see no limit to this power, in slowly and beautifully adapting each form to the most complex relations of life," he is guilty of forgivable optimism. It is a crucial aspect of the stability and limits we find in the ecology of species that evolution itself has limits. Although in the very long term there may not be limits, in the short term it is clear that there are definite limits to evolution which differ in different species. So the ecology of a species, whether we are considering its demographic characteristics, its physiological tolerance, or its interactions with other species, is, in a remarkable way, more an outcome of its evolutionary ability than we have, perhaps, previously suspected.

CHAPTER 11

GENETIC VARIATION WITHIN POPULATIONS

Janis Antonovics

HISTORICAL OVERVIEW

The history of the study of genetic variation in populations has been a sequence of paradoxes. Darwin (1859) fully recognized the importance of genetic variation as the raw material for natural selection and went to considerable pains to document its existence, both in domesticated and wild populations. Yet understanding the origin and maintenance of such variation remained for him a thorn in the side of natural selection. Given his assumption of blending inheritance as the most likely pattern of genetic transmission, he was faced with the paradox of how such variation would not disappear in but a few generations (for discussion, see Fisher, 1958). And this paradox could for him only be resolved by implementing, albeit reluctantly, neo-Lamarckian speculations that environmental variations had concomitant effects on genetic variation. The paradox was seemingly resolved with the rediscovery of Mendel's results and the realization that inheritance was particulate. Yet it was precisely the particulate nature of these Mendelian factors (and the distinct mutants that were used to demonstrate their existence) that led to a conflict between those who ascribed to mutation the major directing role in evolution and those who remained convinced that natural selection could result in emergent novelty by the accumulation of small changes. That such a paradox (whether mutation or selection is more important in evolution) should have retarded evolutionary thinking and generated opposing camps of followers is barely understandable in retrospect (Mayr and Provine, 1980), yet it was the reconciliation of these camps that was hailed as one of the major achievements of the Evolutionary Synthesis. Thus,

Huxley (1942) in *Evolution: The Modern Synthesis* reassured us, in a way that now seems almost patronizing, that "Neither mutation or selection alone is creative of anything important in evolution; but the two in conjunction are creative . . . their interplay is as indispensable to evolution as is that of hydrogen and oxygen to water."

A major component of this synthesis was the growth of population genetics, a field of science devoted explicitly to the analysis of genetic variation in populations. As an area of biology, it was unique in having almost from its inception a strong mathematical basis, more analogous to that found in the physical sciences rather than in biology. It therefore remained a relatively small, influential, yet esoteric field of biology, and as it entered the era of The Synthesis, apart from the almost ritualized sparring of the Wright and Fisher schools on many points of emphasis, it presented at last a relatively paradox-free view of genetic variation. Organisms were generally "wild-types," but Mendelian mutations shuffled by recombination and, more often having quantitative rather than qualitative events, provided the raw materials for adaptation. This period and the following decades were marked by detailed studies on the genetics of natural populations of plants (Clausen and Hiesey, 1958a) and animals (Dobzhansky, 1951; Ford, 1964), and by the growth of a rich theory regarding the maintenance of genetic polymorphism (see Williamson, 1958 for an early review). There was a strong focus and interest in clear-cut single gene (or chromosomal) polymorphisms since their genetic basis was easily understood and their frequency readily quantified. But since such polymorphisms were infrequent, there was little appreciation and little thought given to measuring overall levels of genetic variation in populations. During this time quantitative genetics matured into a highly sophisticated science, but while its techniques could partition and identify genetic effects on the phenotype, these techniques were relatively ineffectual in assessing variation at individual loci: at best one could only talk vaguely about "number of effective factors" (Mather, 1949). However, the commonplace of substantial response to artificial selection in a wide range of quantitative traits led Falconer (1960) to hazard that "natural populations probably carry a variety of alleles at a considerable proportion of loci, even perhaps at virtually every locus." While a brilliant guess, it carried only the weight of impression and not that of precise demonstration.

It was not till the advent of electrophoresis and the discovery by Hubby and Lewontin (1966) of large amounts of genetic variation in natural populations that ideas again fell into disarray and new paradoxes arose. Given that populations were so variable genetically, how did such variation arise, and how was it maintained? And given that an almost infinite number of gene combinations were possible with even a modest level of heterozygosity, why is evolution so often slow and imperceptible. What generates and what constrains this vast reservoir of genes?

Recent discoveries in molecular biology promise, not only to give us more absolute measures in terms of levels of variation at the base-pair level, but to elaborate a classification of variation based on the type of genetic material affected. Variation may occur, not only in structural and regulatory genes, but also in regions of the genome having no overtly functional roles. Structural genes may contain untranslated intervening sequences; they may be interlaced with large noncoding intergenic sequences; there may be "pseudogenes" whose nonfunctional DNA sequences are closely homologous to overtly functional loci; and much of the genetic variation may be caused by insertion, deletion, and modification of transposable elements. Our view of the genome as a string of beads is in its final death throes, as is our view of it as a tightly integrated unit: I myself am tempted to see it not as a sophisticated spaceship but as a temporarily functional drag racer improvised from a junk yard of past efforts.

The discovery of large amounts of variation in populations (for evidence in plants, see Hamrick et al., 1979; Hamrick, 1979), if we are to believe the historical tapestry woven by Lewontin (1974), led to a flurry of explanations which divided us into a "classical school" of neutralists battling with a "balance school" of selectionists. Such controversy not only stimulated further quantification of genetic variation in a large range of organisms but also led to a growth of theories and ideas about its origins, loss, and maintenance. If the appearance of texts is to be used as a criterion, population genetics had finally, in the 1970s, matured into a singular, recognized branch of biology.

Although theoretical population genetics has proffered numerous alternative explanations for the existence of genetic variation, it has also become clear that the application of any particular theory to any particular polymorphism is fraught with difficulties. This has been called "a problem with too many solutions" by Jones et al., (1977) in a review of probably the most extensively studied polymorphism, shell color and pattern in *Cepaea nemoralis*. They suggest that "complex and perhaps unique explanations [of the polymorphism] are needed for almost every *Cepaea* population." Certain electrophoretic variants have now been studied in sufficient detail, so we can clearly understand the causative chain between physiological process, ecological function, and selection differentials. Yet such studies (e.g., Koehn, 1978; Koehn et al., 1980) are probably as remarkable for the time and effort involved as they are for the elegance of results; indeed, it is difficult to see how they can realistically be repeated on numerous, preferably randomly chosen, allelic variants. In other words, unless we spend the next 100 years gathering more and more information on more and more polymorphisms, it is likely that a balanced perspective on the relative importance of the various forces impinging on gene frequency in populations will elude us. This has been formidably argued by Lewontin and is the pessimistic conclusion of his paradox of variation (Chapter 5; Lewontin, 1974). In order to escape from this dilemma, we need to look more closely at its causes, at inadequacies of past studies, and at different new approaches that might finesse us out of our apparent fate of having to move the mountain one handful at a time.

If we survey past studies of genetic variation within populations, we see that they have been characterized by particular approaches and assumptions. Above all, such studies have been largely correlative and descriptive. Gene frequency has been correlated with geography, environmental factors, history, breeding system, etc. Although such studies may give information about forces acting to differentiate populations, they rarely answer hypotheses about maintenance of variants *within* populations. Only relatively recently (Linhart, 1974; Schaal, 1975; Hamrick and Holden, 1979; Watson quoted in Antonovics, 1978; Turkington and Harper, 1979b) have descriptive approaches been attempted at a within-population level.

Even at a within-population level, the dynamics of the forces changing gene frequency cannot be understood by a static cross-sectional view (cf. the problem of estimating competition from descriptive studies of species niche relationships). For example, if there is frequency-dependent selection, there may be weak differential fitness at equilibrium; such selection may only be detectable by perturbing the system. Experimental studies on genetic variation in natural plant populations are extremely few, and they have either examined differential fitness in transplants among populations (and then only rarely in natural conditions: Antonovics and Primack, 1982), or they have been carried out in agricultural situations (Suneson, 1969). In animals, too, with few exceptions (Kettlewell, 1956; Gaines et al., 1971; Jones et al., 1977), experimental studies have been largely confined to laboratory populations. We are reminded of the naturalist–experimentalist dichotomy that characterized evolutionary arguments before the evolutionary synthesis (Allen, 1978): this dichotomy is clearly a legacy that evolutionary and ecological studies have yet to outgrow.

Another characteristic of past approaches has been that they have assumed, albeit tacitly, that within-population genetic variance is appropriately explained in terms of either properties of entire groups (e.g., effective population size) or properties of groups of particular classes of genotypes (e.g., average selection coefficients of particular genotypes). This has resulted in studies that ignore two important questions: What is the relevance of genetically variable progeny to the individual? How important is the genetic and environmental context of a genotype to the fitness of the individual? The first question is the paradox of sexual reproduction: Given the large individual disadvantages associated with sexual reproduction, why is such reproduction so common, so persistent, and in many organisms the only reproductive mode? It is a paradox that has stimulated several books and many papers, yet few experiments. Moreover, it is a question that provides an alternative avenue for developing an overview of the mechanisms whereby genetic variance is generated and maintained in natural populations. If we can understand the *raison d'être* of genetic variance for the individual, then do we need to ascribe reasons to each locus? Group properties such as effective population size, gene frequency, and density may be very important factors impinging on genetic variance in particular instances; there may even be group properties that are disadvantageous to the individual yet that maintain genetic variance (e.g., gene flow between adjacent diverging populations); yet answering the question of why individuals have evolved mechanisms for generating highly variable offspring may well help circumvent our current impasse about the source and maintenance of genetic variation within populations. Our recent approaches can be likened to a study of the immune system where we suddenly have access to all the genetic variants that such a system can produce and have decided that a correct approach is to understand the particular significance of each variant to each population of cells. Although the overall significance of the immune system to the individual is clear, this significance would only be dimly seen (if at all) by a detailed study of each variant. As we shall see, this analogy may not be too farfetched. It can also be pursued further. It was not till monoclonal antibodies were developed that precise study of the mechanisms of their individual interactions and origins was possible. Similarly, we have yet to develop techniques and approaches for studying the fitness of particular individuals rather than of genotypic or phenotypic classes.

Lewontin (1974) concluded that "context and interaction [of genes] are not simply second order effects to be superimposed on a primary monadic analysis. Context and interaction are of the essence." The truth of this assertion has never been tested: We have generally attempted to measure selection on a group of individuals with an average genetic background and in an average environment. The effects of the local external environment and the genetic background of a particular individual are rarely considered. Both are likely to be particularly important in plants: first, because plants do not run around averaging out environmental heterogeneity, and second, because their mixed-mating systems permit both conservation and change of gene interactions. We do not know the answer to many simple questions. How many gene–character combinations lead to equivalent fitness? For any pair of alleles, what is the within-allele genetic variance in fitness due to background, what is the between-allele variance, and what is the interaction effect of allele with background? More plainly, we may wish to ask what is the genetic variance in fitness, either by way of addressing rather grandiose issues such as the applicability of Fisher's fundamental theorem of natural selection, or by way of addressing nuts-and-bolts field biology questions about whether seedling mortality has a genetic component.

GENETIC VARIANCE AND INDIVIDUAL FITNESS

I have just made out my first grass, hurrah! hurrah! I must confess that fortune favours the bold, for, as good luck would have it, it was the easy *Anthoxanthum odoratum*: nevertheless it is a great discovery: I never expected to make out a grass in all my life, so hurrah! . . . It has done my stomach surprising good.

Darwin to Hooker, June 1855

To illustrate how the maintenance of genetic variation in natural populations can be addressed, not as a group property of the population, but as a property of relevance to the ecology of the individual, I will, in this section, outline a few experiments we have been doing with the grass *Anthoxanthum odoratum*. First I will consider briefly some experiments bearing on measuring the fitness gains from producing genetically variable as opposed to uniform progeny. Then I will use some results from these experiments to argue how fitness of individuals can be measured using cloning techniques. And, finally, I will point out how such techniques can be used to map individual phenotypes onto fitness.

Throughout, we have been using as a model system the grass *Anthoxanthum odoratum* (Sweet Vernal Grass) growing in a mown field that has had the same management for over 30 years. Genetically uniform "progeny" have been obtained by cloning tillers from field-collected adults, and genetically variable "progeny" have been obtained as tillers from plants grown from seed produced by those same adults. These tillers are then transplanted as "phytometers" (Clements and Goldsmith, 1924; Antonovics and Primack, 1982) back into the field in formal experimental designs, with a minimum of disturbance to the natural community. The survivorship and fecundity of these transplants is then followed over a number of years and provides an assessment of the fitness of each of the progeny types.

In the first experiment, genotypes were grown both as plots of uniform asexually cloned individuals and as plots of genetically variable half-sib families (tillers from seeds produced by those same individuals). Each plot was planted in a hexagonal fan design (Antonovics and Fowler, in press) of eighty plants to give a range of densities within each plot (Figure 1A). At each of two sites in the field there were four genotypes, represented as variable and uniform "progeny," and each replicated twice to give a total of ca. 2500 tillers. The tillers were preweighed and planted directly into the ground following rooting for four days in water. Individuals were marked with a toothpick and a plastic ring and measured for survival and fecundity for two years, by which time most of them had died. The results (Figure 2) showed that the genetically variable plots outyielded the genetically uniform ones in both sites. The distribution of fitness as estimated by the net reproductive rate ($\Sigma l_x m_x$) over two years was highly skewed, making statistical analysis difficult. However, if we consider the number of plants in each of the inflorescence number

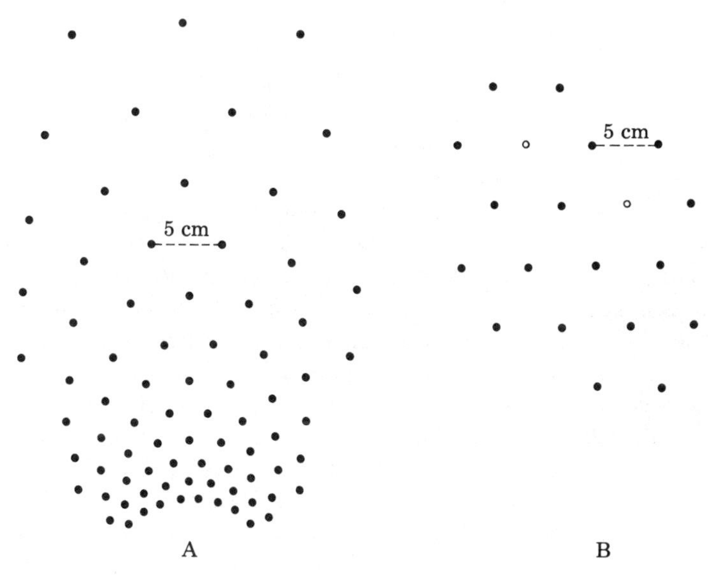

FIGURE 1. Planting designs (A) for experiment to examine effect of density on fitness of genetically variable and nonvariable "progeny," and (B) for experiment to compare fitness of minority and majority genotypes; the minority type is shown in open circles. (See text for further explanation.)

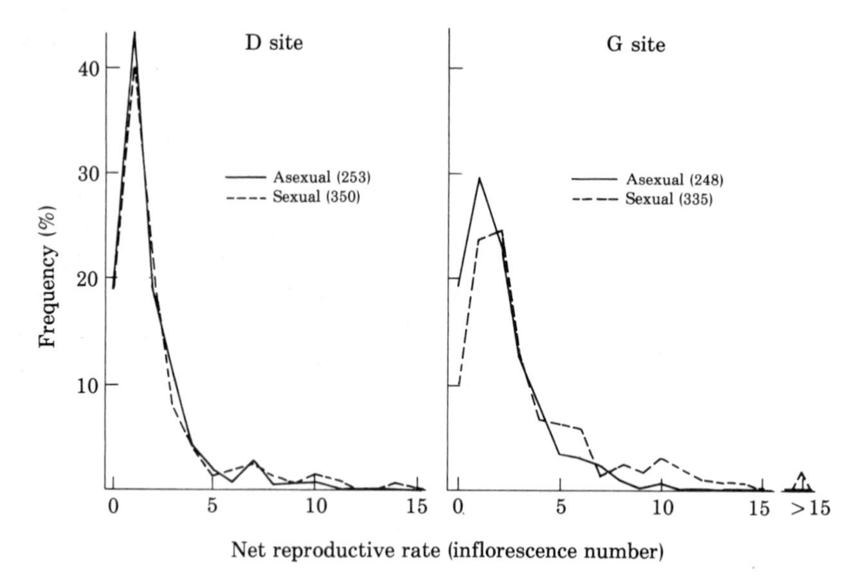

FIGURE 2. Frequency distribution of net reproductive rate over two years for sexually produced half-sibs (genetically variable) and asexual clone (nonvariable) progeny arrays, planted at two sites within a field. Sample sizes are in parentheses.

classes of 0, 1, 2, 3, 4, 5–9, and >10, the two-way interaction of inflorescence class and progeny type (sexual or asexual) was significant ($\chi^2_6 = 16.1$; $P < 0.05$). If we consider the second-year reproductive output only, this interaction was highly significant $\chi^2_6 = 17.1$; $P < 0.01$). The overall means gave a relative fitness of the sexual to asexual progeny of 1:0.97 at the D site and of 1:0.80 at the G site. If we consider the second year of the experiment only, these differential fitnesses were much greater (1:0.63 at the D site and 1:0.42 at the G site). Because of the large variance in the data (not surprising in an experiment carried out in situ, in the field), density did not show any consistent pattern in its effects on this relative fitness. This experiment is now being repeated to simulate more closely the seed dispersal profile around a parent individual and so to enable us to develop more realistic estimates of the relative fitness of uniform and variable progeny.

In a second experiment we have looked more closely at the possible cause of this better performance of genetically variable versus genetically uniform progeny. Parents from eight sites were used in the field, and the experiments were planted back into those same sites. At each site, there were 12 plots, and each plot consisted of two genotypes, a "majority type" and a "minority type" within a hexagonal honeycomb of 20 individuals (Figure 1B). There were six pairwise genotype combinations, each member of a pair being represented as a majority and minority genotype in reciprocal plots. The overall result was that the minority types gave a net reproductive output over three years of 1.89 inflorescences per planted individual, whereas the majority type gave a net reproductive output of 1.26 individuals (for further details, see Antonovics and Ellstrand, in press). Since competitive interaction among individuals were weak, this generalized frequency-dependent selection was probably not mediated by resource partitioning under competition but was probably determined by pathogen effects. Clearly such effects could cause a very large advantage for individuals that are in some sense different from the parent and in a minority around the parent. That frequency dependence is a factor maintaining genetic variance in natural populations is not new; however, its importance as a general mechanism promoting genetically variable progeny has only relatively recently been suggested (e.g., Levin, 1975b; Hamilton, 1980; Lloyd, 1980a; Price and Waser, 1982). Our earlier analogy with the immune system may not be farfetched; genetically variable progeny of an individual may be a pathogen resistance system similar in function to the genetically variable antibody system within individuals of most vertebrate taxa.

However, many other individual advantages of sexual reproduction have been hypothesized (for general discussion, see Williams, 1975; Maynard Smith, 1978a) and this conclusion could well be premature.

Our second issue is the study of the fitness of individual genotypes. Estimating the fitness of an individual requires that we can estimate its contribution to the following generation, that we can ascribe this contribution (in part at least) to a particular phenotype, and that such estimates can be made under natural conditions. To estimate the impact of this contribution on genetic variance, we furthermore need to understand the genetic basis of each phenotype. The problems and difficulties in this process are legion. For example, contribution as measured by schedules of survivorship and fecundity require that we also know male fecundity, that we take into account physiological and genetic quality of the offspring, and that we know how the individual affects the fitness of other related and unrelated individuals in the population. A major problem from a purely methodological standpoint is that a single individual represents an unreplicated unit, whose fitness may be a product of either the local environment or its genotype and whose phenotype and genotype are difficult to determine particularly if, for example, it has died! These problems can be overcome if we use cloned individuals, since then a particular genotype can be replicated and the relative contribution of genotype and environment to fitness can be assessed. We can illustrate this using data from the minority-advantage experiment discussed earlier. In each experimental plot (Figure 1A), the majority genotype was replicated 18 times as a cloned tiller. Because we know these individuals are genetically identical, the variance in individual fitness (Figures 3 and 4) can now be entirely ascribed to environmental variance: Had these been single individuals of unknown genotype, as in a natural population, the genetic and environmental sources of this variation would be totally confounded. Had different cloned genotypes been randomized within these plots, it also would have been possible to calculate the environmental and genetic contributions to overall fitness and to estimate, with appropriate error variances, the finite rate of increase of each genotype (see Lenski and Service, 1982, for these techniques applied to aphid "clones"). However, in the above studies, the experimental unit was the single genotype plot; and in the case of "home"

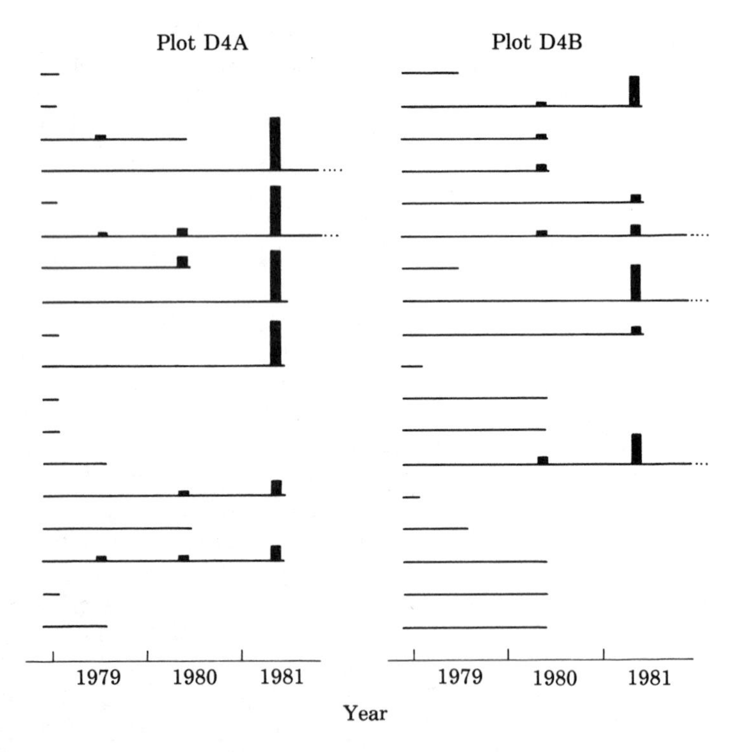

FIGURE 3. Life-histories of genetically identical individuals in two plots (D4A and D4B) of the minority-advantage experiment. Each horizontal line shows the life span of an individual; the vertical bar shows the reproductive output. For scale, the distance between pairs of horizontal lines is equivalent to ten inflorescences.

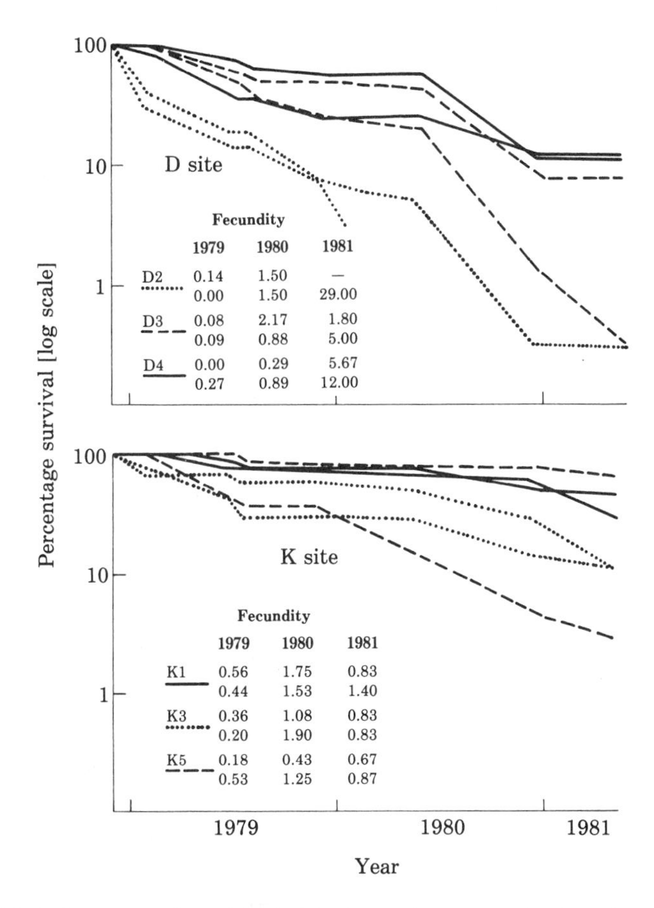

FIGURE 4. Demography of individual genotypes, each cloned as eighteen tillers in two plots of the minority advantage experiment, at two sites (D and K). Similar lines represent replicate plots within each site.

genotypes (i.e., genotypes originating from the same site as that into which the individuals were planted), these were replicated twice. Differences between plots thus also represent environmental variance in fitness, but on a larger (between-plot) scale. Using plot means, it is possible to partition fitness (here measured as net reproductive rate over three years) and its components into variance between sites, between genotypes within site, and between plots within genotype. It can be seen (Table 1) that there are significant differences in survival between genotypes within sites, but that genetic variance in overall fitness with regard to the net reproductive rate is not significant and

TABLE 1. Analysis of variance for differences in survival (arcsin square root transformed), fecundity, and net reproductive rate among genotypes within eight sites (three genotypes per site, each replicated twice).

		Survival to May 1980		Number of inflorescences per flowering plant		Net reproductive rate	
	Degrees of freedom	Mean square[a]	Variance[b] (%)	Mean square	Variance[b] (%)	Mean square	Variance[b] (%)
Site	7	0.553***	60.8	14.16**	25.9	899.6	18.9
Genotypes within sites	16	0.075*	18.3	3.12	0.0	394.1	7.2
Replicate plots within genotypes	24	0.027	20.9	5.25	74.1	329.5	73.9

[a] Significance levels: *, $P < 0.05$; **, $P < 0.01$; ***, $P < 0.001$.

[b] Percentage of total variation accounted for by each effect was calculated from variance components.

accounts for only 7% of the total variance. Within the subpopulations sampled at different sites in the field, genetic variance in fitness is clearly very low relative to the environmental variance in fitness.

This experiment was not specifically designed to measure genetic variance in fitness and I use it only to illustrate an approach that has not previously been attempted. Moreover, such experiments could be made still more realistic if genetically identical individuals are not vegetative clones but are seed progeny. Genetically identical seeds may be produced by apomixis (as in experiments using parthenogenic strains; e.g., Service and Lenski, 1982) or by special techniques. We are currently exploring techniques using crosses among doubled haploids to generate large amounts of genetically identical, but normally heterozygous, seed. Haploids can be generated in a number of ways. In *A. odoratum* we have used tissue culture of postmeiotic anthers, modifying only slightly techniques used in grass and cereal breeding (Kasperbauer et al., 1980). We have also begun screening (and eventually selection) for production of double embryos. Double embryos may include individuals that develop from an unfertilized synergid cell and thus produce a haploid plant (Riley and Chapman, 1957). In *A. odoratum* we have found such "twins" to occur at a frequency of one per several thousand seedlings, but at present we do not know what fraction are double zygotic, what fraction of these are haploid in origin.

Given such "cloned" individuals, it is possible, not only to place them in natural populations in formal experimental designs, but to assess them for morphological and physiological traits in the greenhouse and laboratory. Character states can thus be "mapped" onto fitness. Moreover, the characters in question can be analyzed genetically; or conversely, individuals from particular crossing designs can be cloned into the field. Such procedures could, of course, also be carried out with progeny groups rather than with clones; however, in such cases comparisons are less direct and interactions more diffuse. The potential here is tremendous, yet plant population biologists have never exploited the full experimental potential of plants in studying natural selection in wild populations.

Our conclusions from these considerations are naively Darwinian: A powerful approach to understanding the importance of genetic variation in populations is to ask what are the advantages of such variation for the individual and to make direct estimation of individual fitness in natural populations using experimental methods. It is certainly not the only approach and was not a prerequisite for Darwin himself: Cumulative information gained from less intensive, correlative studies will always bear on particular hypotheses and lead deductively to valid conclusions. However, experimental approaches may provide us with new insights and methods of studying variation in natural populations; and in that these approaches are a direct extension and test of Darwin's ideas, it is for the historians to analyze why at least 100 years have elapsed without thorough studies of differential individual fitness within natural populations.

CHAPTER 12

IMMIGRATION IN PLANTS: An Exercise in the Subjunctive

Donald A. Levin

INTRODUCTION

Plants stand still and wait to have their seeds and pollen dispersed by animals, wind, water, or some combination thereof. Most of these disseminules (namely, pollen and seeds) are deposited within a few canopy diameters of the parent plant (Levin and Kerster, 1974; Levin, 1981). A small proportion are transported tens of canopy diameters or more, some reaching sites beyond the boundaries of the population in which they were produced. That long distance dispersal occurs is evident in the founding of new populations and the spread of species into new areas. The pattern and rate of invasion has been documented for several weeds (Plummer and Keever, 1963; Salisbury, 1961; Ter Borg, 1979; Baker, 1972a).

The introduction of seeds into hospitable vacant sites (namely, colonization) and the introduction of pollen and seeds into existing populations (namely, immigration) have important demographic and genetic ramifications. The arrival of one or a few seeds into a suitable site sets the stage for the development of a new population. Thus, the few propagules may have a significant and immediate effect on the local sociology of a species. The arrival of one or a few seeds or pollen grains into an existing population may be of little demographic importance as the recipient population probably will be producing an enormous pollen and seed rain. Even if the local pollen and seed pools were enriched by, say, 5% immigration, the demographic properties of populations are unlikely to be substantially altered. Populations founded by a few seeds may be less variable and differ substantially from their parental population(s) because they must traverse a genetic bottleneck. Colonization leads to increased gene frequency heterogeneity among populations, especially at those loci where allelic differences are neutral or nearly so. The genetic consequences of immigration are the converse of colonization. The level of variation within populations is likely to be enhanced, and differences in the genetic composition of populations will decline, especially at those loci where allelic differences are neutral.

The rate of seed introduction into a site uninhabited by a given species can be monitored with relative ease for most species. On the other hand, the numbers of pollen grains and seeds introduced into an existing population remains one of the least understood demographic variables in plant populations and is likely to remain such in the foreseeable future because of the difficulty, if not the impossibility, of identifying immigrant pollen and seed. Whereas the dispersal patterns of pollen and seeds suggest that the immigration rate is likely to be much less than 1% (Levin and Kerster, 1974; Levin, 1981), it is important to at least know the order of magnitude of this rate, because even low rates may have an enormous impact on the genetic structure of single populations and the organization of variation among populations. On the other hand, substantial progress is being made in understanding the factors which affect the outcome of a given level of migration. These include the relative fitness of aliens, the spread of alien genes in local populations, the fitness of hybrids, and the pattern of immigration. The manifold effects of gene exchange on the organization of variation among populations and on the level of population divergence are better understood now that theoretical and numerical models have taken into account kin-structured and variable migration and have incorporated colonization into the gene flow dynamics. My purpose in this chapter is to discuss some of the factors affecting the outcome of immigration on genetic variation within and among populations after briefly reviewing what we know about immigration.

SINGLE POPULATIONS

The rate of immigration

The immigration rate may be defined as the ratio of the numbers of alien pollen and seeds to the total numbers of pollen and seeds in a local population. If the input of alien pollen and seeds is constant, then the immigration rate (or level of gene flow) depends upon the local pollen and seed production: the greater the local pollen and seed production, the lower the immigration rate. Accordingly, a population which is small by virtue of youth, contraction, or habitat inhospitality will be influenced more by the pollen and seed rain than a large population (McNeilly and Antonovics, 1968; Antonovics, 1968b). Correlatively, as the size of a population changes, the level of immigration will change in an inverse manner. The inverse relationship between population size and the incidence of interpopulation hybridization has been demonstrated in several crop species (Fryxell, 1956; Williams and Evans, 1935; Bateman, 1947; Crane and Mather, 1943).

The actual level of immigration into natural populations is practically unknown. Based on studies of gene flow within natural populations, the level typically must be very low unless populations are separated by a few meters (Levin and Kerster, 1974; Levin, 1981). In most crops, a distance of 1 km will confer nearly complete isolation from gene exchange via pollen. The isolation distances for 50 crops, as described by Kernick (1961), are plotted in Figure 1. The species with the greatest isolation requirements are cross-fertilizing. Isolation requirements do not differ significantly between insect- and wind-pollinated crops. Unfortunately, crops do not offer insight into seed immigration levels. In most natural species, seed dispersal is likely to be more conservative than pollen dispersal (Levin and Kerster, 1974).

Immigration rates in natural populations are poorly understood because they are too difficult to determine. They are best considered in population systems wherein the donor population has genetic markers which are absent in the recipient populations and detectable in F_1 hybrids. The recipient populations then may be unambiguously assayed for hybrids, and the immigration rate may be estimated from

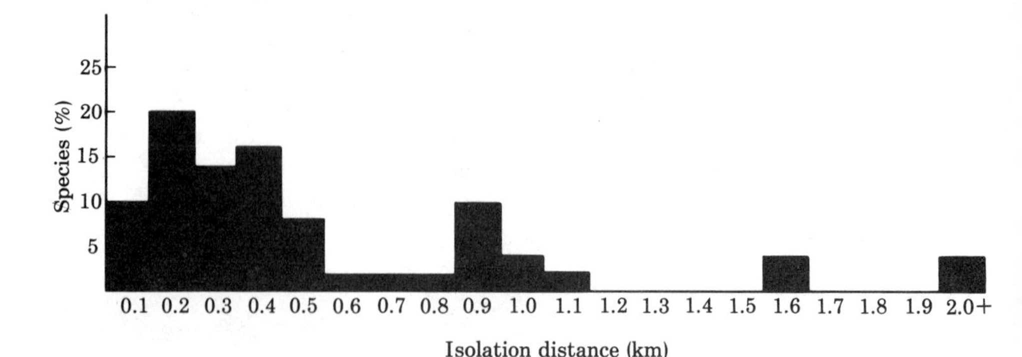

FIGURE 1. Isolation distances necessary to maintain varietal purity in 50 crop species. (After Kernick, 1961.)

the frequency of hybrids and the frequency of the genetic marker in the donor. Slatkin (1981a) conceived a new approach for approximating the level of gene flow from the distribution of neutral gene frequencies among populations. Using computer simulations of interpopulation gene flow, he showed that when allele frequencies were plotted against the proportion of populations in which an allele occurred, the resulting pattern was dependent upon the immigration rate. Accordingly, such a plot of electrophoretic data might provide an approximation of gene flow rates. Slatkin examined data for the plant *Stephanomeria exigua* in this way and concluded that gene flow levels were high. He noted, however, that a plot suggestive of high gene flow could also come from a species in which population number recently had greatly expanded with no gene exchange between populations. The procedure thus has utility for species whose populations have not exploded in response to recent habitat disturbance by man.

Over the long term, gene exchange between populations tends to make populations more similar, so that as the immigration rate increases a given population will be less influenced by selection or genetic drift. The immediate effect of a single immigration episode on the genetic constitution of a population is dependent upon the genetic similarity of the immigrants and the recipient population as well as the immigration rate. If the immigrants had the same gene frequencies as the recipient population, no level of immigration would have an effect on the genetic makeup of the recipient. From a genetic perspective, immigration would be a nonevent. When the genetics of the immigrants and natives are divergent, immigration alters the genetic composition of the native population: the greater the divergence, the greater the impact of a given level of immigration. Accordingly, 1% immigration from a different race, subspecies, or species may have a substantially greater effect on a population than 10% immigration from a population similar to the recipient.

Pattern of immigration

Hybrid seed production is distinctly nonrandom in space in both wind- and insect-pollinated plants. The plants nearest the alien gene source produce the most hybrid seed, and those farthest away the least (cf. Levin and Kerster, 1974). In wind-pollinated crops such as corn and brome grass, the proximal side of the population typically produces five to ten times as much hybrid seed as the distal side (Jones and Brooks, 1950; Knowles and Ghosh, 1968). In insect-pollinated plants, the gradient from proximal to distal rows is not as steep (Green and

Jones, 1953) and may decline as the isolation distance increases (Stringham and Downey, 1978). The proximal row hybridizes more extensively than the others because of its proximity to the alien pollen source and because of a deficiency of local pollen, which characterizes all border rows (Pedersen et al., 1961; Knowles and Baenziger, 1962).

The distribution of the hybrid parents reflects the initial distribution of alien genes. The distribution of hybrids after seed dispersal and plant establishment is apt to be similar to that of their parents, as seeds typically are dispersed within a few canopy diameters of the parent plant (Levin and Kerster, 1974; Levin, 1981). The greater the correlation between the location of parents and their offspring, the more organized the distribution of immigrants and their alien genes in space.

The immigration pattern may have a substantial effect on the rate of increase of a favorable recessive gene and on the equilibrium value achieved by a deleterious recessive gene within a population. This is seen in the computer simulations of Levin and Kerster (1975) who treated gene flow via pollen. When a favorable gene is recessive, the rate of increase and the mean time to fixation is slowest when hybrids are produced only on the proximal border row and fastest when hybrids are produced throughout a population. When the alien gene is deleterious, the equilibrium frequency is the highest when hybrids are produced throughout a population and lowest when they are produced only in one border row. The more restricted the breeding structure of the recipient population, the greater the impact of the immigration pattern.

Given that each population occurs in a unique environment and is adapted to it, immigration will, in general, introduce unfavorable genes and thus lower the fitness of the local population. The frequency of unfavorable genes is dependent, not only upon the selective differential and the immigration pattern, but also upon the gene disseminule. With pollen as the vehicle, selection is delayed until the next generation, and the products are hybrids. With seeds as the vehicle, selection may operate in the present generation, and aliens may be homozygous for unfavorable genes. Under low selection intensities gene flow via seeds has greater impact on the genetic composition of populations than gene flow via pollen; the converse is true when selection is strong (Antonovics, 1968a).

The mode of gene transport affects the level of heterozygosity of alien genes as well as their frequencies. If the vehicle is pollen, there may be a great excess of heterozygotes over Hardy-Weinberg expectation, which might suggest that the alien genes are being favored. If the vehicle is seed, heterozygote frequency for the alien genes will not deviate from expectation since the addition of these genes occurs before selection and random mating (Antonovics, 1968b).

The fitness of immigrants

The consequences of gene flow depend, not only on the frequency and pattern of immigration, but also on the fitness of immigrants relative to locals. The greater the relative fitness of the immigrants, the greater their impact. Because gene flow in plants occurs during the gametophyte and sporophyte generations, immigrants take the form of pollen grains, hybrids, and sporophytes which carry two doses of alien genes. The latter will hereafter be referred to as aliens or immigrants.

Consider first cross-compatibility, that is, the fate of pollen grains on stigmas. In general, stigmas of a population are more receptive to pollen of their own population or a very similar population than to pollen from a divergent system, especially if the latter is another race or subspecies (Levin, 1978b). Thus, the pollen that might have the greatest impact on the genetic constitution of a population usually is the least likely to germinate or will be at a disadvantage relative to local pollen in time to germination or pollen tube growth rate. However, D. A. Levin and K. Clay (unpublished data) found evidence of alien pollen advantage in the self-compatible *Phlox cuspidata*. In a crossing program involving 15 populations, an average of 26% alien pollen germinated on stigmas compared to 20% for local pollen. A significant alien advantage was found in 11 of the 15 populations.

In predominantly outbreeding crop species, the vigor and fecundity of hybrids typically increase as the genetic distance between parental strains increases, until some critical level of divergence is reached when interactions at a few loci counterbalance the effect of heterozygosity. In maize, heterosis is a positive function of the level of divergence of strains except for the most divergent ones (Moll et al., 1965). In *Nicotiana* (Matzinger and Wernsman, 1967) and *Gossypium* (Mariani and Avieli, 1973), heterosis increases with divergence to the level of related species. In natural population systems, maximum heterosis also is associated with moderate levels of divergence. Heterosis has been documented in interracial crosses of Norway spruce (Nilsson, 1974), Douglas fir (Orr-Ewing, 1969), and loblolly pine (Woessner, 1972; Owino and Zobel, 1977). Greater heterosis and higher fertility were obtained in the progeny of crosses between *Liriodendron tulipifera* of different populations (often only a few miles apart) than in crosses within the same population (Carpenter and Guard, 1950). Heterosis in *Mimulus* is best developed in hybrids between populations which have undergone moderate degrees of divergence regardless of their geographical relationships (Vickery, 1978). Beltran and James (1974) demonstrated heterosis in hybrids between chromosomally homozygous populations of *Isotoma petraea*: the more inbred the populations, the greater the vigor of their hybrids.

If populations grow in very different habitats and if hybrids are intermediate to their parents, then interpopulation hybrids may be ill-fit in either population; it is unlikely they will be better adapted than local residents. A prime example of this is seen in the copper tolerance of *Agrostis tenuis* adults and naturally produced progeny at Drws-y-Coed copper mine (McNeilly, 1968). A large proportion of open-pollinated progeny of copper-tolerant plants have nontolerant-pollen parents. They are substantially less tolerant than their egg parents and are strongly selected against in populations growing on copper-rich soils. Conversely, some open-pollinated progeny of low-tolerance parents in adjacent grasslands have tolerant-pollen parents and are selected against in the normal pastures.

Statistically, aliens deviate more than hybrids from the indigenous population, and thus are likely to be less fit than hybrids. Correlatively, immigrants are unlikely to be superior to local plants, unless the population is poorly adapted to its environment. This finding is evident in reciprocal sowing and transplant experiments in natural populations and in garden and greenhouse cultures, experiments which reflect differences present in the field. The relative fitnesses of aliens in several species are summarized in Table 1. Aliens are at a strong disadvantage when they come from habitats very different from that of the recipient population in soil chemistry, grazing, or disturbance. Alien fitnesses of less than 50% are common. When aliens are from sites which differ from the recipient in terms of moderate differences in precipitation or minor differences in temperature and soil composition, they tend to fare somewhat better.

K. Schmidt and D. A. Levin (unpublished data) carried out a two-season study of alien fitness in the annual *Phlox drummondii*, using a complete reciprocal sowing design in eight populations of the species. The populations were distributed throughout the range of the species, which is endemic to central and southeastern Texas. Relative fitness was measured in terms of the finite rate of increase obtained from seed germination, plant survivorship, and seed production. The mean relative fitnesses of aliens were 0.55 in 1979 and 0.58 in 1980 (Table 2). Locals were more fit than aliens in 13 of the 15 comparisons of the finite rate of increase. There is a significant negative correlation between alien fitness and the distance to the alien site of origin ($r = -0.23$) (Figure 2).

The long-term impact of an immigration episode was studied experimentally with two subspecies of *Phlox drummondii* (Levin, 1983b). In 1973, a 25-m line of 1100 ssp. *mcallisteri* seed was sown in a

TABLE 1. Relative fitnesses of immigrants in various species.

Species	Habitats and relative fitness of immigrants	Type of experiment[a]
Agrostis tenuis	Mine 0.05; pasture 0.60	grnh; weight[b]
Agrostis tenuis	Mine 0.45; pasture 0.95	grnh; weight[b]
Agrostis tenuis	Normal soil 0.53	grnh; weight[c]
Agrostis tenuis	Normal soil 0.30	fld; weight[d]
Anthoxanthum odoratum	Mine 0.01; pasture 0.70	grnh; weight[b]
Anthoxanthum odoratum	Unlimed 0.91; limed 0.73	grnh; weight[b]
Anthoxanthum odoratum	Normal soil 0.61	grnh; weight[c]
Anthoxanthum odoratum	Normal soil 0.30	fld; weight[d]
Anthoxanthum odoratum	Limed plots 0.59 Unlimed plots 0.75	rcpr; weight[e]
Anthoxanthum odoratum	Field 0.75; woodland 0.60	fld. seed/transplant[f]
Anthoxanthum odoratum	Pasture 0.78; roadside 0.85	fld; survival[g]
Plantago lanceolata	Normal soil 0.71	grnh; weight[c]
Plantago lanceolata	Normal soil 0.48	fld; weight[d]
Plantago lanceolata	All sites 1.0	fld; survival[h]
Chenopodium album	no atrazine 0.0 0.5 kg/ha atrazine 0.50 1.0 kg/ha atrazine 0.39	grnh; stem weight[i]
Chenopodium album	Northern garden 0.79 Southern garden 0.77	fld; weight[i]
Agrostis stolonifera	Cliff 0.20; pasture 0.50	grnh; weight[b]
Rumex acetosa	Normal soil 0.32	fld; weight[d]
Poa annua	Open field 0.68	grdn; growth rate[j]
Plantago major	Clipped 0.0; unclipped 0.75	grnh; weight[k]
	Mown plots 0.95; roadside 0.24	fld; weight
Bellis perennis	On lawn, after mowing:	fld; fruit number[l]
	heavily grazed popns 0.71	fld; fruit number
	lightly grzd/mwn pop 0.32	fld; fruit number
	ungrazed popns 0.32	
Trifolium repens	Li phenotype 0.69	fld; establishment[m]
Taraxacum (agamospecies)	Upland 0.14; roadside 1.5	fld; number of fruits[n]
Phlox drummondii	Roadside 0.57	fld; finite rate of[o] increase

[a] grnh, Greenhouse; fld, field; rcpr, reciprocal; grdn, garden.
[b] Jain and Bradshaw (1966); [c] Cook et al. (1972); [d] Hickey and McNeilly (1975); [e] Davies and Snaydon (1976); [f] Grant and Antonovics (1978); [g] Kiang (1982); [h] Antonovics and Primack (1982); [i] Warwick and Marriage (1982a); [j] Law et al. (1977); [k] Warwick and Briggs (1980a); [l] Warwick and Briggs (1979); [m] Ennos (1981); [n] Ford (1981); [o] K. Schmidt and D. A. Levin (unpublished data).

TABLE 2. Finite rates of increase of local populations, the mean finite rates of increase of alien populations, and the alien fitness coefficient relative to the local, for 1979 and 1980.

Site	1979			1980		
	Local	Alien	Alien fitness	Local	Alien	Alien fitness
Austin2	34.06	15.10	0.44	0.43	0.19	0.44
Mason	—	—	—	0.72	0.63	0.87
Gonzales	11.94	8.16	0.68	3.87	2.44	0.63
Nixon2	9.88	5.53	0.56	11.94	8.00	0.67
Lytle	1.60	0.90	0.56	9.40	3.46	0.37
Natalia	1.53	1.99	1.30	0.76	1.32	1.74
Raymondville	1.12	0.86	0.77	3.34	1.89	0.56
Sarita	1.57	1.45	0.92	1.34	0.72	0.54
Mean	8.81	4.86	0.55	3.98	2.33	0.58

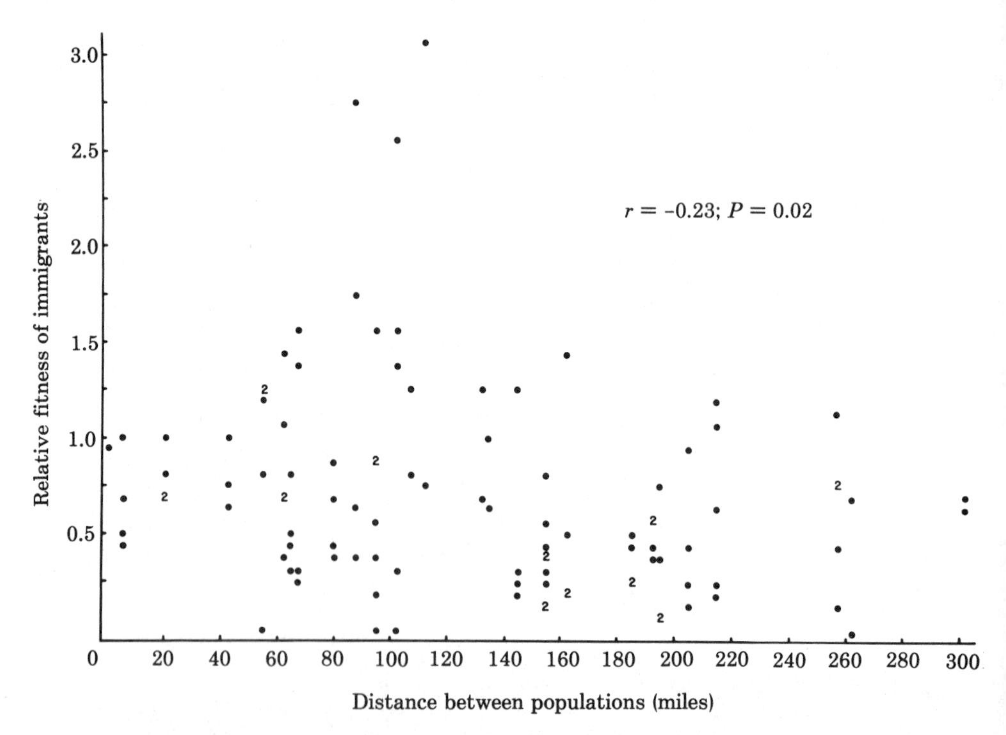

FIGURE 2. The relationship between immigrant fitness and the distance to the immigrant source population in *Phlox drummondii*. The 2's correspond to two data points.

natural population of ssp. *drummondii*. In 1974, 89 immigrants reached flowering, and all were on the sowing line. Immigrants continued to appear on the sowing line from 1975 through 1979 as a result of seed dormancy. From 1975 through 1980, immigrants occurred up to 5 m from the line. These products of immigrant × immigrant crosses were dispersed via explosive capsule dehiscence. The number of immigrants declined from 1974 through 1980; none were present in 1981.

There were 41 magenta-flowered subspecific hybrids in 1975. Their numbers declined gradually through 1981, when only 6 such plants were observed. Their presence over a seven-year period reflects continued hybridization between the immigrants and natives as well as hybrid seed dormancy. The mean distance of hybrids from the sowing line was about 3.8 m. Hybrids have occurred more than 15 m from the line.

IMPLICATIONS OF IMMIGRATION FOR POPULATION SYSTEMS

Differentiation in static population systems

Given some level of interpopulation gene exchange, it is of interest to know the extent to which populations may diverge under the effect of mutation and genetic drift. Theoretical solutions have been formulated with the assumption that immigrants are a random sample of the entire population system (the island model) (Wright, 1931, 1943; Crow and Maruyama, 1971; Latter, 1973) or that migration is between adjacent populations in one- or two-dimensional networks (Kimura and Weiss, 1964; Kimura and Maruyama, 1971). It is quite apparent that neither characterization of migrants is realistic for plants since there is no normalizing rain of migrants from some average source or metapopulation. Nevertheless, these models are useful points of departure.

Consider how much differentiation via genetic drift will occur among populations if the migrants were chosen at random from the population system as a whole, if the migration rate is constant, and if there is no mutation. The amount of differentiation between populations may be expressed as the normalized variance of gene frequencies, F_{ST}, which is the correlation between two gametes drawn at random from each population. Following Wright (1943), $\hat{F}_{ST} = (1 - m)^2/[2N - (2N - 1)(1 - m)^2]$ where m is the migration rate and N is the population size. With $m = 0.10$ and $N = 200$, $\hat{F}_{ST} = 0.010$.

The assumption that the migration rate is constant from one generation to another certainly will be violated in nature as the migration rate has a stochastic element which may be considerable. Wright (1948) first addressed this problem and demonstrated that variation in migration rate increases gene frequency heterogeneity. Within the framework of the island model, Latter and Sved (1981) showed that $\hat{F}_{ST} = [1 + 2(N - 1)(2m - m^2)]^{-1}$, if the variance in migration is from the binomial distribution $[\overline{m}(1 - \overline{m})/2N]$. The level of interpopulation differentiation will be greater with stochastic migration than with deterministic migration. With $m = 0.10$ and $N = 200$, $\hat{F}_{ST} = 0.014$: 40% higher than with deterministic migration. The variance in migration rate may be much greater than that dictated by a binomial distribution. In this case, Nagylaki (1979) showed that the equilibrium gene frequency variance for the island model may be approximated as

$$\sigma_x^2 \approx \frac{\overline{x}(1 - \overline{x}) + \overline{x}(1 - \overline{x})(\overline{m}^2 + \sigma_m^2)}{1 + 2N_e(\overline{m} - \sigma_m^2)}$$

where σ_m^2 is the variance in migration rate. $\hat{F}_{ST}$ is obtained by dividing the variance by pq. If $\overline{m} = 0.10$, but the migration rate was characterized by a three-year cycle with 0 in years 1 and 2 and 0.3 in year 3, $\hat{F}_{ST}$ would be ca. 0.016, or 60% greater than with deterministic migration. If $\overline{m} = 0.10$, but migration occurred only every fifth year, $\hat{F}_{ST} \approx 0.019$.

In the models of differentiation in which genetic drift is counterbalanced by migration, it is assumed that immigrants are unrelated. This assumption is not valid for many plant species, especially when the unit of dispersal is a multiseed fruit. In a multiseed fruit, migrants would be half-sibs or full sibs. If the species were apomictic, migrants would be identical. Given the possibility of kin-structured migration, it is of interest to know how such migration influences divergence by genetic drift, as it introduces an element of sampling error.

In order to assess the effect of migration with kin-structuring on interpopulation differentiation resulting from genetic drift, A. Fix and D. A. Levin (unpublished data) employed a computer program to simulate intergenerational random drift and migration in a linear array of 25 populations. In order to monitor the decay of some existing pattern of gene frequencies as well as to determine the level of gene frequency heterogeneity between populations after the system had equilibrated, the initial array had two alleles, p and q, with gene frequencies alternating between 0.25 and 0.75 as one moved down the array. Population size was set at 200 and remained constant. Each generation a population received ten migrants from each of its two neighboring populations. Intergenerational genetic drift was simulated by choosing a new gene frequency at the beginning of each generation from a normal distribution, with the expected being the original gene frequency and variance equal to $\overline{p}\,\overline{q}/2N$. The effect of migration on population structure was determined after the adjustment for genetic drift.

Migrants are treated as excess individuals whose genotypes are functions of the source population but whose absence has no bearing on the source population. Four cases were considered: the migrants (1) have the same genotype, (2) are full sibs, (3) are half-sibs, and (4) are chosen at random from a population. A deterministic model with migrant gene frequencies equal to the source population gene frequencies serves as a standard for comparison. Populations were run for 50 generations. Each situation was replicated ten times.

F_{ST} declines for 10–15 generations, and then in accordance with theory (Nei et al., 1977) gradually rises regardless of how migrants are chosen (Figure 3). The deterministic run has the lowest F_{ST} values at generation 15; and thereafter the heterogeneity is the result of inter-

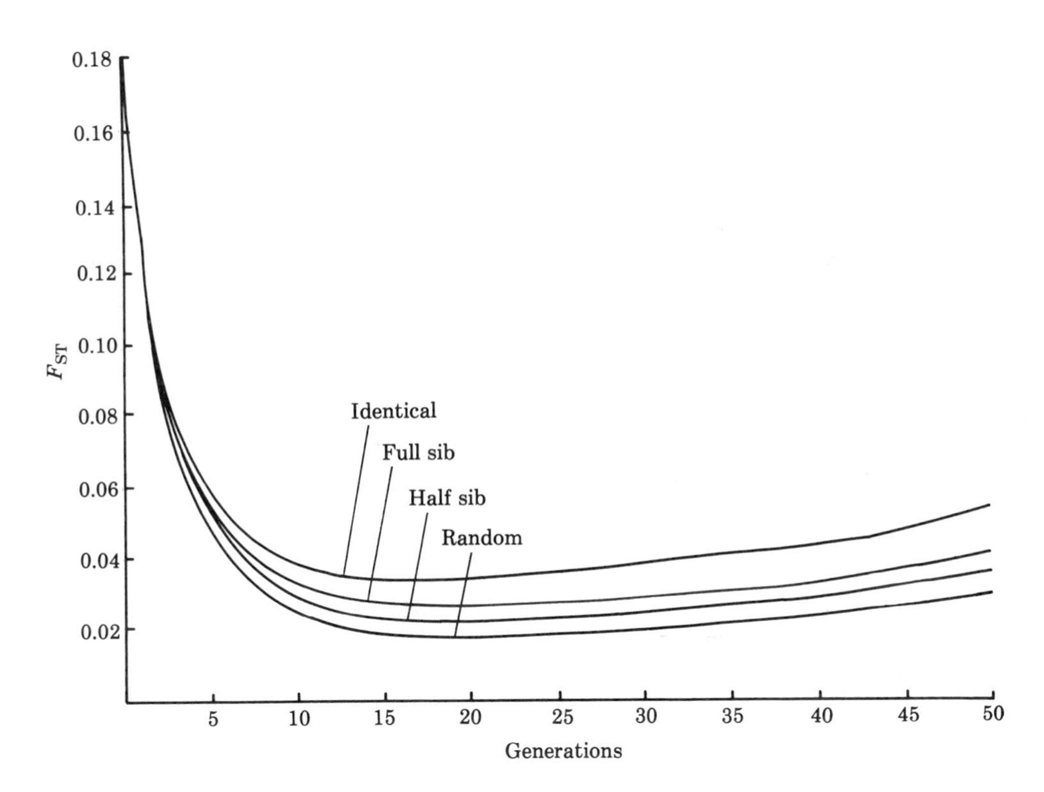

FIGURE 3. The level of interpopulation differentiation over time as influenced by the relatedness of migrants. The curve for the deterministic model (not shown) is very similar to the curve for random migration (see text for further information).

generation genetic drift. With random choice of migrants, some additional sampling error is introduced, but it is small since $N = 20$. At generation 50, F_{ST} is about 0.03, marginally higher than for the deterministic model. With two groups of full sibs entering a population each generation (one from the left and the other from the right), the size of each migrant group effectively is 2.61 and much more sampling error is introduced. After 50 generations of full sib migration, F_{ST} is approximately 0.040. Half-sib migration reduced the effective number of migrants (3.85 per group) and yields an F_{ST} of about 0.039. The maximum increase of F_{ST} over the deterministic model occurs with identical migrants which afford an effective size of 1 per group and correlatively considerable sample error. After 50 generations, F_{ST} averages about 0.055. Thus, with identical migrants in each of two groups, the degree of differentiation is about twice that afforded by deterministic migration and random migration. Thus, a moderate amount of genetic differentiation can develop among populations in the presence of substantial migration when the migrants are related and population size moderate. Were population sizes smaller, the amount of genetic drift per generation would have been greater.

Differences in the relatedness of migrants determine the extent to which genetic drift is counterbalanced. In our simulations, F_{ST} values were determined in each generation before and after migration. The mean differences in F_{ST} in each population before and after migration are dependent upon the relatedness of migrants, as is the degree to which genetic drift counterbalances the effects of migration. With deterministic migration, the effects of genetic drift are reduced an average of 8.5% each generation, whereas with full sibs the reduction is 3.9%, and with identical migrants, 2.0% (Table 3). In other words, with determinsitic migrants as the baseline, the impact of migration is reduced by over 75% when the migrants are identical and over 50% when the migrants are full sibs.

As there is no general stabilizing force (selection) in our model,

TABLE 3. Amount of reduction in the Wahlund variance generated by intergenerational genetic drift due to migration.[a]

Run set	Relationship	Percentage reduction
I	Clone	2.0
II	Full sibs	3.9
III	Half-sibs	4.6
IV	Random individuals	7.2
V	Deterministic	8.5

[a] $M = 0.10$; $N = 200$.

values of F_{ST} will continue to increase. If we allow for some level of migration from the population system as a whole, then an equilibrium will be reached, and approximated as follows: $\hat{F}_{ST} \approx 1/[1 + 4N_e(2m_1m_\infty)^{1/2}]$ where m_1 refers to migration between adjacent populations and m_∞ to long-distance migration (Kimura and Weiss, 1964). If $m_\infty = 0.001$, and $m_1 = 0.10$, and no relationship exists between the migrants, $\hat{F}_{ST} = 0.08$. With kin-structured migration, this value increases. If the two groups of migrants entering a population each were composed of half-sibs, $\hat{F}_{ST} = 0.10$; with full sibs, $\hat{F}_{ST} = 0.11$; and with identical individuals, $\hat{F}_{ST} = 0.14$. If each population received all migrants from one adjacent population in one generation and the other adjacent population the next, the effect of kin-structured migration would increase because there would be groups of relatives entering a population each generation rather than two (Table 4). With half-sibs, $\hat{F}_{ST} = 0.1$; with full sibs, $\hat{F}_{ST} = 0.13$; and with identical individuals, $\hat{F}_{ST} = 0.22$. The greater the relatedness of migrants, the greater the level of interpopulation differentiation as a result of genetic drift.

Differentiation with extinction and recolonization

Having considered differentiation in static population systems, it is of interest to know how extinction and recolonization affect the level of differentiation since population turnover occurs in all species to some degree. The effects of colonization and extinction only recently have received rigorous theoretical treatment. The pioneering study of Slatkin (1977) involves a diploid species with nonoverlapping generations and a single locus with two alleles (p and q), whose frequencies are maintained in a source of migrants at a constant frequency. There is a large number of local populations, each of effective size N and each of which has a fraction m of its individuals replaced by migrants from

TABLE 4. Differentiation in a linear array at equilibrium ($\hat{F}_{ST}$).[a]

Relationship among colonists	One source per generation	Two sources per generation
None	0.08	0.08
Half-sibs	0.13	0.10
Full sibs	0.15	0.11
Identical	0.22	0.14

[a] $m_1 = 0.10$; $m_\infty = 0.001$.

the source. A fraction e_0 of the local populations goes extinct, and the vacant sites are colonized by a propagule with k individuals from one of the populations chosen at random. The newly founded populations grow to size N in the first generation after they are founded. For a population such as this

$$\hat{F}_{ST} = \frac{(1 - e_0/2k)1/2N + e_0/2k}{1 - (1 - e_0/2k)(1 - m)^2(1 - 1/2N)}$$

An equilibrium is achieved because the effects of migration and extinction are in opposition when mutation is absent; migration reduces interpopulation differences, whereas colonization increases them as a result of founder effect.

The effects of migration and extinction on F_{ST} are illustrated with some numerical examples in Table 5. Extinction and recolonization episodes increase the level of differentiation which occurs in the face of gene flow. The sampling error associated with the extinction–recolonization process far outweights the effect of genetic drift in determining the variance, whenever $N > k$, as is typically the case. Recall that migrants are chosen from the population system as a whole. Were they from single populations, F_{ST} would be considerably higher than indicated above.

If colonists in this model were related, the sampling error would be greater than k suggests. The extent to which kin-structured migration amplifies the effect of the extinction and recolonization process is shown in Table 5. The effect may be pronounced. For example, if e_0 = 0.01, m = 0.01, k = 10, and colonists were unrelated, F_{ST} = 0.07. With colonists being half-sibs, F_{ST} becomes 0.08 and increases to 0.11 with full sibs and 0.22 with identical colonists.

Thus far we have considered a model wherein colonists are from a single population and migrants are from the population system as a whole. When colonists are drawn from multiple sources, a second

TABLE 5. Differentiation with migration and colonization.

		F_{ST}			
m	e_0	**Random**	**Half-sibs**	**Full sibs**	**Identical**
0.001	0.001	0.22	0.24	0.26	0.33
0.001	0.01	0.33	0.47	0.55	0.73
0.01	0.01	0.07	0.08	0.11	0.22
0.01	0.10	0.22	0.41	0.50	0.72
0.10	0.10	0.03	0.06	0.10	0.22

derivation of Slatkin (1977) describes F_{ST} at equilibrium as follows:

$$\hat{F}_{ST} = \frac{(1 - e_0)/2N + e_0/2k}{1 - (1 - e_0)(1 - m)^2(1 - 1/2N)}$$

Drawing colonists from the population system as a whole rather than from single populations reduces $\hat{F}_{ST}$ for any set of variables. For example, if $m = 0.01$, the $\hat{F}_{ST} = 0.03$ with multiple sources versus 0.07 when colonists are chosen from single sources.

Slatkin and Wade (1978) extended the aforementioned pair of models to describe the interpopulation variance at equilibrium for quantitative characters under polygenic control. Given that all colonists are drawn from single populations and that migrants are drawn from the population system as a whole,

$$\hat{\sigma}^2 = \frac{vh^2}{2\tilde{N}(s + g)}$$

where h^2 is the heritability; v is the variance within populations and is assumed to be the same in each population; s is the selective advantage of an optimal character state and is the same in each population; g is the migration rate; and $1/\tilde{N} = 1/N_e + e_0/k$. For the sake of discussion, assume that $v = 2.8$, $h^2 = 1$, and $N = 1000$. Treating v and N as constants, the effects of differences in migration and colonization rates on the interpopulation variance at equilibrium is illustrated in Table 6. An increase in extinction–colonization episodes is accompanied by a considerable increase in interpopulation variance. This variance also is increased if the colonists are related. With g and $e_0 = 0.01$, and with ten founders, the variance increases from 0.28, which accrues when

TABLE 6. Interpopulational variance for quantitative characters and gene frequencies.

m	e_0	$\hat{\sigma}^2$ for characters	$\hat{\sigma}^2$ for gene frequencies
0.01	0.01	0.28	0.28
0.01	0.10	1.54	0.88
0.10	0.01	0.03	0.01
0.10	0.10	0.15	0.12
0.01	0.01 half-sib	0.50	0.32
0.01	0.01 full-sib	0.67	0.44
0.01	0.01 identical	1.54	0.88

they are unrelated, to 0.50 for half-sib colonists, 0.67 for full sibs, and 1.54 for identical individuals.

The interpopulation variances for a quantitative character may be compared with interpopulation gene frequency variances in the previous model if we assign values to p and q. Recall that $F_{ST} = \sigma^2/pq$. Accordingly, if $p = q = 0.5$, the variances for various cases of m and e_0 are the product of $F_{ST} \times 0.25$, as shown in Table 6. With $p = q$, the variances in the two models are 0.28, with $m = e_0 = 0.01$ and $N =$ 1000. The effects of colonization and relatedness of colonists are more pronounced in the character model. A tenfold increase in the extinction rate raises the variance 5.4 times, whereas the same increase in the single-locus model increases it 3 times. Exactly the same is true when randomly chosen colonists are replaced by identical colonists. Full-sib colonization in the single-locus model increases the variance 1.5 times over random colonization, whereas in the character model the variance increases 2.4 times.

The effects of migration and colonization on heterogeneity between populations also has been treated in the presence of mutation. Slatkin (1977) and Maruyama and Kimura (1980) derived formulae describing interpopulation divergence via genetic drift and mutation in terms of the probabilities of allele identities. They found that colonization reduces interpopulation divergence in the presence of mutation, which is contrary to the outcome in the absence of mutation.

In order to understand how colonization and migration interact in the presence of mutation, it is useful to consider the Maruyama-Kimura model and the variables therein. Consider a species with a finite number (n) of populations, each of which is subject to extinction at rate e_0. Whenever a population goes extinct, it is immediately replaced by a population derived from founders drawn from a single population. Each population has an equal chance of being the donor. Migration occurs between extant populations at a specified rate, the migrants being drawn from the species as a whole. Mutations occur at a specified rate to a new allele. At equilibrium, the average probability of identity of two randomly chosen genes within populations is $P = 1/[1 + 4Nv + 4Nnmv/(nv + e_0 + m)]$; and the average probability of two randomly chosen genes from different populations being identical is $Q = 1/(1 + 4Nv)[1 + nv/(e_0 + m)] + 4Nmv/(e_0 + m)$, where N is the effective population size, n is the number of populations, v is the mutation rate, m is the migration rate, and e_0 is the extinction rate. Interpopulation differentiation is best described as the ratio of Q to P, which indicates the genetic similarity of two different populations relative to the extent of differentiation within a single population (Spieth, 1974). The larger the ratio, the smaller the level of interpopulation differentiation.

Consider first the effect of extinction and colonization and of migration on P and Q. Assume a species is composed of 100 populations each with an effective size of ten and a mutation rate of 10^{-3}. Migration and extinction rates vary from 0 to 0.10, as shown in Table 7. We see that extinction–colonization cycles increase the level of homozygosity within populations (P) and increase the probability that genes picked from two populations will be identical (Q). On the other hand, migration increases P but decreases Q. Although both variables affect Q in the same way, a change in the migration rate has a larger effect on the probability of identity of alleles from different populations than a change of the same magnitude in the colonization rate. These differences notwithstanding, the ratio Q/P is altered the same amount by both factors. An increase in population replacement elicits the same decline in interpopulation differentiation (increase in Q/P) as an equivalent increase in migration rate (Table 7).

Colonization is a powerful means by which dispersal may retard or preclude interpopulation differentiation and is an integral part of the dispersal dynamics. In areas of rapid range expansions and in which populations are very far apart, colonization may be much more important in species integration than is migration. However, as with migration, the overall effects of colonization do not accrue after one or a few generations. Also, as with migration, the area affected per generation is a function of dispersal distance: the smaller the distance, the smaller the area. If two groups of populations were separated by a zone broader than the maximum dispersal distance that was inhospitable for colonization, dispersal via colonization would not alter the fact

TABLE 7. Probability of allele identity within and among populations.[a]

m	e_0	P	Q	Q/P
0.01	0	0.71	0.06	0.09
0.01	0.001	0.71	0.07	0.10
0.01	0.01	0.73	0.12	0.16
0.01	0.10	0.81	0.43	0.52
0	0.01	0.96	0.09	0.09
0.001	0.01	0.93	0.09	0.10
0.10	0.01	0.34	0.18	0.52
0.001	0.001	0.93	0.02	0.02
0.10	0.10	0.42	0.28	0.67
0	0.10	0.96	0.48	0.50
0.10	0	0.32	0.16	0.50

[a] $n = 100$; $N_e = 10$; $v = 0.001$.

of isolation by distance. In the final analysis, dispersal distance determines the tempo and scope of homogenization.

CONCLUSIONS

The level of immigration via pollen or seeds remains one of the least understood demographic variables. As the genetic consequences may be substantial and are dependent upon the mean and variance of the rate, a quantification of immigration within the same population over time and among populations is a prerequisite for understanding the dynamics of population divergence and its rate and the organization of variation in space. There are numerous factors which affect the outcome of a given level of immigration, and it is the study of these factors which is providing the most information about immigration as an evolutionary force. Genetic markers have yet to be used to actually demonstrate immigration rates in wild plants, but the potential for such exists. I hope they will be employed with whatever ingenuity is required.

CHAPTER 13

MATING PATTERNS IN PLANTS

Mary F. Willson

INTRODUCTION

Random mating is a common assumption for many models in ecology and evolution, yet it is common knowledge that mating is often not random and a number of specific models deal with, for example, the effects of inbreeding or assortative mating on population structure. Deviations from randomness may affect different ecological–evolutionary models to different degrees, depending on the scale of both the models themselves and the actual deviations. Although the importance of deviations from random mating probably varies, analysis of the relative magnitude of the effect is a topic for another essay, which should be written by a more quantitatively inclined author. For present purposes, it may be useful to provide (1) an annotated outline of some of the biological bases of nonrandom mating and (2) a survey of some of the consequences of certain kinds of nonrandomness. This is by no means an exhaustive survey; Levin (1981, 1983b) has recently reviewed much of this material from a slightly different perspective.

SOME SOURCES OF NONRANDOMNESS IN MATING PATTERNS

Absence of cross-breeding

Various forms of parthenogenesis are widespread in the plant kingdom. Some are ameiotic, others are meiotic, and some are initiated only by a stimulus from pollen, which then takes no further part in zygote formation. Furthermore, many plants propagate vegetatively, by a variety of means, and the vegetatively produced "offspring" may be independent of the parent. The frequency of such asexual (or unisexual) modes of reproduction (that is, propagation not involving fusion of paternal and maternal haploid nuclei) often varies among populations, as well as among individuals in a population. Males and females of dioecious species capable of vegetative propagation commonly differ in the capacity for vegetative propagation. Even if such individual differences in asexual reproduction are not directly controlled genetically, they may influence the distribution of associated genes in nonrandom ways (see also Levin, 1978a). Despite its potential importance, the frequency of occurrence of various asexual reproductive modes in different ecological conditions is not well documented.

Perhaps the most commonly considered factor causing nonrandom mating is self-fertilization, a capacity possessed by more plants than animals. The extent of selfing by an individual varies with the degree of self-compatibility (which often varies even within a species), with the behaviors of the pollen vectors (which vary with geographic locale and environmental conditions, in both time and space), and with the capacity for cleistogamy (which enforces selfing and is known to vary greatly among coexisting individuals; references in Willson, 1983). If a chasmogamous-flowered, self-compatible individual produces many flowers at the same time, the potential for geitonogamous (between-flower, within-individual) selfing may be great. This seems to be the case both for many-flowered, single-stemmed plants (e.g., Frankię et al., 1976; Schemske, 1980), although the effect is not universal (e.g., Hodges and Miller, 1981), and for multi-stemmed clonal plants. For example, the frequency of homozygous recessive progeny of the bee-pollinated, clonal *Mimulus guttatus* was greater than expected by random mating, a finding indicating that movement of pollen among the flowers of the clone was very common (Kiang, 1972).

Inbreeding with relatives

Most plants are sessile, and both seed "shadows" and pollen "shadows" are thought to be strongly clumped near the parent in many species (e.g., Levin and Kerster, 1974; Levin, 1981). As a result, most matings may be among relatives. As Levin (1981) has emphasized, however, many factors may mitigate this clumping and open up the population structure at least to some degree. It is not necessary to re-review the material recently covered by Levin, but a few comments are appropriate here.

Levin (1981) noted that seed predation may be a function of seed density and/or proximity to parent; when this is the case, then obviously seed predation would lower the concentration of individuals in any one place and potentially reduce the intensity of inbreeding among relatives (although the survivors are still close together and may inbreed). However, the existing evidence for density- or distance-responsive seed predation is pitifully poor—only some predators respond that way, to some kinds of seeds (references in Wilson, 1983). We know very little about what determines the patterns of predation and why it varies; the variation itself may prove interesting.

With respect to pollination by animal vectors, several studies have shown that nectar-reward levels can affect the foraging behavior at least of bees and birds. Nectar-foraging bees make shorter moves and turn more often when nectar levels are high; this response restricts pollen flow. Lower rewards lead to longer moves and less turning and, presumably, greater pollen flow (e.g., Pyke, 1978; Heinrich, 1979; Waddington, 1981; Zimmerman, 1982). Floral density can influence insect movement patterns (e.g., Levin and Kerster, 1969; Beattie, 1976; Ellstrand et al., 1978; Schaal, 1978) as well as the behavior of birds that carry pollen (Gill and Wolf, 1975; Linhart and Feinsinger, 1980; Schemske, 1980). Nectar levels in flowers may be very uneven on different plants or clusters of plants of the same species and, thus, patterns of pollen flow may differ locally (Pleasants and Zimmerman, 1979); moreover, the response of different pollen vectors to variation in nectar reward levels is not necessarily the same. Furthermore, pollen-collecting bees may not behave the same way as nectar-foraging conspecifics, so that pollen dispersal patterns may differ with the habits of a particular bee on a particular foraging trip (Zimmerman, 1982). Also different species of pollinators may also exhibit different movement patterns on the same flowers (e.g., Schmitt, 1980). Such observations emphasize the necessity of very detailed monitoring of rewards and behavior of all floral visitors. Even then, pollinator movement patterns are not necessarily a good index of pollen movement (e.g., Schaal, 1980a), and such indirect indices are subject to so many sources of error that they should be eschewed, unless preliminary tests clearly show their accuracy (see later).

Moreover, many of the aforementioned studies focus chiefly on the functional responses of individual pollinators and often (but see Schaal, 1978) neglect a possible numerical response: high reward levels are likely to attract *more* pollinators (or, perhaps, more or different *kinds* of pollinators; Schemske, 1983). In certain African sunbirds, very high nectar resources lead to a breakdown of territoriality with an influx of encroaching foragers (Gill and Wolf, 1975); honeybees can recruit fellow workers and a numerical response is likely; bumblebees might gradually discover persistently rich patches. The effect of more vectors might delocalize pollen flow to some degree.

In all of these studies of pollen movement, it seems to be generally assumed that if more pollen is deposited in a particular place, more seeds will result. That is, a dense deposition of pollen near the pollen source leads to an equivalent density of related offspring. This assumes that seed production is constrained chiefly by pollen availability, an assumption that may be commonly untrue (Willson and Burley, 1983). In reality, some pollen may not contribute directly to effective gene flow at all. There is the obvious possibility that pollen is deposited in excess of the number of available ovules. In addition, sibling pollen grains traveling together may not participate directly in fertilization but may sometimes exclude other pollen from a stigma or enhance the growth of pollen tubes from certain of the sibling group (Willson and Burley, 1983). Whether or not the "excess" pollen contributes to the success of the pollen grains that gain access to ovules, there is a clear possibility that dense deposition of pollen does not always lead to gene flow as restricted as the deposition of pollen. Nevertheless, if pollen from nearby sources arrives sooner than pollen from farther away and has an advantage in fertilizing ovules, a high degree of inbreeding among neighbors (assuming they are indeed more closely related than expected by chance) could still result, even if pollen does not limit seed set.

The possible importance of rarer matings

The emphasis on pollen flow among neighbors and on clumped deposition of seeds has tended to obscure the possible importance of rarer events at greater distances. That is, investigators have often neglected the "tails" of the dispersion curve. Two aspects can be considered here.

The first has been called "optimal outcrossing" and has been reported in at least four cases: *Delphinium nelsonii* (Price and Waser, 1979), *Phlox drummondii* (Levin, 1981), *Picea glauca* (Coles and Fowler, 1976), and *Stylidium* spp. (Banyard and James, 1979), although in the last two cases the effect was not equally evident in all populations (see also Levin, 1983). Optimal outcrossing in this sense refers to an observed change in female reproductive success (offspring numbers and fitness) as a function of distance of the pollen source; female reproductive success first increases, then decreases, with increasing distance of pollen source. The scale over which such changes are reported varies from meters to tens of kilometers, indicating perhaps that for different species a given geographic distance has different meanings to the relevant "ecological distance." These observations are generally interpreted to indicate that mating among neighbors may be disadvantageous to some degree (because neighbors are often related) and that matings at too great a distance are also disadvantageous (because of local differentiation of populations). The effect of optimal outcrossing on the mating structure of the population depends on the height and extent of the optimum, the frequency of mating at different distances, and constraining effects of other factors (including the fitness of the offspring in the environment to which they disperse) influencing reproductive success. In addition, it may be worth noting that all these studies have focused on *female* outcrossing—pollen is brought to a maternal parent from various pollen donors. The converse experiment of using the same pollen donor on females at different distances apparently has not been done. There is no a priori reason to suppose that optimal outcrossing effects are identical for male and for female reproductive success. Although the outcrossing itself *may* have similar results because similar haploid genomes are brought together in reciprocal combinations, in natural circumstances the scale and the frequency of male and female outcrossing, as well as the constraints on each, are likely to differ. It is also possible that selection alters gene frequencies in male and female gametes in different ways or to different degrees (Clegg et al., 1978). Therefore, we could expect that gene flow through male and female gametes might have different effects on the mating structure of the population.

The second type of rare event concerns long-distance movements of both pollen and seeds (see Levin, Chapter 12). We are beginning to learn about pollen carryover (references in Levin, 1981) and secondary pollen transfer (K. Garbutt, unpublished data) over relatively short distances, but long-distance events are generally neglected—I presume for the obvious reason that they are very difficult to study. Nevertheless, long-distance movements can and do occur and may be of considerable importance to mating structure of some populations (see also Levin, 1983a).

For instance, wind-borne pollen of *Zizania aquatica* is carried in abundance to a distance at least 10 m from the source, which is equivalent to well over 100 plant diameters (M. F. Willson, unpublished data), and agronomists commonly plant cultivated varieties at least 300 m apart to prevent pollen flow between stands. However, grass pollen is reported to be quite short-lived (Heslop-Harrison, 1979), which would limit effective gene flow through pollen to relatively short distances on an absolute scale. The pollen of *Pinus, Populus, Pseudotsuga, Betula, Ulmus*, and others can be transported by the wind for several to many kilometers (Wright, 1952; Lanner, 1966; Moore, 1976); its viability after the traveling interval should be assayed (see Hoekstra and Bruinsma, 1975, for longevity estimates of pollens of some plant species). Even when most pollen is deposited close to the source, considerable amounts may travel quite far (e.g., Wang et al., 1960). In extreme cases, pollen clouds have been reported hundreds of kilometers from any possible source (references in Proctor and Yeo, 1973), although many or all of these pollen grains are probably dead, and geographic differences in flowering phenology might often doom any survivors. Even some insect-pollinated plants exhibit long, high tails on pollen dispersal curves (Wolfenbarger, 1946). Even when relatively little pollen travels far from its source, it may have important effects on gene flow; low frequency of occurrence does not necessarily imply insignificance.

Long-distance seed transport may also occur. Post-Pleistocene range expansion of *Quercus* and *Fagus* was greater than expected by random movements, and the expansion (at least of *Fagus*) spanned long distances of unsuitable habitat (Davis, 1981; Webb, 1982). Although long-distance pollen movement is apparently well-documented for *Quercus* (Davis, 1981), it is possible that long-distance transport of the large fruits by birds may account for these rapid and/or discontinuous range expansions (Webb, 1982). The westward spread of *Tsuga* in North America may also have been facilitated by long-distance movement of seeds by wind over frozen lakes (M. B. Davis, personal communication). Piñon pine (*Pinus edulis* and its relatives) seeds are harvested, transported, sometimes for many kilometers, and cached by corvid birds, although those carried the farthest are sometimes deposited in apparently unsuitable habitats (VanderWall and Balda, 1977).

Long-distance seed dispersal may be nonrandom either because the propagules are a very small and arbitrary subset of the gene pool of the source population or because particular genotypes are more likely to be dispersed long distances. For example, several cases of size-selectivity are known for scatter-hoarding vertebrates (references in Willson, 1983); much of the seed-size variation in these examples was not genetically controlled, but the studies indicate the *possibility* of nonrandom dispersal. Random or not, long-distance dispersal tends to open up the population mating structure.

Directional dispersal

Seed flow obviously can be strongly directional in space, especially on wind or water currents (e.g., Staniforth and Cavers, 1976; Waser et al., 1982). Even seedlings may travel downstream and become established (e.g., *Acer saccharum*, F. A. Bazzaz, personal communication). Nevertheless, it seems likely that something counters these unidirectional flows, inasmuch as we seldom notice entire plant distributions moving downstream (Staniforth and Cavers, 1976).

Directional pollen flow is obvious for wind-borne pollen, but it can also occur in animal-borne pollen, for bees tend to fly upwind in strong breezes (Woodell, 1978). Trap-lining bees, birds, bats, and butterflies clearly could produce strongly directional movement of pollen (or seeds, if the animals eat fruits), and certain individuals then are probable fathers of offspring on certain other individuals, subsequently visited.

The temporal scale of directionality may differ greatly among systems. The longer the time period, the more likely is a reversal of direction of wind or trapline, though creeks and rivers are less likely to reverse direction on an ecological time scale. Frequent changes of directionality clearly would tend to erase directional flow patterns; in the case of bumblebees foraging on *Aralia hispida*, some parts of individual traplines changed daily (Thomson et al., 1982), and the effect of directionality would be reduced. The possible importance of short-term directionality varies with the question being asked. That is, if one is concerned with the potential for local population differentiation of a perennial species (which probably often takes several generations), short-term directionality may be of little importance. For annuals, short-term events could be more important. Likewise, short-term directionality may have immense and different consequences for the mating success of male and female reproductive functions; directional pollen flow may have greater effects on the mean and variance of male reproductive success (which is often limited by access to females) than on the mean and variance of female reproductive success (which is less often limited by access to males). We know rather little about individual variations in female reproductive success and almost nothing at all about variations in male reproductive success.

Directionality is also evident in the movement of genes between species. Passage of genes via the backcrossing of interspecific hybrids can be strongly differential to one or the other parent stocks (references in Levin, 1975a). Interspecific crosses in *Betula* are directional because one species is a better pollen donor than receiver of alien pollen (Hagman, 1971). This observation emphasizes the earlier comment that gene passage through male and female function may be quite different. Conspecific populations often differ in the degree of outcrossing achieved, and the range of observed differences varies among species (Schoen, 1982b). It would not be surprising to find, within species as well as between species, that between-population gene transmission through male and female function differs and that the magnitude of the differences varies among species. The possible role of interpopulation gene flow in maintaining the genetic cohesiveness of species' gene pools has been emphasized recently by Morris Levy (personal communication); in his case of *Phlox pilosa*, seed dispersal and female function claimed chief attention.

Assortative and disassortative mating

I refer here particularly to matings between like or unlike phenotypes, not to any form of inbreeding. By definition, both of these population patterns of mating are nonrandom. The available evidence for plants is highly variable: Some studies show clear assortative mating for particular traits (plant height, flower color, or odor); some show that some pollen vectors discriminate but others do not; and some show no evidence of discrimination at all (references in Willson, 1983). At this point it seems impossible to discern any general conditions about when these mating patterns occur. Even the possible advantages and disadvantages of these patterns are little explored at the level of the individual plants (except perhaps in the case of heterostyly), as seems to be the case also for animals (Burley, in press). In some instances, mating patterns may be frequency dependent (e.g., Levin, 1972).

Differential contributions to future generations

It is not uncommon for a few parents of a population to account for a majority of offspring in the following generation; sometimes this differential is very large. Examples of enormous differences among individual contributions to future generations in animals include a number of birds that court and mate on leks—one or two males on the lek may account for 75 to almost 100% of the matings (references in Willson, 1984). Similarly, two of ten ground squirrels resident in a certain study area accounted for all resident female descendants after two years (Michener, 1980). Less extreme examples are legion.

In plants, it is well known that seed production often varies enormously with plant size; large plants are commonly very fecund but relatively rare in the population (e.g., Gottlieb, 1977; Levin, 1978a; Leverich and Levin, 1979; Solbrig, 1981; and others). Very often such differences are determined environmentally irrespective of the genotype (Bradshaw, 1965; Gottlieb, 1977; Solbrig, 1981), but the possibility of genotypic differences should be further explored. In either case, the domination of seed production by a few individuals can enhance the potential for inbreeding within the population.

Two related matters may be mentioned. First, it might be instructive to examine a variety of species in terms of the relationship between plant size and fecundity. I have done this for the data of Palmblad (1968) for several weed species, which indicate that for some species fecundity increases rapidly with plant size whereas for others the increase is slower (Figure 1). Unfortunately, the data are averages for groups of plants rather than individual values, and underground plant parts (which probably differ among species in the proportion of total biomass they constitute) are ignored. Therefore, these particular

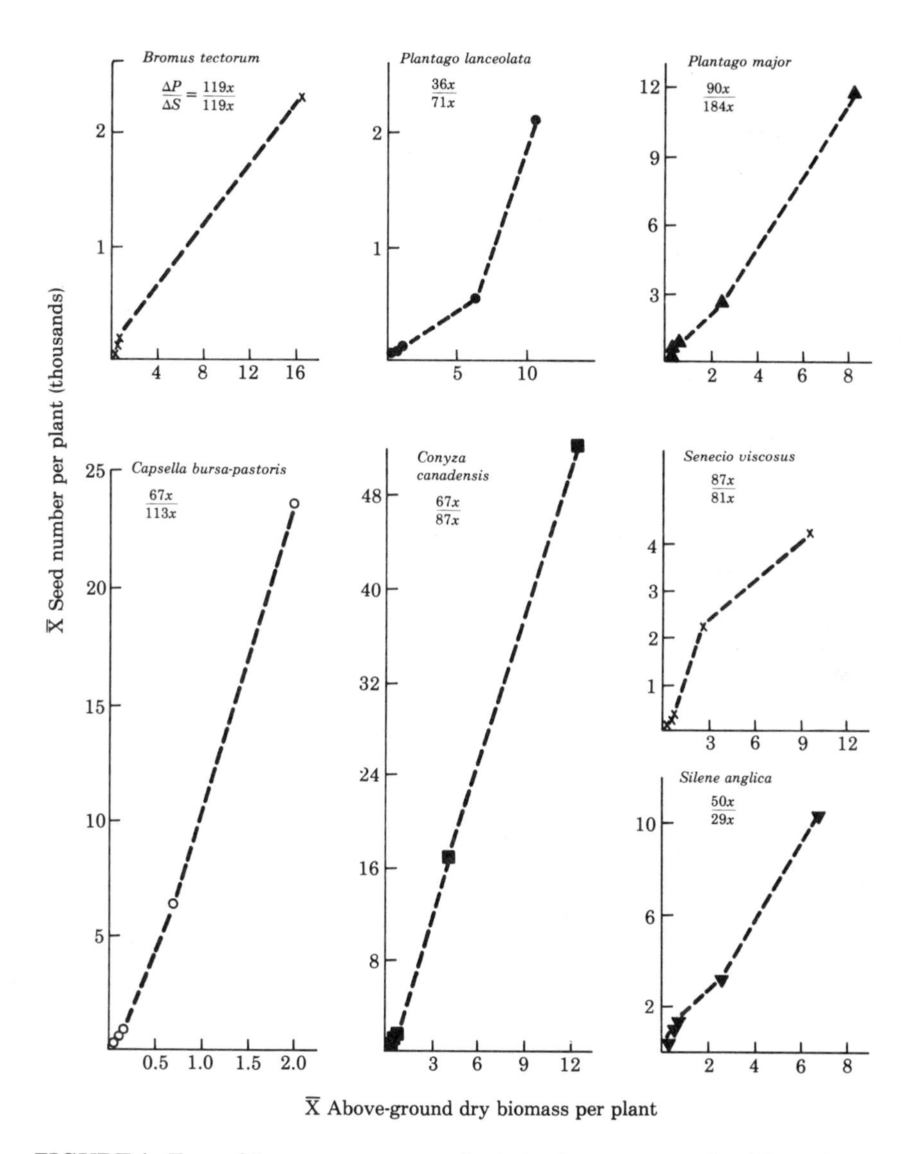

FIGURE 1. Fecundity versus average plant size for seven species. Plant sizes are estimated from size-density relationships given by Palmblad (1968). Of particular interest are the different slopes and perhaps shapes of the curves for different species (see text). ΔP, The change in plant size (g), maximum divided by minimum; ΔS, the change in seed number, maximum divided by minimum.

cases are very imperfect, but they illustrate the kind of analysis that could be done. If species indeed differ in the rate of increase of female fitness gain with plant size, this might be related to the evolution of various forms of sex expression (see Willson, 1983). Second, the relationship between plant size and *male* reproductive success is unstudied. Yet male reproductive success is important, not only because this component of fitness is also related to the evolution of sex expression, but also because the measure of relative domination of the next generations by any hermaphroditic individual (functioning as both male and female) may differ, depending on whether pollen donation or seed production are assessed.

Differential mating ability

I refer here to a differential ability to produce offspring, either maternally or paternally. Not only may some individuals function more successfully as females (pollen recipients) and some individuals more successfully as males (pollen donors), but, in addition, certain combinations may yield more progeny than others. In animals, strong mating preferences and mate choice based on an individual's "personal" traits are well-known, although the consequences of such choices are less well studied, especially for female fitness.

There is also some evidence for differential mating success in plants. I am concerned here, not with the consequences of plant size or other obvious phenotypic differences, but rather with the observation that, at least in some species, pollen donors differ in their ability to deliver pollen and to fertilize ovules and that pollen recipients differ in the array of acceptable pollen donors. A number of species are known to exhibit differential mating ability, here indexed by individual variation in outcrossing success. Listed in Table 1 are 20 species, some cultivated, some wild, for which I have (so far) found reports that different genotypes of pollen recipients (that is, functioning as females) achieve different levels of outcrossing. The genetic marker itself was not necessarily responsible for the observed difference. S-alleles and a prezygotic self-incompatibility (SI) system are implicated in some of these twenty cases, and allelic similarities between donor and recipient seem to account for at least some of the observed variation. In other cases, the strength of the incompatibility system varies among individuals and the SI system perhaps is involved in the observed variation in outcrossing levels, although in most of these cases other factors are involved as well (Barnes et al., 1962; Gabriel, 1967; Sparnaaij et al., 1968; Crowe, 1971; etc.). In fact, in *Medicago sativa*, the degree of self-compatibility and the level of outcrossing were *positively* correlated (Whitehead and Davis, 1954), demonstrating that the SI system cannot explain the observed differentials in this species. More interesting

TABLE 1. Preferential mating in plants.[a]

Species[b]	S-alleles implicated	Reference
Betula spp.	Yes	Hagman (1971)
*Liriodendron tulipifera**	Yes	Wilcox and Taft (1969)
*Asclepias speciosa**	Perhaps	Bookman (1983)
Borago officinalis	Perhaps	Crowe (1971)
*Campsis radicans**	Perhaps	Bertin (1982)
Freesia sp.*	Perhaps	Sparnaaij et al. (1968)
*Lotus corniculatus**	Perhaps	Schaaf and Hill (1979)
*Medicago sativa**	Perhaps	Whitehead and Davis (1954) Barnes and Cleveland (1963)
*Phlox drummondii**	Perhaps	Levin (1976)
*Acer saccharum**	No	Gabriel (1967)
*Euoenothera**	No	Hoff (1962)
Ipomoea purpurea	No	Brown and Clegg (in press)
*Lupinus nanus**	No	Horowitz and Harding (1972a,b)
Lupinus succulentus	No	Harding and Barnes (1977)
Nicotiana rustica	No	Breese (1959)
*Phaseolus lunatus**	No	Harding and Tucker (1964)
*Pinus monticola**	No	Squillace and Bingham (1958) Barnes et al. (1962)
Senecio vulgaris	No	Campbell and Abbott (1976)
Vicia faba	No	Drayner (1959)
*Zea mays**	No	Sari Gorla et al. (1975) Pfahler (1965, 1967)

[a] All species listed have been reported to exhibit individual or genotype-specific differences in female outcrossing.
[b] Species marked with an asterisk exhibit differential male outcrossing.

are the cases in which no SI system is known to be at work; preferential matings may result from genotypic complementarity between particular pollen donors and receivers or pollen acceptability may be determined by the relative competitive ability of pollen from different donors. Some of the potential means of competition between pollen donors and of discrimination by pollen receivers are reviewed by Willson and Burley (1983); they include both prezygotic and postzygotic mechanisms. A complicating factor in at least two cases (*Asclepias speciosa, Euoenothera*) is that crosses occurred between individuals from different populations or races; here the different levels of outcrossing could be as much a function of population differentiation as of individual complementarity.

Still more interesting, perhaps, is that pollen donors (that is, functioning as males) differ in their ability to outcross (* in Table 1; this conclusion for *Ipomoea purpurea* is inferred from patterns of pollinator visitation to different floral morphs). Furthermore, this ability often varies in part with the genotype of the pollen recipients. Interestingly, in *Campsis* and *Freesia*, ability as a pollen donor is inversely correlated with ability as a recipient, indicating the possibility of sexual role specialization even in functional hermaphrodites. Although the importance of the male role in outcrossing has long been known, perhaps reaching as far back as Darwin (Horowitz and Harding, 1972a), it has not received adequate attention, and the emphasis has been more on outcrossing per se than on other aspects of pollen donation.

Differences in levels of outcrossing achieved obviously can be influenced immensely by environmental factors (Clegg, 1980). Thus, the differences evident among individuals of *Clarkia exilis* (Vasek, 1967) may be environmental as much as genotypic, and environmental effects are reported for some of the species in Table 1 also (e.g. *Vicia faba*). Furthermore, some of the studies in Table 1 do not distinguish between differential pollen delivery, which depends on the behavior of pollen vectors, and differential fertilizing ability of pollen that has landed on a receptive surface. For *Ipomoea purpurea*, genetic estimates of female outcrossing paralleled measures of pollinator discrimination among floral color morphs (Brown and Clegg, in press); for *Nicotiana rustica*, female outcrossing varied directly with floral morphology (Breese, 1959); and for *Euoenothera*, outcrossing seemed to be inversely correlated with the capacity for autogamy (Hoff, 1962), but this need not always be so (Horowitz and Harding, 1972a,b).

Of special interest are experiments in which pollen from different sources is applied to the same stigma to test directly the competitive ability of pollen and/or the discriminatory powers of the female; unpublished work with *Campsis radicans* addresses this problem (R. I. Bertin, personal communication). However, even when experiments use only a single pollen donor for each flower used as a pollen receiver, enormous differentials in male success can be seen (e.g., Schaaf and Hill, 1979; Bertin, 1982). While placing different males in direct competition might sometimes alter the order of pollen donors ranked by their success separately, it seems unlikely that the individual differences would be obscured.

SOME CONSEQUENCES OF NONRANDOM MATING IN PLANTS

For convenience in discussion, I will divide the consequences of nonrandom mating that I wish to consider into three categories. Plainly, these categories can interact in determining ecological and evolutionary possibilities, but initially it is easier to discuss them separately.

Population structure

Undoubtedly, the single most discussed consequence of mating patterns in plants (in contrast to animals, see Baker and Marler, 1980) is the genetic structure of populations and the potential for local differentiation of populations in response to selection and random drift. Estimates of effective population size or neighborhood size are numerous and, in many cases, the estimates indicate rather small sizes with relatively high potential for microscale differentiation (e.g. Schaal, 1975; Jain, 1976; Beattie, 1978; Levin, 1978a; Beattie and Culver, 1979; Rai and Jain, 1982; and many others). Levin's (1981) review summarized the problem of the genetic differentiation of populations in relation to gene-flow patterns, especially with respect to random drift (see also Wright, 1940, and later). Local differentiation in response to selection has been studied extensively with respect to microhabitat characteristics (Ledig and Fryer, 1972; other references in Rai and Jain, 1982). Rai and Jain (1982) stress the difficulty of assessing the relative importance of localized selection and of drift when stability of environmental patches is variable or unknown. There are possible ecological consequences of population subdivision, in addition to genetic ones, and these may warrant greater study (Wool and Mendlinger, 1981; see also Antonovics and Levin, 1980). In addition, drift is not necessarily excluded as an agent of evolutionary change even in large populations (e.g., Lande, 1976).

Evolutionary potential

As used in this context, "evolutionary potential" refers not to the facility of speciation (which often may be related to the extent of local population differentiation, just discussed) but rather to the prospects of long-term survival of the population on an evolutionary time scale. The long-term evolutionary prospects of species that reproduce chiefly by asexual means, or that inbreed intensively, have long been thought to be poor, because such species were thought to have much reduced genetic variation by which to respond to changing selection pressures. Likewise, small populations were thought to be genetically uniform at many loci. However, recent work indicates that this need not be so at all. Parthenogenetic animals often exhibit much variability, both genotypically and phenotypically; they can be at least as variable as their closest sexual relatives (e.g., Parker, 1979; Templeton, 1982; M. Lynch, unpublished data). Similar results are reported for plants (e.g.,

Primack, 1980; Ellstrand and Levin, 1982). In some cases variation in typically selfing plants may be maintained in part by occasional events of outcrossing (e.g., Adams and Allard, 1982; but see also Brown and Albrecht, 1980), but mutation rates may also differ in sexual and asexual populations (M. Lynch, unpublished data; Templeton, 1982). Especially in long-lived plants, mutations in polygenic characters may be important in maintaining variation (Lynch and Gabriel, in press; M. Lynch, personal communication). Not to be neglected is the possible importance of somatic mutations that are passed on in the gametes produced by different branches of a single physiological individual (Libby et al., 1969; Whitham and Slobodchikoff, 1981). The high frequency of asexual reproduction and selfing in plants also hints that such reproductive modes may not be without evolutionary potential and that asexually reproducing organisms are not evolutionary dead ends at all (Turkington and Harper, 1979b; Burdon, 1980; M. Lynch, unpublished data; Lynch and Gabriel, in press).

The evolutionary effects of a seed bank in the soil have been explored by Templeton and Levin (1979). Here, however, it is important only to note that whatever the results of nonrandom mating within the spatial dimension of a population, these may be altered in time, depending on the emergence of seedlings from seeds in the bank and on survivorship of seeds.

Life-histories

Mating patterns in animals are commonly discussed in terms of life-histories, and vice versa; competition for mates, size-related changes in fecundity, sex ratios, phenology of breeding, dispersal of siblings, and the evolution of kin-correlated behaviors, and many other features are common elements linking mating patterns and life-histories. Yet this link seems to be less frequently forged in studies of plant reproductive ecology. Perhaps the relationship is actually less close in plants and so deserves less attention, but it surely should not be ignored.

Dispersal of offspring or gametes is a salient feature of life-histories that, at least in theory, may have influenced the evolution of sex expression in plants, of progeny sex ratios in dioecious species, and of resource allocation patterns in functional hermaphrodites (or "cosexes" in the terminology of Lloyd, 1982 and this volume).

Dispersal of male and female offspring often differs, at least in many animals, and several authors have suggested that dispersal differences may lead to biased progeny sex ratios that favor the sex dispersing farther (Bulmer and Taylor, 1980; Taylor and Bulmer, 1980, other references in Charnov, 1982). The nondispersing sex may be more costly to produce if those individuals stay with their parent(s) longer than the dispersing sex (as seems to be true for certain primates and deer, for instance, in which males disperse and females stay at home). When this is true, perhaps parental expenditures on sons and daughters are statistically equal, as argued by Fisher (1930) and later by many others, although this remains to be determined. It may also happen that parents can achieve greater fitness gain through offspring (see also Bull, 1981; Charnov, 1982) of the dispersing sex, by means of reduced sibling competition or increased probability of finding a site for establishment, as Bulmer and Taylor (1980) suggested. If this is true, then we should expect to see correlations of offspring sex ratios with magnitude of dispersal differential between the sexes. At least in vertebrates I have not been able to find that this is so (Willson, 1984), perhaps in part because other factors override any single effect of dispersal distance. Dispersal differences between male and female *seeds* seem to be unknown, however.

Lloyd (1982) has suggested that dispersal differentials between male and female gametes in plants may be one factor influencing sex expression. He argued that in outcrossing species the dispersal of male genomes is invariably greater than the dispersal of the female genome because the paternal genome travels first to the female recipient and then again with the maternal genome in the dispersing seed. The greater the disparity between total dispersal of the paternal genome and that of the maternal one, the stronger the selection for simultaneous functional hermaphroditism, because, for instance, short seed shadows restrict fitness gain through female function and favor addition of the second gender (Lloyd, 1982; see Charnov, 1982, for a review of the general argument).

Given that a plant functions as an hermaphrodite, it is normally expected that there will be equal allocation of resources to male and female function, for essentially Fisherian reasons (Maynard Smith, 1978a; Smith, 1981; Charnov, 1982). However, several factors, including inbreeding and gamete dispersal differentials, may cause this basic 50:50 allocation pattern to shift (Maynard Smith, 1978; Bull, 1981; Charlesworth and Charlesworth, 1981; Charnov, 1982). Selfing shifts the pattern increasingly in favor of female function; the more a plant outcrosses, the more it is expected to spend on pollen and associated male costs (up to the 50:50 ratio). Thus, Lloyd (1972) observed that, among monoecious species of *Cotula*, those with larger flowers are apparently more adapted to outcrossing (but compare Schoen, 1982, for *Gilia achilleifolia*) and produce more male flowers than those with smaller, less outcrossed flowers. However, it is by no means clear that sex allocation patterns in all species conform so easily to expectations (Willson and Ruppel, in press; D. A. Goldman and

M. F. Wilson, unpublished data). In any case, gene passage through male and female functions may influence several aspects of life-history.

It is worth noting here that population size may affect sex ratio (references in Charnov, 1982) and that sex ratio also affects the "effective population size" (references in Falconer, 1960). Selection on clutch size variation also interacts with population size (Falconer, 1960; Gillespie, 1974), and other examples could be provided. Life-history selection apparently preceded mating system selection in *Clarkia xantiana* (Moore and Lewis, 1965), with possible consequences for population structure. The interactions of life-history and population mating structure merit more general consideration.

CHAPTER 14

GENDER ALLOCATIONS IN OUTCROSSING COSEXUAL PLANTS

David G. Lloyd

INTRODUCTION

The sexual strategies of plants are a particularly appropriate topic to discuss in a volume commemorating the centenary of Charles Darwin's death. Darwin wrote three volumes (1862, 1876, 1877) on the adaptive significance of floral features, and he further discussed the sexual strategies of plants and animals in several other books, including *Origin of Species*. In the course of this work, Darwin put forward a novel (and universally accepted) hypothesis to explain floral features that promote cross-pollination. He proposed that the progeny from cross-pollination are generally superior to those from self-fertilization. This hypothesis has provided the foundation of floral biology ever since.

Reproductive characters have continued to occupy a prominent position in studies on the evolution of plants. The types of arguments that have been made have swung dramatically with changing research fashions, however. After the post-Darwinian enthusiasm for pollination mechanisms declined, genetical theories of recombination systems flourished and were incorporated into the synthetic theory of evolution. Most recently, the fusion of ecological and evolutionary streams into modern population biology has focused attention on various strategies for deploying resources, including those concerned with sex.

The central topic of sexual strategies is the relative allocation of resources to male and female offspring or to offspring derived through male and female gametes (Charnov, 1982). Most of the innovative theories and observations, starting with Fisher's (1930) famous argument for equal investment in male and female progeny, have come from zoologists and deal with sex ratios in dioecious organisms. The great majority of seed plants, however, have a single morph that combines male and female functions. The central allocation strategy of such cosexes involves the proportions of sexual expenditure that are directed into paternal and maternal functions. Gender allocations in cosexes were first considered by Maynard Smith (1971), who proposed that a hermaphrodite animal optimally allocates equal resources to the two parental modes—the analogue of Fisher's sex ratio theorem.

Several factors have since been shown to modify Maynard Smith's conclusion. Factors causing departures from equal allocations include nonlinear fitness curves in general (Charnov, 1979; Charlesworth and Charlesworth, 1981), local mate competition (Charnov, 1980), self-fertilization (Charlesworth and Charlesworth, 1981), and variable conditions (Lloyd and Bawa, 1983). The latter two articles and some sections of Charnov's (1982) book deal specifically with plants.

In this chapter I will examine gender allocations in outcrossing seed plants. After the small amount of information on gender allocations in outcrossing natural populations is reviewed, the theoretical effects of several factors operating in a homogeneous environment are analyzed. The results suggest how unequal allocations may be selected in angiosperms.

DATA ON GENDER ALLOCATIONS

There are only a few estimates of the fraction of reproductive resources that are spent on maternal and paternal functions. To obtain such estimates, we must first decide what constitutes maternal and paternal investments. In addition to the direct costs of producing pollen grains and seeds, there are various indirect costs of reproduction. Some indirect costs contribute to both maternal and paternal success, including organs of support (pedicels and peduncles), protection (sepals), attraction (petals, etc.), and reward (nectar). These costs need not contribute equally to maternal and paternal fitness, but in the absence of any theoretical basis for partitioning them, the indirect costs are considered to be bilateral costs that are necessary for both functions and are separate from the gender allocations. The bilateral costs simply reduce the resources available to be allocated to the unilateral (maternal or paternal) functions.

Gender allocations are concerned with the proportions of the unilateral investments directed to structures that contribute to paternal or maternal fitness. In general, the paternal and maternal allocations are invested respectively in the androecia and gynoecia, but several complications may arise. The costs of any nongynoecial structures that are incorporated into fruit should be considered as maternal costs. On the other hand, parts of the stamens or carpels that produce nectar (such as the stylopodia of *Smyrnium olusatrum*) are treated here as bilateral costs. The contribution of pollen grains to attracting or rewarding floral visitors, which enhances both maternal and paternal fitnesses, is a more difficult matter. In the absence of any appropriate theory, the only procedure that is possible at present is to naively treat pollen costs as unilateral paternal investments. The partitioning of costs is considered further in a later section.

Another difficult procedural problem involves the currency that the allocations should be measured in. The proportions of the unilateral expenses that are allocated to pollen versus fruits and seeds are not always identical when measured in different ways, as dry weights, calories, or quantities of nitrogen, phosphorus, or potassium (Lovett Doust and Harper, 1980). This problem is not unique to gender strategies. It is common to all allocation measurements and has not been satisfactorily answered. In the case of gender strategies, a decision is forced on us since the only comparative data that are available use dry weights.

Comparisons of the dry weights of structures contributing to paternal or maternal fitness are further complicated by the difference in timing between paternal and maternal expenditure (Lloyd, 1980b) and by contributions that the fruits of many species make to the photosynthates they utilize (Bazzaz et al., 1979; Herrera, 1982a). Again, we can at present only take the final dry weights at face value as legitimate estimates of expenditure.

The dry weights of structures contributing to unilateral paternal or maternal fitness have been calculated for natural populations of six species of cosexual angiosperms (Table 1), excluding species or flowers that are regularly self-fertilized. The gender allocations vary widely among the six species, which are all herbs. In only one species, *Microlaena polynoda* (chasmogamous flowers), are the allocations to paternal and maternal functions almost identical. In the other five species, there is a pronounced emphasis on maternal investment. The two wind-pollinated species have higher paternal allocations than any of the four insect-pollinated species.

The information on gender allocations in *Gilia achilleifolia* is particularly instructive. The paternal allocation in Table 1 (9.3% of the combined unilateral costs) is an extrapolation to no self-fertilization

TABLE 1. Estimates of paternal allocations in outcrossing cosexual angiosperms.

Species	Pollinating agent (wind or insects)	Paternal allocation (%)	Frequency of outcrossing (%)	Source
Microlaena polynoda[a]	W	51.2	?	D.J. Schoen (personal communication)
Zizania palustris	W	16.0–33.3	ca. 90	M. Willson (personal communication)
Impatiens biflora[a]	I	9.9	? (Protandrous)	Schemske (1978)
Impatiens pallida[a]	I	9.6	? (Protandrous)	Schemske (1978)
Gilia achilleifolia	I	9.3	Extrapolation to 100	Schoen (1982a)
Smyrnium olusatrum	I	4.0	ca. 100	Lovett Doust and Harper (1980); D. Lloyd (personal observation)

[a]Chasmogamous flowers only, in species with cleistogamous flowers as well.

from a regression based on allocation estimates of six populations with average frequencies of self-fertilization between 0.20 and 0.85 (Schoen, 1982a). The linear correlation between the paternal allocation and the frequency of self-fertilization is very high: 0.99. The precise linear relationship agrees well with theoretical expectations. Ross and Gregorius (1983) derive an equation for the evolutionarily stable strategy (ESS) to paternal allocation in the absence of inbreeding depression, $a = \frac{1}{2}(1 - s)$, where s is the frequency of self-fertilization (cf. Charlesworth and Charlesworth, 1981). In the *Gilia* estimates, the costs of nectar, petals, and sepals were excluded, and pollen was considered as a strictly paternal cost (as discussed earlier). The high correlation between the breeding system and paternal allocations therefore provides considerable reassurance that these conventions and final dry weight measurements do indeed provide meaningful and accurate estimates of relative gender allocations.

Although the available data on gender allocations in outcrossed flowers is meager, they are highly informative. Diverse degrees of inequality are evident among the six species, and the figures suggest that the diversity is related to at least one ecological factor, the pollinating agency. Data on diverse pollen:ovule ratios in outcrossing angiosperms (e.g., Cruden, 1977; Cruden and Jensen, 1979) also suggest lower paternal allocations in some species. Moreover, wind-pollinated species tend to have higher pollen:ovule ratios (Pohl, 1937), in parallel with the apparent difference in paternal:maternal allocations. The interpretation of pollen:ovule ratios is more complicated, however, since they reflect not only the gender allocations but also variation among species in the resources invested in single seeds and pollen grains; moreover, they do not incorporate post-fertilization events such as seed abortion.

GENERAL PROCEDURE FOR THEORETICAL MODELS

The quantitative theory of sex ratios and sex allocations is not intrinsically difficult, although the literature rarely gives this impression. One of the sources of complications arises from the means of relating maternal and paternal fitnesses and finding a combined quantity that is maximized. Several of the procedures used are abstruse to all but the most numerate biologists. The methods used include fitness set analysis (which gives the simplest arguments a hint of the miraculous), the Shaw-Mohler equation, and the counterintuitive maximization of the product of maternal and paternal fitnesses (see Charnov, 1982, for explanations of all three). As a result, even the most powerful results are often lost to biologists who are not mathematically inclined, slowing progress in the subject.

The simplest possible procedure is adopted here. The method rests on the elementary observation that the genetic contributions of male and female gametes to sexual offspring are mutually interdependent and equal (sex-linked genes apart). Hence the combined fitness of any parent or grandparent from the two sources is *always* additive. If m_i, p_i, and w_i are respectively the maternal, paternal, and combined fitnesses of individual i, then $w_i = m_i + p_i$.

The equality of the male and female contributions to each zygote can be combined with Bateman's (1948) observation that paternal fitness is commonly limited by success in fertilizing eggs. The paternal fitness of an individual can then be equated with the maternal fitness of those of its mates' eggs (including its own) that it is able to fertilize in competition with other paternal parents. If a father has access to an "eligible" fraction, e_i, of the eggs of each mate (j) and succeeds in fertilizing a "competitive share," c_i, of that fraction of each of K mate's eggs, then

$$w_i = m_i + \sum_{j=1}^{K} m_j e_i c_i \qquad (1)$$

In some situations (e.g., partial self-fertilization), a cross-fertilizing pollen parent is not eligible for all of a mate's eggs ($e_i < 1$). But in all the situations treated below, a pollen parent has access to all eggs that each mate produces; hence $e_i = 1$ and can be dropped from Equation (1). Furthermore, all mates of a pollen parent may have equal maternal fitnesses, and their eggs are competed for on the same basis. Equation (1) then simplifies to

$$w_i = m_i + Km_jc_i \tag{2}$$

A pollen parent's share of a mate's gametes may often be equated with the proportion that its pollen grains make up in the pollen pool from all competing pollen parents. This assumes that all pollen grains in the pool have equal probabilities of achieving fertilization. That assumption is applicable to populations in which maternal fitness is pollinator-limited as well as to those in which maternal fitness is limited by resources, provided that the paternal fitness of every pollen donor is proportional to the number of pollen grains it produces. If a parent that allocates all its unilateral reproductive resources to producing pollen grains has g grains, an individual that allocates a fraction a of its resources to paternal functions produces ag pollen grains. We assume that an invariable fraction, r_p, of pollen grains remain and compete for fertilization after losses from predation, weather damage, being dislodged by flower visitors, and so forth. Then if individual i competes with a total of K pollen parents, including itself, for a mate's eggs, the individual's competitive shares of zygotes,

$$c_i = ga_ir_p \Big/ \sum_{k=1}^{K} ga_kr_p = a_i \Big/ \sum_{k=1}^{K} a_k = a_i/ka_k \tag{3}$$

if all a_k are the same.

All that remains is to write down precise expressions for m_i, m_j, and c_i in terms of gender allocations, the composition of pollen competition pools, and the fate of seeds and to find the allocation that maximizes the total fitness of an individual. This simple procedure has been used to obtain sex ratios and allocations where local mate competition is involved (Hamilton, 1967; Charnov, 1980). The method can be used generally to derive ESS sex ratios and allocations for a wide variety of ecological factors. The gender allocations in several situations in homogeneous environments are considered below. The effects of heterogeneous environments are described elsewhere (Lloyd and Bawa, 1983).

UNLIMITED OPPORTUNITIES FOR MALE AND FEMALE GAMETES

The simplest possible assumption about paternal and maternal fitnesses is that an increment of investment in either function confers a constant gain in fitness, for all proportional allocations, a_i and $1 - a_i$. That is, the fitness curves obtained by graphing the paternal or maternal fitnesses against their allocations are linear (Figure 1A). The paternal fitness curve is linear if male gametes compete on such a wide basis that the paternal allocations of an individual have a negligible effect on the probability that each of its pollen grains will fertilize an egg. This requires that the number of mates K is very large. The maternal fitness curve will be linear if the success of a seed is independent of the number produced by its parent. Then, if a plant that allocates all unilateral resources to maternal reproduction ($a = 0$) produces n seeds, a constant proportion r_m of these survive all hazards, germinate, and grow to maturity, $m_i = n(1 - a_i)r_m$.

Consider a population of individuals of phenotype 1, with allocations a_1 and $1 - a_1$ to paternal and maternal functions. Suppose there is a single individual of phenotype 2, with slightly different allocations

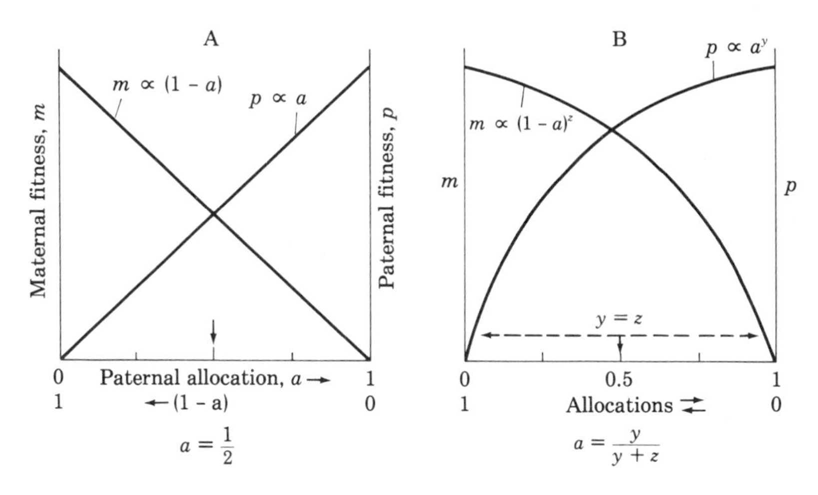

FIGURE 1. Graphs of paternal allocations (a) versus paternal fitness (p) and of maternal allocation ($1 - a$) versus maternal fitness (m). A. Linear fitness gains. The ESS (arrow) is at $a = 0.5$. B. Diminishing fitness gains. The ESS may vary widely (dashed line at bottom) and is at $a = 0.5$ when $y = z$.

a_2 and $1 - a_2$. The evolutionarily stable strategy may be found most simply as the allocation that confers the highest fitness on the mutant of phenotype 2 (Maynard Smith, 1974). The mutant has a competitive share of its mates' eggs, from Equation (3), $c_2 \cong a_2/Ka_1$, since K is large. Then from (2),

$$w_2 = n(1 - a_2)r_m + Kn(1 - a_1)r_m[a_2/Ka_1]$$

The rate of change in the mutant's fitness as its allocation changes,

$$\partial w_2/\partial a_2 = nr_m[-1 + (1 - a_1)/a_1]$$

The maternal and paternal fitness gains cancel each other out when

$$\frac{\partial m_2}{\partial a_2} + \frac{\partial p_2}{\partial a_2} = 0 \qquad \text{that is, } \frac{\partial w_2}{\partial a_2} = 0$$

Then,

$$\frac{a_1}{1 - a_1} = 1 \text{ or } a = 1/2 \tag{4}$$

Here $\partial^2 w_2/\partial a_2^2 = 0$. That is, the equilibrium is neutrally stable. A population in which the average allocations are unequal can be invaded by a mutant with a deviation in the opposite direction (cf. Maynard Smith, 1971). But if the population as a whole allocates equal resources to maternal and paternal functions, the mutant has the same fitness regardless of its gender allocations. The result is analogous to the well-known neutral equilibrium at an equal sex ratio in dioecious populations (Shaw and Mohler, 1953).

The ESS method used here is an abbreviation of the method used by Hamilton (1967) to find a strategy that is "unbeatable" at any proportions of the two phenotypes. The location of an unbeatable strategy occurs when the fitness advantage ($w_2 - w_1$) of phenotype 2 is maximized. The two methods give the same result (Lloyd, 1983). The analysis of unbeatable strategies is more rigorous and more general but it involves an additional step, so the simpler method, maximizing the fitness of a rare mutant, is used here.

The model analyzed above assumes that the success of progeny is density independent. Suppose instead that regulation of the number of plants is density dependent and that the progeny of N seed parents compete for each of x opportunities that are available for every parent in the generation under consideration. If all seeds have an equal chance of competing successfully, the maternal fitness of an individual is equal to the number of opportunities open to its progeny times the fraction of its seeds in the seed pool competing for these opportunities. Then, using (2) and (3), if N is very large,

$$w_2 \cong Nx \left[\frac{n(1 - a_2)r_m}{Nn(1 - a_1)r_m}\right] + \left[\frac{KNxn(1 - a_1)r_m}{Nn(1 - a_1)r_m}\right]\left[\frac{ga_2r_p}{Kga_1r_p}\right]$$

When $\partial w_2/\partial a_2 = 0$, $a_1/(1 - a_1) = 1/1$, as in (4).

Now 1:1 is the ratio of total opportunities available to the male and female gametes of the mutant, since every zygote contains a set of genes from both sources. Hence the ratio of paternal to maternal allocations is equal to the ratio of opportunities available through the two gametes. This result is a special case of that obtained for multiple strategies (situations where one function is performed by different structures on the same individual at one time) in general (Lloyd, in press).

Equation (4) does not contain n, g, r_m, or r_p. The ESS of equal allocations is very robust and does not depend on the size or reproductive effort of plants, the sizes and costs of either pollen grains or seeds, or the fraction of random losses that they experience (their "efficiency"), so long as all these parameters are independent of the allocations. This result comes about because the levels of competition among pollen and seeds, and hence their success rate, are the same for all competing phenotypes and therefore do not affect their relative fitnesses. A number of authors have argued that less efficient functions, such as pollen transfer by agents that cause much of the pollen to be lost, should result in a greater emphasis on that function to compensate for the inefficiency (e.g., Altenburg, 1934; Cruden, 1977; Cruden and Miller-Ward, 1981; Givnish, 1980). Conversely, particularly efficient pollen donation has been thought to require lower paternal allocations. Equation (4) shows, however, that *consistent* levels of efficiency or economy do not affect the ESS allocations. Greater efficiency (or rather, proficiency) leads to more competition, not to a lower investment.

NONLINEAR FITNESS CURVES

The general effects of nonlinear fitness curves on ESS allocations can be obtained by assuming that paternal fitness is proportional to a power of a, a^y, and that maternal fitness is proportional to $(1 - a)^z$ (cf. Charnov, 1979, 1982; Charlesworth and Charlesworth, 1981). Then

$$m_i = n(1 - a_i)^z \text{ and } c_i = a_i^y/\Sigma a_k^y$$

and the fitness of a mutant of phenotype 2 in a population of phenotype 1,

$$w_2 = n(1 - a_2)^z + \frac{K(1 - a_1)^z a_2^y}{K a_1^y}$$

When $\partial w_2 / \partial a_2 = 0$, and $a_1 = a_2 = a$,

$$\frac{a}{1 - a} = \frac{y}{z} \text{ or } a = \frac{y}{y + z} \tag{5}$$

Here

$$\left.\frac{\partial^2 w_2}{\partial a_2^2}\right|_{a_1 = a_2 = a} = n \left[z(z - 1)(1 - a)^{z-2} + \frac{y(y - 1)(1 - a)^z}{a^2}\right]$$

If both the maternal and paternal fitness curves decelerate, that is, the gains for a given increment diminish as the allocation increases ($y < 1, z < 1$; Figure 1B), the second derivative is always negative. Equation (5) then specifies a fully stable equilibrium. Any mutant that deviates from the equilibrium is selected against, whether or not the population as a whole is at the equilibrium point. If one curve is linear (y or $z = 1$) and the other decelerates, the ESS is still stable, as Charnov (1979) pointed out. If one curve decelerates and the other accelerates, the condition for stability, obtained by substituting for a and $1 - a$ in the second derivative, is $y + z > 2yz$. The stability of the allocations depends on the sum and the product of the powers. The stability of the allocations, and of cosexuality itself, is diminished by an acceleration of either function, but one accelerating function is not necessarily sufficient to destabilize cosexuality and select for separate sexes (cf. Bawa, 1980; Givnish, 1980).

The ESS allocations depend on the ratio of the powers, not their sum or product. More resources are invested in the parental function which decelerates least quickly. Various biological factors may influence the curvature of either fitness curve. For example, if a greater amount of pollen makes a plant more attractive to pollinators and increases the relative rate of pollen donation, y increases and the paternal allocation increases as well. Conversely if an increase in pollen makes pollen donation less "efficient" ($y < 1$), the proportional allocation to seeds is increased. If the "efficiency" of a function is dependent on the level of investment in that function, the ESS allocation depends on *the way in which the "efficiency" changes*, and not on the overall "efficiency." We may note, however, that very strong curvatures are needed to cause large deviations from equal allocations. Even when $y = 1/2$, a dramatic deceleration factor for pollen donation, the equilibrium pollen allocation is $a = 1/3$. The deviation is still insufficient to explain most of the species listed in Table 1.

Variations in y and z show the general effects of nonlinear fitness curves, but it is difficult to devise biologically plausible values for the powers. To evaluate the effects of nonlinear factors more precisely, we must find explicit functions to describe the fitness curves.

LOCAL POLLEN AND SEED COMPETITIONS

If a pollen parent fertilizes only a limited number of other plants, an increase in pollen production brings diminishing gains because the plant's pollen grains increasingly compete with each other for the restricted number of gametes available in the mates' flowers (Figure 2B). Hence there is a form of limited mate competition comparable to that in dioecious animal populations which are subdivided each generation when a few fertilized females colonize a site and their offspring fertilize each other before they disperse. Hamilton (1967) showed that local mate competition in such species selects a sex ratio with female predominance. In sessile hermaphrodite animals, local mate competition leads to an analogous emphasis on maternal allocation (Charnov, 1980). In plants, pollen transfer depends on a mediating agent, usually wind or insects, and is almost inevitably limited in extent, often to surprisingly short distances (Levin, 1979). Hence pollen competition is virtually always localized, although to varying degrees.

Assume initially that a cosexual population is self-incompatible.

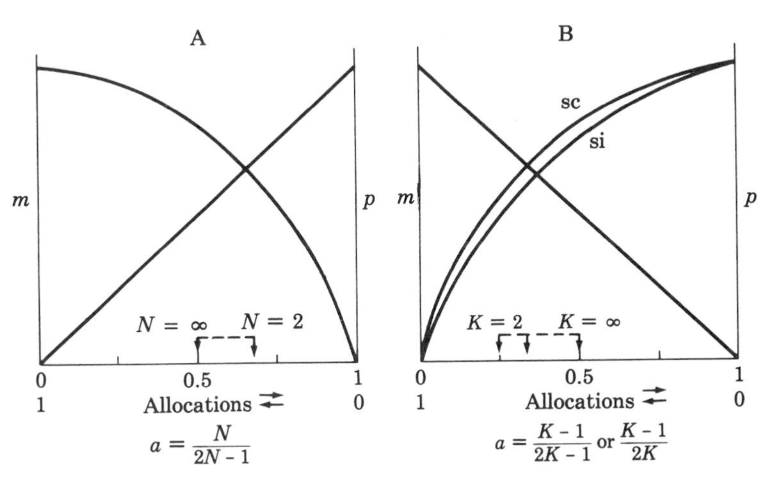

FIGURE 2. Graphs of paternal and maternal allocations against the appropriate fitness, as in Figure 1. A. Local resource competition. B. Local mate competition for self-incompatible (si) and self-compatible (sc) populations.

Pollen is transferred from each plant to K mates equally, where $K < \infty$. A single mutant phenotype 2 in a population of phenotype 1 competes in a pollen pool,

$$\sum_{k=1}^{K} ga_k r_p = gr_p[(K - 1)a_1 + a_2]$$

and

$$c_2 = a_2/[(K - 1)a_1 + a_2]$$

If the survival of seeds is density independent,

$$w_2 = n(1 - a_2)r_m + Kn(1 - a_1)r_m \left[\frac{a_2}{(K - 1)a_1 + a_2}\right]$$

When

$$\left.\frac{\partial w_2}{\partial a_2}\right|_{a_1 = a_2 = a} = 0$$

then

$$\frac{a}{1 - a} = \frac{K - 1}{K} \tag{6}$$

the result Charnov (1980) obtained for barnacles. The second derivative is always negative, so the equilibrium is fully stable. The deviation toward an emphasis on maternal investment is slight unless K is very small. Even for $K = 5$, $a = 0.44$. Note that it is the number of competing individuals, not the number of competing male gametes, that determines the ESS.

If plants are self-compatible, paternal fitness diminishes slightly faster as a increases because a plant has fewer (of its own) female gametes to fertilize (Figure 2B). Here c_2 is the same as for self-incompatible plants, but if a pollen parent competes for the eggs of K mates, including itself

$$\sum_{k=1}^{K} m_j = [(K - 1)(1 - a_1) + (1 - a_2)]nr_m$$

Proceeding as before gives,

$$\left.\frac{\partial w_2}{\partial a_2}\right|_{a_1 = a_2 = a} = 0$$

when

$$\frac{a}{1 - a} = \frac{K - 1}{K + 1} \tag{7}$$

The ESS allocation emphasizes maternal investment slightly more in self-compatible populations. For example, when $K = 5$, $a = 0.40$. The result is formally equivalent to the sex ratio in dioecious organisms with local mate competition (Hamilton, 1967).

There has been considerable controversy about the nature of local mate competition. To achieve local mate competitions in mobile animals, a population must somehow be subdivided each generation into spatially delimited subpopulations, with mating (but not subsequent dispersal) occurring within each subpopulation. Thus, temporary "groups" are set up, within which there is considerable inbreeding among sibs. Local mate competition in such specially structured animal populations has been attributed to inbreeding or sib-mating (Hamilton, 1967; Maynard Smith, 1978a; Borgia, 1982) or group selection (Colwell, 1981).

The preceding models for plants reveal the nature of local mate competition clearly. Plants are sessile and require an external agent for fertilization. Hence local mating is a normal event in plants, even in a completely homogeneous population. The models of local mate competition in plants do not contain any element of population subdivision, even for a single generation, or any element of inbreeding or sib-mating. Consequently the deviations in gender allocations cannot be caused by any of these factors. Instead, local mate competition operates through a restriction in the number of matings available to pollen donors, which causes pollen grains from the same plant to compete with each other. The paternal fitness curve therefore decelerates, whereas there is no such restriction on maternal fitness. Thus, the allocation at which paternal and maternal fitness gains balance out ($\partial m_2/\partial a_2 = -\partial p_2/\partial a_2$) is reached at a paternal allocation which is less than one-half. Taylor (1981), Charlesworth and Toro (1982), and Charnov (1982) have expressed similar views, primarily from reconsiderations of Hamilton's model.

It can readily be shown that Equations (6) and (7) also apply to density-dependent regulation of offspring numbers if the seeds from very many parents compete for each site. Then

$$m_1 = Nx \left[\frac{n(1 - a_1)r_m}{Nn(1 - a_1)r_m}\right] = x$$

$$m_2 \cong Nx \left[\frac{n(1 - a_2)r_m}{Nn(1 - a_1)r_m}\right]$$

and neither N nor x affects the ESS. In all real plant populations, however, seed dispersal occurs over finite and often very restricted distances (Levin, 1979), a situation producing a form of unilateral "local resource competition" (Clark, 1978) among the maternally derived offspring of a parent. Such local seed competitions are virtually universal among plants, although the degree of localization varies. We assume here that seeds that survive noncompetitive hazards are dispersed from their mother evenly over an area such that they compete with the seeds of N other parents at every site. (The assumption is not entirely realistic, but it is the simplest formulation that shows the basic nature of local seed competition.) Then, $m_1 = x$ as before, and

$$m_2 = Nx \left[\frac{n(1 - a_2)r_m}{n[(N - 1)(1 - a_1) + (1 - a_2)]r_m} \right]$$

If we assume in the simplest case that the various seed shadows of the mates of a pollen parent do not overlap and that pollen competition is widespread ($K \rightarrow \infty$) (Figure 2B), then $c_2 \cong a_2/Ka_1$. By substituting in (2), we find when

$$\left. \frac{\partial w_2}{\partial a_2} \right|_{a_1 = a_2 = a} = 0$$

$$\frac{a}{1 - a} = \frac{N}{N - 1} \tag{8}$$

The second derivative is negative. When a mate's seed shadows do not overlap, the effects of local seed competitions are exactly the reverse of the effects of local pollen competitions in self-incompatible populations. Local seed competitions are similarly due to limited opportunities and the consequently diminishing returns for one function only, not to group selection or inbreeding.

Since local pollen and seed competitions are both virtually universal, we should consider their joint effects. These depend on the extent to which various seed shadows overlap. Only the simplest situation is considered here, where the seed shadow of a plant and those of the plants it fertilizes do not overlap at all. Then m_1 and m_2 are as above for local seed competition alone, and $c_2 = a_2/[(K - 1)a_1 + a_2]$, as above for local pollen competition by itself. The same procedure as before finds that the optimal allocations occur when

$$\frac{a}{1 - a} = \left[\frac{K - 1}{K} \quad \frac{N}{N - 1} \right] \tag{9}$$

The effects of the two local competitions are multiplicative when they operate quite independently of each other (cf. Lloyd, in press).

In natural plant populations, local seed and pollen competitions will tend to cancel each other out. If $N = K$, the paternal and maternal allocations are equal (with nonoverlapping shadows), regardless of how restricted pollen and seed dispersal are. The local competitions are not likely to cause a wide deviation from equal allocations in many natural populations. Nevertheless, they are important factors in cosexual populations because they stabilize the gender allocations.

INTERACTIONS BETWEEN MATERNAL AND PATERNAL FUNCTIONS

The models up to this point have assumed that the paternal and maternal functions on a plant operate independently of each other, but this need not be so. We suppose here that either function interferes with or facilitates the other to a degree that depends on the gender allocations. More particularly, we assume that an interaction increases as the allocation to the activating function increases and that to the affected function decreases.

Interference between paternal and maternal functions is considered first. The presentation of pollen may interfere with seed production in several ways; for example, by clogging the stigmas with self-pollen so that access for cross-pollen is reduced (Bawa and Opler, 1975; Lloyd and Yates, 1982). Assume that an increasing pollen allocation causes a linearly increasing interference with the maternal function, that is, $m_i = 1 - Ia$, where $I > 1$ (Figure 3A). If pollen and seeds are dispersed widely and the success of the seeds is density independent, the fitness of a rare mutant,

$$w_2 = n(1 - Ia_2)r_m + Kn(1 - Ia_1)r_m[a_2/Ka_1]$$

Following the usual procedure, the ESS allocation is found to be

$$\frac{a}{1 - a} = \frac{1}{2I} \text{ or } a = \frac{1}{2I} \tag{10}$$

The allocation to the function that causes the interference is decreased in proportion to the degree of interference. If, for example, $I = 2$ (the maternal fitness decreases twice as fast with an increase in a as it would without interference), then $a = 1/4$.

Conversely, the maternal activities of a flower may interfere with the removal of pollen. If the structures contributing to maternal fitness interfere with the paternal function of a plant so that the effective pollen production is $g[1 - J(1 - a)]$, the ESS is

$$a/(1 - a) = (2J - 1)/1 \text{ or } 1 - a = 1/2J \tag{11}$$

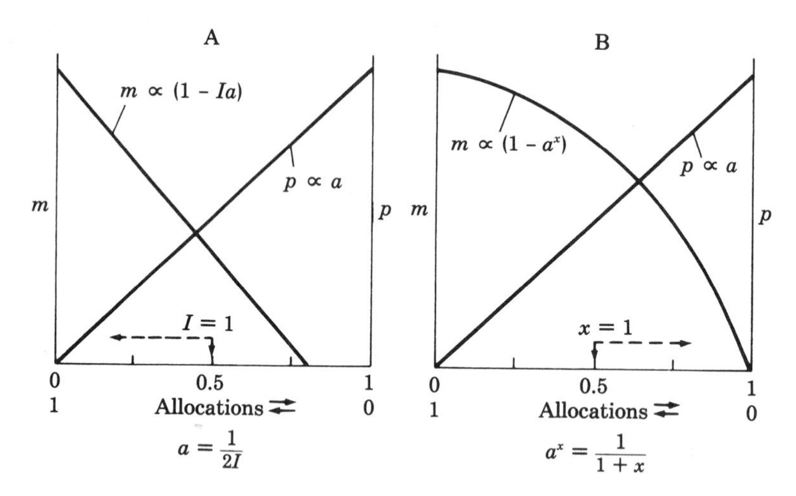

FIGURE 3. Graphs of paternal and maternal allocations against the appropriate fitness, as in Figure 1. A. The paternal allocation interferes with maternal fitness. B. The paternal allocation facilitates maternal fitness.

If both functions interfere with each other, the ESS is

$$a/(1 - a) = (JI + J - I)/(JI - J + I) \tag{12}$$

Unlike the joint effect of local pollen and seed competitions considered earlier, the effects of mutual interference are not the product of the two single effects because the two types of interference do not vary independently. They both depend on the gender allocations.

Any interference between paternal and maternal activities decreases the total fitness of the parent. Plants have evolved a variety of ways of reducing interference, including monoecy, dichogamy, herkogamy, and possibly heterostyly (Bawa and Opler, 1975; Lloyd and Yates, 1982). Consequently interference may have only a minor effect on gender allocations in most angiosperms.

Either parental function may facilitate the other. The paternal allocation may assist maternal fitness if pollen is an attractant or reward for visitors (see Coleman and Coleman, 1982; Thomson et al., 1982), particularly if maternal fitness is pollinator-limited. The reverse interaction, a maternal investment that boosts pollen fitness, is probably uncommon since the bulk of the expenditure on maternal fitness usually occurs when fruits mature, after pollen donation by a flower is completed (Lloyd, 1980b).

To examine the effect of pollen facilitating maternal fitness, suppose that $m_i = n(1 - a_i^x)$, where $x > 1$ (Figure 3B). (Note that $x < 1$ gives an alternative formulation of interference.) If pollen and seeds compete very widely and the success of offspring is density independent, the fitness of a rare mutant,

$$w_2 = n(1 - a_2^x)r_m + Kn(1 - a_1^x)r_m[a_2/Ka_1]$$

Fitness is maximized when

$$\frac{a^x}{1 - a^x} = \frac{1}{x} \text{ or a}^x = \frac{1}{1 + x} \tag{13}$$

The function that assists the other is slightly emphasized. For example, if $x = 2$, $a = 0.58$. The deviation occurs because an increase in the maternal allocation lowers the paternal investment and hence the amount of facilitation. As a result, the maternal fitness curve decelerates and the paternal and maternal gains cancel each other out while the maternal investment is less than one-half.

The model for facilitation provides one approach to the difficulty raised earlier of partitioning partly bilateral attractants and rewards, particularly when these are the pollen grains (or ovules or seeds) which contribute directly to parental fitness.

VARIOUS COSTS OF REPRODUCTION

Several types of costs may be distinguished on the basis of how fitness increases as the costs are outlaid and whether the costs are unilateral or bilateral.

Production costs

Some unilateral costs are constant for each offspring invested in, regardless of the size of the appropriate allocation. The models presented in the preceding sections have incorporated these "production costs" by specifying the number of structures (n or g) that are produced at the maximum allocation. Production costs may include indirect costs that are not part of the seeds or pollen grains themselves, such as anther walls or fruit tissues, provided these costs are proportional to the number of pollen grains or seeds produced.

We have already seen that production costs do not affect the allocation strategy that is selected.

Fixed costs

Unilateral fixed costs were first recognized by Heath (1977). He considered costs that are incurred once only before any reproduction is possible and that are not added to subsequently. The reproductive ducts of animals are examples of fixed costs. Fixed costs may be unilateral maternal costs F_m or paternal costs F_p or bilateral costs F_b (such as those of peduncles and pedicels, prior to anthesis). Suppose that a cosexual plant has total resources R available for reproduction and that each seed costs k_m units to produce. Then $n = [R - F_b - F_m - F_p]/k_m$. Similarly, if each pollen grain costs α_p units to produce, $g = [R - F_b - F_m - F_p]/k_p$. Since neither n nor g modify the ESS allocations, it is apparent that fixed costs do not alter the allocations. Unilateral fixed costs reduce the stability of cosexuality, however, as Charnov (1979) pointed out.

Plants characteristically produce many reproductive structures (inflorescences, flowers, and fruits), whereas animals usually have only one or a few sex organs. Fixed costs are, therefore, less important in plants generally than in animals. There may, however, be considerable fixed costs in plants with a single large inflorescence, such as some palms.

Recurrent costs

In plants, recurrent costs that are repeated more than once but less often than for every offspring (e.g., every flower or fruit) largely replace the fixed costs that occur in animals. Recurrent costs, like fixed costs, must be outlaid in full before any reproduction is possible in the structure concerned. For example, fruit structures such as capsule walls or the fleshy tissues of squashes, apples, or citrus fruits may be required in full before even one seed can be dispersed. If the fruits contain more than one seed, these costs are recurrent in the sense used here. Recurrent paternal costs are unlikely to be high since paternal investment ends at anthesis and the costs of filaments and anther walls are relatively slight.

Bilateral recurrent costs (such as those of floral defenses against predators or pedicel costs prior to anthesis) have no effect on gender allocations. Like bilateral fixed costs, they simply reduce the resources available to be partitioned between maternal and paternal functions.

To consider a unilateral recurrent cost of paternity, assume that there is a lower limit of paternal allocation a_r below which there is no paternal fitness and above which paternal fitness increases linearly (Figure 4A). For any $a_i > a_r$,

$$c_i = \frac{g(a_i - a_r)r_p}{\sum_{k=1}^{K} g(a_k - a_r)r_p}$$

If seed and pollen compete widely and the success of seeds is density

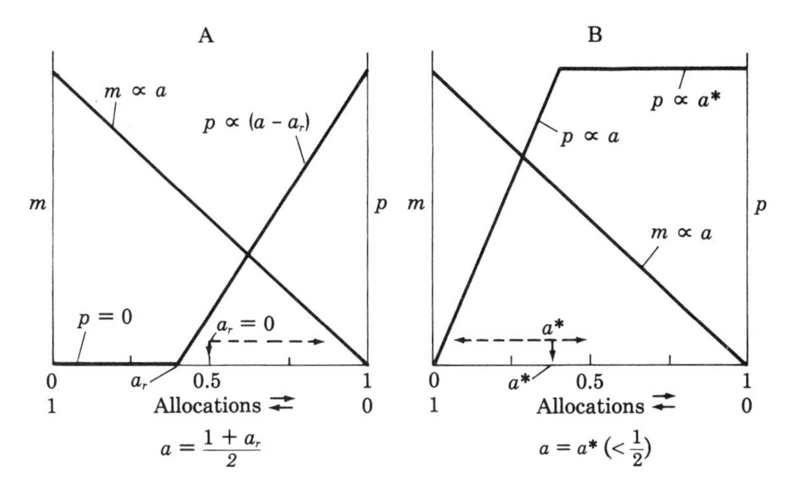

FIGURE 4. Graphs of paternal and maternal allocations against the appropriate fitness, as in Figure 1. A. Recurrent paternal cost a_r. B. Upper limit to paternal fitness is reached at a^*.

independent, the fitness of a rare mutant,

$$w_2 = n(1 - a_2)r_m + Kn(1 - a)r_m \left[\frac{g(a_2 - a_r)}{Kg(a_1 - a_r)}\right]$$

Fitness is maximized when

$$\frac{a}{1 - a} = \frac{1 + a_r}{1 - a_r} \text{ or } a = \frac{1 + a_r}{2} \tag{14}$$

Recurrent costs, in contrast to fixed or production costs, affect the gender allocations. The allocation to the function that incurs a recurrent cost is increased by half the cost.

Note that if $a_i \leq a_r$, an individual is functionally unisexual, succeeding only as a mother. Moreover, $\partial w_1/\partial a$ is then negative, so selection will oppose an increase in the paternal allocation. Hence a high recurrent cost can destabilize cosexuality and cause selection for unisexuality, in the same way as a unilateral fixed cost can (Heath, 1977; Charnov, 1979).

If there is a recurrent cost for maternal activities such that there is no maternal fitness until the maternal fitness rises above $1 - a_s$ and maternal fitness then increases linearly, the ESS allocation can be shown to be at

$$\frac{a}{1 - a} = \frac{1 - a_s}{1 + a_s} \text{ or } a = \frac{1 - a_s}{2} \tag{15}$$

If both parental functions incur unilateral recurrent costs, the ESS allocation is given by

$$\frac{a}{1 - a} = \frac{1 + a_r - a_s}{1 - a_r + a_s} \tag{16}$$

Whichever function has the higher cost receives the larger allocation.

AN UPPER FITNESS LIMIT

Here we consider the reverse of recurrent costs, an upper limit of an allocation beyond which fitness does not increase. For simplicity, fitness is assumed to increase linearly to the limit, although a gradual slowing of fitness gains would probably be more realistic in most instances.

An upper limit to paternal fitness could occur, for example, if a wasteful (inefficient?) pollinator picks up only a restricted amount of pollen but dislodges the rest, no matter how much is present in a flower. Let the limit in paternal fitness occur at the allocation a^*. Then, if pollen is dispersed widely, when $a_i, a_k < a^*$,

$$c_i = \frac{g a_i r_p}{\sum_{k=1}^{K} g a_k r_p} = \frac{a_i}{\sum_{k=1}^{K} a_k}$$

and selection proceeds normally, to $a = 1/2$ when fitness gains are linear.

When $a_i, a_k \geq a^*$,

$$c_i = \frac{a^*}{\sum_{k=1}^{K} a^*} = \frac{1}{K}$$

If pollen and seeds are dispersed widely and there is no density regulation, the fitness of a rare mutant is then

$$w_2 = n(1 - a_2)r_m + \frac{Kn(1 - a_1)r_m}{K}$$

$$\therefore \partial w_2 / \partial a_2 = -nr_m$$

As $a_2 (\geq a^*)$ increases, the fitness of the mutant decreases. That is, selection drives the allocation back toward a^*.

When the upper limit for paternal fitness a^* occurs at a paternal allocation less than one-half, the derivative $(\partial w_2/\partial a_2)$ is positive if $a < a^*$ and negative if $a > a^*$. That is, there is a stable allocation at

$$a = a^* \tag{17}$$

A unilateral upper limit to fitness does not tend to destabilize cosexuality, as fixed and recurrent costs do. An upper limit to fitness has a stronger effect than a recurrent cost on the ESS allocation. There is no theoretical limit to how low an allocation can be held in this way.

Now suppose there is an upper limit on maternal fitness at $(1 - a)^*$. By parallel reasoning, when $(1 - a)^* < 1/2$, there is an ESS at

$$1 - a = (1 - a)^* \tag{18}$$

While this chapter was in preparation, an identical argument for the effect of an upper maternal limit was proposed by Charnov (1982), formalizing the hypothesis of Heath (1979) to explain hermaphroditism in animals which brood offspring and have limited space in the brood pouch.

It would probably be more realistic in most instances to suppose that an upper limit to paternal or maternal fitness is approached by a gradual slowing down of fitness gains, rather than being reached abruptly. A model for this situation would be considerably more complex, and none will be considered here. It may be noted, however, that a gradual approach will cause an even greater deviation from equal allocations because the ESS will occur at the point where the paternal and maternal fitness gains cancel each other. In the case of a paternal limit, the ESS allocation will occur at a paternal allocation less than a^*, the point where there is *no* further increase in paternal fitness.

DISCUSSION

The diverse factors that modify the position and stability of gender allocations in cosexual plants may be divided into two general groups. Some factors affect the relative number of opportunities to be gained from expenditure on paternal and maternal activities, represented by the height or average slope of the fitness curves. The only "height" parameter considered here is the almost ubiquitous feature of sexual reproduction, that male and female gametes contribute genes equally to each zygote. In the absence of other factors, the equal contributions cause equal allocations (Maynard Smith, 1971). This provides the reference point for evaluating deviations caused by other factors. The occurrence of self-fertilization, which decreases the paternal allocation at the ESS (Charlesworth and Charlesworth, 1981), can also be formulated in terms of height parameters, the relative opportunities obtained from expenditure on seeds and outcrossing pollen (D. G. Lloyd, unpublished model). Charnov (1982) gives a rather different interpretation.

The remaining factors considered above affect the shape of the paternal or maternal fitness curve, influencing either the curvature over the full range of allocations or causing a restriction in the range over which fitness increases. We must now assess the importance of the various "shape" factors as determinants of unequal allocations in natural populations of angiosperms. The meager information on natural allocations that is available at present, supported somewhat uncertainly by data on diverse pollen:ovule ratios in outcrossing species (e.g., Pohl, 1937; Cruden and Jensen, 1979), indicates equal allocations or deviations of varying extent toward a greater emphasis on maternal expenditure.

Among the shape factors described earlier, local pollen and seed competitions are not likely to cause wide deviations from equal allocations, and they frequently tend to cancel each other out. In theory, interference of one function by the other can cause any degree of bias in the allocations, but it lowers the total fitness of individuals, and a variety of mechanisms have evolved to reduce interference (Bawa and Opler, 1975; Lloyd and Yates, 1982). Facilitation probably acts most frequently when pollen attracts visitors to flowers and thereby facilitates maternal fitness as well as contributing to paternal fitness. The ESS bias that results is toward a heavier paternal allocation and therefore does not help to explain the observed deviations.

Unilateral recurrent costs are probably most frequently incurred in the production of fruit and the maintenance of ancillary structures such as pedicels during seed maturation. The resulting allocation deviates toward a higher maternal investment, as observed in several species. But only one-half the cost is recouped and the costs are not likely to be high except in fleshy fruits, which occur in a considerable proportion of angiosperms.

An upper limit on paternal fitness offers the most promise of explaining the observed deviations emphasizing maternal expenditure. There is no limit to the deviation possible. Paternal fitness limits can be imposed (or overcome) in three general ways:

A limited number of visits to each flower that result in pollen pickup. In the process of picking up a fraction of the pollen on a flower, many animal visitors cause much of the remaining pollen to be lost from the surfaces where it is presented, either because they eat pollen (e.g., syrphid flies) or collect it or because they dislodge it and scatter it wastefully (e.g., "mess and soil" pollinators; Faegri and van der Pijl, 1979). On the other hand, very precise pollinators may remove so little pollen that even numerous visits require only a limited paternal expenditure. Hence both grossly "inefficient" and supremely "efficient" pollinators may limit the quantity of pollen that is profitably produced. Again it is evident that it is not the efficiency of pollination as such that controls gender allocations. It is the way in which pollinator actions affect the shape of the paternal fitness curve that is important.

Other pollinators may remove only moderate amounts of pollen during a visit but render a flower unattractive to subsequent visitors because nonrenewable rewards are exhausted or the integrity of the floral pattern is damaged (many beetle or bird visits or explosive blossoms). The number of visitors to still other flowers may be restricted by the time a flower can remain open, either because of the theft of pollen or nectar by nonpollinating visitors or because changes occur in floral parts after pollen is deposited on the stigmas.

Any of these factors may limit the number of visits during which pollen is removed from a flower. These forces are evidently widespread because flowers possess a variety of mechanisms which increase the number of pollen-donating visits, including antitheft devices (Faegri and van der Pijl, 1979) and the parsimonious presentation of pollen in small packages (staggered dehiscence of anthers, poricidal anthers, etc.; Lloyd and Yates, 1982).

Limits to the number of fertilizations resulting from one visit. In a few plants the pollen grains are removed *en masse* while they are held together by viscin threads or in pollinia (Willson, 1979; Cruden and Jensen, 1979). Willson postulated that intrasexual selection in these species had promoted an increase in the number of pollen grains transferred at one time. This strategy will be prohibited when it adversely reduces the number of visits during which pollen can be removed. Moreover, if the number of pollen grains that are transported together exceeds the number of ovules to be fertilized in an ovary, a further increase in pollen will often result only in more intense competition between pollen grains from the same donor and will not increase the fitness of the parent (assuming there is no pollen carryover). There will be an upper limit to the number of pollen grains that are profitably removed in one mass, even in species with many ovules per fruit, as in the Orchidaceae and Onagraceae. The strategy is unlikely to succeed elsewhere.

Limits on the number of polliniferous flowers. Where there are limits on the number of pollen grains that can succeed from one flower, the paternal fitness may be improved by increasing the number of flowers that donate pollen. Andromonoecy and low-fruiting hermaphroditism, both of which allow more pollen flowers than fruit, attest to this means of increasing paternal fitness. There may be limitations imposed on the number of pollen flowers, too, for example, by increased structural expenses or a rise in the frequency of geitonogamous self-fertilizations.

In total, there are many factors that may impose a limit on profitable paternal allocations in animal-pollinated species. There is insufficient information at present to determine how far these limits are responsible for the low paternal investment observed in some angiosperms.

The hypothesis that paternal fitness limits have selected low paternal allocations in some angiosperms gains support from the pattern of observed gender allocations shown in Table 1. All the paternal fitness limits described earlier operate only in animal-pollinated species. In the process of wind pollination, pollen is simply released into the surrounding air and travels to stigmas in an undirected manner. Pollen is not lost by the operation of the pollinating agency itself, and the wind is not limited in the amount of pollen it can carry. Hence one would not expect pollen fitness to become saturated, as it frequently may do in animal-pollinated species. The fact that wind-pollinated species appear to have higher paternal allocations (Table 1) and higher pollen:ovule ratios (Pohl, 1937; Cruden 1977) conforms with the hypothesis that low paternal allocations may be selected in animal-pollinated species because there is an upper limit to pollen fitness. More data are needed, however, before we will be able to evaluate the importance of paternal fitness limits and other factors, such as recurrent maternal costs, as causes of unequal gender allocations.

The imposition of paternal fitness limits will generate selection pressures to overcome these limits and increase paternal fitness. It will be proposed elsewhere that intrasexual selection to increase the proficiency of pollen donation, particularly *the number of visitors that can remove pollen from a flower*, is the major selective force guiding floral evolution. Selection for paternal proficiency may explain the ecological trend toward increasing precision of pollination (described in Faegri and van der Pijl, 1979) and morphological trends toward reduction and fusion of floral parts and zygomorphy, as well as reversions or halts in these trends.

SECTION III
PLANTS AS INTEGRATED ECOPHYSIOLOGICAL UNITS

INTRODUCTION TO SECTION III

Science seems to proceed by recurrent processes of convergence to and radiation from points of interrelation among related or complementary disciplines. This book, and particularly this section, states repeatedly the need to intersect the fields of plant population ecology, genetics and ecophysiology. Yet, one can trace the origins of ecophysiology to the earliest attempts at formal synecological studies of Schimper, Du Rietz and others where very detailed phytogeographic information was set against the backdrop of several components of the environment—particularly climate—in an effort to relate and explain different morphological traits of plants. The comparison of plant formations in similar climates but geographically distant areas, or in highly contrasting climates, was a usual procedure to explore this area of broad ecological-physiological relations.

The seminal work of Clausen, Kieck and Hiesey, on the experimental study of variation of plant populations in relation to the different environments they experience, went to the core of the problem that the early phytogeographic approach was trying to address. These studies inspired many researchers in the fields of population genetics and population ecology. However, on the whole the new wave of studies stimulated by the Carnegie Institution group became too punctilious in the attempt to explain the underlying bases of plant variation in ever finer detail, losing the valuable broader approach of the original studies.

The three chapters that constitute this section coincide in stressing both (1) the need to understand the physiological basis of individual responses in survivorship, growth and reproduction that constitute the demographic features of plant populations; and (2) the need to study physiological responses at the level of whole organisms—being concerned not only with how a given physiological mechanism operates but also with how this determines the efficiency with which an organism performs in a given environment.

Up until now the understanding of environmental and evolutionary constraints on photosynthesis has been derived by research done at levels of organization ranging from the molecule to the whole leaf, mostly in highly idealized environments. Mooney and Chiariello (Chapter 15) stress that there are important variations in performance within and between plants in a 24-hour period, and these cannot be dealt with by studies on single leaves. These variations create important problems in attempts to link existing plant physiological information to population biology. Physiological models based on single-leaf behavior are unsatisfactory when dealing with the situations confronted by a whole plant immersed in the complex matrix of factors which constitute its natural environment.

Mooney and Chiariello show the plant as an integrated and balanced system in which certain mechanisms, such as those responsible for the acquisition of resources, can be considered over the entire range of levels, from the molecule to the whole plant. Processes that need to be considered in isolation can be nested to provide a description of whole plant responses. Resource (carbon) balance and optimization models are proposed as tools to understand whole plant responses. Patterns of resource allocation, especially between growth and reproduction, are still largely unknown.

The application of plant ecophysiology to population biology is enormously rich in approaches. The use of leaf life tables to assess leaf contribution to the growth of the population of modules that constitute an individual plant (Bazzaz, Chapter 16) is a representative example. The study of interdependence between plant parts and ramets is particularly important in order to assess the internal structure and integration of the plant unit.

It is clear that the same physical conditions determine sharp differences in the performances of individuals of different species, particularly in regard to demographically meaningful performance. Bazzaz shows that these responses may even alter sexual proportions in populations and consequently have a considerable impact on the genetic structure of a population. The differences between species' responses to environmental factors which affect demographic parameters may also be present at the intrapopulational level. The impressive degree of polymorphism present in many species need not be restricted to morphological and behavioral traits; very likely it includes similar variability of physiological responses to factors such as light level, water deficit, mineral availability, etc.

Jeffries (Chapter 17) shows that the plastic responses of individual organs in plants are often regulated by the long distance transport of metabolites and growth substances within the plant. He also draws our attention towards assessing the effects of the environment on the control of phenotypic expression, which is still very poorly understood. This also applies to the definition of the ontogenic constraints limiting the boundaries of phenotypic expression.

Plant demography has reached a point where the differences in responses encountered at the individual level cannot be explained if one cannot account for them in physiological terms and assess how much of a given physiological trait is genetically determined.

CHAPTER 15

THE STUDY OF PLANT FUNCTION—THE PLANT AS A BALANCED SYSTEM

Harold A. Mooney and Nona R. Chiariello

INTRODUCTION

Plant function is studied at levels ranging from the molecule to the whole plant. Physiologists generally investigate function at levels ranging from molecules to organs. Physiological ecologists often extend this scope of inquiry to organ systems or even to groups of individuals. The physiological ecologist is concerned not only with how a particular mechanism operates but also with the efficiency of a given mechanism in a particular environment. The evolutionary biologist, on the other hand, is concerned principally with assessing the fitness of individuals. Often information derived from lower levels of integration is used to infer fitness or to assess controls on fitness.

An interchange of concepts and techniques among biologists operating at these different levels of integration is obviously important in the development of a comprehensive understanding of the basis of the success of a plant in a given environment. The thesis we develop in this chapter is that although it is essential to utilize information from many levels in order to derive a mechanistic understanding of plant function, this information must be integrated into a whole plant context in order to avoid inappropriate conclusions regarding the functional significance of a given trait. The viewpoint presented is that of a physiological ecologist. Particular emphasis is placed on the approaches utilized in physiological ecology and on their possible application in population biology.

A great amount of work in physiological ecology the past two decades has focused on leaf-level phenomena. For example, much of the progress in our understanding of environmental and evolutionary constraints on photosynthesis has been at the whole-leaf level (as well as at the molecular level). The development of energy balance analysis (Gates, 1962; Raschke, 1956) has emphasized the interaction between the environment and the leaf, rather than the whole plant, and a very detailed and robust set of models has been used to examine leaf shape (Vogel, 1970; Balding and Cunningham, 1976), color (Mooney et al., 1977), pubescence (Ehleringer et al., 1976), orientation (P. C. Miller, 1967), size (Smith, 1978), and outline (Gottschlich and Smith, 1982). Recently, the interaction between mass and energy exchange has been explored by linking biophysical models of energy balance with biochemical, biophysical, and empirical models of photosynthesis (Cowan and Farquhar, 1977; Ehleringer and Mooney, 1978; Williams, 1983) (Figure 1). From careful measurements made on single leaves, it is now possible to predict fairly accurately the effect of varying leaf microclimate on leaf temperature, photosynthesis, and transpiration. Moreover, we can quantitatively assess the effects on carbon gain of specific leaf characters and can compare their effects with those of hypothetical, alternative characters. These two applications of physiological models greatly expand comparative ecology to include variation that cannot be studied through experimental manipulation. From models of light extinction in canopies (Monsi and Saeki, 1953), leaf photosynthesis can also be translated into canopy production (Duncan et al., 1967).

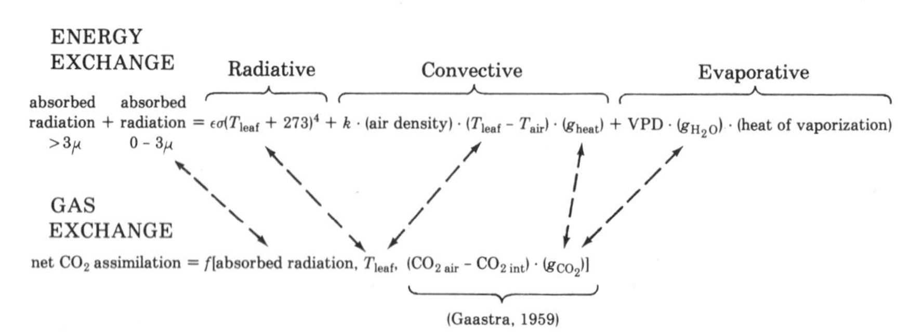

FIGURE 1. Direct linkages between the determinants of gas exchange and energy exchange at the leaf level. Many additional interactions exist (e.g., between vapor pressure deficit and conductance).

The principal questions of population biology, however, generally cannot be answered by single leaf measurements, no matter how elegant or detailed they may be. Each leaf on a plant has a unique history and lives in a distinctive microclimate. Large differences occur both within and between plants in the daily course of physiological responses of leaves. This variation creates problems in attempts to link physiology and population biology. It confounds surveys of genetic variation based on single leaf measurements, and it introduces error in whole-plant models based on leaf behavior. Yet these two kinds of information (levels of genetic variation and the whole-plant consequences of genetic differences) are essential for understanding the physiological basis of individual differences that are ecologically and evolutionarily important.

Obtaining this information requires, we believe, a framework from which we can understand the integrating principles in whole plant function and adaptation to suggest evolutionary trade-offs, much in the same sense that leaf models have linked energy, water, and carbon dioxide exchange to explain photosynthetic efficiencies. To achieve these purposes, whole-plant models linking different levels of organization are needed. A goal of this chapter is to suggest how patterns of integration might be derived from models nesting various levels of organization. By integration we mean the consequences of time-integrated, interacting processes for whole-plant responses. We propose that central features of integration are likely to derive from the constraints imposed by one level of organization on another and from the changes in these constraints through time. These models may provide a framework for identifying physiological traits in plants that are diagnostic for different patterns of integration and compromise and may provide greater insight into the types of variation that may be most critical in a population.

THE PLANT AS A BALANCED SYSTEM

Integration among leaf-level processes includes direct coupling between energy and gas exchange and also feedback between these two processes. Similarly, whole plants have been viewed as systems balancing two interacting processes: carbon fixation by the shoots and uptake of water and nutrients by the roots. Patterns of coupling and feedback have been described for whole plants in studies on growth and allometry (Troughton, 1956; Ledig et al., 1970; Thornley, 1972; Wareing and Patrick, 1975). Although conceptually related, our focus is specifically on the study of balance in plants through the integration of patterns at different levels of organization. Leaf energy balance again provides an illustration. At the leaf level, there is feedback control of transpiration through the effects of water loss on leaf temperature and the effects of temperature on transpiration. Additional feedback occurs, however, if this leaf-level process is nested in whole-plant water transport. A leaf's transpiration rate can affect whole-plant water status, and this may modify the leaf-level feedback process.

Many types of acclimation and adaptation are characterized by a suite of interacting processes, occurring at different levels, rather than a single trait. Because acclimation at the leaf level is often embedded in responses at higher levels, the whole-plant response cannot be predicted purely from the response at the leaf level. In order to illustrate the range of levels and interactions contributing to whole-plant responses such as growth rate and seed set, we examine two cases of acclimation: adaptation to light and responses to grazing. In both cases, photosynthetic changes stand out as an important feature of the overall response. However, responses at higher levels (stem architecture, resource allocation) may be of equal magnitude. To understand the photosynthetic responses from the viewpoint of integrated plant function, these higher level responses must be considered.

Light acclimation

Plants adjust to their light environment through a whole syndrome of interacting processes (Boardman, 1977), involving mechanisms at all levels of organization (Table 1). The overall pattern results in a correlation between photosynthetic capacity and light intensity. In moving from low-light to high-light environments, chlorophyll content per unit area and electron transport capacity increase, enabling the plant to use the higher radiant energy. The extra chemical energy generated by the splitting of water in the light reaction is utilized by the increased production of ATP and NADPH, which facilitate ribulose bisphosphate (RuBP) regeneration in the dark reactions. Increased levels of RuBP carboxylase allow consumption of the increased RuBP. As more light energy is trapped and the potential for CO_2 fixation is enhanced, the CO_2 supply rate is increased via higher stomatal density and higher conductance of gases.

Increases in radiation alter the leaf energy balance and tend to increase leaf temperature, which can increase the ratio of water lost to carbon fixed. This tendency is counteracted by adjustments in leaf thickness, which reduce the ratio of leaf area to leaf volume (Mooney et al., 1977). Under high light, leaves may also be oriented at steeper angles, forsaking carbon gain at midday (when water-use efficiency is low) for carbon gain earlier in the day, when water-use efficiency is greater (Mooney et al., 1977). Decreases in leaf size and increases in

TABLE 1. Common adjustments at different levels of organization to an increase in the greater availability of radiation.

Level	Adjustment
Molecular	High electron transport capacity High RuBPase activity High RuBP concentration
Organelle	Smaller and fewer chloroplasts More stomata
Organ	Thicker leaves Angled leaves
Branch	High leaf turnover Short nodes
Plant	High root:shoot ratio High reproductive output

leaf angle affect the penetration of light through the plant canopy and are coupled to reductions in branch internode length. The combined changes result in a dramatically different branching pattern and canopy structure in sun and shade plants (Figure 2). Greater resource

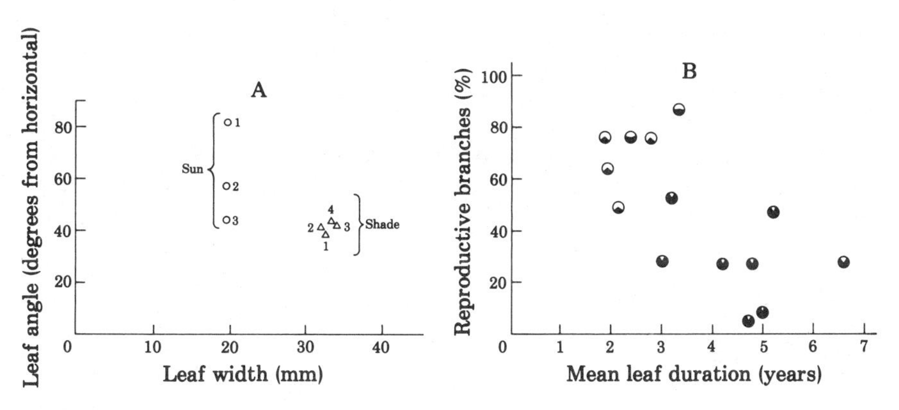

FIGURE 2. Leaf and reproductive characteristics of branches of the Californian evergreen shrub, *Heteromeles arbutifolia*, naturally growing in the sun or shade. Leaf duration and angle differ greatly among sites. A. Numbers give the age class (years) represented by each point. B. Each point represents the mean for five branches on a single plant. The unshaded fraction of the circle indicates the percentage of sky seen by each plant, as determined from fish-eye photographs. (From N. R. Chiariello, C. Field, H. A. Mooney, R. Pearcy, R. Robichaux, J. Keeney, and S. Tilley, unpublished data.)

acquisition by the sun plant may also result in a greater reproductive output.

The integrated outcome of adjustment to a particular light environment is likely to vary among species according to the manner in which a plant's morphology shapes the microenvironment of its leaves. Growth form responses to environment may be as important as the environment itself in determining the distribution of microsite types occupied by leaves. The joint analysis of leaf response to microenvironment and plant control over microenvironments offers a promising means of quantifying whole-plant patterns of acclimation. This kind of integration has been examined in a sun plant whose leaves have a life-history in which they are initiated under high light intensity but experience progressively lower light levels with time due to self-shading. In this case, leaf adjustments in photosynthetic capacity parallel changes in the light microenvironment, with the result that carbon gain per day exceeds that of a hypothetical nonadjusting leaf (Field, 1981).

Because net photosynthetic rate responds to light both instantaneously and developmentally, many ecological questions concerning carbon gain require a nesting of several levels of analysis. As discussed earlier, the instantaneous rate of carbon gain by a leaf can be predicted quite accurately if certain physiological characteristics are known. These characteristics reflect the integration of processes occurring at the various levels of organization that we have discussed. Although our understanding of adaptation to light has been based mainly on comparisons of plants grown in different light environments, the same adjustments occur when the light environment of a single plant is changed. Perturbation experiments such as this are more difficult to interpret because the responses occurring at different levels each may have a particular time course of readjustment. Examining only a single level of organization in isolation from its full network of interactions can lead to inappropriate conclusions, as can an analysis based on too short a time frame.

Defoliation and photosynthesis

Several studies indicate that grazing may enhance plant productivity (McNaughton, 1979). This contradicts simple models of plant growth, which assume that growth rate is a linear function of biomass or photosynthetic capital (Monsi, 1968). One hypothesis that accounts for this finding is that grazing frees plants from internal bottlenecks on growth-related processes, particularly by increasing sink strength relative to source strength. Several studies have, in fact, found that the photosynthetic rate of a leaf remaining on a plant increases after adjacent leaves are removed (Carmi and Koller, 1979; Ericsson et al.,

1980b; Painter and Detling, 1981; Detling et al., 1979). This response has been interpreted as a release from the limits on photosynthesis due to product accumulation. It has also been suggested that grazing improves the water-use efficiency of plants (McNaughton, 1979)—another mechanism that might increase productivity. These hypotheses are key issues in understanding the coevolution of plants and herbivores, of which grasses and grazers represent a special case that has received considerable attention (see Owen and Wiegert, 1981).

There are alternative explanations, however, for increases in photosynthesis following defoliation. Viewing herbivory as a disruption of the plant's integration provides a different framework for analysis, one which emphasizes the coordination of resource use among levels of organization. For example, one alternative explanation for photosynthetic responses to defoliation is that leaf removal disrupts the normal balance between the amount of canopy developed and the rate of nutrient uptake and supply [or hormonal supply from the roots to shoots (Carmi and Koller, 1979)]. Removal of leaves may result in a greater supply of nitrogen to the remaining leaves and an increase in the activity of photosynthetic enzymes. By a similar mechanism, partial defoliation is likely to relieve water stress to some degree, increasing photosynthesis by decreasing stomatal limitation (Ericsson et al., 1980b). These mechanisms may explain the compensatory growth observed in short-term experiments. Simultaneous regrowth from stored reserves may also contribute to the compensatory response (Ericsson et al., 1980b).

This framework also allows us to extrapolate from short-term responses at the leaf level to the effects on long-term productivity. In a long-term accounting of the effects of herbivory, the duration and degree of compensatory response must be weighed against the resources, especially nitrogen, lost through herbivory. Under herbivory regimes comparable to grazing, declines in productivity due to resource loss may well exceed the compensatory response (Wallace et al., 1982). Two mechanisms, however, could operate to increase productivity: an increase in the light available to photosynthetic tissues and an increase in the ratio of carbon gained to water lost. If dead or senescing tissue intercepts a significant fraction of incident light, then defoliation may cause an increase in the total light available for photosynthesis and an increase in productivity. This is likely to occur in grasses whose meristems are intercalary, rather than terminal, resulting in the shading of younger tissue by older tissue.

While defoliation may increase light availability, little evidence supports the hypothesis that the second mechanism (increased water-use efficiency following defoliation) is significant. In *Lolium multiflorum*, one of the few grasses that have been investigated, the short-term response to defoliation includes a decrease in stomatal resistance and a corresponding increase in photosynthesis, relative to undefoliated controls (Gifford and Marshall, 1973). Because water-use efficiency decreases monotonically with decreasing stomatal resistance, this short-term response to defoliation increases carbon gain but decreases water-use efficiency. This can be seen by calculating an index of water-use efficiency from stomatal and residual resistances:

$$\text{WUE} = \text{photosynthesis/transpiration}$$

$$\text{WUE} \propto \frac{r_{H_2O,\ gas}}{r_{CO_2,\ gas} + r_{CO_2,\ residual}}$$

In *Lolium multiflorum*, water-use efficiency decreased in the short-term response but then gradually rose to control values (Figure 3).

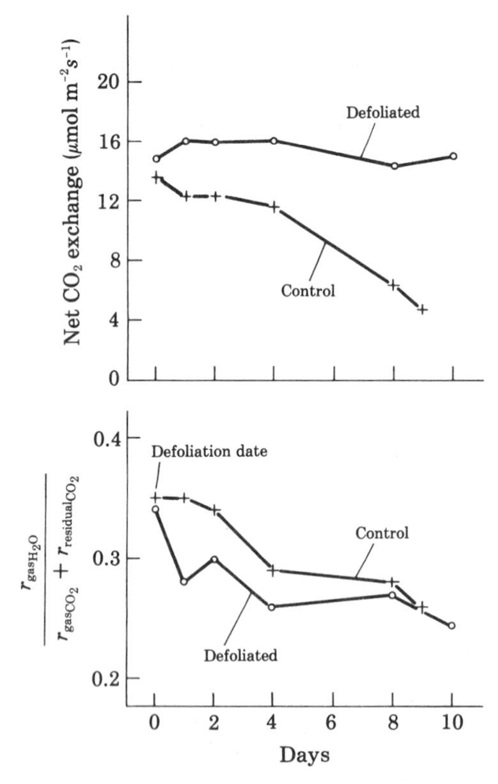

FIGURE 3. Changes in gas exchange characteristic of control and defoliated plants of *Lolium multiflorum*. Residual resistance represents nonstomatal leaf resistances; gas phase resistance includes stomatal and boundary layer resistances but is primarily stomatal resistance under the measurement conditions. (After Gifford and Marshall, 1973).

During this period, photosynthetic rates of control leaves decreased by two-thirds but remained constant on the partially defoliated plant. This suggests that the effect of partial defoliation was to arrest the effects of aging on photosynthesis but not the effects on water-use efficiency.

This pattern is only a part of the picture, even at the level of leaf gas exchange. Leaf age affected both the time span of the photosynthetic response in *L. multiflorum* and the movement of photosynthate produced by remaining tissues. Similar patterns are seen in crop species, which constitute the bulk of the literature on photosynthetic responses to defoliation and other forms of source–sink alteration. Increasing sink–source strength rapidly affects the movement of photosynthate produced by remaining tissues without altering photosynthesis in the short term (Fellows et al., 1979; Fondy and Geiger, 1980), but in some cases modifying it over the course of days (Thorne and Koller, 1974). The response is also affected by leaf age (Peet and Kramer, 1980).

Although the literature is weighted toward crop species, especially legumes, the combined results suggest that responses to defoliation occur at many levels of control and function and that the importance of photosynthetic responses can be assessed only within this larger context. For example, recent studies of differential grazing tolerance identify "rejuvenation" of leaf tissue as a significant factor but assign primary importance to responses in growth form and biomass allocation (Caldwell et al., 1981; Painter and Detling, 1981; R. Nowack, personal communication). These studies also emphasize the importance of different time frames of adjustment in different processes.

Carbon balance and growth

Studying whole-plant function as the integrated result of many levels of organization requires methods of integrating and quantifying feedback between levels. The carbon balance approach has been used at both the level of the whole plant and the level of the leaf and provides a meaningful currency for linking these two levels. Carbon gain can be measured with high precision, carbon allocation can be measured over both short and long time scales, and the carbon cost of building and maintaining various tissues can be calculated through biochemical pathway analysis (Penning de Vries et al., 1974), compositional analysis (McDermitt and Loomis, 1981), or gas exchange characteristics (Kimura et al., 1978).

Quantitative models linking carbon acquisition, allocation, and tissue costs can be used to predict the growth of plants through time.

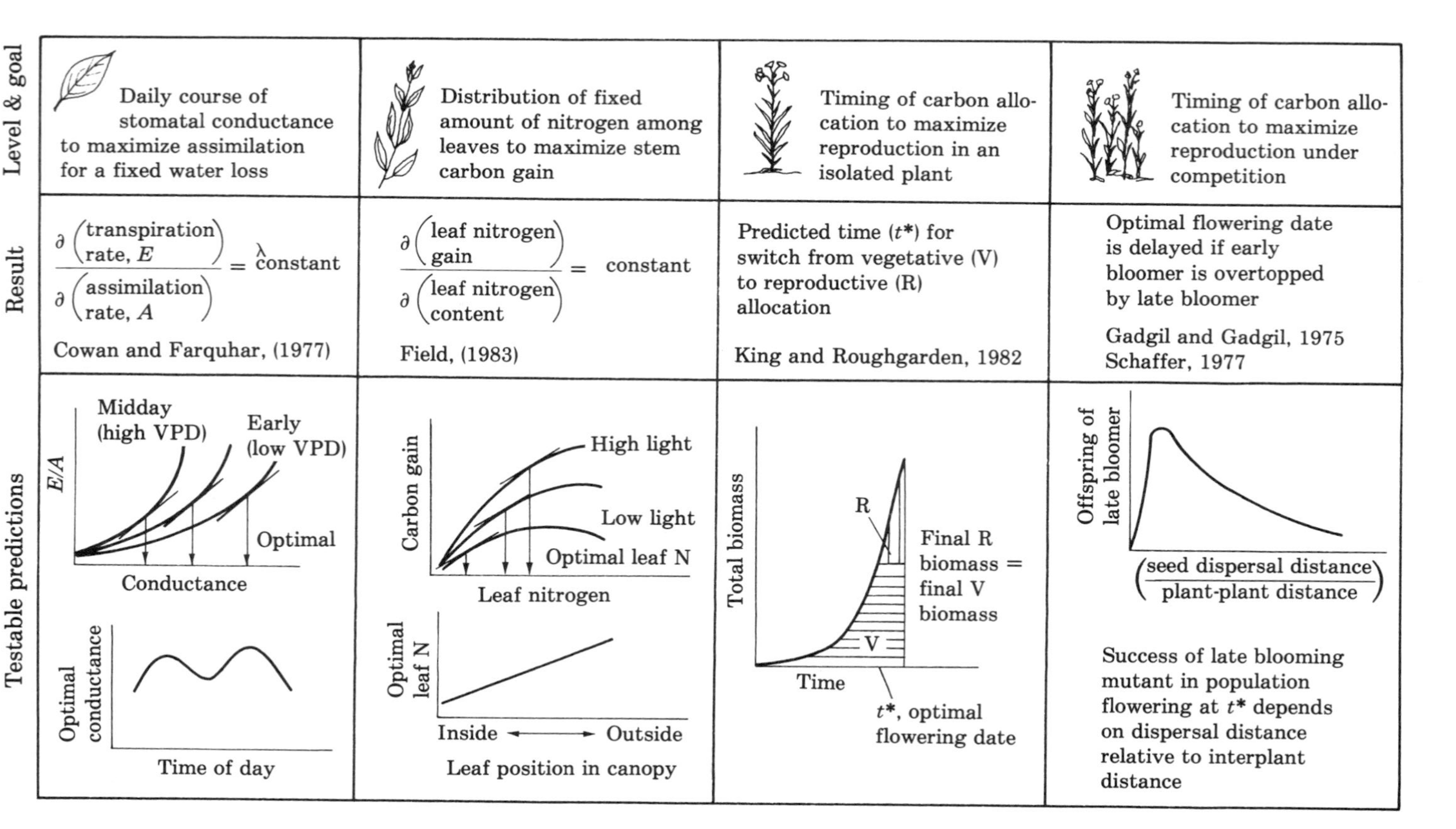

FIGURE 4. Levels-of-optimization models of plant adaptation.

Growth models have been derived for a variety of crop plants, including cotton, soybeans (Wann and Raper, 1979), sorghum (Vanderlip and Arkin, 1977), tobacco (Wann et al., 1978), and sugar beets (Fick et al., 1973). Growth models are needed in ecology and, in particular, could add necessary structure to optimization models of carbon gain, seed set, or other possible correlates of fitness. Cohen's (1966, 1971) optimization models for allocation pattern have stimulated a series of optimization models at levels ranging from the leaf to the whole plant (Figure 4). To population biologists, models at higher levels of organization will be most useful since they contain information on plant size and reproductive outputs. Whole-plant optimization models, however, generally have not contained enough physiological information to suggest the mechanistic basis of results or to provide the links between physiology and features of the life table. Combined with more realistic models of plant growth, as developed for crops, optimization models can be a framework for nesting models at various levels of organization and can provide tests for the whole-plant consequences of function at lower levels. Tools such as these may help elucidate the adaptive nature of different growth forms and the relationship between growth form and life history.

One principal limitation in model building is the derivation of rules for allocation patterns in plants. The importance of allocation pattern in determining plant productivity is clear from both experimental and theoretical studies. For example, compare the growth of commercial and wild radish, which have the same photosynthetic rate per unit of leaf area but differ in allocation pattern. After five weeks of growth, the wild radish accumulates nearly twice the dry weight of the cultivated radish (Figure 5). This results from greater allocation of carbon to leaves in the wild form. The converse may also be true: Selection for higher photosynthetic rates often does not translate into increased productivity as a result of accompanying changes in allocation pattern. As yet, there is little systematic information on the controls of allocation patterns in plants. Growth models have used empirically derived allocation patterns or have assumed allocation priorities in plants.

Existing growth models have generally been successful only during the vegetative phase of growth. A particularly important aspect of allocation is the nature of allocation changes during the phenological switch from the vegetative to the reproductive phase. Little is known about the controls on this allocation switch, yet it has major consequences for both competitive ability and reproductive success. Optimization studies have explored the potential trade-offs between

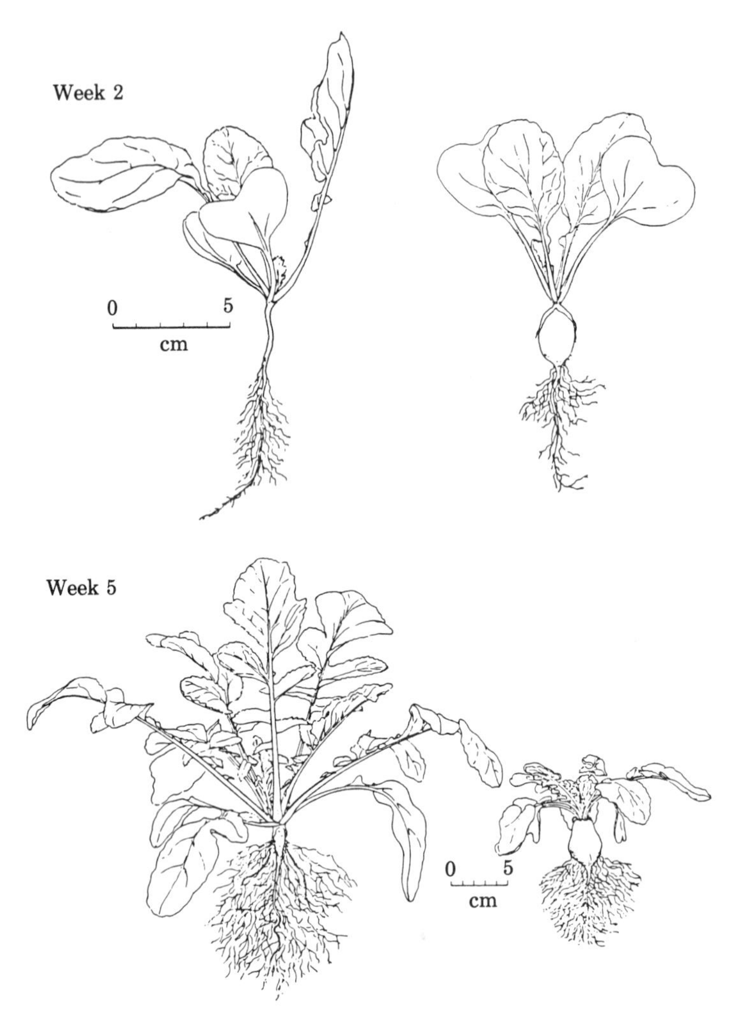

FIGURE 5. Even-aged wild (left) and cultivated (Cherry Belle, right) radish with different allocation patterns after two weeks and five weeks of growth.

reproductive activity and vegetative persistence, but this question needs further work that incorporates physiological information on resource costs. As outlined in the next section, a multiresource accounting may be very important to understanding the constraints underlying allocation rules, particularly rules for phenological switches.

Carbon and nitrogen interactions

The carbon balance approach has been criticized on the grounds that resources other than carbon may also limit plant growth and may be important evolutionary constraints on integration in plants. Nitrogen is a primary candidate because it appears to be limiting in most habitats. Nitrogen and carbon are actually linked constraints because leaf nitrogen content is a primary determinant of the light-saturated photosynthetic capacity (Figure 6).

Because carbon and nitrogen are harvested by separate spheres of the plant (shoot and root), the size and activity of one sphere must be commensurate with the size and activity of the other. Variation in nitrogen availability affects components of productivity at all levels. In the Californian shrub *Diplacus aurantiacus*, increased nitrogen availability results in increased nitrate uptake, leaf nitrogen content, and photosynthetic capacity (Gulmon and Chu, 1981); biomass allocation to roots decreases, as predicted by the model of Thornley (1972) (Figure 7). These responses to nitrogen availability can be analyzed in a carbon balance context with simple models that compare growth rates under hypothetical levels of adjustment (Figure 8). Similarly, the effect of leaf nitrogen content on the costs and benefits of defensive compounds for leaf protection can be calculated (Mooney and Gulmon, 1982). Models of crop growth have in some cases coupled the dynamics of carbon and nitrogen (Wann and Raper, 1979). These models, in a sense, translate nitrogen availability into a cost of acquiring nitrogen, measurable in carbon units.

Several features of plants may require a different type of analysis that emphasizes the changes in the cost of acquiring nitrogen through-

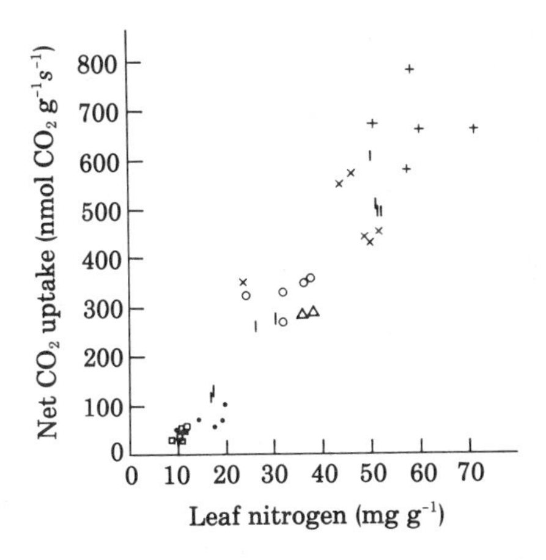

FIGURE 6. Leaf nitrogen versus net photosynthesis at light saturation for a wide range of plant species. (From H.A. Mooney and C. Field, unpublished.)

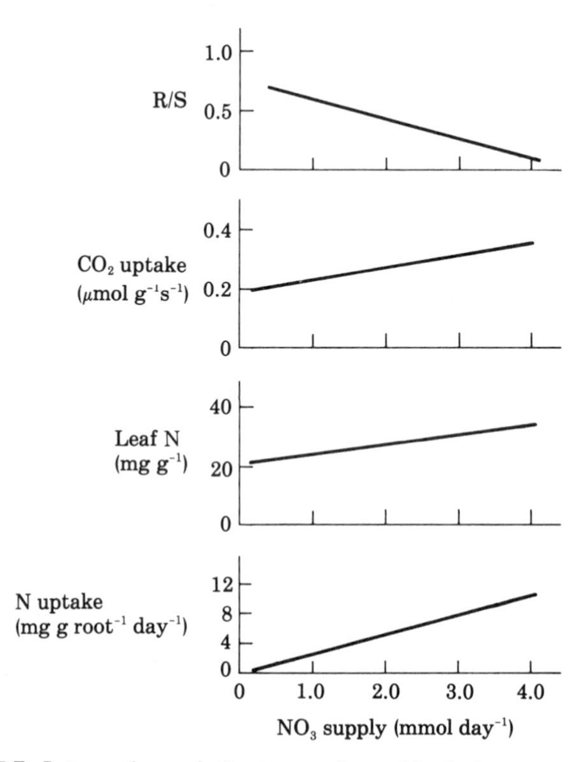

FIGURE 7. Interactions of nitrate supply and leaf nitrogen content, leaf CO_2 uptake, and root:shoot ratio for plants of *Diplacus aurantiacus*. (From Gulmon and Chu, 1981.)

out the lifetime. This type of analysis may be applicable at both the leaf and the whole-plant level. Early in their development, leaves receive an initial aliquot of nitrogen, which is subsequently diluted by the addition of carbon to the leaf, and then is further reduced by translocation out of the leaf. Movement of nitrogen out of the leaf to leaves developing in higher light maximizes whole plant carbon gain (Field, 1983). Later, nitrogen is diverted from the leaves to developing reproductive tissue. Thus, during the life of a leaf there appears to be a carbon-poor and a carbon-rich period, and a nitrogen-poor and a nitrogen-rich period. The life cycle of an annual plant shows similar phases (Figure 9).

The relative availability of carbon and nitrogen is determined by both the environment and the plant, in the same sense that the environment and the plant interact to create leaf microclimates. Reproductive structures may be self-supporting in terms of carbon, but this is achieved at the cost of nitrogen export from the leaves, possibly because root growth and activity diminish when flowering is initiated (Noodén, 1980). Thus, changes in allocation entail interactions be-

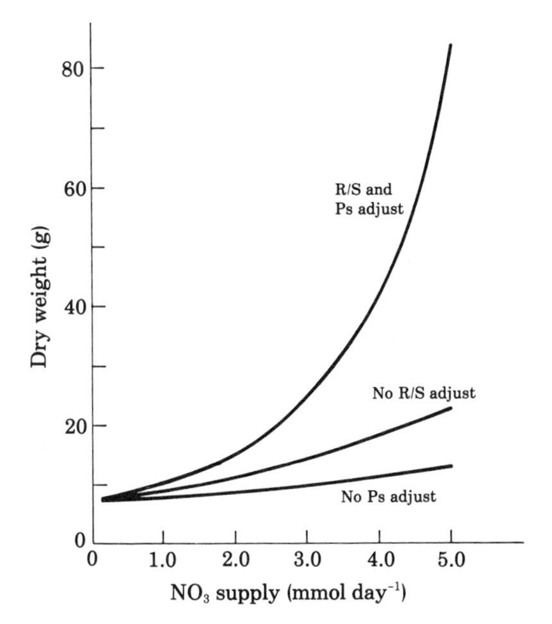

FIGURE 8. Simulated plant growth during three weeks where photosynthetic capacity and root:shoot ratio are functions of nitrate supply, as in Figure 7. These simulations indicate the carbon gain benefits resulting from altered root:shoot ratio and photosynthetic capacity in relation to increased resource availability.

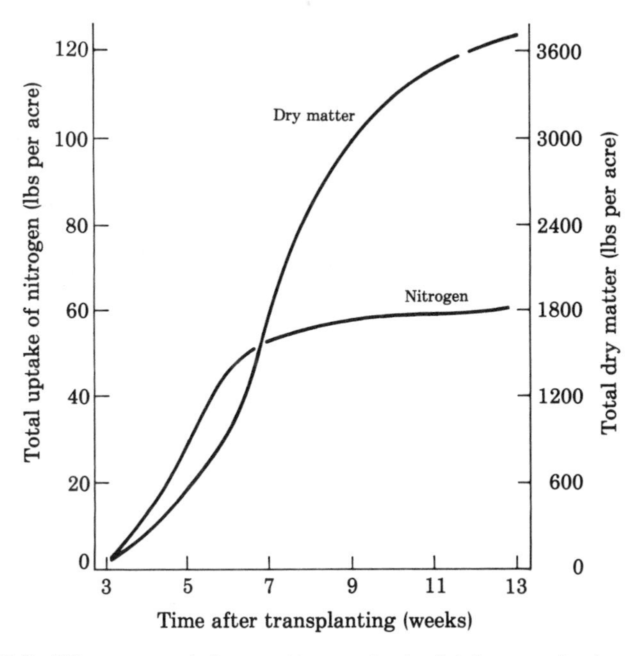

FIGURE 9. Nitrogen and dry matter content of tobacco plants versus age. (From Tso, 1972.)

tween the environment and many levels of organization in the plant. Extending carbon balance models to include changing relative costs of carbon and nitrogen may elucidate the resource constraints from which allocation rules can be derived. For example, monocarpic perennials must reach a threshold size before flowering is induced (Werner, 1975; Hirose and Kachi, 1982). This may reflect a threshold requirement for a resource due to the overhead costs of reproduction. Or, for example, coupling the activities of starch metabolism and flowering (Fawzi and El-fouly, 1979; Lercari, 1982) may reflect an alternative to photosynthetic reproductive structures which requires greater starch storage and a lowered growth rate in the vegetative phase.

Predicting competitive interactions

One of the major goals in population biology is predicting the outcome of competitive interactions among genotypes and species. Several models have been developed that predict the dynamics of competition (Watkinson, 1981). However, each parameter in the competition model combines information on many aspects of physiology and many types of resources. By expanding each parameter in terms of the underlying physiology, we may be able to construct models that provide a mechanistic understanding of competitive outcomes (Figure 10).

We have emphasized that individual growth rate can be derived from carbon balance considerations that account for the cost of necessary resources. To link this analysis with predictive models of competition, we need to quantify and predict a resource matrix in a habitat. The resource matrix may result from environmental heterogeneity or neighborhood structure. In habitats where light is the principal factor limiting growth, techniques are already available and demonstrate the predictive potential of this approach. Fish-eye photographs from the leaf or plant's perspective can be coupled with predicted solar tracks for any day of the year to determine the temporal pattern of light availability (Anderson, 1964). Under conditions of light limitation, growth can be predicted from the amount of sunlight a plant captures (Figure 11). Where competitive outcomes depend on the unequal growth of individuals, because of differential light capture, we should be able to predict competitive dynamics using models of light-limited growth.

More typically, growth is limited by resources that are difficult to quantify, vary seasonally, and interact. As with our understanding of nitrogen, a first step in identifying other critical resources is to examine changes in plant nutrient balance through time (Abrahamson and Caswell, 1982). Models can then be constructed that will enable us to predict plant growth as a function of various resource levels. At the same time, studies are needed that systematically consider the effect of competition on modifying the optimization criteria appropriate to understanding plants grown in mixtures. Austin's (1982) studies pre-

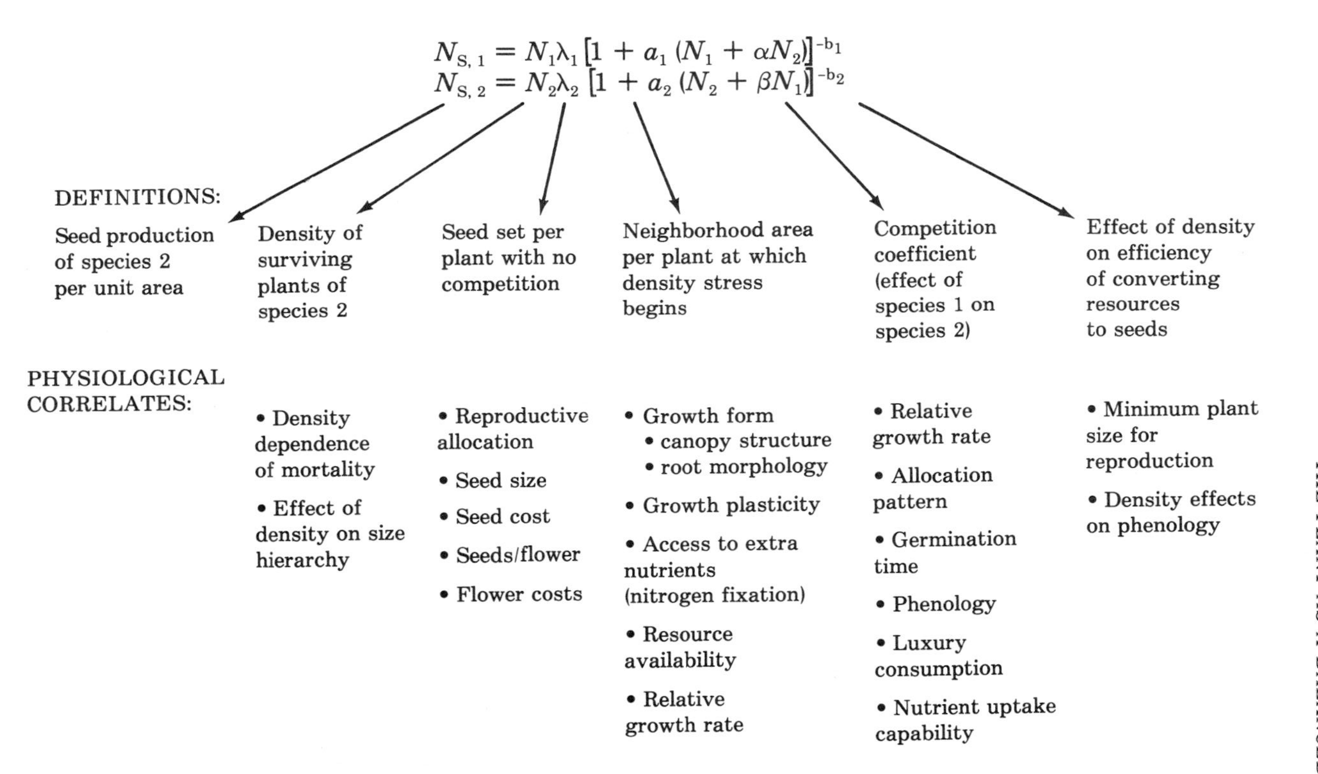

FIGURE 10. The governing equations for Watkinson's (1981) model of two-species competition. For each term, physiological traits which potentially contribute to the term are listed.

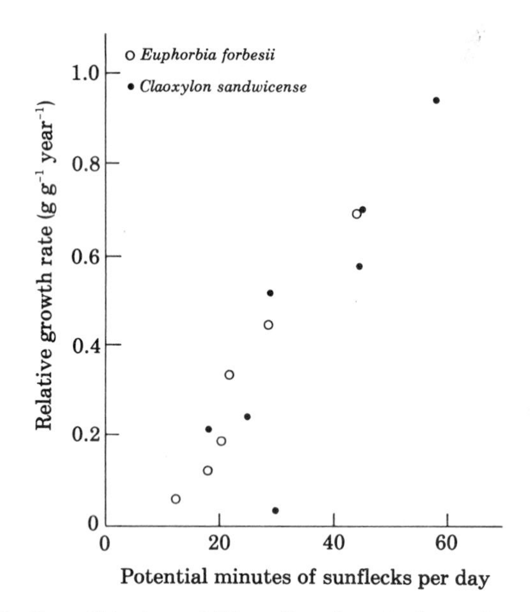

FIGURE 11. Growth rates of Hawaiian forest plants in relation to photosynthetically active radiation received. (From R. Pearcy, unpublished.)

dicting yields in mixtures from monoculture yields imply that under some circumstances, ecological and physiological optima are the same. This finding should stimulate considerably more attention to linking physiological performance with plant interaction and distribution.

CONCLUDING REMARKS

1. Progress in plant physiological ecology over the past two decades has been principally at the leaf level because of the predictive power of energy balance (e.g., Gates, 1962) and gas diffusion models (e.g., Gaastra, 1959), which have been principally leaf-based.
2. Whole plant approaches are needed in physiological ecology in order to link up with the field of population biology.
3. The whole plant is a balanced system:
   a. Mechanisms to enhance resource acquisition may be seen from the molecular to the whole-plant level.
   b. The significance of adjustments to the removal of plant parts should be assessed at the whole-plant level and through time.
   c. Optimization models generally consider isolated processes but can be nested to yield whole-plant information.
4. Consideration of budgets of resources (e.g., carbon) provides a whole plant perspective.

5. The carbon balance approach permits a quantitative assessment of dry matter accumulation, based on the following considerations:
   a. Carbon acquisition can be measured easily, given certain precautions, or can be estimated from leaf parameters.
   b. Carbon costs of maintenance and construction of tissues, which are necessary to construct growth models, are easily calculated. As yet, however, we have no rules to predict allocation patterns, which are also essential in these models.
6. Resource balance approaches further offer the possibility of making cost–benefit analyses of various tissues and processes.
7. A particularly crucial unsolved issue is what determines the switch time from the deployment of resources to growth versus reproduction, since this time affects the amount of biomass accumulated and the final reproductive output.
8. Carbon balance models can be linked to nitrogen balance models, offering new understanding of resource interaction and controls on allocation.
9. Knowledge of physiological interactions controlling relative growth rates offers tools to predict competitive outcomes, although additional information is required.
10. Information about the resource base of a microsite is readily obtainable for certain resources (e.g., light) but not others (e.g., nutrients).
11. Models of plant competition may be modified to include information on physiology in order to give a mechanistic basis to competitive outcomes.

Plant physiological ecology offers promise of providing information on the mechanisms controlling plant response to a given resource base and on the mechanisms controlling plant–plant and plant–animal interactions. However, more focus is needed on whole-plant responses rather than leaf-level phenomena. Resource balance and optimization models offer the tools to accomplish this. Important problems which need to be solved are those related to controls on resource partitioning within the plant through time.

CHAPTER 16

DEMOGRAPHIC CONSEQUENCES OF PLANT PHYSIOLOGICAL TRAITS: Some Case Studies

Fakhri A. Bazzaz

INTRODUCTION

The fields of physiological ecology and population ecology originated and developed quite independently. Physiological ecology had its roots in plant geography, as early plant geographers and collectors sought to discover what physical factors of the environment governed the distribution of species and what morphological and physiological adaptations these species had developed to cope with their physical environment. Interest has been strong in the study of convergent physiological responses of taxonomically unrelated species occurring in quite geographically separated areas. Much emphasis was placed on plants from extreme environments, perhaps because of the higher probability of discovering plant adaptations to them. There was initially special emphasis on the water economy of desert plants and on the germination, growth, and carbon economy of plants in cold environments. The advent of infrared gas analysis techniques, especially field instruments, has revolutionized research in physiological ecology, and within the last few years our understanding of physiological responses of plants to their environment has increased considerably. Physiological ecology has advanced much by the infusion of ideas and techniques from plant physiology, micrometeorology, biophysics, and biochemistry. Physiological ecologists have begun to make use of theories developed in other fields, such as optimal foraging and cost-benefit analysis in studying resource allocation patterns and engineering theory to construct organism models for energy budget considerations.

Until recently, the principles of population ecology were almost exclusively based on the behavior of animal populations. Much progress in plant population ecology has been made by plant ecologists and by ecologists who were trained as zoologists but found plants to be more suitable than other organisms for testing particular theories and predictions. The renewed interest in plant-animal interactions, for example, herbivory and defense (Dirzo, Chapter 7), pollination (Willson, Chapter 13), and dispersal (Levin, Chapter 12), has also contributed to this situation. In particular, the recent surge in tropical biology research has been influential because of the fascination with plant-animal coevolutionary interactions, which are especially vivid in seasonal tropical forests. The field of population ecology gained much from advances in and application of mathematical models in ecology and from population genetics. Furthermore, plant population ecology has been significantly advanced by researchers with a strong interest in agricultural ecosystems. Much of the theory of plant population biology is now based on the behavior of agricultural (including weedy) species under field and glasshouse conditions. Models of competition, responses to density, patterns of allocation, and survivorship are prominent examples (see Harper, 1977).

The fields of population ecology and physiological ecology have advanced a great deal over the last two decades. They have done so quite independently, however, and each now has its own gurus, revolutionaries, and renegades. Whereas physiological ecology has a strong mechanistic and experimental approach, population ecology is more theoretical, with a strong genetic and evolutionary bent. Physiological ecology emphasizes the response of individuals to their physical environment, whereas population ecology concerns the behavior of individuals with respect to each other and to the rest of the biological environment. Physiological ecology mostly concerns species distribution, whereas population ecology mostly concerns species abundance.

There have been some attempts in the past to combine the philosophies and approaches of both fields. Outstanding among these have been the works of Clausen, Keck, and Heisey at the Carnegie Institute of Washington at Stanford. F. E. Clements was also aware of the link between population ecology and the physiological responses of plants, and some of his early experimental work had elements of both.

Recently there has been increased interest, especially among young ecologists, in working at the interface between physiological ecology and population ecology. Already the work has very convincingly shown that it is advantageous, and even necessary, to approach ecological problems in this fashion. Many of the questions before ecologists are complex, and their answers require much breadth and sophistication in the way they are approached and in the techniques that have to be brought to bear on them.

My associates and I have been working on the problem of species replacement in successional plant communities. We have been asking questions about the life-history characters of the species and their role in colonization, establishment, growth, competition, and the evolution of resource-sharing mechanisms. We have used physiological, demographic, genetic, and statistical techniques in various combinations to test our hypotheses and predictions about these events. In this chapter I will give some examples from this work to illustrate these points and to show some of the demographic consequences of plant physiological behavior. I will discuss two winter annuals, one spring annual, and one long-lived clonal perennial.

DEMOGRAPHY IN RELATION TO GERMINATION BEHAVIOR AND PHOTOSYNTHETIC RESPONSES IN TWO *ERIGERON* CONGENERS

Erigeron annuus and *E. canadensis* are winter annuals of early successional habitats in many places in temperate regions. Their seeds mature in summer and are dispersed by wind. The seeds have no obvious dormancy and germinate readily on moist ground. Germination and emergence in both species is asynchronous, and emergence bursts occur in summer and fall. The resulting small seedlings grow slowly and form rosettes that overwinter.

Seeds of *E. annuus* are shed earlier than those of *E. canadensis*. Whereas emergence of *E. annuus* seedlings is usually completed by early fall and more or less shows a peak in late summer, emergence of *E. canadensis* is usually spread over a longer time in the fall and continues until the onset of frost (Figure 1). Emergence in this species seems to be controlled by the timing of rainfall and the resulting in-

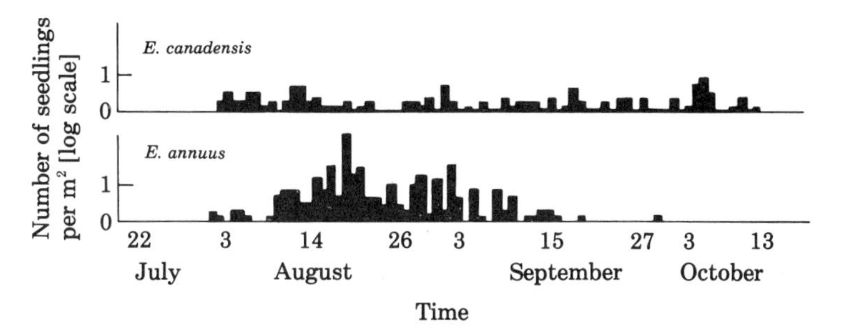

FIGURE 1. Pattern of emergence of seedlings of *Erigeron canadensis* and *E. annuus* in the field.

crease in soil moisture (Regehr and Bazzaz, 1979). Furthermore, additional germination and emergence of *E. canadensis* may occur in the following spring.

One consequence of earlier germination of most *E. annuus* seeds is that the rosettes of the species are larger than those of *E. canadensis* prior to the onset of frost in November. The average rosette of *E. annuus* has a longer time to photosynthesize and grow in warm weather than does that of *E. canadensis*. The size class distributions of the rosettes of the two species in November are quite different (Figure 2). Many more large size classes are represented among *E. annuus* than among *E. canadensis* rosettes. Rosette size is also influenced by density. In high density arrays, growth of individual rosettes is restricted and the rosettes are usually small when they enter winter.

The rosettes of both species show a considerable degree of acclimation to temperature. The optimum temperature for photosynthesis in winter is ca. 15°C, whereas in summer it is 25°C. Winter maximum photosynthetic rates are only slightly lower than summer rates, and there are no significant differences between species in these rates. However, the two species differ in the degree of decline of photosynthetic activity at lower temperatures (Figure 3). For example, at 0°C photosynthetic rate is 55% of maximum in *E. annuus* whereas in *E. canadensis* it is only 30% of maximum (Regehr and Bazzaz, 1976). This wider range for *E. annuus* may further contribute to a more favorable carbon budget for the rosettes of this species relative to the other.

Under sunny conditions in winter, leaf temperatures of both species may be significantly higher than air temperature. For example, at ca. 1°C air temperature, leaf temperature may be 10°–15°C which is near the optimum temperature for photosynthesis. The light compensation

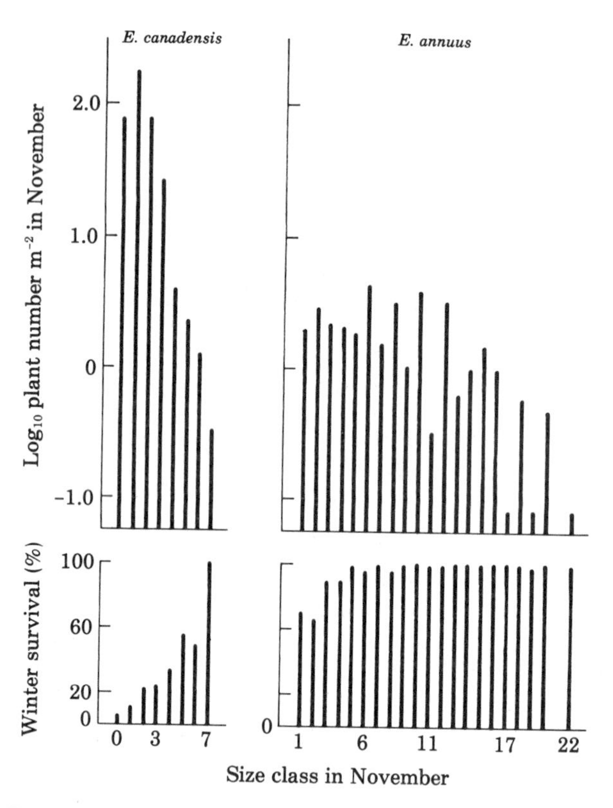

FIGURE 2. Frequency distribution of rosette diameter in November and percentage survival over winter in *E. canadensis* and *E. annuus*.

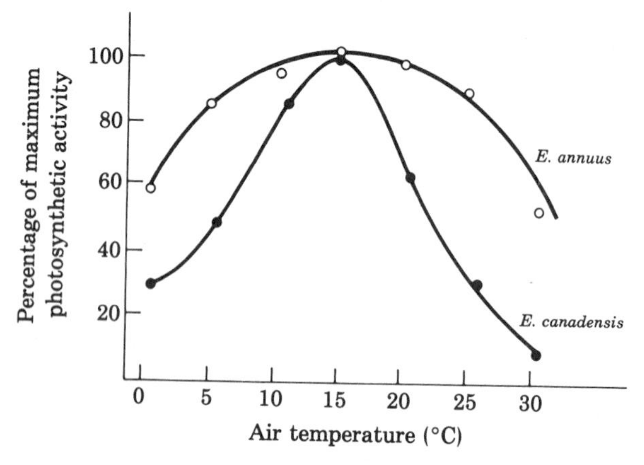

FIGURE 3. Photosynthetic response to temperature of winter rosettes of *E. canadensis* and *E. annuus*.

points of the rosettes of both species shift in winter and under low temperature to very low levels. Photosynthetic start-up time after dark is also very short in winter.

Thus, although the basic strategies of carbon gain of these species are somewhat similar, they differ in ways that could make *E. annuus* rosettes larger and better established than those of *E. canadensis*.

The rosettes of both species experience considerable mortality in winter, mainly from uprooting due to frost heaving. In *E. canadensis*, smaller rosettes suffer disproportionate mortality; the probability of death declines with increased size (Figure 2). Larger *E. canadensis* rosettes and most *E. annuus* rosettes are well anchored because of the former's larger taproot and the latter's extensive fibrous root system. *Erigeron canadensis* populations experience further mortality in early spring, primarily as a result of the weakened condition of many partially uprooted rosettes. Later on in the season, competition among larger individuals of both species and from summer annuals contributes to further mortality. Spring-germinated individuals suffer severe mortality because of their relatively small size.

The large rosettes of *E. annuus* begin rapid growth in spring. A shift in optimum temperature for photosynthesis from ca. 15° to ca. 25°C occurs. Undoubtedly, translocation of stored reserves from underground parts aids in the fast growth of the rosettes, enabling them to overtop those of *E. canadensis* and of summer annuals as well (Figure 4). Bolting occurs early, and flowering peaks in June. In con-

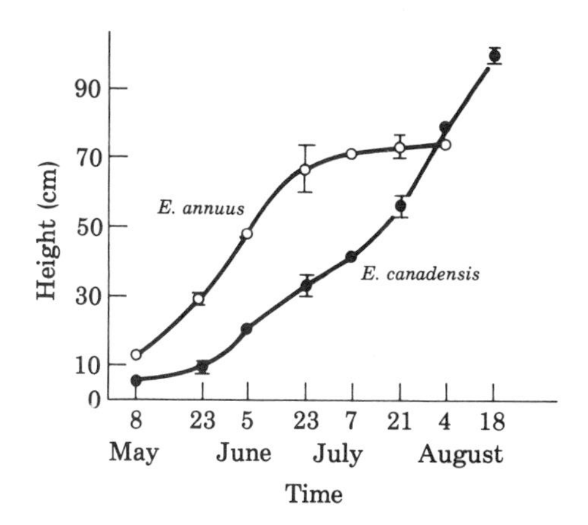

FIGURE 4. Mean height (± SE) of *E. annuus* and *E. canadensis* in plots during the first growing season. (All means between species are significantly different.)

trast, the small rosettes of *E. canadensis* grow slowly in spring, bolt in early summer, and flower in late summer (Figure 5). Thus, there is only a little overlap in flowering time of the two species. The height extension growth rate in *E. canadensis* increases markedly after the flowering of *E. annuus* (Figure 4).

There are at least three consequences of delayed growth and flowering of *E. canadensis* relative to *E. annuus*. First, *E. canadensis* individuals suffer competition from *E. annuus* and from summer annuals that may be present in the field [e.g., *Ambrosia artemisiifolia, Polygonum pensylvanicum, Amaranthus retroflexus*, and *Setaria faberii* (Raynal and Bazzaz, 1975)]. In contrast, *E. annuus* individuals attain considerable height before the rapid growth phase of summer annuals and consequently suffer less competition from them. Second, *E. canadensis* populations suffer considerable mortality and reduced seed production because of severe infection with aster yellows, a mycoplasma disease that is transmitted to a wide range of host plants by a leaf hopper. This insect is usually transported from the southern United States by strong winds in the summer. *Erigeron annuus* plants escape infection by early growth and flowering. Third, competition for pollinators between the two species is reduced because of their different flowering phenologies. The two species use the services of the same pollinators to a large extent and are likely to compete strongly for them if the flowering times overlap.

The differences in phenology of growth and reproduction in the two species and the resulting patterns of mortality and competitive interactions are therefore greatly influenced by the timing of germination, the patterns of photosynthesis and growth, and the morphology of the root system of the rosettes.

EMERGENCE TIME AND CANOPY STRUCTURE IN RELATION TO PHOTOSYNTHETIC ACTIVITY AND DEMOGRAPHY OF INDIVIDUALS AND LEAVES IN *AMBROSIA* POPULATIONS

Ambrosia trifida L. (Compositae) is a fast-growing annual common in disturbed ground in the eastern and midwestern United States. In the spring, *A. trifida* seedlings emerge prior to those of other species in annual communities in Illinois. *Ambrosia trifida* seeds are capable of germinating at lower temperatures than those of all other species usually associated with it (Figure 6). Seeds of *A. trifida* are the largest among associated annuals and produce large seedlings with green, photosynthetically active cotyledons. As a result of early germination and rapid growth, the species dominates the annual community early in the season, even when its density is relatively low. The fast growth of the plant is aided by its high photosynthetic rate (Bazzaz and Carlson, 1982) and by its relatively high tolerance to low night temperatures

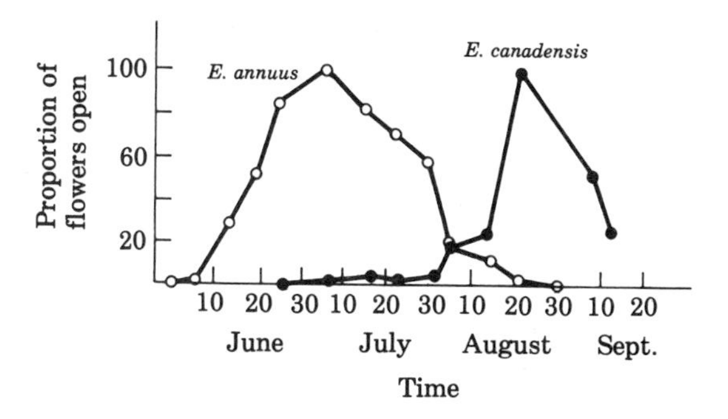

FIGURE 5. Seasonal flowering patterns of *E. annuus* and *E. canadensis*.

early in the growing season (Figure 7). On fertile soils a stand of *A. trifida* may develop to ca. 3 m in height and produce ca. 1600 $g \cdot m^{-2}$ of biomass in one growing season. In some stands, up to 98% of the stand's biomass is contributed by *A. trifida*. When the seedlings of the species are removed early in the season, the other annuals are released from severe competition and the species diversity of the community increases greatly, but total plant production declines sharply (Abul-Fatih and Bazzaz, 1979a). The species is therefore a very strong competitor and dominates the annual communities in which it is present. It acts like an "organizer" or a "keystone" species (*sensu* Paine, 1969).

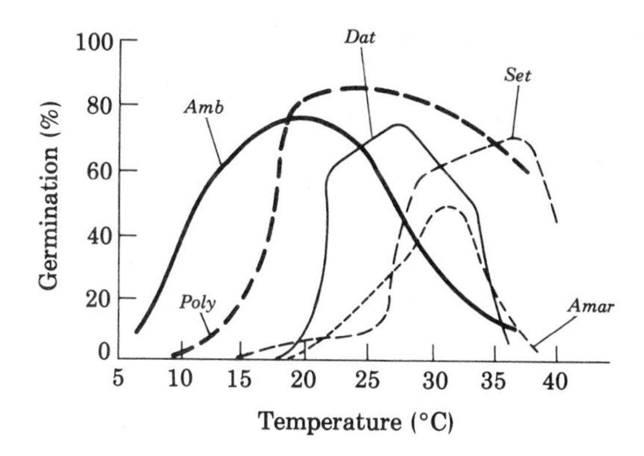

FIGURE 6. Response of seed germination to temperature in *Ambrosia trifida* (*Amb*), *Polygonum pensylvanicum* (*Poly*), *Datura stramonium* (*Dat*), *Amaranthus retroflexus* (*Amar*), and *Setaria faberii* (*Set*).

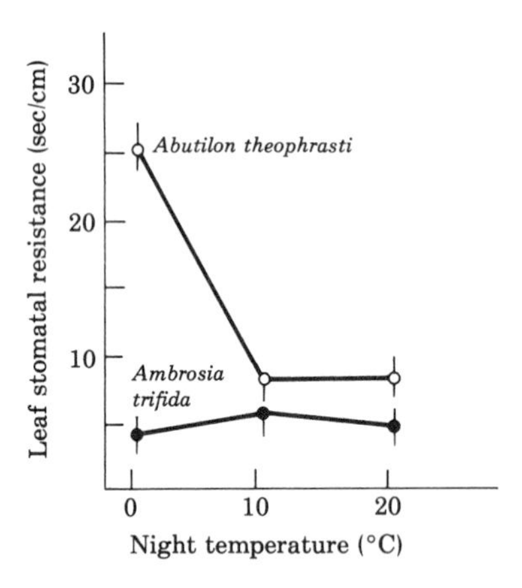

FIGURE 7. Leaf stomatal resistance (± SE) in *Ambrosia trifida* and *Abutilon theophrasti* following nights of various temperatures. (Data from Drew and Bazzaz, 1982.)

The aim of this discussion is to show how physiological responses interact with demographic patterns at both the individual and leaf levels to produce such a remarkable behavior.

There are several factors that control the emergence time of *A. trifida* seedlings in the field. Seed depth and differential response to the rising temperature of the soil contribute substantially to emergence time. Seeds germinate under a wide range of temperatures (8°–41°C), though germination is slower at the lower end of the temperature gradient (Goloff, 1973). Although under experimental conditions germination is essentially completed in six days at the optimal temperature of 24°C, it is only 30% complete at 10°C. Thus, under field conditions, emergence time will be influenced by the rapidity of changes in soil temperature. In the field, emergence time in a given location may be spread over 20–30 days (Figure 8). A hierarchy of plant heights usually develops, with early-emerging seedlings being taller than late-emerging ones. Because there is a tendency for late-emerging seedlings to come from lower depths (Abul-Fatih and Bazzaz, 1979b), their subordinate position in the canopy is caused both by their late emergence per se and by the loss of reserve material expended by the seedling to reach the soil surface. Seedlings that emerge first have the highest probability of survival; this declines with delayed emergence. Furthermore, earlier emergence results in taller, heavier individuals and higher seed production (Figure 9).

The position and fate of an individual in a plant population will be further influenced by its response to the physical environment, its competitive interactions with its neighbors, and its response to the

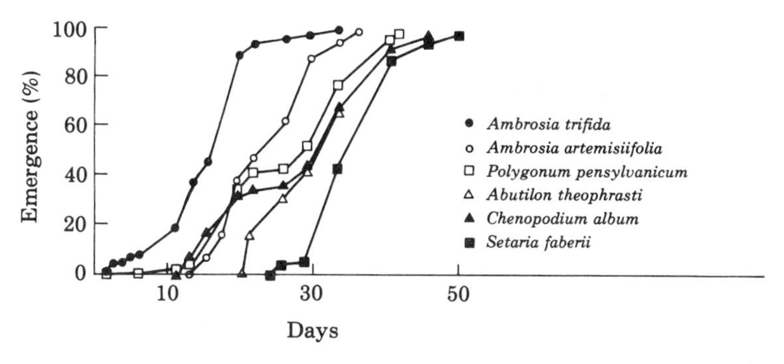

FIGURE 8. Seedling emergence in the field of six annuals in Illinois.

rest of the biological environment. Plants respond to crowding by mortality, usually of suppressed individuals, and by differential reduction of the growth rates of all individuals (Abul-Fatih et al., 1979). Relative leaf production rate and relative leaf-area ratio also decline with crowding. The growth responses of *A. trifida* are very plastic, and crowding causes changes in plant architecture as well. For example, plants in high-density arrays produce straight, slender, and unbranched stems,

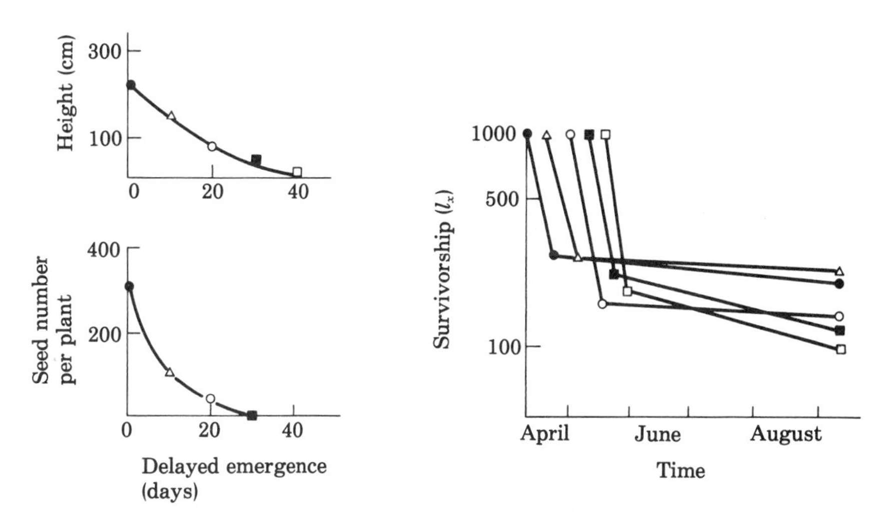

FIGURE 9. Height, seed number per plant, and survivorship of five cohorts of *Ambrosia trifida* plants in relation to delayed emergence in the field.

whereas plants in low-density arrays produce numerous branches and leaves. This flexible growth response results in differences in canopy structure, light interception patterns, microenvironment within the canopy, and other parameters. These will, in turn, influence CO_2 fixation, water use, material allocation patterns, and several other aspects of the physiological responses of the plants. Furthermore, plant-height distributions and canopy architecture are dynamic and change during the season. The distributions of leaf biomass and leaf area in the canopy change during the season as well. In high-density populations, the plants usually accumulate a massive leaf area early in the season. Later on, the plants discard lower, older, less active leaves as they develop new ones above. Thus, *A. trifida* canopy essentially develops to near-maximum leaf area and then is raised up as the season progresses (Hunt and Bazzaz, 1980). In very high-density populations, maximum leaf area is reached early in the season and declines later (Figure 10). Apparently, leaf production rate is so fast that it exceeds the optimal light interception capacity initially, and the plants discard lower, extremely shaded leaves and adjust the rate of production of new leaves to produce a canopy structure optimal for photosynthetic carbon fixation.

At the individual plant and leaf levels, populations of *A. trifida* respond to density in different ways. In lower-density stands, there is little or no death of individual plants, but there are high rates of leaf birth and death. In contrast, in high-density populations, individual plant mortality occurs, and leaf birth and death rates are low (Figure 11).

Regardless of the causes that determine the status of an individual in a population, there are several consequences for the individual plants in a subordinate position in the canopy, in addition to depriva-

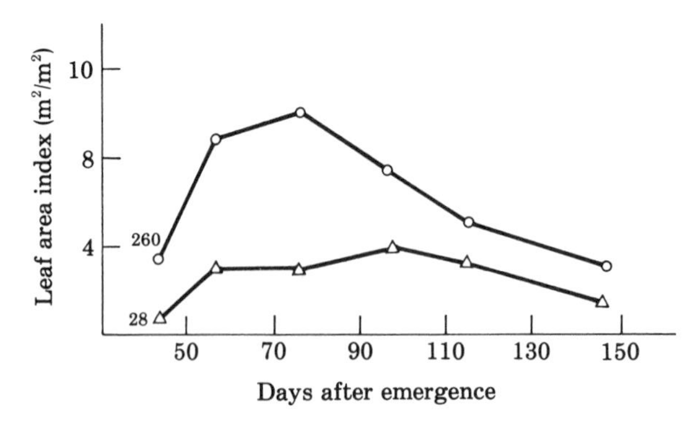

FIGURE 10. Trends in leaf area index of *Ambrosia trifida* stands in two densities: 260 plants per m^2, and 28 plants per m^2.

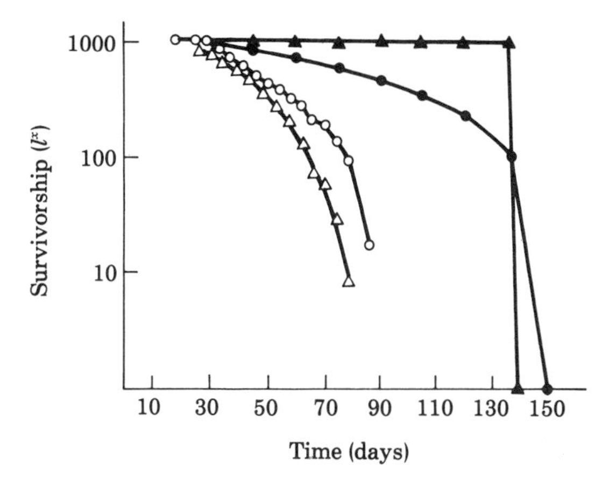

FIGURE 11. Plant (solid symbols) and leaf (open symbols) survivorship curves in *Ambrosia trifida* plants grown at low density (triangles) and at high density (circles).

tion of resources, for example, light and water. Two such consequences in *A. trifida* are sex expression and the level of seed predation. The plant is commonly monoecious, producing male flowers terminally and female flowers at the leaf axils. However, suppressed individuals in high-density stands usually produce female flowers only (Figure 12). These plants have a higher seed weight:plant weight ratio than do bisexual tall plants (10% versus 3%). The short plants apparently invest

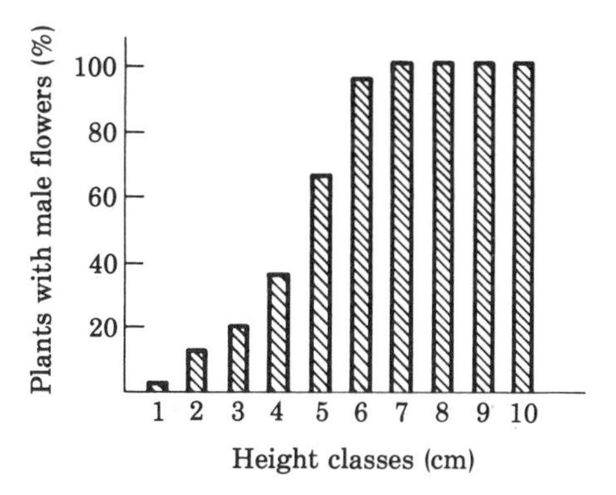

FIGURE 12. Relationship between flower maleness and plant height in a crowded population of *Ambrosia trifida*.

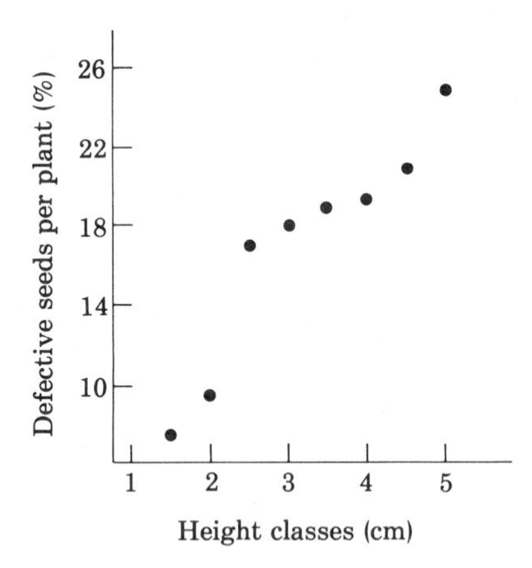

FIGURE 13. Relationship between seed predation (defective seeds) and plant height in *Ambrosia trifida*.

their reproductive energy in seeds, whereas tall plants partition it between male and female functions. Although this is a plastic, environmentally induced response (the plants derived from seeds of the females produce bisexual plants), it must have some consequences to the genetic structure of the progeny in that none will result from selfing.

The degree of seed predation is greater in tall individuals than in short ones (Figure 13). The tall plants may be more apparent to insects whereas short ones perhaps are protected against insect invasion from above by the dense canopy of their taller neighbors. The different physical microenvironment and the nutritional value and defense capabilities of the short individuals may also be involved in this situation. This response is not generalizable, however, because in the congener *Ambrosia artemisiifolia* L. suppressed individuals suffer more seed damage by insects than do taller individuals (Raynal and Bazzaz, 1975).

Leaves may carry out three functions for plants: photosynthesis, storage of nutrients, and shading of neighbors. Undoubtedly, the relative importance of these functions varies among species, individuals, and environments. In *A. trifida* leaf birth and death rates vary considerably with density and nutrient levels, as discussed earlier. The life spans of leaves on an individual plant vary considerably as well. Generally, leaves produced early and late in the season have short life spans, whereas those produced at the middle of the season live much longer (Figure 14). There is also a tendency of these individual leaves to be larger than those produced early or late.

In *A. trifida* and other fast-growing mesophytes, especially when resources are in good supply, a leaf that is born in the sun finds itself deep in the canopy shortly thereafter, perhaps spending as much as 60% of its life in the shade. The contribution of a given leaf to the carbon economy of the plant is a complex phenomenon, as leaves also become older as they are buried deeper in the canopy. The decision on the part of the plant as to when to discard a particular leaf is further influenced by the ease of moving nutrients out of the leaf, cost of defense of that leaf, and availability of other resources to replace it with a younger, more efficient, well-lighted one above.

The relationship of photosynthetic rate to leaf age at a given density of *A. trifida* individuals is shown in Figure 15. Initially, of course, a young leaf is a sink for carbon, as carbon is required to build the appropriate machinery (Mooney and Gulmon, 1982). *Ambrosia trifida* leaves become sources of carbon at about day 7 after their unfolding and reach peak photosynthetic activity around day 20, declining to become a sink again at ca. day 30. The area of an individual leaf also increases with age; thus, an individual leaf's contribution to the carbon economy of the plant increases quickly with age at first and then declines slowly after the peak. Although leaf demography has been examined in some species and life-table analysis has been applied to leaves (e.g., Bazzaz and Harper, 1977; Abul-Fatih and Bazzaz, 1980), there have not been many attempts to relate leaf demography to a leaf's life-long contribution to the plant's photosynthesis. Applying principles of population biology to *A. trifida* data, we can calculate the contribution of the leaves to the increase in the dry weight of the plant. Populations of leaves on a plant have demographic features (e.g., age structure) which permit the use of life-history tables. All that is re-

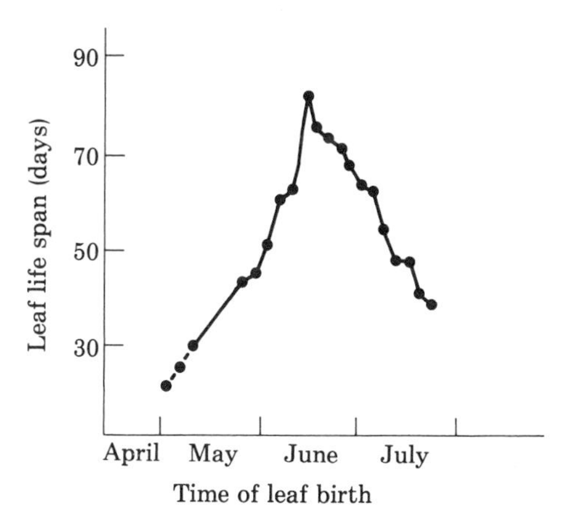

FIGURE 14. Leaf life span in relation to time of leaf birth in *Ambrosia trifida.*

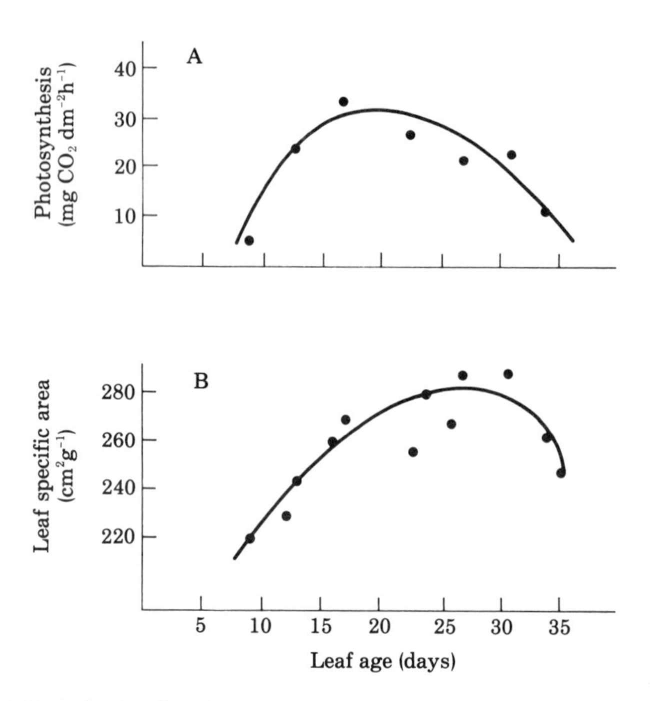

FIGURE 15. Relationship between leaf age and (A) photosynthetic rate and (B) leaf specific area in *Ambrosia trifida.*

quired for these calculations are leaf birth and death schedules and age-specific leaf photosynthetic rates. The whole plant has an age-specific reproductive value V_x which is the relative lifetime contribution by an individual of age x to growth of the population. The reproductive value of that individual is given by:

$$V_x = \frac{e^{rx}}{l_x} \cdot \sum_{y=x}^{n} e^{-ry} l_y b_y$$

where l_x is the number of individuals entering age class x, b_y is the fecundity of individuals in age class x, and r is the intrinsic rate of population growth. V_x is a measure of the reproductive contribution of individuals of age x to the next generation (Wilson and Bossert, 1971). Similarly, leaves of age x have an analogous reproductive value V_x representing their value in terms of their relative contribution to the total assimilation of the plant (Table 1). D. C. Hartnett and F. A. Bazzaz (unpublished data) have used this approach to examine leaf dynamics in *Solidago canadensis* and have suggested that these parameters of the leaf photosynthetic activity and other demographic features may be important factors influencing the selection of host leaves by phytophagous insects.

It is well established that sun- and shade-adapted leaves differ from each other in their structure and photosynthetic activities (Björkman, 1968). Shade leaves are usually greener, flatter, and thinner than sun leaves; they have a low leaf specific weight, low respiration rates, and low light-compensation points. Sun leaves usually have higher photosynthetic rates on a per area basis. They light-saturate at high light intensities, have higher light-compensation points, and typically have a higher chlorophyll a:b ratio than do shade-adapted leaves. Sun leaves also have higher ribulose bisphosphate (RuBP) activity than do shade leaves. Plant species differ widely in their ability to acclimate to light conditions, with species from open, early-successional habitats being more responsive than species from closed, late-successional ones (Bazzaz, 1979; Bazzaz and Carlson, 1982). In fast-growing mesophytes such as *A. trifida*, the leaves are born in the sun but spend a significant amount of their life in the shade. It would be advantageous if these leaves could switch, later in their life, from sun-adapted to shade-adapted behaviors.

The most relevant changes for leaves as they become buried in the canopy would be to increase surface area or decrease thickness, in-

TABLE 1. Cohort life table for leaves of *Ambrosia trifida*.

x^a	$l_x{}^b$	$L_x{}^c$	$b_x{}^d$	L_xb_x	e^{-rx}	$e^{-rx}L_xb_x$	V_x	V_x/V_0
0	1000	1.0	0.0	0.0	0.88	0.0	29.1	1.0
5	1000	1.0	0.0	0.0	0.69	0.0	37.4	1.3
10	1000	1.0	0.7	0.7	0.54	0.4	48.2	1.7
15	1000	0.99	11.2	11.1	0.42	4.7	61.5	2.1
20	980	0.89	26.3	23.4	0.32	7.5	56.7	1.9
25	810	0.77	37.8	29.1	0.25	7.3	67.0	2.3
30	730	0.67	35.0	23.4	0.20	4.7	44.7	1.5
35	620	0.58	14.0	8.1	0.15	1.2	13.4	0.5
40	540	0.49	0.0	0.0	0.12	0.0	0.0	0.0
45	450	0.39	0.0	0.0	0.09	0.0	0.0	0.0
50	340	0.27	0.0	0.0	0.06	0.0	0.0	0.0
60	210	0.17	0.0	0.0	0.04	0.0	0.0	0.0
70	130	0.29	0.0	0.0	0.02	0.0	0.0	0.0
80	46	0.03	0.0	0.0	0.01	0.0	0.0	0.0
90	19	0.01	0.0	0.0	0.01	0.0	0.0	0.0
100	0	0.00	0.0	0.0	0.00	0.0	0.0	0.0

[a] x = Age (days).
[b] l_x = Number of survivors entering age class x.
[c] L_x = Mean number alive during age class x.
[d] b_x = Fecundity (photosynthetic rate per leaf) for leaves of age x.
$r_m = 0.05$ leaves • leaf^{-1} • day^{-1}

crease chlorophyll concentrations, and change the chlorophyll a:b ratio, perhaps at the cost of the RuBP pool. An increase in the slope of the initial curve of light saturation would also be advantageous. Since many of these activities are related to nitrogen concentration (Mooney and Gulmon, 1982), this probably involves exporting nitrogen from these shaded leaves.

In *A. trifida* both leaf age per se and the degree of shading (determined by the position of the leaf in the canopy), influence leaf photosynthetic characters. Although there is some reduction of dark respiration with age, the reduction is much more pronounced with shading. For example, dark respiration of mature leaves that remained in the sun declined to 95% of initial rate, but dark respiration of leaves of the same age that were in the shade declined to 39%. Light compensation points do not change during the same period for leaves that remain in the sun but decline appreciably for those that are shaded. There is no significant difference in the patterns of maximum rate of photosynthesis; the rate declines with age to about the same level in both the sun and shade leaves.

Leaves of *A. trifida* may be subjected to severe damage by herbivores, and the level of damage may be a determining factor in the timing of leaf senescence and fall. At the time of flowering and seed set in the field, most of the leaves are either dead or severely damaged. The plants, however, produce large numbers of small, bractlike, lanceolate leaves which subtend the female flowers. These leaves are produced at very high rates (19.4 leaves$\cdot$day^{-1} in isolated individuals) and remain photosynthetic during seed filling. Together with the green flowers and fruits, they contribute significantly to the reproductive effort of the plants. Reproductive assimilation accounts for 41 and 57%, respectively, of the carbohydrate required to produce the male and female inflorescences. Individual flowers and their subtending leaves can produce some seed even in individuals from which all regular leaves have been removed. Moreover, individual flowers retain their ability to elaborate seed even if they are removed from the parent plant at the time of pollination (Table 2).

Photosynthetic capacity of fruits may be an important factor in reducing the impact of herbivores. For example, repeated 50% defoliation reduced seed production by only 30% under field conditions.

The contribution of flower and fruit photosynthesis to seed weight varies considerably among species (Bazzaz et al., 1979). In some species, green flowers and fruits may be sufficiently active photosynthetically to pay a substantial portion of the energetic and carbon costs of their own and of the resulting seed. One implication of this is that the energetic cost of an animal-pollinated plant with showy, nongreen flowers will be twofold, the cost of making the structures and the cost of sacrificing its potential photosynthetic contribution (Bazzaz et al., 1979).

TABLE 2. Reproductive characteristics of control and defoliated plants and of the excised fertilized female flowers of *Ambrosia trifida*.[a]

Plant class	Total seed weight (g)	Seed number	Weight of single seed (mg)	Reproductive ratio[b] (%)
Control	1.5±0.5	39±16	41± 6	27± 6
Defoliated	0.5±0.2	10± 0.2	55±18	36± 7
Excised flowers (in bracts)	0.2±0.1	7± 2	30±11	34±12

[a] All data are means ± SE. From Bazzaz and Carlson (1979).
[b] Ratio of reproductive to total biomass.

The decisive competitive success of *A. trifida* in annual communities is based on a combination of several physiological and demographic features:

1. Largest seeds among the plants in the community.
2. Earliest emergence of seedlings in the spring.
3. Largest cotyledons that are photosynthetically active.
4. High tolerance of seedlings to cold nights.
5. Rapid growth rates, heavy initial commitment to leaf production, and rapid leaf area growth rate.
6. Very high leaf area index, which is achieved early in the season.
7. High flexibility in biomass allocation to various plant parts and in plant architecture and sex expression.
8. Rapid adjustment of individuals to changes in environmental resources by death of individuals and/or changes in leaf birth and death rates.
9. High rate of birth of bract leaves subtending and supporting female flowers and seeds.
10. Significant amount of photosynthesis in flowers and seeds.

DEMOGRAPHIC CONSEQUENCES OF PHYSIOLOGICAL INTEGRATION OF CONNECTED RAMETS IN *SOLIDAGO CANADENSIS*

The clonal growth habit is common in herbaceous and woody plants in successional communities. Among these species, *Solidago canadensis* (goldenrod) has a long residency time in old fields in parts of the midwestern and eastern United States and Canada. The species usually becomes dominant during the third or fourth year after disturbance and persists in large clumps until it is outcompeted by trees or other late successional plants many years later. The established genets increase in size as new ramets are produced via rhizomes. The new shoots usually emerge in late March and grow through spring and summer. Ramets begin to produce inflorescences and new rhizome buds in September. After seed set, some rhizomes continue to grow until early November. Mature rhizomes overwinter, then resume growth in the following spring.

The colonization of successional fields by *S. canadensis* is episodic. Large numbers of seeds are blown from adjacent fields into newly disturbed sites. These established populations grow by genet expansion and, depending on the initial density, the population may lose some genets with time. Although the seed rain increases greatly after the first flowering season of the established populations, the probability of seedling survival declines sharply soon after establishment of the genets in the field. Thus, *S. canadensis* recruitment is possible only for a short period of time after a field is disturbed. The ramets may maintain rhizome connections for up to four years, and a single genet will be composed of one to several units of about 10 to 20 interconnected ramets.

Within the growing season, the interdependency among ramets within a clone decreases with time. The newly developing ramets are strongly dependent upon their parental clone during emergence and establishment, but become progressively less dependent as they grow larger. When ramets are separated from the remainder of the clone by severing the underground connections early in the growing season, the isolated ramets suffer high mortality. Delaying severing time to later in the growing season, when the individual ramets are large, reduces the probability of ramet death. Severed ramets also suffer reduced shoot extention growth, reduced leaf number, reduced growth rate, and reduced total dry weight (Hartnett and Bazzaz, 1983). The proportion of ramets that flower and set seed and the number of new rhizomes produced are also reduced (Figure 16).

The specific resources for which the new ramets depend upon their parental clones are unknown. In daughter ramets, root development lags behind shoot extension, so there may be a dependency upon the parent clone for water and nutrients. It is also possible that photosynthates are translocated from the parental clone to support the growth of the daughter ramets. This notion was experimentally tested by growing pairs of ramets under different light environments.

Shaded and unshaded ramets were either interconnected or independent of each other as follows: one shaded and one unshaded interconnected ramet; one shaded and one unshaded severed ramet; or two unshaded interconnected ramets. Height extension rates, survivor-

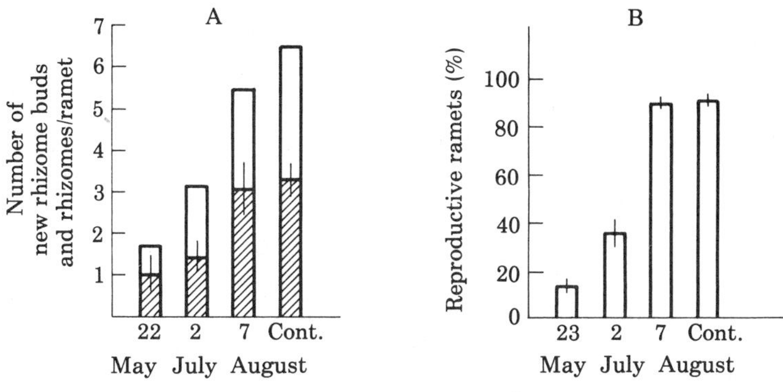

FIGURE 16. A. Number of new rhizome buds (open bars) and number of rhizomes per ramet in relation to the time of severing from the rest of the clone. B. Percentage reproductive ramets in relation to the time of severing from the rest of the clone.

ship, leaf population dynamics, and photosynthetic rates, prior to and after shading, were measured (Hartnett and Bazzaz, 1983).

The most striking result was that, following the application of shade to one of two connected ramets, the rate of photosynthesis of the unshaded ramets increased by approximately 20% (Figure 17). This

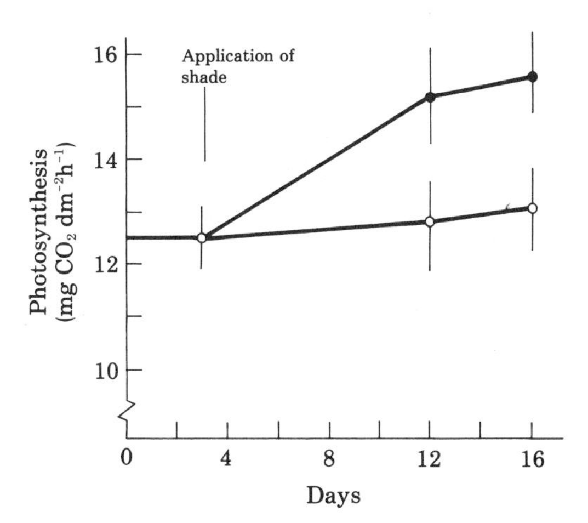

FIGURE 17. Photosynthetic rates (mean ± SE) of *Solidago canadensis* ramets. Unshaded ramets connected to unshaded siblings, ○; unshaded ramets connected to shaded siblings, ●.

could have resulted from an increased assimilate demand on, and increased rate of export of assimilates from, the unshaded ramets to support the growth of the shaded ones. This response was evident within a few days following the imposition of shade on the sibling ramets. Unshaded ramets connected to unshaded siblings experienced no mortality during the 100 days following the beginning of the shading experiment. The unshaded ramets with severed rhizomes also experienced no mortality. The shaded ramets that had no intact connections to unshaded siblings experienced high mortality within a few days after the imposition of shade. By contrast, the shaded ramets connected to unshaded siblings experienced no mortality for about 50 days following the treatment (Figure 18).

It appears, therefore, that intraclonal ramets are physiologically integrated while they remain attached and are able to transport resources from resource-unlimited to resource-limited parts.

Integration of resource acquisition and use among ramets over a given area enables a genet to integrate local environmental heterogeneity and, hence, sample resources in a more fine-grained manner than would physiologically independent ramets. Within a clone some ramets may be in shaded patches, while others are located where light is abundant but soil moisture and nutrients are limiting. As a result of interconnection, each ramet acts differentially as source or sink for different resources such that all resources are more-or-less equitably distributed among ramets of a clone. As a result, physiologically in-

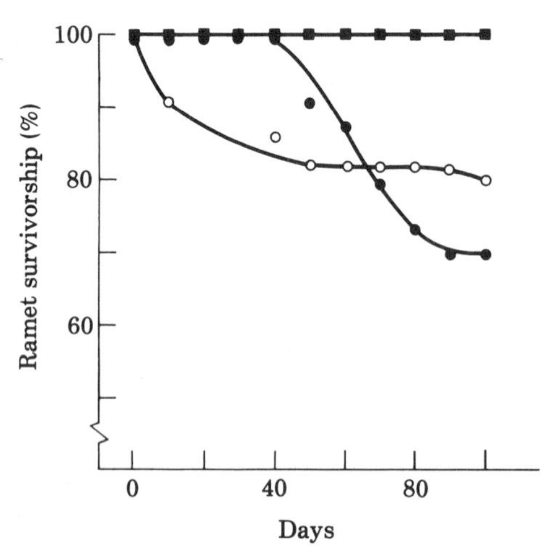

FIGURE 18. Percentage of survivorship of *Solidago canadensis* ramets: connected ramets, ■; shaded ramets but connected to unshaded siblings, ●; shaded ramets alone, ○.

tegrated ramets may respond less sensitively to differential resource limitations imposed by neighbors, herbivores, or other factors than would a population of independent ramets. An individual ramet, therefore, may be able to successfully occupy a resource-poor patch where an independent ramet would be unsuccessful.

Field experiments in which interconnected and severed ramets were located in the same or in different neighborhoods confirm this prediction (D. C. Hartnett and F. A. Bazzaz, unpublished data). Interconnected ramets of *S. canadensis* were located either in *Poa pratensis*, *Aster pilosus*, or *S. canadensis* neighborhoods, or in a combination of the three. The performances of *S. canadensis* ramets in these situations, with respect to survivorship, growth, leaf dynamics, height, final weight, and number of daughter rhizomes, were compared. The responses of a ramet to one neighbor species differed depending upon the neighbors of the adjacent ramets to which it was connected. In addition, the variation in responses to different neighbors was greatly reduced when interconnected ramets were each located in a different neighborhood than when all were located in a single neighborhood (Figure 19). Thus, there is indeed a high level of physiological integration and movement of resources between ramets of a clone. In addi-

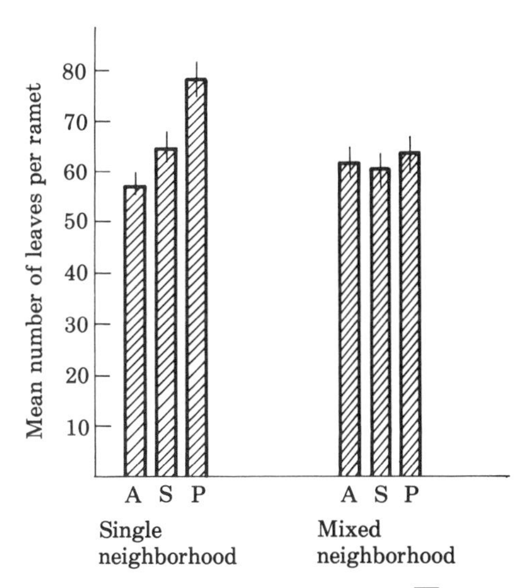

FIGURE 19. Performance as number of leaves ($\overline{X} \pm 1$ SE) produced on *Solidago canadensis* ramets grown in single neighborhood and mixed neighborhoods of *Aster pilosus* (A), *Solidago canadensis* (S), and *Poa pratensis* (P).

tion, formal mathematical weight:density relationships characteristic of dense populations were exhibited by independent ramets grown at a range of densities, but not by dense populations of interconnected ramets. This further supports the notion that the genet and not the ramet is the ecologically functional individual and that intraclonal ramets are highly integrated.

It is likely that different genets in a population will become more ecologically equivalent as they expand and integrate the local environment. Genet mortality, which does occur during succession, will be more equitably distributed among genotypes. The integration of ramets over a large area may also buffer a given genet against very localized patch-specific selection pressures and, therefore, prevent selective filtering out of genotypes through successional time. As a result, genotypic diversity of *S. canadensis* populations is maintained over time, and in any given field this diversity is mainly determined by the identity of the initial invaders of the field.

Plants in nature, more often than not, are found as populations and communities of interacting genotypes and species. The biology of any individual is the product of its genotype–environment interaction. As individuals, plants respond to the sum total of their neighborhood environment, both physical and biological. It is highly unlikely that a plant will respond to one aspect and not to the other at any given point in its life. If the aim of plant ecology is to understand the behavior of plants in nature, then physiological, demographic, and genetic approaches have to be used to achieve this understanding. These approaches individually contribute substantially to this, but their combined application would give us more insight into the workings of individuals, populations, communities, and, ultimately, ecosystems. Demographic features of populations are based on individual responses which have physiological bases. These, in turn, have a genetic basis. A thorough knowledge of the plants' biology will have to eventually consider all of the fields.

The examples discussed in this chapter were chosen to illustrate the value of such a multidisciplinary approach. It is hoped that they will stimulate more ecologists to adopt the philosophy espoused therein.

CHAPTER 17

THE PHENOTYPE: Its Development, Physiological Constraints, and Environmental Signals

Robert L. Jefferies

INTRODUCTION

Phenotypic traits appear at every stage of the life cycle of plants. Examples of discrete plasticity, such as seed polymorphisms (Thurston, 1957; Williams and Harper, 1965), the production of dimorphic leaves (Bradshaw, 1965), and changes in sex expression (Policansky, 1982) occur widely in plants. However, many phenotypic responses involve changes in the size and numbers of plant organs (continuous plasticity) (Harper, 1977). Modifications of plant form and structure and changes in the number of plant parts may affect not only the fate of individuals but also their reproductive fitness, topics which are very much the concern of the plant population biologist.

As indicated by Wright (1931), a single genotype may produce an array of environmentally dependent phenotypes (phenotypic plasticity). Although events in the genome provide no way of predicting phenotypic behavior, the ability to produce different phenotypes uncouples the gene pool from selection and buffers the genotype against its effects (Grant, 1963; Stearns, 1982). Alternatively, different genotypes may produce the same phenotype. On the whole it appears that the greater the developmental flexibility of the phenotype, the better a species can cope with environmental uncertainty without genetic reconstruction (Johnson and Cook, 1968). A number of persistent and nagging questions concerning the ability of a genotype to produce different phenotypes remain unresolved. They include the following:

1. What are the mechanisms which control phenotypic expression?
2. What are the ontogenetic constraints that limit and define the boundaries of phenotypic expression?
3. Is the ability of the genome to produce different phenotypes itself subject to selection?
4. Are the environmentally induced phenotypes adaptive?
5. What is the evolutionary significance of phenotypic expression?

Although there are a number of studies that provide insights into these questions, overall the phenomenon of phenotypic plasticity is poorly understood.

CLASSIFICATION OF PHENOTYPIC TRAITS, FITNESS, AND ENVIRONMENTAL HETEROGENEITY

Most schemes that have been proposed during the last three decades to describe and classify phenotypic traits are based on that of Schmalhansen (1949). One such scheme places the emphasis on the ecological and evolutionary significance of different structures. For example, Bradshaw (1965) recognizes three major groupings of morphological traits: those structures which show continuous plasticity, either as new forms or else as modifications of existing structures in response to environmental stimuli (e.g., size of vegetative parts, number of shoots, leaves and flowers); discrete structures which represent alternative forms depending on the nature of the stimulus (e.g., heterophylly in *Potamogeton natans*) and lastly, structures which appear to show no plasticity, either because the appropriate stimulus is missing or because the form is conserved (e.g., shapes of inflorescence and floral characters).

Another scheme (Wareing, 1977) emphasizes the physiological processes involved in phenotypic plasticity and the role of environmental stimuli in triggering responses. Two major divisions are recognized: (1) autonomic traits in which, for example, development of organs is controlled by internal factors (e.g., the development of a leaf or a flower), but their size and form may be modified by environmental conditions;

and (2) environment-determined traits which require a specific environmental signal for the initiation of their development (e.g., in day-length-sensitive species in which the onset of flowering depends upon changing photoperiods).

The various schemes are not mutually exclusive; they represent different approaches to describing the same phenomenon. As indicated by the authors, the boundaries between the different groupings are not discrete—a reflection of the integration of ontogenetic events at all levels of organization within a plant.

Continuous changes occur in the availability of resources (e.g., water, nutrients) which modulate allometric growth of plants and bring about changes in reproductive fitness of the individual. The identification and measurement of environmental factors and the determination of their effects on the reproductive fitness of a phenotype constitute an unresolved problem in assessing selection pressures in unpredictable environments. Frequently, plant population biologists regard the fitness of an individual in a given environment as the result of integration over time of the effects of environmental conditions on growth and development. This approach results in poor definition of both the selection forces in the environment and the mechanistic basis of phenotypic plasticity and in loss of information on the control of phenotypic expression.

In contrast, physiologists have used other methods to study the responses of plants to environmental factors in order to obtain information on the adaptation of plants to different habitats. In comparative studies, bioassay techniques have been used to monitor the responses of individual plants from contrasting environments to a particular environmental factor (see Mooney and Chiariello, and Bazzaz, this volume). The responses of plants from various habitats to a factor are interpreted in relation to the status of that factor in the different habitats, but rarely is reproductive fitness measured. The technique has been particularly effective in examining variations in physiological traits among individuals from different populations. However, information on the physiological behavior of plants, based on large differences in environmental conditions, cannot readily be useful to the population biologist in determining the underlying causes of change in the number of individuals. The physiological data represent short-term responses, often obtained under atypical conditions, where realism has been sacrificed for precision. The spatial scales are frequently too large for the demographer, who is concerned with changes in the fate of individuals over short distances but long periods of time. In addition, biotic influences, such as plant competition (Donald, 1963), pattern and process (Watt, 1947), and the effects of neighbors (see Turkington, this volume) are topics which, although of concern to agronomists and population biologists, rarely involve the physiologist. However, in the case of environmentally determined traits, such as flowering in day-length-sensitive species, major changes in reproductive fitness of individuals may occur if the appropriate signal is received. The recognition of the signal and an understanding of the subsequent developmental changes, which are often separated in time from the signal, represent an area in which close cooperation between the physiologist and population biologist is essential.

Overall, in a number of studies field measurements are incomplete, selection agents' pressures are poorly defined, and as a result, correlations between reproductive fitness of individuals and the prevailing environmental conditions are weak.

EXPERIMENTAL METHODS USED IN STUDYING VARIATION IN PLANT POPULATIONS IN RELATION TO ENVIRONMENT

Four general approaches have been used in studies of both genetic variation in populations and phenotypic plasticity of individuals. All of them complement one another and involve the use of well-established experimental methods (Clausen et al., 1940; Clausen and Hiesey, 1958b; Harberd, 1957; Wilkins, 1959; Heslop-Harrison, 1964; Bradshaw, 1965; Grant and Antonovics, 1978).

They are:

1. The use of reciprocal transplants.
2. Growth of plants under controlled conditions, or in a "neutral" environment.
3. Responses of plants to perturbations of the environment.
4. Responses of plants in ecologically marginal or geographically peripheral conditions.

When these approaches are coupled with demographic methods, together they provide evidence of both the phenotypic response and the reproductive fitness of individuals in different environments. The genetic and physiological bases of the differential responses of individuals both within and between populations can then be examined in relation to prevailing environmental conditions.

One example of the use of these four procedures and of the use of demographic methods is based on studies of the genus *Salicornia*. This genus consists of a number of annual species which grow in coastal and inland saline areas. Electrophoretic evidence has been used to examine genetic variability in two nearly indistinguishable diploid annual species, which are cleistogamous and which grow around the coasts of England and elsewhere (Jefferies and Gottlieb, 1982). All of the 800 individuals of the two species were homozygous at all loci examined. Each species, at least in England, appears to consist of a single completely homozygous lineage, and there is no evidence of genetic variation within a species, as far as can be determined by electrophoresis. The electrophoretic mobilities of 24 of the enzymes were identical in all plants, but for the other 6 enzymes each plant possessed one of two alternate sets of alleles, depending on the species. *Salicornia ramosissima* is common in the upper levels of marshes, whereas *S. europaea* is common in open areas at the seaward end of marshes. Although seeds of both species germinated in late winter or spring, upper marsh seedlings (*S. ramosissima*) grew little until July, whereas continuous growth characterized individuals (*S. europaea*) from the lower marsh. These differences in phenology were maintained when reciprocal transplants were made or when nutrients or sea water were added to permanent plots.

Because neither species showed any evidence of genetic variation, based on electrophoretic data, responses of individuals to environmental conditions are a measure of phenotypic plasticity. We are taking seeds from known parents and studying the growth and development of the offspring under different environmental conditions in order to assess the degree of phenotypic plasticity in each species.

As a first step, reciprocal transplants have been made of seeds of the two species between the upper (landward) and lower (seaward) levels of a marsh. Transplants of seeds were also carried out within each level to provide controls. In this way, the survivorship of individuals was followed over the whole cycle between the seed and the development of a mature, seed-bearing plant (A. J. Davy, Smith, and R. L. Jefferies, unpublished data). In all cases studied, no more than 10% of the seeds developed into plants of reproductive age. In the case of *S. europaea*, there was little difference between the survivorship of individuals in the lower and upper marshes. However, mortality among individuals of *S. ramosissima* at the seedling stage was higher in the lower marsh than in the upper marsh. Observations showed that much of the mortality in the lower marsh was caused by wave action washing seedlings of *S. ramosissima* out of the mud. Complementary field studies carried out in Denmark on these two species indicated that mean relative growth rate of the root at the seedling stage was lower in *S. ramosissima* than in *S. europaea* (A. Jensen and R. L. Jefferies, unpublished data). Because of the slow growth of the roots, seedlings of *S. ramosissima* were very vulnerable to wave action, and this accounted for the low survivorship of this species in the low marsh. Plants of *S. ramosissima* were found in the lower marsh, but they usually grew in a closed sward of other vegetation, which offered the seedlings protection from the waves.

This study of the fate of plants under field conditions involved the use of genecological, demographic, and physiological methods to evaluate the survival and reproductive fitness of individuals. The results indicate that it is possible to link poor survival with a particular physiological attribute, in this case the low relative growth rate of the root. Individuals of *S. ramosissima* were unable to survive in the low marsh, because the growth rate of the root did not exhibit sufficient plasticity.

As indicated earlier, one of the experimental field methods which may be used to study both phenotypic plasticity of individual genotypes and genotypic variation in populations is to examine the responses of plants in ecologically marginal and/or geographically peripheral habitats (Grant and Antonovics, 1978). There have been few studies involving determination of phenotypic expression in different environments (Clausen and Hiesey, 1958b; Abbott, 1976a,b).

Recently we have examined the biology of a marginal population of *S. europaea* agg. on the shores of the Hudson Bay, the northernmost limit of this species in coastal environments in Canada (Jefferies et al., 1983). The plants are confined to south-facing slopes of shallow drainage channels, above the upper limit of spring tides. The angle of slope of the banks of these channels was 5° or less. The difference in temperature of the surface sediments between south- and north-facing slopes was as much as 7°C, particularly early in the season. Although most seedlings emerged in June, germination continued throughout the summer, but plants that appeared late in the season failed to set seed. The breeding behavior was strongly cleistogamous. Because there was no tidal inundation of the site, groups of seedlings emerged in the immediate vicinity of the dead remains of the parent plant. Although electrophoretic variation within and between groups of individuals has not been examined, the cleistogamous behavior and the clumping of individuals indicated that, within groups of plants, genetic variation was likely to be very low. Seeds or seedlings from within a group of plants from a south-facing slope were transplanted within the same site and to a north-facing site. Few or no seeds germinated at the north-facing site. When seedlings were transplanted to a north-facing slope, only a few of the plants died, but the seed production of mature plants was low compared with that for individuals transplanted within south-facing sites.

The movement of individuals of this species just two or three meters across open flat terrain was sufficient to result in poor germination and a fall in seed output of the experimental plants. The near-absence of this species from north-facing slopes indicates that the individuals, which are likely to be of one or a few genotypes, lack the necessary phenotypic plasticity to exploit this niche. The results indicate the usefulness of marginal habitats in experimental studies of the limits to natural selection in plant populations. As Pigott (1982) has pointed out, experiments conducted on vegetation are not only indispensable but are the definitive characteristic of experimental ecology.

Perturbation of the environment in which plants are growing offers an alternative approach to determining plant responses to changes in the environment. Generally, this approach has been underutilized, and in particular the use of endogenous growth substances to modulate the growth of natural vegetation has not been widely used as a technique in ecological studies. The addition of these compounds to natural vegetation may bring about a substantial modification of the architecture of individual plants, a modification which results in changes in competitive ability and fitness of individuals.

In an example quoted by Wareing (1977), apical dominance in birch seedlings is affected by mineral deficiencies. Axillary buds develop actively under conditions of high light and high nitrogen; but if nitrogen is in short supply, apical dominance is strong and lateral bud growth strongly inhibited. When exogenous cytokinins are applied to the shoots of nitrogen-deficient birch seedlings, lateral buds commence rapid growth within a few days. The significance of this study is that it demonstrates the possibility of uncoupling the environmental stimulus from the modulator of the growth process, a procedure resulting in important demographic changes in the number of plant organs. Although it is well known that each type of plant growth substance has a wide spectrum of physiological effects, and that quite different effects may be produced by the same substance in different species, there appears to be considerable scope for modifying plant architecture and bringing about demographic changes with the use of exogenous growth substances.

Growth substances may also be used to modify the breeding system of individual plants. For example, *Turnera ulmifolia* (Turneraceae) is distylous and self-incompatible. However, application of GA_3 to flower buds prevented style elongation in the "pin" flowers (long-styled form). In this state the flowers were self-compatible (J. S. Shore and S. C. H. Barrett, unpublished data). This method clearly has considerable potential for examining the genetic structure of different populations of *T. ulmifolia.*

ONTOGENETIC PROCESSES AND PHENOTYPIC TRAITS

It is important to recognize that the powers of selection to mold phenotypic traits are limited by ontogenetic and morphological characteristics of organisms. These constraints affect allometric growth at different levels of organization in plants. For example, at the cellular level, the presence of a tonoplast and a cell wall is universal in higher plants. Because of the nature and design of the tonoplast membrane, unless both the osmotic potential and the hydrostatic potential (turgor pressure) are the same in the cytoplasm as they are in the vacuole, the membrane will rupture (Dainty, 1979). Likewise, it can be shown that the development of maximum turgor pressure in a cell is affected by the elasticity and thickness of the cell wall and the radius of the cell. These three physical parameters interact to determine the maximum turgor pressure that can develop in the cell. In a number of plants, under conditions of low external water potential, cell size is reduced and cell wall thickness increases in new leaves. These physical changes enable the cell to maintain turgor under conditions of high internal osmotic pressure without rupturing. Alterations in the elasticity of the wall may occur, alterations which involve changes in cell wall composition and the orientation and packing of fibers within the wall. The constraints imposed by the physical properties of cellular structures and the dimensions of the cell determine the degree of plasticity which cells can attain with respect to turgor pressure.

An example of the possible effect of these changes in turgor pressure on development is seen in the aquatic angiosperm *Callitriche heterophylla*, which can develop two different leaf types with distinctive morphological characteristics (Deschamp and Cooke, 1983). The elongate, often dissected, leaves that originate on submerged apices are called water forms and the shortened, broad leaves from emergent apices are designated land forms. Cellular turgor pressure appears to be the mechanism responsible for the determination of leaf form under controlled conditions. The land form of leaf can be experimentally induced in submerged apices by various treatments which mediate cell expansion through their effects on turgor pressure and wall extensibility.

Leaves in different classes of vascular plants show heteroblastic development (i.e., a progressive elaboration of leaf size and shape). Allsopp (1954a,b) indicated that heteroblastic manifestations in different and unrelated species can usually be attributed to the size and nutritional and hormonal status of the apex. The process of development may be reversed when well-developed plants are subjected to unfavorable conditions. Wardlaw (1965) has demonstrated that both nitrogen and sugar sources must be present in large amounts for the frond of *Marsilea* to become large and multipinnate. Although the mechanisms that account for these changes are poorly understood, the modifications are examples of continuous adjustment of allometric growth in relation to environmental changes. The subapical region emerges as one which shows great plasticity under the influence of factors of different kinds.

At another level of organization, Raschke (1960), Gates (1965), and Givnish (1979) have examined the consequences of changes in leaf size and shape in relation to the maintenance of leaf temperature of plants growing in different environments. The linear or needlelike leaf shape of plants from arid environments is of selective advantage because it increases the efficiency of convective heat transfer between the leaves and the surrounding environment. It is possible to predict the effects of changes in leaf size and shape on the energy exchange of leaves and water flow in plants. These physiological consequences of alterations in the morphology in turn affect the survival and fitness of individuals.

Long-lived perennial plants subject to environmental changes may be expected to show phenotypic plasticity in response to the changes. An example from our own investigations provides evidence of heteroblastic development of leaves of *Triglochin maritima* (Juncaginaceae) in relation to salinity (Rudmik, 1983).

Populations of this circumboreal species are widespread in both calcareous and saline environments. Individuals are long-lived and over 80% of the biomass is below-ground. Field observations indicate that individuals growing at sites of high salinity ($Na^+ > 0.5$ *M*) and low water potential (below −2.5 MPa) have small leaves. Experimental results showed that the plants exhibited considerable phenotypic plasticity in response to salinity. Two-year-old plants from a clone were grown in sand cultures and watered with a dilute seawater solution. The births and deaths of leaves were recorded, so that after a complete turnover of leaves the age structure of the leaf population on each plant was known. From this time on, 50% of the plants were watered with undiluted seawater. Subsequently, differences in leaf demography were observed between plants which received diluted seawater (1:20) and those which received undiluted seawater. There was a shift in the age structure of the population of leaves in plants which were grown in seawater. The birth rate of leaves per shoot did not alter but the death rate fell, resulting in a decrease in the rate of turnover of leaves. The mean leaf length of all age classes was less than half of that of leaves treated with diluted seawater, the percentage volume occupied by lacunae was reduced, and cell sizes were also much reduced. When the external salinity was changed back to diluted seawater, the existing leaves were replaced by the original type. Although the detailed ontogeny has yet to be described, it appears that the type of leaf produced depends upon the environment to which the plant is exposed.

This capacity of individuals of *T. maritima* to adjust to a wide range of external salinities is a characteristic shared by only a few plant species.

The reduction in leaf length and cell size of plants grown in seawater results in leaf dimensions similar to those of juvenile leaves from the control plants. However, the physiological characteristics of the cells are very different in the two groups of plants. Osmotic adjustment of plants growing in seawater resulted in a substantial increase in the concentration of proline and sodium and chloride ions in the leaves as well as small cell sizes. There were indications that the photosynthetic rates of the leaves of plants grown in seawater were higher, irrespective of the units of expression.

Plants of this and other halophytic species exhibit a high level of physiological plasticity in response to environmental variability. Continuous morphogenetic changes occur in response to shifts in external salinity.

PHENOTYPIC RESPONSES TO GRAZING

Recently, Owen and Wiegert (1981) have suggested that grazers maximize the fitness of the plants they eat. Their hypothesis is based on the effects of grazing on grasses. As they indicate, it is important to distinguish between a ramet and a genet. Clonal propagation in perennial grasses can result in populations of grasses consisting of few genets (genetic individuals) and many ramets (physical individuals). The selective advantage of ramet formation is that the probability of extinction of the genotype decreases as more and more ramets are formed (Cook, 1979a). The genet spreads over a larger and larger area, so that although individual ramets may be lost, the genotype becomes very old. Owen and Wiegert (1981) postulate that ramet-forming, long-lived grasses have evolved in response to grazing (see, however, Herrera, 1982b and Silvertown, 1982a). Grazing tends to inhibit flowering, so short-term fitness is reduced, but as the genet spreads over a large area, long-term fitness is enhanced because a small number of seeds is produced each season.

We have examined the effects of grazing on salt-marsh vegetation at a site on the Hudson Bay coast. Subarctic salt marshes along the southern shores of Hudson Bay are important breeding grounds for populations of the lesser snow goose (*Anser caerulescens caerulescens*). At La Perouse Bay, Manitoba, there is a breeding population of approximately 4000 pairs of lesser snow geese. Grazing by geese of two perennial, prostrate, graminoid species, *Puccinellia phryganodes* and *Carex subspathacea*, increased annual net above-ground primary production by between 30 and 80% compared with that of ungrazed sites, in two years of study (Cargill, 1981). Grazing also resulted in an increase in the average total nitrogen content of grazed plants. The increase in production was achieved by sustained growth of grazed plants until late in the season, whereas little growth occurred in ungrazed sites after mid-July. Demographic data on the stoloniferous grass *P. phryganodes* indicated that the number of axillary shoots produced per main shoot in grazed areas was approximately double that of shoots in ungrazed sites. Under the influence of grazing, the development of these axillary stolons was delayed until later in the season. In ungrazed plants, early senescence of leaves of both main and axillary shoots took place, whereas a large number of leaves of grazed plants were still alive in September. These results are consistent with the sustained growth of grazed plants until late in the season. The ability of a potential forage plant to maintain growth throughout the season when grazed (by the development of axillary shoots) is of considerable selective advantage in the survival of the clone.

If the grazing pressure is removed, rapid successional development occurs. In exclosures set up three years ago, 12 species are present, compared with only 4 in adjacent grazed areas (D. Bazely and R. L. Jefferies, unpublished data). *Puccinellia phryganodes* and *C. subspathacea*, the major forage species, are dominant in the grazed marsh but have largely been replaced by other, unpalatable species in the ungrazed areas. The colonial feeding behavior of the goose population appears to be a prerequisite for the maintenance of suitable grazing habitats.

Plants of *P. phryganodes* have never been known to set seed, and individuals undergo extensive clonal growth. However, electrophoretic evidence indicates widespread genetic variation both within and between populations (Jefferies and Gottlieb, 1983). It is plausible that somatic mutation may be at least partly responsible for this variation (see Whitham and Slobodchikoff, 1981). The different electrophoretic patterns in individuals provide an opportunity to record the survival of genets of *Puccinellia* in the presence and absence of grazing. In addition, the fate of individual ramets can be followed using demographic methods. The approach is similar to that used by Cahn and Harper (1976a,b), who examined the effects of sheep grazing on *Trifolium repens* in relation to leaf-mark polymorphisms.

The increase in the number of axillary shoots per main shoot in grazed plants suggests that, although the apex of the main shoot may have ceased meristematic activity, the removal of the shoot activates the development of axillary shoots. Whether changes in the types and/or amounts of growth substances upon removal of the apex account for this development remains to be established.

An example of such an interaction is the effect of auxin, gibberellin, and cytokinin on the development of axillary buds of *Solanum andigena* (Kumar and Wareing, 1972; Woolley and Wareing, 1972). If an aerial shoot is removed and a mixture of IAA and GA_3 is applied to the cut end, the uppermost lateral bud develops as a stolon. If only IAA is applied, the lateral bud fails to develop; and if GA_3 alone is placed on the cut end, the lateral bud develops into an orthotropic leafy shoot (Booth, 1959). Thus, the development of an axillary bud, either as a leafy shoot or as a stolon, can be controlled by the appropriate application of growth substances. The effects of these growth substances on plasticity and the development of ramets may be substantial.

CONCLUSIONS

Although the discussion has centered on the plastic responses of individual organs of plants, the control mechanisms which govern the responses frequently involve long-distance transport of metabolites and growth substances within plants. The implication is clear, that the whole plant must be considered in examining the mechanisms which govern the development of individual phenotypes. The demographer is primarily concerned with the ecological behavior of individual phenotypes and with their survival and reproductive fitness. If these demographic events in individuals and in plant populations are to be interpreted in relation to environmental changes, much more attention will have to be given to the effects of the environment on the control of phenotypic expression.

SECTION IV
AGRONOMIC IMPLICATIONS OF PLANT DEMOGRAPHY

INTRODUCTION TO SECTION IV

A major component of present-day plant population ecology owes its existence to studies derived in the context of man-managed ecosystems, particularly agricultural systems.

From its very beginning, the invention of agriculture by early man was embedded in the manipulation of ecological traits of the "embryonic" cultivars and the new ecosystems in which they were induced to grow. Indeed, the success of agricultural systems depended almost totally on how well man understood several basic principles of plant yield, and how intelligently he managed them. A relentless game of natural selection dictated the rates of success or failure in this process of plant domestication: If man was unable—through his agricultural system—to replace those calories spent in the food-raising effort, he almost certainly perished.

The contribution of plant population ecology to the management of weeds has been very restricted. This is a point clearly made both in Chapter 18 by Mortimer and Chapter 19 by Snaydon in this last section on the agronomic implications of fundamental research on plant population ecology. Integrated control of weeds contains an element of the management of the weeds and the agricultural system, but certainly could benefit from an ampler ecological knowledge of the system as a whole.

Mortimer points out that despite the fact that "weedy" species have been some of the organisms which originally attracted the interest of plant population ecologists, very few detailed studies have been carried out on injurious species with long-term agricultural interests in mind. The crop protection literature is conspicuously poor in examples of natural regulatory agencies of weed populations.

Using case studies as examples (Snaydon) or referring to the general literature (Mortimer) reference to different agricultural managements is made, but it becomes quite apparent that empiricism has been more the basis of manipulation of the agroenvironment and the crop than the application of knowledge derived from fundamental research.

In addition to the questionable efficiency of chemical methods of weed control, the very complex problem of herbicide resistance by certain weeds (Mortimer) suggests that probably certain highly technological, short-term, cash-oriented agricultural systems may be trapped in a blind alley. Perhaps their reorientation to economically less attractive (in the short term) but ecologically (and hence economically in the long term) more sustainable systems is necessary.

Although both chapters in this section deal mostly with aspects of the biology of weeds or the application of demographic studies to the management of agrosystems, there are other very important concepts in plant population ecology that need to be mentioned.

The introduction of knowledge about plant form and efficient use of resources (e.g. by being compatible with other crops) has been even more limited than the application of demographic studies. A considerable degree of plant-form manipulation goes on automatically in plant breeding efforts but it is not a purposeful selection or design of ideotypes. Characters other than those concerned with the efficient capture of resources by given plant forms have tended to dominate plant breeding efforts.

There is a clear need to explore genetic traits that maximize ecological combinatory ability of crop plants. This is counter to what has happened up to now. Crop plants have been selected and bred to become highly "selfish" rather than combinatory organisms. On the other hand, weeds have become extremely good "sharers" (opportunist users) of the agricultural system. No wonder many present day cultivars originate from "weedy" species once associated to primeval crops.

Snaydon's claim for the need to study the agroecosystems as integrated systems is very important to understanding them properly. Lack of this vision, for example, does not allow agronomists or ecologists to fully perceive whether what we call weeds are really injurious, neutral or beneficial components of the agroecosystem.

It is intriguing that the flow between the fundamental knowledge in plant population ecology and the applied research in agronomy has been highly unidirectional, with plant population ecology being the principal recipient (Snaydon). But there are good reasons why this may be happening.

The full development of the industrial revolution, the discovery of fossil sources of energy, and the social changes that occurred during the nineteenth century produced profound changes in the basic patterns that governed the agricultural systems in the newly industrialized colony-dependent countries in Europe and America. A general trend to conform the agricultural systems to the economic framework of these countries was established. No longer was their economic organization largely a result of—or at least highly dependent on—an agricultural system governed by ecological constraints. Thus, economic constraints largely started dictating the characteristics of these man-managed, food-producing systems.

This change in "governing patterns" resulted in stereotypes of agricultural systems: monocultures, weed-free crops, plants and agrosystems designed for maximum applicability of mechanization, plant breeding efforts highly directed to satisfy marketing demands, etc. Some components of these stereotypes made more sense than others, but all were exported almost wholesale to many areas of the world where they became nearly official standards of agricultural systems, to be achieved at all costs.

Within this context of "ecologically alienated" agricultural systems, it should not come as a surprise that fundamental research in plant population ecology has contributed little to generate criteria for the better management of these systems.

It is more logical to expect a greater impact of fundamental research on those agricultural systems that are structured on ecological bases rather than on short-term economic constraints. A need to investigate more ecologically based, traditional systems of food production, both in and out of the tropics, would surely help us learn how more integrated, ecologically harmonious systems can be devised or bettered. At least, these systems are the carefully shaped product of centuries (and often millennia) of empirical but intimate knowledge of crops and their associated flora (including some injurious species) in an integrated and very long-term view of the system.

With the concern for the availability of fossil sources of energy for various inputs in the agricultural systems, and for the environmental effects of some of those inputs (e.g. pesticides and herbicides), a new look at more ecologically compatible systems may not be too far, not at least in the under-industrialized world. A truly "promised land" of contributions from population ecology for the better management of agricultural systems lies in the future for those who would devote efforts to research within the integrated concept of agricultural systems. This involves applying knowledge coming from plant demography and population genetics, and requires thorough understanding of morphological and physiological traits of crops and their associated companions, the weeds.

CHAPTER 18

POPULATION ECOLOGY AND WEED SCIENCE

A. Martin Mortimer

INTRODUCTION

The impact of weeds upon the activities of man constitutes a paradigm of a natural hazard of longstanding. Kates' (1970) definition of such a hazard[1] is clearly appropriate to the complaint about unwanted plants (weeds) made by civilizations dating into early times (Godwin, 1960), facets of which surface in their vernacular. An early English language usage of the word *weed* has been traced to the Anglo-Saxon *woed*, and it is tempting to contemplate that these people in Britain equated weeds with problems at harvest, for their name for the month of August was *woed-month* (Wall et al., 1977). Then and now in temperate agroecosystems, the phenology of many unwanted species is such that they may conspicuously betray their presence by flowering in late summer and indeed contribute to problems at harvest. Whether it was perceived, however, that by this stage other damage to the crop had already occurred is perhaps conjectural. But Holzner (1982) attests that it was well known at least to the peoples of the Fertile Crescent.

Damage to crops by animal pests, microbial pathogens, and weeds on a worldwide basis was estimated by Cramer (1967) to be in the region of 35%, and recent estimates (Pimentel, 1976) have confirmed this level despite developments in agricultural technology. There exists a belief that animal pests, in causing a greater fraction of total

[1] An interaction of people and nature governed by the coexistent state of adjustment in the human use system and the state of nature in the natural events system.

crop damage, contribute in a proportionally similar way to financial loss, yet this may be questioned when overall growers' budgets are considered. In 1960 the cost of land tillage was 16% of crop value, and approximately one-half of cultivation was made necessary by the presence of weeds (Shaw and Loustalot, 1963). Taking this into account, Wall et al. (1979) cited in the case of forage and vegetable crops in the state of California that the cost of tillage for weed control alone far exceeded the losses due to animal pests or plant diseases. Clearly, however, any attempt to derive comparisons of gross estimates (whether state, national, or global) is fraught with difficulty and prone to error, and this example is out of date in more ways than technologically. However, as Parker and Fryer (1975) have commented, there is a lack of contemporary information; and moreover, it is often that documented estimates concern loss of production (yield loss × price) alone and ignore the important resource allocation and market effects caused by weed infestations and their control (Vere and Auld, 1982). In attempts to maximize world food production, the pressing need to equate the costs of pest control in the broadest sense to the proven benefits gained from control (Geier, 1978) has been particularly highlighted in the last decade (Auld et al., 1979a). As the cost of raw materials has escalated, being powered by the price of fossil fuel (see Pimentel, 1979, for commentary on the relative "values" of fossil fuels), so too has concern for the environment (Anon., 1979; Pimentel et al., 1980; Robertson, 1981). To conceive and write the equation(s) governing cost–benefit for a crop production system and to replace the algebraic terms with reliable numerical estimates requires at the very least the integration of the skills and knowledge of the ecologist, the economist, and the grower. This review explores the role of the plant population biologist in the context not only of the governing views of economics but in a broader ecological perspective. However, it is not my intention to review a vast literature (see Holzner and Numata, 1982) but rather to concentrate on salient points that are apposite. The definition of the word *weed* is narrowed to injurious species in plant communities which are managed for food production.

THE CONCEPTUAL FRAMEWORK— THE EQUATION OF COST–BENEFIT

Within most sociopolitical systems, the declared aim of crop protection is the maximization of desirable biological production for the least costs, in essence a process of optimization subject to both internal and external constraints (Norton and Conway, 1977). Cost may be evaluated with differing scales and units if aesthetic and moral arguments are invoked, but comparative assessments are conveniently handled in financial terms.

Southwood and Norton (1973) argued that a primary objective of insect pest control was to maximize the function

$$Y[A(S)] \times P[A(S)] - C(S)$$

and as such this relationship (an equation of cost-effectiveness) has equal applicability to weed control. The control function (level of attack) $A(S)$ or the weed infestation resulting from a strategy of control (S) determines through the quantity (crop yield) $Y[A(S)]$ and quality (price) $P[A(S)]$ damage functions the monetary value of the crop which is produced for the cost $C(S)$ of the control strategy. While pertinent to all cropping systems in describing the relationship between biological production and economic constraint in financial terms, this function belies the wide range of options available to the producer of a crop and the interactive nature of variables governing the decision-making process during the life of the crop.

Most cropping systems involving the harvesting of vascular plants are phasic by nature. This is most evident in forestry, arable farming, and horticultural enterprises where there is often entire removal of plants at harvest; but it is also illustrated by perennial cropping systems, such as grassland, which requires controlled management for optimal use (Anslow and Green, 1967). This phasic nature of agroecosystems results, on the one hand, from the tendency for periodicity of biomass production determined by seasonal climatic changes as well as the act of harvest and, on the other, from the desire of the grower to maximize productivity of the enterprise by optimizing biomass production. Slash-and-burn agriculture and systems of intensive and extensive land use are some of the strategies which have evolved in response to this desire; and one which has in many cases (grain, fruit, and forage crops) been approached by the growing of monocultures demanding a specific schedule of husbandry practices. The logical extension of this simplicity, the concept of a weed-free cropping environment (Elliott and Boyle, 1963), appeared as a practical reality by 1970 in the western hemisphere with the progressive use of herbicides, discovered over two decades earlier.

In many modern crop production systems, a herbicide-based technology has replaced the traditional cultural one (for example, *Beta vulgaris*; Gunn, 1977), and the dependence of many world food production systems on herbicides is paramount (Fryer, 1981; Haas and Streibig, 1982). This change has arisen for a variety of reasons [not the least of which are the cost of labor and the desire to prevent soil erosion (Robinson, 1978)] and has gone hand in hand with herbicide development. The virtual weed-free cropping environment in such cases is a concomitant, desirable travelling companion if not a goal, despite the specter of herbicide resistance (Harper, 1957). Nevertheless, Sagar (1974) argued that the proposition underlying the concept of a weed-free cropping environment was not necessarily tenable. Tolerance of weeds by crops in terms of lack of economic injury, the tacit assumption of the unilateral detrimental value of weeds, and the maintenance of reinfestation potential of cropped land were cited as serious issues for the weed scientist. For many cropping systems they remain unexplored, and it is only very recently that they have become a focus of attention. The motivation is mainly economic and not ecological even though interspecific selection through changed management practices has resulted in altered weed floras (Fryer and Chancellor, 1970). In 1981 British agriculture spent in excess of £100m on herbicides to exclude weeds; and although well employed, much of this expenditure occurred in the absence of both defined objectives and performance criteria (Elliott, 1982). With these comments in mind, it is appropriate to assess the objectives of weed control in relation to quantity and quality damage functions.

Discerning the components of crop damage and the causal agents and measuring their effects is more readily achieved for crops which require whole-plant destruction at harvest. At the opposite end of the spectrum, in rangelands and grassland pastures the transition between trophic levels makes assessment complex. Weeds in these agroecosystems may have lethal effects on the production units, with considerable financial penalties (Nielsen, 1978), but the effect of nontoxic indigenous species on livestock production is a matter of debate (Auld et al., 1979a). The elements in the chain of relationship—prehension, palatability, digestibility, conversion efficiency—between plant biomass on offer and herbivore weight on hoof have been substantially investigated for cultivars of many species (Allden and Whittaker, 1970; Reed, 1972; Stobbs, 1973), but detailed studies on the impact of weeds in pastures are rare. Productivity that requires transference across a trophic level has a rigorous predictive base still in its infancy, particularly as far as weeds are concerned (Noy-Meir, 1975; Harris, 1978).

The comments that follow, therefore, relate specifically to homogeneous monocultures of crops where mixtures of species are undesirable, although some parallels may be deduced for other simple cropping systems (intercropping). Complex heterogeneous crops (pastures and rangelands) are well discussed by Snaydon (1980a, 1982) and Wilson (1978).

THE AIMS OF WEED CONTROL

Homogeneous cropping systems are characterized by uniformity of crop species, defined planting arrangements, and resource additions to maximize yield. The habitat is commonly predictable, with an inherent periodicity (seasonality) due to harvesting and a changing resource spectrum—not the least of which are radiant energy at the ground surface as canopy relations change and nutrient availability as fertilizer budgets are consumed. Crops with annual life cycles typify such systems but so do perennial trees and shrub crops in which the range of influence of weeds may be just as great (Atkinson and White, 1981). Where weed control is practiced frequently, the above-ground injurious community is a poor reflection of dormant reserves below-ground (Thompson and Grime, 1979), which may well be represented by a very few species giving a "characteristic" weed flora.

Damage functions

The damage caused in the diversification of a crop by weed infestations may be assessed in relation to existing and future cropping cycles. Within a cropping cycle (planting, growth, and harvest), damage to the crop occurs through loss of yield and a lowering of harvest efficiency (quantity damage components) and by determining the price of the crop (quality damage). The nature of this damage to the "current" crop will depend on the components of yield that are harvested and their response to interference from weeds. The recurrence of damage to "future" cropping cycles from a weed infestation will in turn relate to these quantity damage components—reflecting in fitness parameters of the weed species. Overlaying these direct effects of the presence of weeds are those indirect ones that are more difficult to assess—the role of weed populations in determining pest and pathogen incidence (Way and Cammell, 1981). Table 1 illustrates the easily envisaged damage relationships for a weed in a cereal crop.

The recognition of damage functions and their relationship to the cropped plant community is not a new concept to agricultural ecology. For insect pests (Conway, 1976) and plant pathogens (Zadoks, 1971), it is well documented and forms an integral part of control rationale (e.g., Krause et al., 1975; Croft et al., 1976). For weed species, however, the literature displays a dearth of information that enables both the current and the future damage functions to be precisely described, although they may well exist in the minds of some successful growers. Although the functions describing arable crop yield response to the density of weed infestation at crop harvest have been extensively cataloged (see Dew, 1972), the reciprocal effect of the crop on the weed has received far less attention. Yet a crop is often a potent agent of biological control (extending the definition of Huffaker, 1957), which through competitive processes may reduce the fitness of individuals in

TABLE 1. Causes and relationships in the damage function for a weed infestation in a cereal crop.[a]

Damage	CROPPING CYCLE	
	Current	Future
Quantity (causes of crop yield loss)	Competition for growth resources Predation from animal pests and pathogens migrating from host weeds Harvest inefficiency Application of control measures	Contamination by reproductive propagules Survival of adult plants Maintenance of hosts for pests and pathogens
Quality (causes of commodity price loss)	Grain and straw contamination	

[a] See text for further details.

weed populations dramatically. In theory, measurement of the quantity damage function enables perception, action and economic damage thresholds to be identified in a cropping cycle and in so doing leads to a rationale for the cost effective use of control procedures.

These quantity damage functions bear an intrinsic relationship to the life cycles of weed species which in life-history may be constrained in the realized agricultural niche (Mortimer, 1983). Notionally, at least, they may be considered in relation to the relative contributions to damage that individuals in injurious populations make. For monocarpic weed species with an age-distributed population in an annual cropping system, individuals with a similar or greater age than the crop are likely to cause disproportionate damage in terms of crop yield loss in comparison to younger members of the weed infestation. Conversely, these late-emerging individuals may make the largest contribution to inefficiency at crop harvest because they are delayed in maturity and, in consequence, constitute proportionally the bulk of living biomass at harvest and interfere with harvesting operations. Human-applied discriminatory control procedures against injurious plants may also interact with age or perhaps more appropriately its partial correlate: the stage of development. Those individuals which by virtue of appropriate size and physiological stage of development become the targets of control may protect suppressed individuals in a hierarchy, individuals who are released only with the demise of the elements of the population conspicuous to the control measure. Contributions to future damage may also be distributed disproportionately according to age-state. The act of harvesting may prohibit late-developing individuals from dispersing reproductive propagules to the ground through death or by removal with the crop at harvest. Clearly, comments such as these are specific to weed crop associations and any generality is restrained to particular crop production systems.

Control functions and costs

A control function in the context of an equation of cost-effectiveness measures the level of weed infestation after the application of a given strategy of control. However, the design of a control strategy invokes decisions at two levels—intrinsic and extrinsic to the crop.

Plasticity of plant parts and mortality of entire individuals constitute intrinsic control measures that are density-related in both crop and weed populations. The decision at the start of a cropping cycle to plant a given variety of a crop species at a chosen density and to follow specific husbandry procedures (for plant establishment and growth) which favor the predominance of one component of the community reflect a choice made with hindsight of a predetermined community response strongly tilted toward maintenance of a particular low-diversity structure. If any generalizations are useful, the agricultural significance of yield of harvestable product per unit area and its relation to choice of planting or sowing density (Donald, 1963; Willey and Heath, 1969) results in spatial heterogeneity of crop biomass (even if regularly arranged) during part, if not all, of the cropping cycle. Thus, there is the opportunity for invasive increase of weeds at densities at which natural intraspecific regulation may be important. However, the part played by self-thinning as opposed to plasticity in regulation of weed populations before appreciable crop loss occurs remains largely unestablished.

The second level of decision-making invokes extrinsic control measures that perturb the plant community and reflect the grower's desire to manipulate the course of events within it and the trajectories of individual plant populations. Traditionally these have been cultural, and only relatively recently have they become dominated in the developed world by chemical means (Fryer, 1981). Ease of application and costs vary according to the assemblage of species present and the stage of development of the plant community. For example, the suppression of a weed species generically similar to the crop in a well-established community requires specificity of herbicidal action and precision in application of the control and has high attendant costs. Conversely, selection against dicotyledonous species in the presence of monocotyledonous species may exploit gross morphological and physiological differences, require less precision in application, and have low cost.

The distinction between prophylactic and therapeutic measures at the second level is, however, dependent on time scale and a priori–posteriori decision-making. A commitment at the start of the cropping cycle to plough—to deliberately bury weed seeds—may be considered prophylactic to the coming cropping cycle but therapeutic in the long term. Alternatively, chemical control as a selective postemergence herbicide is therapeutic to the current crop but prophylactic in the long term, if successful. The distinction and its time scale has obvious economic implications.

The gamut of control measures available for weed control (Table 2) suggests the subtleties with which integrated weed management systems may be devised. While testifying to the ingeniousness of man, if not to his ecological ingenuousness (May, 1975), all seek to regulate the size of weed populations with a view either to eradication of weeds from a cropping system or to containment to a level satisfactory to economics, visual attractiveness, or pride.

Implicit to the foregoing discussion is the notion that the financial costs of the components of weed control practices bear acute consideration. Although these are not as readily attainable as first appearances might suggest (namely, fixed versus variable costs; Nix, 1978), it is perhaps salutory to consider the acceleration of costs. In this respect

TABLE 2. Components of weed control strategies in homogeneous cropping systems.[a]

Class of weed control measures	Component
Cultural	Land cultivations during seed bed preparation Mechanical removal during crop growth Minimal disturbance procedures: vehicular access points and tramlines Resource control: fertilizer, irrigation Timing of husbandry operations Crop rotation
Chemical	Pre- and post-crop-emergence herbicides Crop growth regulators Crop safeners
Biological	Choice of crop and variety Microbial herbicides Biological control (sensu stricto)

[a] See Roberts (1982) for further details.

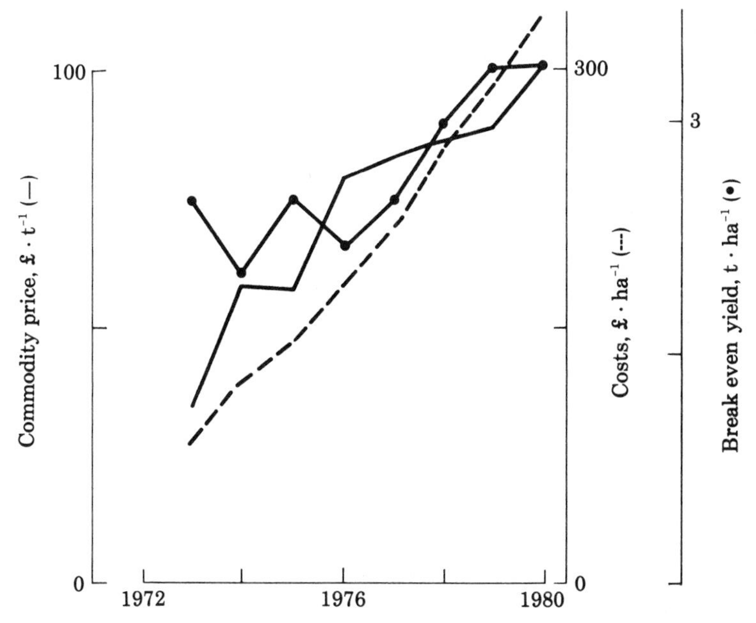

FIGURE 1. Trends in the production of winter wheat in the United Kingdom. (Data from North, 1981.)

(Figure 1 and Table 3), the data speak for themselves. McCarl (1981) gives similar data for agriculture in North America.

A cost less overt to the ecologist but of significance to the agriculturalist is that associated with changing weed control programs because of resistance to control measures in weed populations. Both ecological and evolutionary changes in weed populations are well docu-

TABLE 3. Comparison of winter wheat production in the United Kingdom.[a]

Production parameters	1969	1980
Price of seed ($£ \cdot kg^{-1}$)	0.025	0.105
Price of nitrogen fertilizer ($£ \cdot kg^{-1}$)	0.10	0.28
Price of machine fuel (trade) ($£ \cdot l^{-1}$)	0.0176	0.202
Mean nitrogen fertilizer rate ($kg \cdot ha^{-1}$)	90	145
Mean yield ($t \cdot ha^{-1}$)	3.52	5.89

[a]Modified from Fowden (1982).

mented (Baker, 1974), but the significance of population response to selection pressures posed by herbicides has only begun to be fully appreciated and understood (Gressel et al., 1982).

REGULATION OF INJURIOUS SPECIES IN PRACTICE

In 1976 Sagar and Mortimer concluded that description of the natural regulatory agencies of weed populations was largely absent in the crop protection literature and that little attempt had been made to seriously consider density-related agents. In stark contrast is the substantial agrochemical literature on density-independent crop protection measures. In part this arises from the variability shown by many homogeneous crops to the effects of weeds (Zimdahl, 1980) and the fact that prevention of economic damage is characteristically cost effective only after logarithmic-scale changes in weed density (Cussans, 1980). The desire for a clean cropping environment and the undoubted success of chemical control measures in reducing weed floras to a few pernicious species has done much to delay this line of inquiry into the regulation of weeds in an ecological setting.

Injurious species which infest annual crops often mimic the crop (Baker, 1974) in some characteristics (for example, relative growth rate, reproductive allocation) and display attributes typical of pioneer species in secondary plant succession. Such reflections led Bunting (1960) to argue that in many instances crop production constituted managed secondary succession in perpetuity; but, as Snaydon (1980a) remarks, this comment masks the immense variation that the weed flora may display. Nevertheless, in cropping systems punctuated by periodic planting and harvesting, judicious choice by man of the timing of events is a primary means of regulation. These events comprise the planting of the crop, the onset and duration of interference between crop and weed, and the time of harvest. These arrangements circumscribe the temporal backdrop upon which other elements may influence the outcome.

Land preparations

The act of "seed" bed preparation for the coming crop involves a perturbation of the soil or its immediate surface chosen to precede favorable climatic conditions for plant establishment. Acquisition of a clean "seed" bed requires manipulation of the agencies controlling the fates of seeds and more particularly their incorporation into the soil seed bank and their longevity in the soil (Sagar and Mortimer, 1976). This subject has recently been reviewed by Cook (1980), who comments on the paucity of demographic data on aspects of seed dynamics that may elucidate the adaptive significance of characters of seed populations, features of which are now well documented, particularly for weed species (Harper et al., 1970). From an ecological and managerial point of view, a considerable data base is available for the prediction of the loss in situ of seeds in the soil profile in a range of environments (e.g., Roberts and Ricketts, 1979; Roberts and Neilson, 1981). Characteristically, total soil seed losses may be large, with seed exposed to a constant probability of death which differs among species. What happens to seed populations on the soil surface in agroecosystems is, however, less defined; but changes, as might be expected, are dramatic. Table 4 compares levels of mortality visited on wild oats populations in the United Kingdom through husbandry practices and natural causes. Clearly, although straw burning in autumn reduced by one-third the size of seed populations where it was effective (Wilson and Cussans, 1975), natural processes in uncultivated land (given time) may lead to similar overall reductions. Moreover, by breaking seed dormancy, straw burning also increased the seedling densities present in the autumn, as did stubble cultivations, which necessarily aided seed burial. Timing of autumn cultivations for seed bed preparations then may have significant effects on future wild oat infestations by lessening the risk of mortality to seed by burial within the soil profile (see Wilson, 1978, for long-term effects under cropping). Yet a strategy of delaying cultivations until required for a spring crop brings with it alternative costs—ingress of other species and often the lowered yields associated with a shorter growing season.

TABLE 4. Influence of cultural practices on the fates of seeds of *Avena fatua.*[a,b]

	Survivorship in surface seed population				**Proportion of total seed population present as seedlings in October**	
Cultural practice	**Sept.**	**Oct.**	**Nov.**	**Dec.**	**After stubble cultivation**	**After no cultivation**
Stubble burnt	0.589	0.515	0.490	0.234	0.118	0.035
Stubble not burnt	0.870	0.669	0.630	0.286	0.032	0.007

[a]Mean data from Wilson and Cussans (1975).
[b]Distribution of seeds after harvest (September):
proportion on soil surface, 0.87;
proportion within soil surface, 0.13.

To a large degree, elucidation of the natural causes of mortality in weed seed populations remains to be attempted. Although for the practitioner, the influence of tillage has been established, at least for some important weeds (Cussans et al., 1979; Moss, 1980), the effects of control practices on the agents that undertake seeds and shape the survivorship curve of a seed population on the soil surface has been investigated in only a few instances, often with emphasis on rodents (Masselink, 1980; Bochert and Jain, 1978). Invertebrates, too, may have equally significant regulatory influences (McRill and Sagar, 1973; and Figure 2). Herbicides do not appear to have any direct action on the soil fauna (Way and Cammell, 1981), although there is evidence that the magnitude and timing of soil disturbance does (Edwards, 1975; Altieri, 1981). Polymorphisms determining seed shape and size, together with dormancy, have much to do with the regulation of rates of seed burial and the partition of a seed population to the alternate and sometimes lower constant risks of mortality within the soil profile as opposed to those on the surface. They are worthy of further study, particularly in relation to changing tillage practices. Minimum tillage and, particularly, direct drilling techniques have brought with them the creation of niches for new weeds (Fryer, 1981).

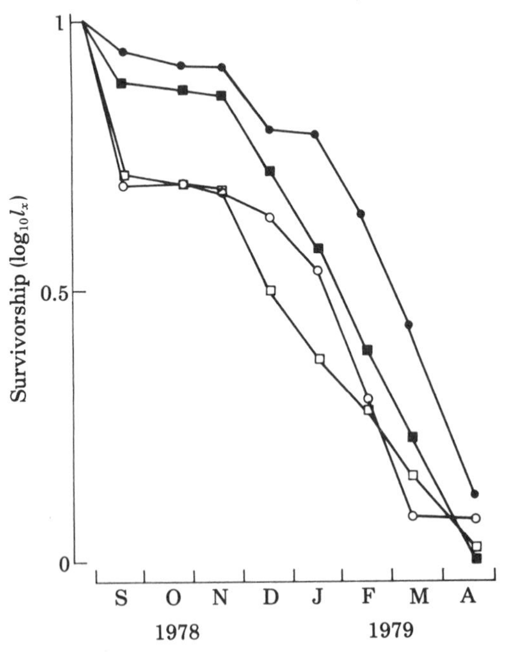

FIGURE 2. Survivorship of *Plantago lanceolota* seeds on the soil surface in grass swards (solid symbols) and on bare ground (open symbols) where invertebrate soil fauna was present (circles) or chemically excluded (squares).

Interference in crop–weed communities

The agronomic and horticultural literature is extensive in its description of the effects of weeds on crop yield, but attempts at more than trivial generalizations appear impossible (Zimdahl, 1980). In part this would appear to stem from a lack of appreciation of the theoretical basis originated by Bleasdale and Nelder (1960) and Bleasdale (1966), more recently extended by Watkinson (1980, 1981), a model which enables mortality of whole plants and plasticity of yield components to be assessed in relation to density using additive experimental designs. The state of the art of controlling weeds by mediating crop growth has been critically reviewed by Snaydon (1982a), who exposed the current level of ignorance of this mechanism of regulation of injurious species. Four points deserve comment.

The first is that, from an ecological point of view, density-independent factors may disturb the intensity of reciprocity of interference between a crop and a weed by magnifying the absolute difference in emergence time (often by displacement of weed seed lower in the soil profile). In populations of wild oats, individuals that emerge with spring barley or soon after make the major contribution to yield loss, later ones having little effect (Peters, 1978). Reliance on differences in emergence time, however, assumes equivalence in relative growth rates and resource demands and partition. Differences in the former, at least, and the desire to know when effort is most effectively expended in the removal of weeds has led to the empirical concept of "critical periods for competition" (see Figure 3, following Nieto et al., 1968; Roberts, 1976). Yet though of practical value for specific (constrained) situations, its empiricism underlines its limitations. The weed flora in toto is presumed injurious, and no distinction is made between the nature of the competitive processes occurring. In the early stages of the interaction, exploitation-competition may predominate, to give way later on to interference-competition as the components of the mixture change. Little is known of the significance of variation in limiting resources during the cropping season. Preemptive occupation of biological space and consequent disproportionate interception of light may be the single dominant factor.

Manipulation of resources and crop density at first sight may seem a powerful tool for weed control. For arable crops, a number of workers (e.g., Felton, 1976) have shown that increasing crop density reduces yield loss due to weeds, as does the application of nitrogenous fertilizer (Scott and Wilcockson, 1976), but this is not always the case (Appleby et al., 1976). Although crop rotation is a well-established control mechanism exploiting competitive mechanisms in the broad sense, the

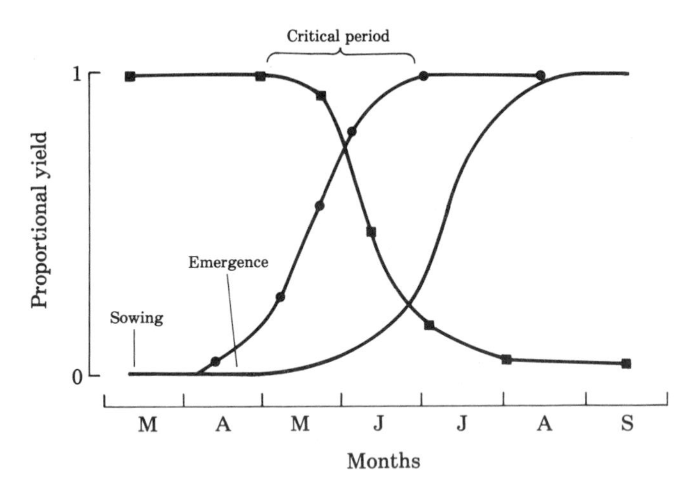

FIGURE 3. The "critical period of competition" illustrated for a crop of *Allium cepa.* ——, Changes in crop dry weight from sowing to harvest; ■, yield response resulting from delaying the start of continuous weed removal; ●, yield response occurring from delaying the termination of weed removal. (After Roberts, 1976.)

paucity of information on spatial and temporal availability of resources capable of limiting growth in agroecosystems and an exact description of the realized niche of weeds and crops precludes definitive comment on resource manipulation for weed control (McWhorter and Shaw, 1982). Indeed, in some cropping systems (wide-row crops), at least, governing constraints imposed by high resource inputs and harvesting techniques predetermine crop density and spatial arrangement of crop and leave few alternatives.

Third, the mechanisms of resource deprivation (and identification of limiting resources) in crop–weed associations remains in many instances to be experimentally explored (Hall, 1974). Below-ground interactions appear in some instances (*Hordeum vulgare* and *Polygonum lapathifolium*; Aspinall, 1960) to be as important as those involving light interception in leaf canopies (Donald, 1963). As Harper (1977) illustrates, discovering causal relationships is often an immense challenge, and couch grass *Elymus* (*Agropyron*) *repens* provides a good example (Bucholtz, 1971). This species is an extravagant consumer of nutrients (Werner and Rioux, 1977) and is a persistent weed of arable crops (Håkanson, 1975) because of its regenerative powers from buds on underground rhizomes. Clones of *E. repens* comprise a series of tillers morphologically interrelated at two levels: proximally within the shoot complex of primary shoot (from erected rhizome apex) and subtending tillers and distally among shoot complexes inter-

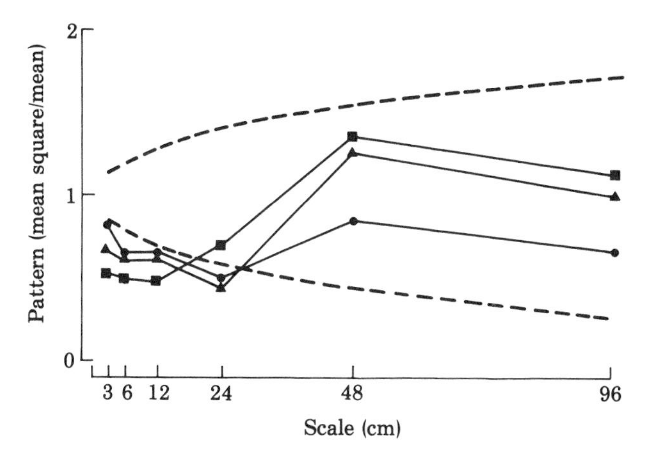

FIGURE 4. The spatial pattern of shoot complexes in *Elymus repens* populations at high densities. ●, Pure stand (668 m^{-2}); ■, on the boundary of a crop (720 m^{-2}); ▲, within a crop of wheat (586 m^{-2}). Dashed lines are 95% confidence limits.

connected by rhizomes. In a growing season, a plant arising from a single bud in unrestricted conditions has prodigious fecundity and may give rise to a clone bearing over 2000 buds, which, if separated by fragmentation, have a high probability of establishment (McMahon, 1982). In minimal till agriculture, clonal integrity is often retained as rhizomes are undisturbed by soil cultivations, and expression of clonal architecture may occur. In pure stand and in the presence of a crop (Figure 4), this results in an even packing of shoot complexes at several scales of pattern, which in contrast to other species (Harper and Bell, 1979) appear only partly explicable by application of rules of morphology (Mortimer and McMahon, 1982, and unpublished data). Plagiotropic primary rhizomes arising from the base of a shoot complex grew at random in a two-dimensional orientation, whereas lateral rhizomes branching from parental rhizomes at small acute angles served to amplify the direction of colonization of a primary. Computer simulation led to the conclusion that only regulation of small scales of pattern (3–6 cm) was a consequence of morphology (Figure 5). Mortality of shoot complexes in the field in the absence of husbandry practices was slight, and population flux with distance-dependent mortality was insufficient to account for observed patterns. In *E. repens*, rhizomes constitute a historical record of growth; and hence, resource availability and internodes showed conspicuous variation in length

(1–80 mm; McMahon, 1982). Moreover, oscillations in internode length were found to correlate with deficits in soil moisture, a resource known to be critical for growth in this species (McIntyre, 1967). This finding suggests that rhizome apices may sense the soil environment in a fine-

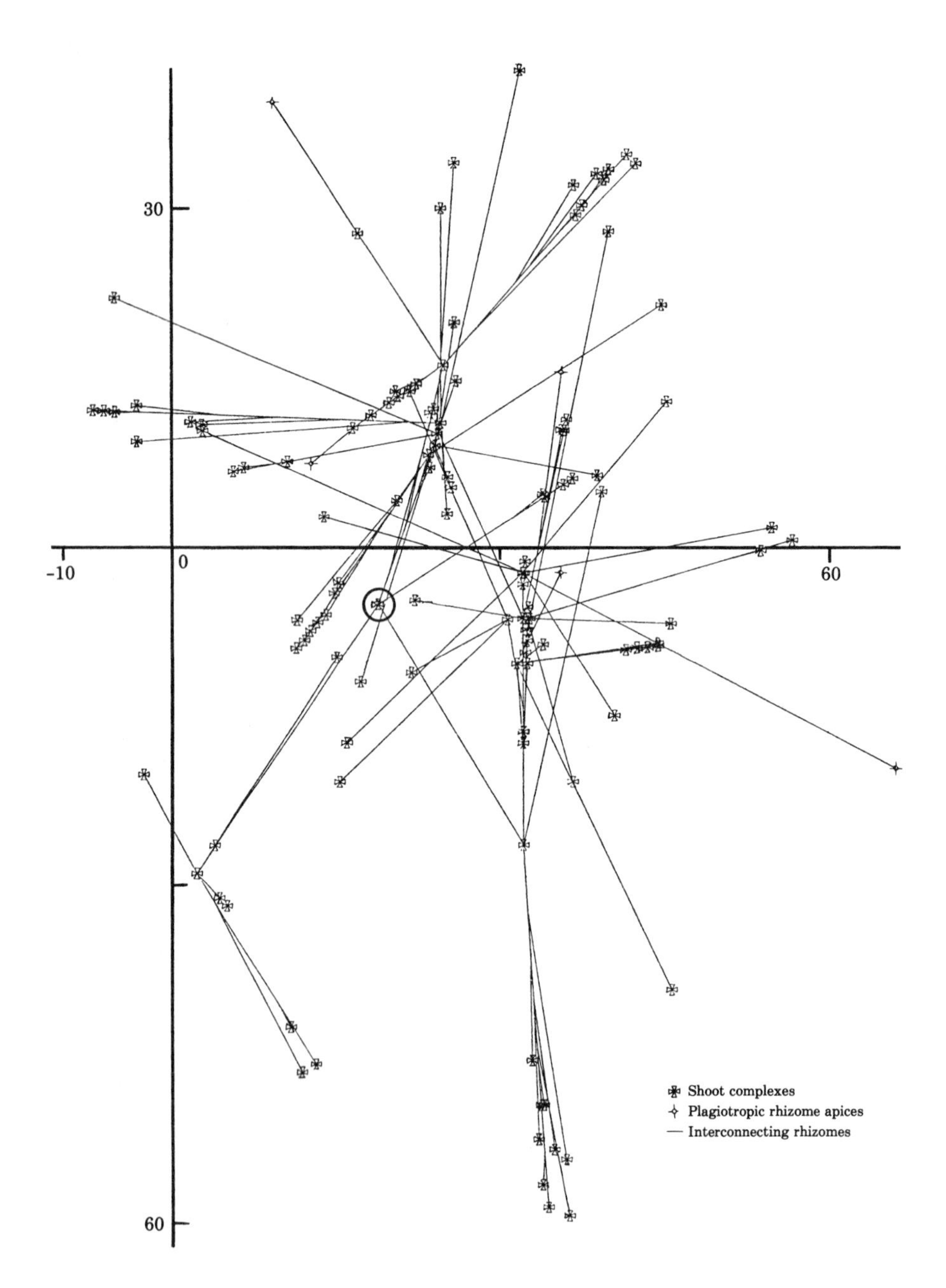

grained manner, possibly through the extensive rooting systems that occur at nodes and thus behave in an individualistic fashion. Orthotropic expression in rhizomes and, hence, position of recruits to the aerial population may in part be determined by resource levels in the soil (possibly nitrogen), which may become locally depleted by extant shoot complexes. If threshold resource levels are a prerequisite for rhizome apex erection, such a mechanism may contribute to an evenness in distribution of shoot complexes at scales beyond morphological determination. Moreover, depending on time and sequence of rhizome erection, a phalanx of shoots may interfere with light interception by the crop. Systematic two-dimensional exploitation of resources proffers an additional explanation of the intense competitive pressure this species displays toward a crop (Welbank, 1961). The hypothesis remains to be tested.

A final comment on interference relates to the economic return from the usage of herbicides to ameliorate the competitive pressure exerted by weeds. Snaydon (1982) argues on the basis of collated yield data (e.g., Wallgren, 1980; Niemann, 1980) that average crop yield response to herbicides is poor (2% or less), although for annual grass weeds, control may give yield responses of 10–20% (Baldwin, 1979). Although these data appear disappointing, carefully disentangling the deleterious effects of herbicides on the crop, on target weeds, and, more importantly, on the process of weed crop interference has not been a major endeavor of weed scientists (but see Hawton, 1980). Moreover, the results of field trials for comparative herbicide performance are frequently evaluated with fitness parameters appropriate to future damage—mature inflorescences, seed number. Paterson (1977) has shown the importance of considering fitness as dry weight in assessment of population–herbicide interactions. The results of a pot experiment showing the relative magnitudes of the ef-

◀ **FIGURE 5.** The clonal architecture of *Elymus repens*. A computer simulation of the growth form of a clone from a single rhizome bud (circled) over three seasons. The placement of shoot complexes, plagiotropic rhizome apices, and interconnecting rhizomes were determined by stochastic driving variables derived from field measurements: seasonal rate of primary rhizome production by shoot complexes, angle and rate of primary and secondary (lateral) rhizome growth, rhizome bud fate—dormancy, extension or mortality, chance of rhizome erection. The simulation illustrates the means by which evenness in small scale pattern is generated (see text for details). Scale is in unit internode lengths (mean: 2.14 cm).

fects of crop interference and a herbicide are shown in Figure 6 (R. J. Manlove, A. M. Mortimer, and P. D. Putwain, unpublished data). *Avena fatua* and *Hordeum vulgare* (cv. Athos) were sown successively to give different emergence times and a selective post-emergence herbicide (1-flam-prop-isopropyl) applied after one month's growth in a greenhouse. To mimic field conditions in which loss of herbicide efficiency occurred as a result of uncontrollable variables, two-thirds of the manufacturer's recommended dose was applied. Wild oats were found to decrease the growth of barley by up to 46% of that in a pure stand but were in turn suppressed by the barley. The influence of the herbicide was not strictly additive. Proportional depression of wild oats fitness in monoculture by herbicide alone occurred irrespective of plant size. However, a very similar reduction was achieved also when the weed emerged together with the crop in the absence of chemical control. Addition of herbicide reduced fitness further but significantly less than might have been predicted from direct observation of herbicide performance alone. When wild oats were younger than the crop, little advantage was gained by herbicide spraying, the converse being true if ages were reversed. Variable amounts of spray reaching the target and crop damage due to herbicide alone were not detected. It is more likely that changing allometric relationships in the mixture and their interaction with the herbicide were the cause. Moreover,

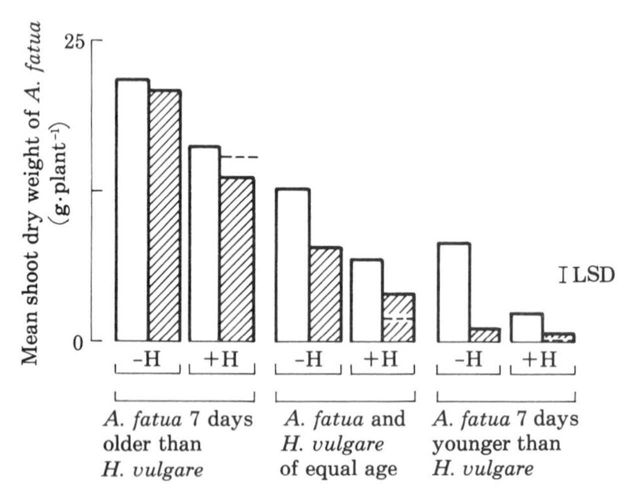

FIGURE 6. The interaction of post-crop-emergence herbicide, competitive pressure, and growth differential on fitness of *Avena fatua* in pure stand (white bars) and in mixture with *Hordeum vulgare* (black bars). The herbicide (1-flam-prop-isopropyl) was applied foliarly after 28 days growth from the first planting (+H). Dashed lines indicate additive expectation. LSD ($P \leq 0.05$) appropriate for pairwise independent comparisons.

significantly, no fitness (dry weight) advantage to barley was accrued by herbicide application.

Harvest

Community composition at the point of harvest, reflecting the accumulation of the effects of past control practices, influences the means of harvesting as well as the quality of the crop. Elliott (1980) has pointed to the significance of matter other than grain in cereal harvesting, and he argued the need for identification of those species that grow late in the season in the life of a crop. In cereals in Britain, contributors to harvesting losses include *Polygonum lapathifolium*, *Galium aparine*, and *Elymus repens*. An analysis of community dynamics in weed-crop associations, however, is needed for an ecological understanding of the interaction between harvester—mechanical (Bell, 1977) or man (Moody and De Datta, 1980)—and injurious flora. This community, for which there are few quantitative descriptions, may well comprise species which probably have contributed little to yield loss through interference but whose niche is determined by changing resource availability as crop maturation and senescence occurs. The use of nonselective phytotoxic herbicides, when virtual crop death has occurred but before crop harvest (O'Keefe, 1980), is a novel development of chemical control in such circumstances.

HERBICIDE RESISTANCE—
EVOLUTIONARY AND ECOLOGICAL CHANGE

A substantial body of evidence now exists showing the adaptive significance of life-history characteristics and morphological traits in weed species (e.g. Gadgil and Solbrig, 1972; Law et al., 1977; Grant and Antonovics, 1978; Bradshaw, Chapter 10). They lead to the inevitable conclusion that the evolutionary heritage of many species that are present-day weeds is one of prolonged adaptation in the face of selection pressures which arise, in the main, from periodic catastrophic habitat disturbance and competitive pressures from other species and from the attack of herbivores (Baker, 1974). Weeds may be opportunistic adventives, being preadapted as well as undergoing further evolution—for example, toward crop mimicry (Yabuno, 1966). Progenitors of cultivars of *Oryza* species illustrate this particularly well (Morishima, 1978), displaying a wide range of ecological attributes that may be fitted in part to the classifications either of MacArthur and Wilson (1967) or of Grime (1977).

In contrast to the time span over which evolutionary responses to cultural practices might occur, the time span for response to chemical control is exceedingly short (40 years). Nevertheless, to date 30 common annual weed species in 18 genera (23 dicots and 7 monocots) have now been shown to be resistant to the triazine group of herbicides (Lebaron and Gressel, 1982), and intraspecific variation has been reported for herbicide tolerance to a range of other groups [phenoxy herbicides (2,4-D); carbamates; uracils (Bandeen et al., 1982)]. Yet, overall, the rate of appearance of herbicide resistance is characteristically much less than might be expected by (superficial) comparison with other pesticides (Georghiou and Taylor, 1977). This has prompted several contemplative reviews (e.g., Gressel and Segel, 1978) and the development of models for the prediction of occurrence and increase in herbicide resistance (Gressel and Segel, 1982).

Genetic variability to herbicide resistance occurs in natural populations, and there is evidence that weed species possess a wide diversity of modes of inheritance—monogenic, polygenic, and maternal (Cormstock and Anderson, 1968; Schooler et al., 1972; Warwick and Black, 1980; Scott and Putwain, 1981). While each has specific implications for weed control practices and the breeding of herbicide-resistant crops (Faulkner, 1982), the intensity of selection, the fitness of resistant genotypes, and the potential for ecological change have significant effect.

The subtlety in ecological and evolutionary change is particularly highlighted in the case of triazine resistance in *Senecio vulgaris* (Putwain et al., 1982). In soft-fruit orchards in the United Kingdom, simazine is used extensively for control of weeds between crop rows; residual herbicide persists in the surface soil layers for several months after spring application. In *S. vulgaris*, escape from herbicide phytotoxicity occurs in susceptible populations by the switch from a summer to a winter annual phenology. Plants emerging in spring have a short life expectancy when sprayed (mean, 14 days), and it is only those individuals that germinate in late summer, survive winter, and escape further simazine applications through deep rooting that produce progeny. In the absence of herbicide spraying, however, seed production is only achieved by plants that germinate early in spring (February, March); intense competition from other weed species prohibits the success of later recruits. This change in phenology appears not to be a consequence of selection for dormancy but rather a modification of seasonal availability for microsites for germination. Putwain et al. (1982) argued that in unsprayed sites cultivation practices in autumn for weed control ensure both a large pool of buried viable seed which is returned to the surface and microsites for germination by the activities of the soil fauna. In sprayed sites, lack of cultivation and a characteristic bryophyte flora on the soil surface restricts seed incorporation within the soil profile and exposes seeds and seedlings to prolonged climatic hazards on the surface. The winter annual characteristic in sprayed sites is then a direct consequence of a hazard-free period for seedling establishment in late summer.

When alleles for simazine resistance occurred in populations, the summer annual habit was restored in herbicide-treated sites. Additionally, seed shed during the summer also germinated the following autumn, and there were two partially overlapping generations per year, one behaving as a summer annual and the other as a winter one. In populations comprising a mixture of susceptible and resistant genotypes, however, an equilibrium may be maintained despite rapid selection for predominantly resistant populations through maternal inheritance. K. R. Scott and P. D. Putwain (personal communication) have recorded a change in genotypic frequency from 2 to 90% in one season. This equilibrium arises partly from spatial escape from the herbicide in practice but also partly as a result of the relative greater fitness of susceptible genotypes over resistant ones (Conrad and Radesovich, 1979). Thus two niches, temporally separated, exist. In the first (spring), simazine concentrations result in the relative fitness of susceptible genotypes being minute; whereas in late summer through to winter (niche 2), although climatic hazards induce mortality, susceptible genotypes survive and by virtue of greater seed output ensure escape from absolute elimination from the mixture.

These studies expose the complexities that changes in agronomic practices may bring. Although evolution of herbicide resistance may occur, ecological changes to modifications in physical and chemical environment can be expected. Putwain (1982) has pointed to further areas of study.

WEED DEMOGRAPHY AND THE DESIGN OF WEED CONTROL STRATEGIES

Traditionally, the disciplines of pest control have evolved separately, and agroecosystems have in consequence been considered from discrete viewpoints. Yet the need for a holistic approach is borne out by the demonstration that the ecological assemblage of entire cropping systems may be manipulated to encourage interactions beneficial to the farmer (e.g. *Eleusine indica* and *Leptochloa filiformis* around *Phaseolus vulgaris* in repelling the leafhopper *Empoasca kraemeri*; Altieri et al., 1977). Highly pertinent to management policies that advocate utilizing plant species other than the crop is the means of arresting community development in subsequent cropping cycles. This

TABLE 5. Infestation rates of *Elymus repens* under various control regimes.[a]

Management	Fecundity		Survivorship				Finite rate of increase
					Shoot complex		
	Seed	Rhizome buds	Rhizome buds	Seedling	Flowering	Barren	
A monoculture of couch	13.7	460	0.96	0.25	0.95	0.95	15.95
In winter wheat with tine cultivations	0.63	170	0.96	0.01	0.5	0.5	6.69
In winter wheat with tine cultivations and a contact herbicide applied in autumn (October/November)	0.63	85	0.96	10^{-5}	0.05	0.0	4.37
In winter wheat with tine cultivations and a systemic herbicide applied in autumn (October/November)	0.63	85	0.009	10^{-5}	0.05	0.0	0.73

[a]Demographic statistics (per annum) are collated from experimental and literature sources. See Mortimer (1983) for details.

is merely the question addressed previously (in a wider context), but it is a question which underpins the design of integrated weed management systems and the design of weed control strategies for individual persistent weeds in otherwise largely clean crops. Surprisingly few detailed studies have been conducted on injurious species with long-term agricultural interests in mind (Sagar and Mortimer, 1976), but, ironically, "weedy" species initially attracted the interests of plant population biologists (Harper, 1977).

Studies on couch grass and wild oats in cereals in particular illustrate, on the one hand, the regulatory influence that the crop itself may exert on weed species but, on the other, emphasize the need for intensive control even if containment of weed infestations, and not total eradication, is the aim. These conclusions are hardly novel in themselves, but it is in few instances that they have been carefully measured.

In monoculture, a finite rate of increase of 16 was recorded for couch grass in the field (Table 5); this value was more than halved by the growing of a companion wheat crop. This reduction arose by the depression of rhizome bud and seed fecundity and seedling survivorship (interspecific interference) together with shoot complex survivorship (primarily by soil cultivations); but it also involved a phenological response in rhizome apex erection by the weed. Birth pulses of shoot complexes in *E. repens* conspicuous in midsummer in monoculture were precluded by the presence of wheat and only occurred at crop establishment and maturity (McMahon, 1982). Responses to resource limitation in couch are complex (Courtney, 1977), and capture of radiant energy by parental shoot complexes within a crop canopy is an important determinant of rhizome apex behavior underground (Williams, 1970). Moreover, bidirectional transport of assimilates along rhizomes is a noticeable feature of intact clones (Smith and Rogan, 1980) and may afford a mechanism by which the effects of spatial heterogeneity in available light within the developing plant canopy are offset. Certainly the observed low rates of natural mortality in shoot complexes in pure stand (5% per annum) and in winter wheat (3%, during crop growth) suggest strong internal homeostatic control, which may extend in rhizomes up to a meter. Not surprisingly, eradication of couch became feasible if all demographic statistics were reduced by orders of magnitude.

In age-distributed populations of wild oats, elimination of the majority of cohorts is a requisite for containing infestations. Figure 7 illustrates the contributions to annual rates of increase in populations arising from large seed reservoirs on the surface of the soil ($500 \cdot m^{-2}$).

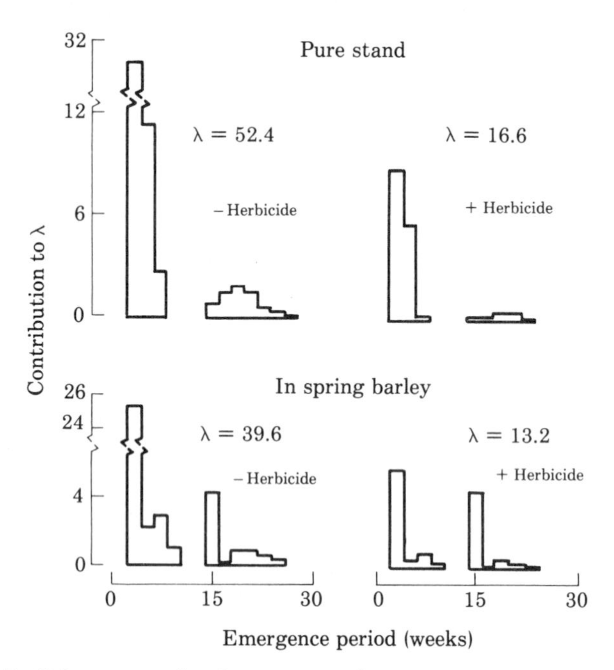

FIGURE 7. Cohort contributions to population increase (λ) in *Avena fatua* alone and infesting spring barley. Cohorts are grouped in 14-day emergence periods, the first being present in November 1980. Postemergence herbicide (see text) was applied in week 26.

In monoculture, these contributions were distributed inversely according to age, being discontinuous as a result of lack of seedling emergence in winter (December–January). The influence of a postemergence herbicide applied in late spring was to reduce mean seed fecundity of all cohorts rather than to influence mortality. The net result of herbicide application (a reduction in λ by one-third) was also achieved on wild oats infesting spring barley, the rate being further suppressed by the crop. Quite clearly containment of the size of this infestation requires intensive control, especially when up to one-third of the seed population in the soil may survive into the subsequent cropping season. Density-dependent regulation of plant survivorship was not observed in these field trials, and it was only in high total plant densities in winter wheat that seed fecundity in wild oats showed evidence of density regulation (R. J. Manlove, A. M. Mortimer, and P. D. Putwain, unpublished data). The cost of reducing annual infestation rate in *Avena* was, however, paid for by yield losses in *Hordeum*.

These data illustrate the utility of describing flux in weed populations and begin to fulfill some of the criteria necessary for prediction of the size of weed infestations (Mortimer, 1983). Integrating models that forecast yield loss in the crop together with long-term dynamic trends in injurious populations becomes the necessary next step. Although, from a systems analysis point of view, parallels exist in other crop protection disciplines (animal pests and pathogens), they are distinguished from the approaches needed for weed control by considerations of scale both spatially and temporally. The grower of homogeneous crops is confronted in a cropping cycle with infestations that may well originate from in situ sources on a predetermined dispersal pattern, migratory events during the cropping cycle probably being of less importance than with other pests (but see Sagar and Mortimer, 1976). "Spot treatments" then become a feasible tactic within a control strategy. Differences in temporal scale are largely a reflection of generation times in weeds in comparison to other pests and for chemical control necessitate prediction of the transient dynamics of a weed flora. Selective chemical measures frequently exploit differences in growth stage between crop and weed, and the window of crop safety may often be small. Moreover, in grass weeds at least, age and stage distributions often show poor relation (A. M. Mortimer and R. J. Manlove, unpublished data), and it may well prove necessary to invoke both characters in forecasting optimal periods for application of control. Additionally, if crop compensatory responses are to be harnessed as a component of a weed control program that relies on post-emergence therapeutic methods, the relationship between time of weed removal and time-dependent allometric responses in crops requires further study (see Barnes, 1977; Aikman and Watkinson, 1980). Despite the vast body of knowledge of qualitative relationships and empirical quantitative descriptions in the literature, putting "system" into weed science will almost certainly require the detailed censusing methods of the "pure" plant population ecologist and agriculturally a greater interest in the individual in weed populations.

CONCLUSIONS

"Will the writer of the highly remarkable article on weeds in your last number have the kindness to state why he supposes that 'there is too much reason to believe that foreign seed of an indigenous species is often more prolific than that grown at home' . . . I have no doubt that such an acute observer has some good reason for his belief" (Darwin, 1857). "We can see that when a plant or animal is placed in a new country amongst new competitors, the conditions of life will generally be changed in an essential manner, although the climate may be exactly the same as its former home" (Darwin, 1859). That adventive weeds were of interest is beyond question and the extension of this interest into competitive interactions so easily observed in plants is one of the crucial roots of modern agricultural science. In removing the barriers to crop productivity, botanical concerns have rested in the disciplines of physiology and genetics (Olsen, 1982) and ecological interests have remained dormant. The techniques of population ecology and, in particular, those which identify transitions in a life-history to which population growth rate is most sensitive (Varley and Gradwell, 1960; Sarukhán and Gadgil, 1974; Caswell and Werner, 1977) are an obvious pressing application. Yet this would be only a small repayment of the debt that "ecology" owes pest control (Geier and Clark, 1979). As with phytopathology (Browning, 1981), the agroecosystem–natural ecosystem dichotomy has persisted to the detriment of weed science, but less so because of the relative youth of plant population ecology. It would be gratifying to hope that future research will satisfy the concerns of Southwood (1981) and that the translation of theory into practice is not too protracted. Darwin surely would have been sympathetic to this.

CHAPTER 19

PLANT DEMOGRAPHY IN AN AGRICULTURAL CONTEXT

R. W. Snaydon

INTRODUCTION

Darwin and applied studies

It is apt that this centennial volume to celebrate Charles Darwin should include at least two chapters on applied topics. Darwin was greatly influenced by his studies of cultivated plants and domesticated animals. Reading his two volumes on *Variation of Animals and Plants under Domestication* (Darwin, 1868) it is easy to see how these studies underpinned *Origin of Species*. It is not by chance that the first chapter of *Origin* was entitled "Variation under Domestication," since Darwin states in his Introduction that "At the commencement of my observations it seemed to me probable that a careful study of domesticated animals and of cultivated plants would offer the best chance of making out this obscure problem. Nor have I been disappointed; in this and in all other perplexing cases I have invariably found that our knowledge, imperfect though it be, of variation under domestication, afforded the best and safest clue."

Darwin's interest in cultivated plants was obviously exceptional among botanists of his time. He wrote in the first volume of *Variation of Animals and Plants under Domestication* that "Botanists have generally neglected cultivated varieties as beneath their notice." To a considerable extent, the same could be said today.

Pure and applied studies

It is interesting to note that, just as Darwin's studies of cultivated plants led to the development of his ideas on the mechanisms of evolution, so studies of cultivated plants have more recently laid the foundations of plant population ecology, as I have argued previously (Snaydon, 1980a). Indeed, the literature is so full of such applied studies, relevant to plant population ecology, that it would be impossible to do justice to it in one chapter, or even in one volume. Instead, I have selected two contrasting case studies for consideration.

Before considering these case studies, it is useful to recognize some important differences between pure and applied studies, not only in population ecology, but more generally. In comparing studies of agricultural and natural populations, or agricultural and natural communities, botanists tend to focus on the uniqueness and "unnaturalness" of cultivated species and agricultural environments. I have argued elsewhere (Snaydon, 1980a,b) that these differences are usually not as large, or as important, as is often thought. Darwin wrote in *Origin* "It has been argued that no deductions can be drawn from domestic races to species in a state of nature. I have in vain endeavoured to discover on what decisive facts the above statement has so often and so boldly been made. There would be great difficulty in proving its truth." The same could still be written today. By contrast, the reasons for carrying out pure and applied studies, the ways in which the studies are carried out, and the interpretation placed upon the results are usually very different.

Understanding and prediction

Perhaps the most important difference between pure and applied studies is that, whereas the objective of pure studies is solely to increase knowledge and understanding, the objective of applied studies is to manipulate biological systems to increase the output of products useful to man or, more often, to increase the profitability of manipulating those systems. To achieve these objectives, it is necessary to be able to predict accurately the outcome of any proposed manipulation (Snaydon, 1980b). It might be argued that this predictive ability in manipulating agricultural systems results from greater knowledge and understanding of the system and hence springs from fundamental studies. However, this is not necessarily so. Increasing understanding in science is usually considered to occur through increasing knowledge at progressively lower levels of biological organization; that is an analytical approach. Thus, increasing understanding at the population level has been sought by studies at the level of the individual, plant organs (e.g., Bell; White, this volume), or at even lower levels (e.g.,

Brown, 1979). This analytical approach has certainly increased knowledge at lower levels of organization and sometimes has shed fresh light on population processes, but it has not been particularly successful in improving predictions of the outcome of management practices at the level of the population, community, and ecosystem in agriculture. As we shall see, advances in agriculture have usually been made empirically, and by extensive studies at the relevant level of organization, rather than by extrapolation from lower levels of organization.

Manipulation of agricultural systems

The manipulation of agricultural systems concerns both biological and environmental components (Snaydon, 1980a). Manipulation of the biological components involves not only control of the species present (e.g., crop species planted and weed species removed) but, in most crops (as opposed to pastures), also involves the control of the density of sown species. In most crops the genotype is also controlled, both by selection of a particular cultivar and by the breeding of cultivars.

The control of the environmental component mainly concerns soil factors, especially mineral nutrients and soil water, though also, to some extent, soil physical conditions. There is little direct control over climatic conditions, though the conditions experienced by the crop can be varied to some extent by varying the sowing date (e.g., autumn versus spring sowing).

Predicting responses

It is extremely difficult to predict the outcome of manipulating either the biological or the environmental components of agricultural systems because of the complexities caused by biological and environmental diversity and by interactions between and within the biological and environmental components. Biological diversity (i.e., differences between species, between cultivars and genotypes, and between the developmental stages of a genotype) makes it difficult or impossible to predict the response to manipulation without specific experimental evidence. Similarly, environmental diversity, both in space and time, makes prediction more difficult. Prediction is made even more difficult by interactions between environmental factors; at its simplest this is represented by Blackman's "law of limiting factors" (Blackman, 1905), but it often takes much more complicated forms. Prediction is also made more difficult by differences between genotypes and cultivars in their response to environmental factors, that is, genotype–environment interactions (Hill, 1975). With these various complexities, it is hardly surprising that prediction from a lower level of organization to the next is difficult, and sometimes impossible (Passioura, 1979, 1981).

CASE STUDIES

Since I cannot hope to consider more than a very small part of the relevant literature on the many aspects of plant demography in agricultural systems, I shall consider two contrasting case histories. The first concerns annual cereals, which are genetically uniform and in which there is usually fairly close control over both the biological and environmental components of the system. The second concerns pasture species, which are genetically heterogeneous and in which there is much less control of biological and environmental components. In addition, pasture species are usually grown in species mixtures, though most of the data I shall use here refers to single-species stands, that is, to plant populations rather than plant communities.

CEREAL CROPS

Features of cereal production

There is usually very close control of the biological components in cereal production. In developed countries, cereals are grown in pure stands; they are essentially free of weeds and are genetically uniform. Even with outbreeding species (e.g., corn), the use of hybrids gives considerable genetic uniformity. The density of the crop is also under close control.

There is generally less control of environmental components, and control is more variable. There is essentially no control over climatic factors, and control of mineral nutrient supply varies greatly between countries and also, to some extent, between crops.

Causes of increased yield

Before considering some of the demographic implications of these controls, it would be sensible to consider the relative contribution of the various controls to increasing crop yield. In a few cases (e.g., Cardwell, 1982), attempts have been made to assess the effects of the separate biological components (e.g., control of weeds, crop density, and improved cultivars) and of the separate environmental components (e.g., fertilizer use, drainage, cultivation and drilling techniques) on the yield of a crop. However, evidence is usually inadequate, and most studies have simply assessed the contribution of new cultivars to the overall increase in yield (e.g., Evans, 1981). Such studies indicate that the contribution of new cultivars to yield increase ranges from 11% of the total (for wheat in Sweden) to 60% (for wheat in Britain). However, these figures are probably overestimates since there is an assumption that the greater yields of cultivars obtained in trials are also achieved under farm conditions. This assumption is incorrect and probably overestimates the effects of new cultivars by at least twofold (e.g., Davidson, 1965; Walker and Simmonds, 1981).

Although the contribution of plant breeding to increasing crop yields may have been overestimated, there is little doubt that new cultivars have made an important contribution. To what extent can that contribution be interpreted in demographic terms?

Reproductive allocation

The most important change in cultivars that has led to increases in crop yield has been in reproductive allocation. Whereas the yield of shoot material by cultivars of most crops has not significantly changed over the past 100 years, the yield of grain has increased exponentially. In the case of barley in Britain (Figure 1), the grain yield of cultivars has increased but total shoot weight has not, so the ratio of grain yield to total shoot yield (i.e., the harvest index) has increased

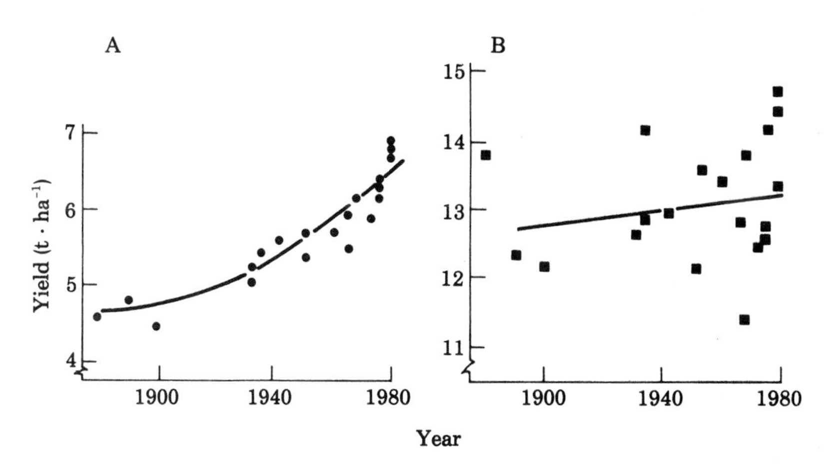

FIGURE 1. The grain yield (A) and total shoot yield (B) of barley cultivars released at various times between 1880 and 1980. (Data from Riggs et al., 1981.)

from 0.35 for cultivars released in the late 1800s to 0.48 for recently released cultivars. Similar increases in harvest index have occurred in wheat (e.g., Evans, 1980a, 1981) and in other cereals (Donald and Hamblin, 1976).

This increase in reproductive allocation has taken cereal crops outside the range normally found in wild annual species (Harper, 1977). Presumably, the large allocation to seed production must carry some ecological penalty, associated with reduced allocation to root, stem, and leaf production; there is some evidence of this. For example, Hamblin and Rowell (1975) found that barley genotypes that yielded the most grain were least competitive in mixed genotype stands. However, since the genotypes (i.e., cultivars) are normally grown in pure stands, where all are equally uncompetitive, this is no disadvantage. Indeed, Donald (1981) has argued that breeders should select for poorly competitive genotypes: his "ideotype" is a plant with a single culm, having a short stem, few small erect leaves, and a large harvest index. Donald contrasts such "communal plants" with the competitive plants that are likely to result from natural selection.

In the wild, there is probably an optimum balance between seed production and vegetative growth. The ratio is greater in annuals than in perennials (Harper, 1977) but, even for annuals, there are presumably limits to the amount of resources that can be safely diverted to seed production from vegetative growth, which largely determines competitive ability. The greater yield of recent cultivars has, therefore, been dependent upon releasing the crop from natural selection. This has been achieved by using single-genotype stands, by resowing the crop each year, and by maintaining a weed-free crop.

Although annual crops have been released from competition, competition still occurs between genotypes within segregating families, in the early stages of the breeding program. Hamblin and Rowell (1975) have argued that there is danger of selecting for competitive, but ultimately low-yielding, genotypes in F_2 and F_3. An alternative method will be considered later.

The importance of breeding for changes in resource allocation, including reproductive allocation, is not restricted to cereals. For example, most of the success in potato breeding has been achieved by changing the ratio of marketable to unmarketable tubers (Harris, 1980). Sugar production, from both beet and cane, has been increased largely by increasing the sugar content. In timber production, the main contribution of breeding has been to increase the size of the harvested trunk, relative to the size of the branches (J. D. Matthews, personal communication). By contrast, breeding has had little effect on the productivity of pasture species (see later) because there is little opportunity for the reallocation of resources, since all the above-ground parts are used.

In animal breeding, most successes have also involved changes in resource allocation, especially in reproductive allocation. For example, breeding has considerably increased egg production in poultry, milk production in cattle, and the number of offspring per litter in pigs, sheep, and cattle. It has also increased wool production in sheep and changed the "conformation" of meat-producing animals. By contrast, improvements in live-weight gain have been modest; it is easy to increase the overall size of animals, but not to change the important "conversion ratio," that is, the ratio of live-weight gain to food consumed. Natural selection seems to have already maximized biomass production and efficiency of conversion in both plants and animals.

Crop density

The most obvious way in which agriculturalists can modify and manipulate the demography of crops is through regulating crop density. However, it is extremely difficult to disentangle the effects of crop density from those of other factors, under normal agricultural practice. This difficulty occurs, first, because various environmental factors (e.g., soil water content, temperature, and depth of sowing) and biotic factors (e.g., pests and pathogens) affect germination and the establishment of seedlings, and so affect density. Second, crop density interacts with many factors in determining the yield of crops; for example, crop density interacts with nitrogen supply (see later) and water supply. In addition, the yield of harvested components (e.g., grain) usually respond in a different way than total plant weight (Willey and Heath, 1969).

Total dry matter yield usually increases asymptotically with increasing density, whereas grain yield, and most other harvested components, decreases at higher plant densities (Willey and Heath, 1969). As a result, it is important to sow crops so as to achieve the optimum density at which the yield of the harvested component is maximized. This optimum varies greatly among species but is also affected by environmental conditions (see earlier).

Differences between species and cultivars. The optimum density for various annual crops has been determined empirically. It ranges from about 1 plant • m^{-2} for crops such as cassava and sugar cane up to 300–400 plants • m^{-2} for cereals (e.g., wheat and rice) and vegetables (e.g., carrots and onion) (Willey, 1982). Cabbage, corn, and sugar beet have optima of 5–10 plants • m^{-2}; peas, field beans, and soybeans have optima of 50–100 plants • m^{-2}.

Is it possible to define the factors which determine the differences in optimum density between species? For example, are these optima determined by maximum plant size (achieved at very low density) or by competitive ability? At the moment we know little about the determining factors, and there is obviously a need to study the features which determine the response of crops to density.

There are also smaller differences in response to density within species. In particular, cultivars of corn differ in response to density, and the differences appear to be associated with ability to tolerate shading (Stinson and Moss, 1960), though nitrogen nutrition may also be involved (Hageman et al., 1961).

Even greater differences between cultivars in response to density are apparent when more extreme densities are used. For example, Syme (1972) and Fischer and Kertesz (1976) found little correlation ($r = 0.10$ and 0.31, respectively) between yield at normal crop densities and yield of spaced plants when they compared 49 and 40 cultivars of wheat, respectively. Performance in noncompetitive conditions is, therefore, not related to performance under competitive conditions in dense stands.

Density and reproductive allocation. The fact that grain yield declines at high densities, while total plant yield remains constant (see earlier), indicates that reproductive allocation declines at high densities. In general, harvest index tends to decline progressively with increasing density (Donald and Hamblin, 1976) though, in a few cases, it may increase with increasing density at low densities.

There seems to be less cultivar–density interaction for reproductive allocation than for grain yield per se. For example, Syme (1972) and Fischer and Kertesz (1976) found closer correlations between spaced plants and normal crop densities for harvest index ($r = 0.85$ and 0.81, respectively) than for grain yield (see earlier). As a result, they suggested that the harvest index of spaced plants might prove a valuable selection criterion in the early stages of a breeding program, when there is insufficient seed to sow plots. The use of spaced plants at this stage also has the advantage of overcoming the problem of competition between genotypes in segregating families at F_2–F_4 (Hamblin and Rowell, 1975) discussed earlier.

Nitrogen fertilizer

We have already seen that it is difficult to quantify the effects of new cultivars and crop density in increasing crop yields over the past 40 years. It is even more difficult to disentangle the relative importance of the many other biotic and environmental factors which have contributed to increasing yields. However, there seems little doubt that fertilizer use, and especially nitrogen (N) fertilizer, has contributed substantially to the increase, especially in developed countries but also to the so-called "green revolution" in developing countries. There is a close correlation between fertilizer use and grain yield both over time in a given country and between countries (Evans, 1980b).

At first sight, the use of N fertilizer seems to have few consequences in terms of plant demography but, on closer examination, there are several important implications; some of these will be discussed briefly here.

Nitrogen and density. The response to density of both individual plants within the crop and the crop as a whole is crucially dependent upon the supply of limiting resources, since it is competition for limiting resources that determines the performance of individuals and hence the crop. Nitrogen is a major limiting factor and, hence, a commonly used input in many agricultural systems (see earlier); it therefore interacts strongly with density.

The optimum density, and as a result the maximum grain yield, is greater with increasing applications of N (e.g., Lang et al., 1956). In other words, when supplies of the major limiting resource are made nonlimiting, more plants can be crammed into a unit area of land. Indeed, it would seem that the main effect of N applications is to enable a higher density and hence a greater grain yield.

Nitrogen and reproductive allocation. One effect of N is to extend the length of the vegetative phase, that is, postpone flowering and seed production, in many crops. While this increases the shoot weight, it generally reduces the harvest index (Donald and Hamblin, 1976). The outcome of these two conflicting responses depends on environmental conditions, especially water supply. Where water supply is adequate, the increase in shoot yield exceeds the reduction in harvest index and grain yield is increased by N applications; where water supply is inadequate, especially at the end of the growth period, harvest index is severely reduced and grain yield may actually be reduced by N applications (Donald and Hamblin, 1976).

The effects of N on seed production depend, in particular, on the stage at which N is applied (e.g., Langer and Liew, 1973). Applications of N before ear development increase the number of grains by increasing the number of spikelets per ear; these early applications also increase grain size. Applications of N between floret initiation and ear emergence increase the number of grains per spikelet and increase grain size. Applications after ear emergence have no effect on grain number or size but do increase the percentage of N in the grains. The slight effect of N after ear emergence probably reflects the fact that N uptake almost ceases after ear emergence.

Most studies on reproductive allocation in cereals, and also in wild species (Harper, 1977), have concentrated on the production and allocation of photosynthates (see reviews by Thorne, 1974; Evans and Wardlaw, 1976; Gifford and Evans, 1981). However, there is considerable evidence that photosynthesis and its products are not as limiting as the uptake and utilization of nitrogen. For example, there is normally no correlation between photosynthetic rate and crop yield (Elmore, 1982; Gifford and Jenkins, 1981) and breeding has tended to lead to a decrease in photosynthetic rate (Evans and Wardlaw, 1976). In addition, remobilization of photosynthates into the grain generally only accounts for a small proportion of the final grain weight. The proportion ranges from 5% in some studies of wheat to a maximum of 40% in some studies of rice (Evans and Wardlaw, 1976).

Nitrogen, on the other hand, is a major limiting factor for crop production in most areas where water is not the main limiting factor, and it greatly affects grain production (see earlier). In addition, there is usually a strong negative correlation between grain yield and percentage of N in the grain (Lupton and Pushman, 1975), indicating that nitrogen is a limiting resource within the plant. The uptake and utilization of nitrogen are therefore likely to be important components of reproductive allocation (Fischer, 1981) and deserve further study. Initial studies (e.g., Halloran and Lee, 1979) indicate considerable variation between cultivars in N allocation, at least under spaced plant conditions.

Empirical or analytical methods?

To what extent have the increases in yield achieved by plant breeding—largely by increasing harvest indices—been achieved predictively? Most evidence indicates that the advances were made empirically, perhaps in part by chance, with only post hoc recognition of the modus operandi. Although the idea of a "harvest index" was mooted by Beavan (a plant breeder) as long ago as 1914 (Donald and Hamblin, 1976), there seems to have been no further recognition of its importance or of its use as a selection criterion until the mid-1970s. Probably the main factor leading to an increase in "harvest index" over the past 40 years has been selection for shorter straw length. This was carried out, not to increase harvest index per se, but to prevent "lodging" under conditions of high N inputs; Donald and Hamblin (1976) concluded that "any increase (in harvest index) has been an unplanned secondary effect of breeding for grain yield, shorter straw, and earliness."

Over the past 30 years, crop physiologists have suggested various, often conflicting, criteria for increasing grain yield. These have included reduced (or increased) tillering, more upright leaves, increased leaf photosynthesis, and greater leaf area duration. None of these seem to have proved successful, and, as Passioura (1981) observed, "Overt attempts by physiologists and breeders to collaborate have led, almost without exception, to failures"; this is probably due to the negative associations that exist between attributes (Adams and Grafius, 1971). As a result, crop physiologists seem more uncertain of their ability to define selection criteria (e.g., Evans, 1981), and plant breeders continue to use empirical methods.

PERENNIAL PASTURE SPECIES

Features of pasture production

The biotic components of pastures are less closely controlled than those of annual crops. Most pastures consist of mixtures of species, often with only slight control of the species composition, though the composition of recently resown pastures is usually more controlled. Most pasture species are outbreeders, with the result that cultivars are complex mixtures of genotypes. The genetic composition of these cultivars can change quite rapidly after sowing (Snaydon, 1978b). Although the initial density of pastures can be quite closely controlled, by the seeding rate, thereafter there is little control of density. Indeed, it is difficult to define the density of some pasture species because of their clonal spread. In these cases it may be more suitable to measure the density of modules (e.g., tillers or stolons) rather than discrete plants.

Environmental conditions in pastures are usually less controlled and more variable, both in space and time, than those in crops (Snaydon, 1980a). In particular, the effects of selective grazing and treading and, perhaps most of all, the effects of dung and urine lead to extreme spatial variation (Beckett and Webster, 1971). As a result, most studies of pastures, especially those considered here, have been carried out in the absence of grazing animals and usually with single species or simple swards.

New cultivars

While plant breeding has led to considerable increases in the grain yield of cereals (see earlier), there has been less success in breeding for increased yield in pasture species. For example, the best of the currently available cultivars of *Lolium perenne* in Britain yield only 5–7% more than those available 50 years ago (N.I.A.B., 1982).

There are several possible reasons for this slow progress: (1) the outbreeding habit; (2) the limited scope for changes in resource allocation; and (3) the effects of natural selection. Most attention will be focused on (2) and (3) since there is little evidence that (1) is an important factor. Although annual crops tend to be inbreeders and pasture species tend to be outbreeders, there are examples of both breeding systems among both crops and pasture species. There is little evidence that success in plant breeding is related to breeding system within either group. For example, many crop plants are outbreeding and yet have been improved by breeding (e.g., maize, rye, *Phaseolus* bean, field bean, sugar beet, safflower, sunflower, carrots, miscellaneous cabbages, and radish), whereas various pasture species are inbreeding, or even apomictic (e.g., *Trifolium subterraneum, Trifolium fragiferum, Melilotus alba, Bromus mollis*, and *Agropyron cristatum*) but have not been improved more than outbreeders.

Resource allocation

There is less opportunity to manipulate the resource allocation of pasture species than that of crop species because most of the aboveground parts of pasture plants are utilized; however, some reallocation is possible. In theory, it might be possible to change the ratio of shoot to root weight, though reducing the size of the root system might have detrimental results (see later). Another possibility is to increase the proportion of the shoot that occurs above the cutting (or grazing) height. For example, cultivars of *Trifolium repens* have different leaf:stolon ratios, where leaves grow above cutting height and stolons grow below. Rhodes and Harris (1979) showed that, in the first year, cultivars of *T. repens* with large leaves, that is, long petioles and large laminae, gave 50% more harvested yield than those with small leaves. This difference was almost exactly matched by contrary differences in stolon weight; harvested yield and stolon weight were negatively correlated ($r = -0.9$): the total shoot weight of the cultivars was similar.

Resource allocation and survival

Although such an allocation of resources to leaf material may give an increase in harvested yield in the short term, it may not be advantageous in the long term if it reduces competitive ability or survival. Since the continued survival and spread of plants of *T. repens* depends on stolon growth, it is not surprising that Baines et al. (1982) found a negative correlation ($r = -0.9$, Figure 2) between the survival of clover populations, after two years in grass swards, and their "morphological index" (i.e., leaf weight:stolon length). Similarly, Rhodes and Harris (1979) found a negative correlation between the harvested yield of cultivars in pure stands and their competitive ability in mixtures with ryegrass, even in the first year.

There is indirect evidence that the same phenomenon also occurs within grass species. For example, data for *L. perenne*, *L. multiflorum*, and their hybrids (N.I.A.B., 1982) indicates that there is a negative correlation ($r = -0.64$) between harvested yield of cultivars in the first harvest year and their persistence (Figure 3). The higher yielding, less persistent cultivars have a tall, upright habit, with a few large tillers and a greater reproductive allocation.

Bred cultivars of *Lolium* do not represent the full range of variation present in agricultural pastures. They have been selected for various agronomic attributes, such as yield, quality, and disease resistance, but also, to some extent, for seed production, since relatively cheap supplies of seed are needed for resowing pastures. Seminatural populations of *Lolium*, from agricultural grasslands which have not been resown, contain plants which produce fewer flowering heads and little seed (Breese, 1966). Populations from the most heavily grazed grasslands produce the least seed (Hayward, 1967) and are generally more

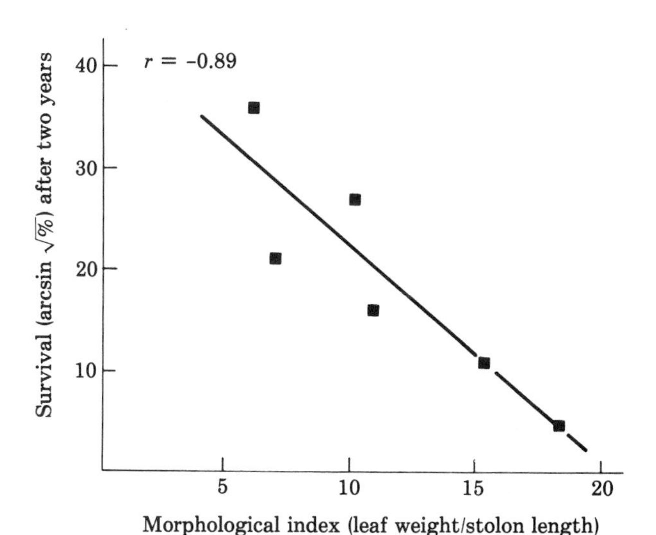

FIGURE 2. The relation between the survival of six populations of *Trifolium repens*, two years after being transplanted into swards, and their "morphological index" (see text). (Data from Baines et al., 1982.)

prostrate. These populations, which are apparently adapted to long-term survival in grazed swards, may appear less productive in cut plots.

All of the grass species normally sown in Britain (i.e., *L. perenne*, *L. multiflorum*, *Dactylis glomerata*, *Phleum pratense*, and *Festuca pratensis*) have an upright tufted habit and little capability for lateral spread. This perhaps represents an unconscious selection for a large "harvest index" (i.e., the ratio, harvested material:total shoot material). However, as we have seen for *T. repens*, this probably carries the penalty of poorer persistence, since the sown species are progressively displaced by other grass species with the capability for lateral spread (e.g., *Agrostis* spp., *Holcus lanatus*, and *Festuca* spp.) (Morrison, 1979).

Density

Although the initial density of perennial pasture species can be controlled to a considerable extent by varying the seeding rate, thereafter density is largely uncontrolled; this contrasts with the close control of density in annual crops.

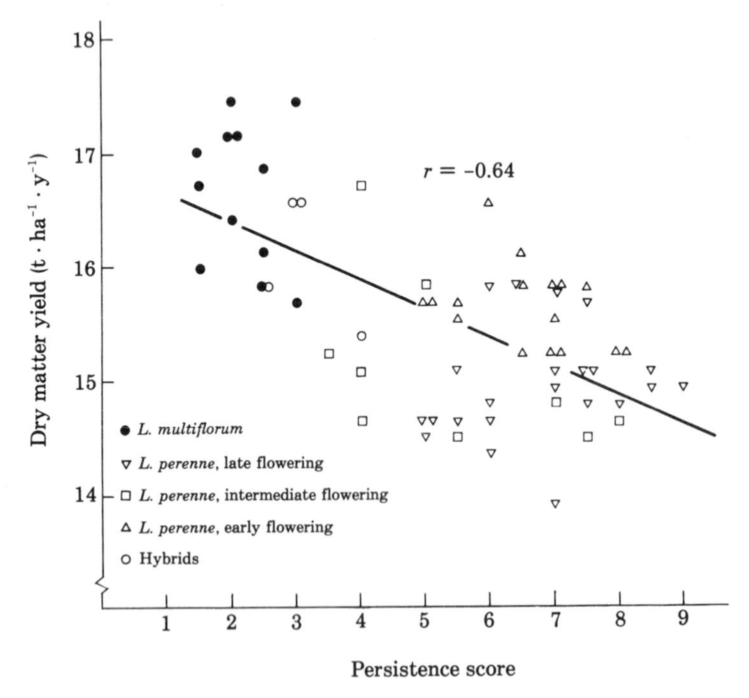

FIGURE 3. The relation between the annual dry matter yield ($t \cdot ha^{-1} \cdot y^{-1}$) and persistence of cultivars of *Lolium multiflorum*, *L. perenne*, and their hybrids. (Data from N.I.A.B., 1982.)

The other important differences between most perennial pasture species and crops is the ability of many pasture species to spread vegetatively, making the definition of individual plants difficult or sometimes impossible. As a result, many of the numerous studies of density in pastures are based on tillers and stolons, rather than on plants.

Plant density after sowing. Langer et al. (1964) have studied changes in plant and tiller density of *Phleum pratense* and *Festuca pratensis* swards for a three-year period, and Kays and Harper (1974) have studied changes in *Lolium perenne* swards for a six-month period. Langer et al. varied the cutting regimes, whereas Kays and Harper varied the sowing rate and light intensity.

Plant density declined with time in both studies; the decline continued for all three years in Langer's study. Density declined more rapidly with infrequent cutting and after three years was only 5% of the original density in the case of *P. pratense* (Langer et al., 1964). Density declined rather more rapidly with dense sowing, but light intensity had only a slight effect (Kays and Harper, 1974). In the long-term study (Langer et al., 1964), most of the plant mortality occurred during the period of maximum growth (April–June). This is a seasonal pattern similar to that detected for several dicotyledonous perennial herbs (Harper, 1977) and probably reflects the more intense competition for scarce resources (e.g., nitrogen and light) that occurs during the period of most rapid growth.

Rapid initial reductions in the plant density of sown pasture have also been recorded by Charles (1961). He found that mortality was not random but that when cultivars were sown in mixtures large changes in the proportions of the cultivars occurred in the first few years. It therefore seems likely that the reductions in density (described earlier) would be accompanied by biotype depletion and changes in the genetic structure of populations (Snaydon, 1978b). It would be interesting to know how long these processes continue and to what extent there is later recolonization by seedlings. Currently, little is known about seedling establishment in closed pasture (e.g., after disturbance), though observation indicates that it is quite rare for perennial species. On the other hand, clonal spread by indigenous species is common (see earlier).

Tiller density after sowing. In the studies by both Langer et al. (1964) and Kays and Harper (1974), the number of tillers initially increased for four months and then declined to an equilibrium level irrespective of sowing density. Tiller density was slightly reduced by shading. Tiller density of *P. pratense* was greatly reduced by infrequent cutting, but *F. pratensis* was less affected. Other studies have shown that frequent cutting (or grazing) often reduces tiller density, though the effect differs among species.

Tiller density, in the long-term study (Langer et al., 1964), varied almost twofold within each year. Tiller density was generally greatest in early spring (March–April) and least in June. As in the case of plant mortality, tiller mortality was greatest when growth was most rapid, during late May. However, in the case of tiller mortality, the cause may not be only more intense interplant competition at that time, but also more intense intraplant competition. Flowering occurs at that time, and it is known that many of the smaller vegetative tillers die during the period of flower head development. The peak of tiller appearance occurred in June–July, slightly after the peak of growth and flowering.

Tiller size and number. The herbage yield of a pasture depends on both tiller number and tiller size, since stand yield is the product of the two. Various management factors can change either, or both, size and number. For example, in a two-year study of *Lolium* spp. by Wilman and Mohamed (1980), cutting frequency greatly affected tiller size but had little effect on tiller number (Figure 4). The difference (2.5 times) in tiller size was due to the fact that the interval between cuts was twice as long, with the result that the annual yield was only 30% greater under infrequent cutting; this is a normal difference between such treatments (e.g., N.I.A.B., 1982).

Increasing nitrogen fertilizer applications, from 0 to 525 kg N $\cdot$ ha^{-1} $\cdot$ y^{-1}, increased both tiller number and tiller size about twofold (Figure 4); as a result, annual yield was increased fourfold.

Cultivars differed considerably in tiller number and size (Figure 5). However, the greater tiller density of Melle was offset by smaller tiller size, so that there was no difference in annual dry matter yield between Melle and Sabrina, though the greater tiller density of Melle probably contributes toward its greater persistence.

Thus, we see that several management variables affect tiller density and tiller size. However, neither of these attributes is under direct control, and the management variables are varied empirically to control total yield rather than tiller attributes.

CONCLUSION

These two contrasting case studies—annual cereals and perennial pasture species—clearly indicate that the control of demographic phenomena in agricultural systems varies widely. In cereals, there is close control of species composition (elimination of weeds), genotypic composition (genetically uniform cultivars), and crop density (resown annually). By contrast, these attributes are all poorly controlled, or not under direct control, in pastures.

Although many of the results obtained from studies of agricultural systems have formed the basis on which the principles and concepts of plant demography have been built (Harper, 1977), there is as yet little evidence that these principles can be used predictively to define management regimes for crops or pastures. I have argued that the present management and manipulation of crops and pastures is based almost entirely on empirical studies. This is true not only of the manipulation of the environment (e.g., the use of N fertilizer) but also of the manipulation of the crop itself (e.g., control of density or breeding new cultivars).

The analytical approach, which is the foundation of "pure" biology, is not a satisfactory basis for defining management regimes for agri-

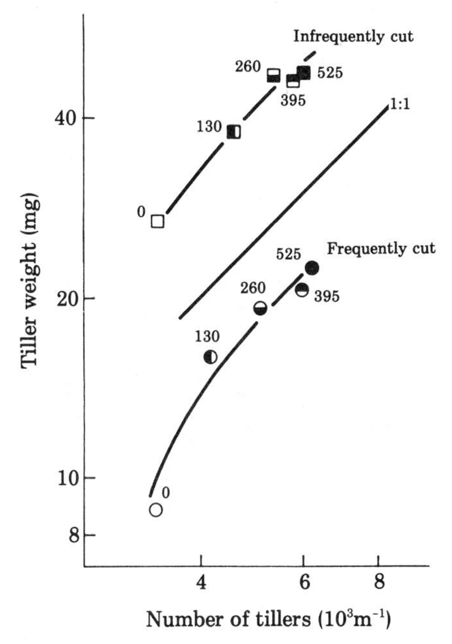

FIGURE 4. The effects of nitrogen fertilizer (0–525 kg N • ha^{-1} • y^{-1}) and cutting frequency (four or eight cuts per year) on the number and weight of tillers of *Lolium*. Each value is the mean of four cultivars (two of *L. perenne*, and one each of *L. multiflorum* and the hybrid), and of two years. Note the logarithmic transformation of both axes. The numbers by the symbols correspond to nitrogen levels (K • ha^{-1} • y^{-1}). (Data from Wilman and Mohamed, 1980.)

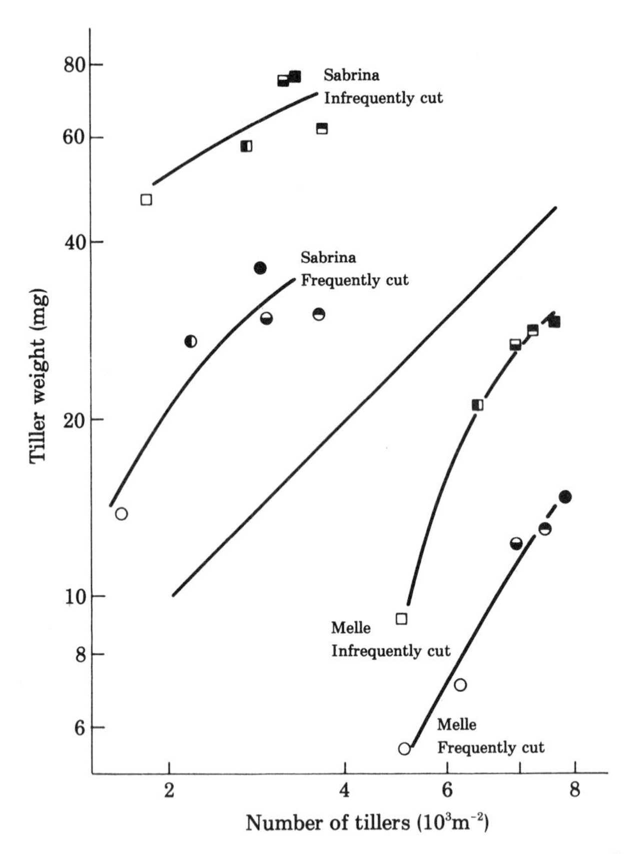

FIGURE 5. The effects of nitrogen fertilizer (0–525 kg N • ha^{-1} • y^{-1}) and cutting frequency (four or eight cuts per year) on the number and weight of tillers of *L. perenne* (cv. Melle) and *L. perenne* × *multiflorum* (cv. Sabrina). Each value is the mean of two years. Note the logarithmic transformation of both axes. The symbols for nitrogen levels are the same as in Figure 4. (Data from Wilman and Mohamed, 1980.)

cultural systems. It is largely inadequate because of the severe limitations in extrapolation from one level of biological organization to the next higher level. Thus, it is difficult, and often impossible, to extrapolate from the level of the individual to that of the population. Although the population is the relevant level of organization for most arable crops (e.g. a wheat crop), many agricultural crops are mixtures of species (e.g. pastures and mixed cropping), and so are communities, while grazed pastures must be viewed as ecosystems. Most of the information and most of the principles and concepts needed to define management regimes in each case must be obtained at the relevant level, though sometimes information at the next lower level may be helpful. Information, principles, and concepts at much lower levels (e.g., the molecule or cell) have no direct value in agriculture. This is not to argue that such analytical studies should not be carried out; they are equally as valid as any other in "pure" botany. However, we should not be deceived into believing that they have value in any predictive or functional context in agriculture.

Finally, I must question the generally accepted view that applied science and technology are spin-offs from pure science. If we study the history of the physical sciences and industrial technology through the eighteenth and nineteenth centuries (e.g., Hounshell, 1980) and the history of biology and agriculture in the past 100 years, it is apparent that pure science is far more often a spin-off from applied science or technology. The recent history of plant population biology is a good example of this phenomenon. Darwin leaves us in no doubt about his own debt to applied studies.

LITERATURE CITED

(Numbers in parentheses at the end of each reference indicate the chapter or chapters in which the work is cited.)

Aarsen, L. W. 1983a. *Interactions and Coexistence of Species in Pasture Community Evolution.* Ph.D. thesis, The University of British Columbia. (5)

Aarsen, L. W., 1983b. Ecological combining ability and competitive combining ability in plants: Towards a general evolutionary theory of coexistence in systems of competition. *Amer. Nat.* 122, 707–731. (5)

Aarsen, L. W., and R. Turkington. 1983. What is community evolution? *Evol. Theor.* 6, 211–217. (5)

Abbott, R. J. 1976a. Variation within common groundsel, *Senecio vulgaris* L. I. Genetic response to spatial variations of the environment. *New Phytol.* 76, 153–164. (17)

Abbott, R. J. 1976b. Variation within common groundsel, *Senecio vulgaris* L. II. Local differences within cliff populations on Puffin Island. *New Phytol.* 76, 165–172. (17)

Abrahamson, W. G., and H. Caswell. 1982. On the competitive allocation of biomass, energy, and nutrients in plants. *Ecology* 63, 982–991. (15)

Abul-Fatih, H. A., and F. A. Bazzaz. 1979a. The biology of *Ambrosia trifida* L. I. Influence of species removal on the organization of the plant community. *New Phytol.* 83, 813–816. (16)

Abul-Fatih, H. A., and F. A. Bazzaz. 1979b. The biology of *Ambrosia trifida* L. II. Germination, emergence, growth and survival. *New Phytol.* 83, 817–827. (16)

Abul-Fatih, H. A., and F. A. Bazzaz. 1980. The biology of *Ambrosia trifida* L. IV. Demography of plants and leaves. *New Phytol.* 84, 107–111. (16)

Abul-Fatih, H. A., F. A. Bazzaz, and R. Hunt. 1979. The biology of *Ambrosia trifida* L. III. Growth and biomass allocation. *New Phytol.* 83, 829–838. (16)

Adams, M. W. 1967. Basis of yield component compensation in crop plants with special reference to the field bean, *Phaseolus vulgaris*. *Crop Sci.* 7, 505–510. (1)

Adams, M. W., and J. E. Grafius. 1971. Yield component compensation—alternative interpretations. *Crop Sci.* 11, 33–35. (19)

Adams, W. T., and R. W. Allard. 1982. Mating system variation in *Festuca microstachys*. *Evolution* 36, 591–595. (13)

Agnew, D. 1981. *The Structural Dynamics of Beech* (Fagus sylvatica *L.*) *Saplings.* M.Sc. thesis, National University of Ireland. (1)

Aikman, D. P., and A. R. Watkinson. 1980. A model for growth and self thinning in even aged monocultures of plants. *Ann. Bot.* 45, 419–427. (18)

Aker, C. L. 1982. Regulation of flower, fruit and seed production by a monocarpic perennial, *Yucca whipplei*. *J. Ecol.* 70, 357–372. (4)

Allard, R. W. 1960. *Principles of Plant Breeding.* Wiley, New York. (10)

Allard, R. W., and J. Adams. 1969a. Population studies in predominantly self-pollinating species. XIII. Intergenotypic competition and population structure in barley and wheat. *Amer. Nat.* 103, 621–645. (5)
Allard, R. W., and J. Adams. 1969b. The role of intergenotypic interactions in plant breeding. *Proc. XII Intern. Congr. Genet.* 3, 349–370. (6)
Allard, R. W. and P. E. Hansche. 1964. Some parameters of population variability and their implications in plant breeding. *Advan. Agron.* 16, 281–325. (5)
Allard, R. W., G. R. Babbel, M. T. Clegg, and A. L. Kahler. 1972. Evidence for coadaptation in *Avena barbata. Proc. Nat. Acad. Sci. USA* 69, 3043–3048. (5)
Allden, W. G., and I. A. McD. Whittaker. 1970. The determinants of herbage intake by grazing sheep: The interrelationship of factors influencing herbage intake and availability. *Aust. J. Agric. Res.* 21, 755–766. (18)
Allen, G. E. 1978. *Thomas Hunt Morgan: The Man and His Science.* Princeton University Press, Princeton. (11)
Allsopp, A. 1954a. Experimental and analytical studies of pteridophytes. 24. Investigations on *Marsilea.* 4. Anatomical effects of changes in sugar concentration. *Ann. Bot. NS.* 18, 449–461. (17)
Allsopp, A. 1954b. Juvenile stages of plants and the nutritional status of the shoot apex. *Nature* 173, 1032–1035. (17)
Altenburg, E. 1934. A theory of hermaphroditism. *Amer. Nat.* 68, 88–91. (14)
Altieri, M. A. 1981. Crop-weed-insect interactions and the development of pest-stable cropping systems. In *Pests, Pathogens and Vegetation,* J.M. Thresh (ed.). Pitman, London. pp. 459–466. (18)
Altieri, M. A., A. Van Schoonhoven, and J. D. Doll. 1977. The ecological role of weeds in insect pest management systems: A review illustrated with bean (*Phaseolus vulgaris* L.) cropping systems. *PANS* 23, 185–206. (18)
Anderson, M. C. 1964. Studies of the woodland light climate. I. The photographic computation of light conditions. *J. Ecol.* 52, 27–41. (15)
Angevine, M. W., and B. F. Chabot. 1979. Seed germination syndromes in higher plants. In *Topics in Plant Population Biology,* O. T. Solbrig, S. Jain, G. B. Johnson, and P. H. Raven (eds.). Columbia University Press, New York. pp. 188–206. (5)
Anon. 1979. Agriculture and pollution. *Royal Commission on Environmental Pollution Seventh Report.* (Cmnd. 7644) London: HMSO. (18)
Anslow, R. C., and J. O. Green. 1967. The seasonal growth of pasture grasses. *J. Agric. Sci. (Camb.)* 68, 109–122. (18)
Antonovics, J. 1968a. Evolution in closely adjacent plant populations. V. Evolution of self-fertility. *Heredity* 23, 219–238. (5)
Antonovics, J. 1968b. Evolution in closely adjacent plant populations. VI. Manifold effects of gene flow. *Heredity* 23, 507–524. (12)
Antonovics, J. 1971. The effects of a heterogeneous environment on the genetics of natural populations. *Amer. Sci.* 59, 593–599. (5)
Antonovics, J. 1976a. The nature of limits to natural selection. *Ann. Mo. Bot. Gard.* 63, 224–247. (5,8)
Antonovics, J. 1976b. The input from population genetics: "The New Ecological Genetics." *Syst. Bot.* 1, 233–245. (5)
Antonovics, J. 1978. The population genetics of mixtures. In *Plant Relations in Pastures,* J. R. Wilson (ed.). CSIRO, Melbourne. pp. 233–252. (5,11)
Antonovics, J., A. D. Bradshaw, and R. G. Turner. 1971. Heavy metal tolerance in plants. *Adv. Ecol. Res.* 7, 1–85. (5,10)

Antonovics, J., and N. C. Ellstrand. in press. Experimental studies on the evolutionary significance of sexual reproduction. I. A test of the frequency-dependent selection hypothesis. *Evolution.* (6,11)

Antonovics, J., and N. L. Fowler. in press. Analysis of frequency and density effects in mixtures of *Salvia splendens* and *Linum grandiflorum* using hexagonal fan designs. *J. Ecol.* (11)

Antonovics, J., and D. A. Levin. 1980. The ecological and genetic consequences of density-dependent regulation in plants. *Ann. Rev. Ecol. Syst.* 11, 411–452. (4,6,13)

Antonovics, J., and R. B. Primack. 1982. Experimental ecology and genetics in *Plantago.* VI. The demography of seedling transplants of *P. lanceolata. J. Ecol.* 70, 55–75. (4,8,10,12)

Appleby, A. P., P. D. Olsen, and D. R. Colbert. 1976. Winter wheat yield reduction from interference by Italian ryegrass. *Agron. J.* 68, 463–466. (18)

Arber, A. 1946. Goethe's botany. *Chron. Bot.* 10, 63–126. (1)

Arber, A. 1950. *The Natural Philosophy of Plant Form.* Cambridge University Press, Cambridge. (1)

Arnold, S. J., and M. J. Wade. in press. On the measurement of natural selection. *Evolution.* (8)

Arthur, A. E., J. S. Gale, and K. J. Lawrence. 1973. Variation in wild populations of *Papaver dubium.* VII. Germination time. *Heredity* 30, 189–197. (8)

Arthur, W. 1982. The evolutionary consequence of interspecific competition. *Adv. Ecol. Res.* 12, 127–187. (5)

Aspinall, D. 1960. An analysis of competition between barley and white persicaria. II. Factors determining the course of competition. *Ann. Appl. Biol.* 48, 637–654. (18)

Aston, J. L., and A. D. Bradshaw. 1966. Evolution in closely adjacent plant populations. II. *Agrostis stolonifera* in maritime habitats. *Heredity* 21, 649–664. (5)

Atkinson, D., and G. C. White. 1981. The effects of weeds and weed control on temperate fruit orchards and their environment. In *Pests, Pathogens and Vegetation,* J. M. Thresh (ed.). Pitman, London. pp. 415–428. (18)

Auld, B. A., K. M. Menz, and R. W. Medd. 1979a. Bioeconomic model of weeds in pastures. *Agro-ecosystems* 5, 69–84. (18)

Auld, B. A., K. M. Menz, and N. M. Monaghan. 1979b. Dynamics of weed spread: Implications for policies of public control. *Prot. Ecol.* 1, 141–148. (18)

Austin, M. P. 1976. Performance of four ordination techniques assuming three different non-linear species response models. *Vegetatio* 33, 43–49. (3)

Austin, M. P. 1980. Searching for a model for use in vegetation analysis. *Vegetatio* 42, 11–21. (3)

Austin, M. P. 1981. Permanent quadrats: An interface for theory and practice. *Vegetatio* 46, 1–10. (3)

Austin, M. P. 1982. Use of a relative physiological performance value in the prediction of performance in multispecies mixtures from monoculture performance. *J. Ecol.* 70, 559–570. (3,15)

Austin, M. P., and B. O. Austin. 1980. Behavior of experimental plant communities along a nutrient gradient. *J. Ecol.* 68, 891–918. (3)

Austin, M. P., and I. Noy-Meir. 1971. The problem of non-linearity in ordination: Experiments with two gradient models. *J. Ecol.* 59, 763–773. (3)
Austin, M. P., O. B. Williams, and L. Belbin. 1981. Grassland dynamics under sheep grazing in an Australian Mediterranean type climate. *Vegetatio* 7, 201–211. (3)
Ayala, F. J. 1970. Competition, coexistence, and evolution. In *Essays in Evolution and Genetics in Honor of Theodosius Dobzhansky,* M. K. Hecht and W. C. Steere (eds.). Appleton-Century-Crofts, New York. pp. 121–158. (5)
Ayala, F. J. 1972. Competition between species. *Amer. Sci.* 60, 348–357. (5)
Baines, R. N., J. H. Grieshaber-Otto, and R. W. Snaydon. 1982. Factors affecting the performance of white clover in swards. In *Efficient Grassland Farming,* A. J. Corrall (ed.). British Grassland Society, Hurley. (19)
Baker, A. J. M., and D. H. Dalby. 1980. Morphological variation between some isolated populations of *Silene maritima* within the British Isles with particular reference to inland populations on metalliferous soils. *New Phytol.* 84, 123–138. (1)
Baker, H. G. 1953. Race formation and reproductive method in flowering plants. *Symp. Soc. Exp. Biol.* 7, 114–145. (5)
Baker, H. G. 1965. Characteristics and modes of origin of weeds. In *The Genetics of Colonizing Species,* H. G. Baker and G. L. Stebbins (eds.). Academic Press, New York. pp. 147–172. (5)
Baker, H. G. 1972a. Migration of weeds. In *Taxonomy, Phytogeography and Evolution,* D. H. Valentine (ed.). Academic Press, New York. pp. 327–347. (8,12)
Baker, H. G. 1972b. Seed weight in relation to environmental conditions in California. *Ecology* 53, 997–1010. (8,12)
Baker, H. G. 1974. The evolution of weeds. *Ann. Rev. Ecol. Syst.* 5, 1–24. (18)
Baker, H. G., and G. L. Stebbins. 1965. *The Genetics of Colonizing Species.* Academic Press, New York. (10)
Baker, J. R. 1938. The evolution of breeding seasons. In *Evolution, Essays Presented to E. S. Goodrich,* G. R. De Beer (ed.). Oxford University Press, New York. pp. 161–177. (7)
Baker, M. C., and P. Marler. 1980. Behavioral adaptations that constrain the gene pool in vertebrates. In *Evolution of Social Behavior: Hypotheses and Empirical Tests,* H. Markl (ed.). Verlag Chemie CmbH Weinheim. (13)
Balding, F. R., and G. L. Cunningham. 1976. A comparison of heat transfer characteristics of simple and pinnate leaf models. *Bot. Gaz.* 137, 65–74. (15)
Baldwin, J. H. 1979. The chemical control of wild oats and blackgrass. *A.D.A.S. Quart. Rev.* 33, 69–101. (19)
Bandeen, J. D., G. R. Stephenson, and E. R. Cowett. 1982. Discovery and distribution of herbicide-resistant weeds in North America. In *Herbicide Resistance in Plants*, H. M. Lebaron and J. Gressel (eds.). Wiley, New York. pp. 9–30. (10,18)
Banyard, B. J., and S. H. James. 1979. Biosystematic studies in the *Stylidium crassifolium* species complex (Stylidae). *Aust. J. Bot.* 27, 27–37. (13)
Barkham, J. P. 1980. Population dynamics of the wild daffodil *(Narcissus pseudonarcissus).* I. Clonal growth, seed production, mortality and the effects of density. *J. Ecol.* 68, 607–633. (4)
Barkham, J. P., and C. E. Hance. 1982. Population dynamics of the wild daffodil

(Narcissus pseudonarcissus). III. Implications of a computer model of 1000 years of population change. *J. Ecol.* 70, 323–344. (2)

Barnes, A. 1977. The influence of the length of the growth period and planting density on total crop yield. *Ann. Bot.* 41, 883–895. (18)

Barnes, B. V., R. T. Bingham, and A. E. Squillace. 1962. Selective fertilization in *Pinus monticola* Dougl. II. Results of additional tests. *Silvae Genet.* 11, 103–111. (13)

Barnes, D. K., and R. W. Cleveland. 1963. Genetic evidence for nonrandom fertilization in alfalfa as influenced by differential pollen tube growth. *Crop Sci.* 3, 295–297. (13)

Barrett, P. H., D. J. Weinshank, and T. T. Gottleber. 1981. *A Concordance to Darwin's Origin of Species* (1st Ed.). Cornell University Press, Ithaca. (1)

Barthou, H. 1979. Analyse des correlations interorganiques au cours du dévelоppement de l'axe végétatif du *Mirabilis jalapa* L. 1. Evolution spatio-temporelles des correlations dimensionnelles locales. *Ann. Sci. Nat. Bot. Paris, Ser.* 13, 117–127. (1)

Baskin, J. M., and C. C. Baskin. 1974. Germination and survival in a population of the winter annual *Alyssum alyssoides*. *Can. J. Bot.* 52, 2439–2445. (4)

Bateman, A. J. 1947. Contamination of seed crops. I. Insect pollination. *J. Genet.* 48, 257–275. (12)

Bateman, A. J. 1948. Intra-sexual selection in *Drosophila*. *Heredity* 2, 349–368. (14)

Bawa, K. S. 1980. Evolution of dioecy in flowering plants. *Ann. Rev. Ecol. Syst.* 11, 15–39. (14)

Bawa, K. S., C. R. Keegan, and R. H. Voss. 1982. Sexual dimorphism in *Aralia nudicaulis* L. (Araliaceae). *Evolution* 36, 371–378. (4)

Bawa, K. S., and P. A. Opler. 1975. Dioecism in tropical forest trees. *Evolution* 29, 167–179. (14)

Bazzaz, F. A. 1979. The physiological ecology of plant succession. *Ann. Rev. Ecol. Syst.* 10, 351–371. (16)

Bazzaz, F. A., and R. W. Carlson. 1979. Photosynthetic contribution of flowers and seeds to reproductive effort of an annual colonizer. *New Phytol.* 82, 223–232. (16)

Bazzaz, F. A., and R. W. Carlson. 1982. Photosynthetic acclimation to variability in the light environment of early and late successional plants. *Oecologia* 54, 313–316. (16)

Bazzaz, F. A., R. W. Carlson, and J. L. Harper. 1979. Contribution to reproductive effort by photosynthesis of flowers and fruits. *Nature* 279, 554–555. (14)

Bazzaz, F. A., and J. L. Harper. 1977. Demographic analysis of the growth of *Linum usitatissimum*. *New Phytol.* 78, 193–208. (7,16)

Bazzaz, F. A., D. A. Levin, and M. R. Schmierbach. 1982. Differential survival of genetic variants in crowded population of *Phlox*, *J. Appl. Ecol.* 19, 891–900. (4)

Beattie, A. J. 1976. Plant dispersion, pollination and gene flow in *Viola*. *Oecologia* 25, 291–300. (13)

Beattie, A. J. 1978. Plant–animal interactions affecting gene flow in *Viola. Linn. Soc. Symp. Ser.* 6, 151–164. (13)
Beattie, A. J., and D. C. Culver. 1979. Neighborhood size in *Viola. Evolution* 33, 1226–1229. (13)
Beckett, P. H. T., and R. Webster. 1971. Soil variability: A review. *Soils and Fertilizers* 34, 1–15. (19)
Beeftink, A. in press. Interaction between *Plantago maritima* and *Limonium vulgare* under natural conditions. *Vegetatio.* (3)
Beeftink, W. G. (ed.). 1980. *Vegetation Dynamics.* Delta Institute for Hydrobiological Research. Junk, The Hague. (3)
Bell, A. D. 1974. Rhizome organization in relation to vegetative spread in *Medeola virginiana. J. Arnold Arb.* 55, 458–468. (1)
Bell, A. D. 1976. Computerized vegetative mobility in rhizomatous plants. In *Automata, Languages and Development*, A. Lindenmayer and G. Rozen (eds.). North-Holland, Amsterdam. pp. 3–14. (2)
Bell, A. D. 1979. The hexagonal branching pattern of *Alpinia speciosa* L. (Zingiberaceae). *Ann. Bot.* 43, 209–223. (2)
Bell, A. D., D. Roberts, and A. Smith. 1979. Branching patterns: The simulation of plant architecture. *J. Theor. Biol.* 81, 351–375. (1,2)
Bell, A. D., and P. B. Tomlinson. 1980. Adaptive architecture in rhizomatous plants. *Bot. J. Lin. Soc.* 80, 125–160. (1,2)
Bell, G. 1982. *The Masterpiece of Nature: The Evolution and Genetics of Sexuality.* Croom Helm, London. (1)
Bell, W. 1977. The cost of combine breakdowns and delayed harvesting. *Farm Management Rev.* 10, 29–40. (18)
Beltran, I. C., and S. H. James. 1974. Complex hybridity in *Isotoma petraea.* IV. Heterosis in interpopulational hybrids. *Austral. J. Bot.* 22, 251–264. (12)
Bennett, E. 1964. Historical perspectives in genecology. *Scot. Pl. Breed. Stn. Rec.* 49–115. (5)
Bentley, S., and J. B. Whittaker. 1979. Effects of grazing by a chrysomelid beetle, *Gastrophysa viridula*, on competition between *Rumex obtusifolius* and *Rumex crispus. J. Ecol.* 67, 79–90. (4)
Bentley, S., J. B. Whittaker, and A. J. C. Malloch. 1980. Field experiments on the effects of grazing by a chrysomelid beetle *(Gastrophysa viridula)* on seed production and quality in *Rumex obtusifolius* and *Rumex crispus. J. Ecol.* 68, 671–674. (4)
Berendse, F. 1981. *Competition and Equilibrium in Grassland Communities.* Ph.D. thesis, University of Utrecht. (3)
Bergh, J. P. van der, and W. G. Braakhekke. 1978. Coexistence of plant species by niche differentiation. In *Structure and Functioning of Plant Populations*, A. H. J. Freysen and J. W. Woldendorp (eds.). North-Holland, Amsterdam. pp. 125–138. (5)
Bernard, R. L. 1972. Two genes affecting stem termination in soybeans. *Crop Sci.* 12, 235–239. (1)
Bertin, R. I. 1982. Paternity and fruit production in trumpet creeper *(Campsis radicans). Amer. Nat.* 119, 694–709. (13)
Bishop, J. A., and L. M. Cook (eds.). 1981. *Genetic consequence of man-made change.* Academic Press, London. (10)

Björkman, O. 1968. Carboxydismutase activity in shade-adapted and sun-adapted species of higher plants. *Physiol. Plantarum* 21, 1–10 (16)

Blackman, F. F. 1905. Optimal and limiting factors. *Ann. Bot.* 18, 281–295. (19)

Bleasdale, J. K. A. 1966. Plant growth and crop yield. *Ann. Appl. Biol.* 57, 173–182. (18)

Bleasdale, J. K. A., and J. A. Nelder. 1960. Plant population and crop yield. *Nature* 188, 342. (18)

Boardman, N. K. 1977. Comparative photosynthesis of sun and shade plants. *Ann. Rev. Plant Physiol.* 28, 355–377. (15)

Bochert, M. I., and S. K. Jain. 1978. The effect of rodent seed predation on four species of California annual grasses. *Oecologia* 33, 101–113. (18)

Bocher, T. W. 1949. Racial divergencies in *Prunella vulgaris* in relation to habitat and climate. *New Phytol.* 48, 285–324. (10)

Bock, W. J. 1977. Toward an ecological morphology. *Die Vogelwarte* 29, 127–135. (1)

Bock, W. J. 1980. The definition and recognition of biological adaptation. *Amer. Zool.* 20, 217–227. (1)

Bogaert, J. van. 1974. The evaluation of progenies of meadow fescue (*Festuca pratensis* L.) in monoculture and in mixture with perennial ryegrass (*Lolium perenne* L.). *Euphytica* 23, 48–53. (5)

Böker, H. 1935–7. *Einführung in die Vergleichende Biologische Anatomie der Wirbeltiere.* G. Fischer, Jena. (1)

Bonaparte, E. E. N. A., and R. I. Brawn. 1975. The effect of intraspecific competition on the phenotypic plasticity of morphological and agronomic characters of four maize hybrids. *Ann. Bot.* 39, 863–869. (1)

Bonnier, G. 1895. Recherches experiméntales sur l'adaptation des plantes au climat alpin. *Ann. Sci. Nat. Bot. Biol. Veg.* 7th ser. 20, 217–358. (10)

Bonnier, G. 1920. Nouvelles observations sur les cultures éxperimentales à diverses altitudes. *Rev. Gen. de Bot.* 32, 305–326. (10)

Bookman, S. S. 1983. Costs and benefits of flower abscission and fruit abortion in *Asclepias speciosa. Ecology* 64, 264–273. (13)

Booth, A. 1959. Some factors concerned in the growth of stolons in potato. *J. Linn. Soc. Bot.* 56, 166–169. (17)

Borgia, G. 1982. Female-biased sex ratios. *Nature* 298, 494–495. (14)

Braakhekke, W. G. 1980. *On Co-existence: A Casual Approach to Diversity and Stability in Grassland Vegetation.* Ph.D. thesis, University of Wageningen. (3)

Bradshaw, A. D. 1959. Population differentiation in *Agrostis tenuis* Sibth. I. Morphological differentiation. *New Phytol.* 58, 208–227. (5,10)

Bradshaw, A. D. 1960. Population differentiation in *Agrostis tenuis* Sibth. III. Populations in varied environments. *New Phytol.* 59, 92–103. (5,10)

Bradshaw, A. D. 1965. Evolutionary significance of phenotypic plasticity in plants. *Adv. Genet.* 13, 115–155. (4,5,17)

Bradshaw, A. D. 1971. Plant evolution in extreme environments. In *Ecological Genetics and Evolution,* R. Creed (ed.). Blackwell, Oxford. pp. 20–50. (10)

Bradshaw, A. D. 1972. Some of the evolutionary consequences of being a plant. *Evol. Biol.* 5, 25–47. (5,10)

Bradshaw, A. D. 1975. The evolution of heavy metal tolerance and its significance for vegetation establishment on metal contaminated sites. In *Heavy Metals in the Environment,* T. C. Hutchinson (ed.). Toronto University Press, Toronto. pp. 599–622. (10)

Bradshaw, A. D. 1983. The importance of evolutionary ideas in ecology—and vice versa. In *Evolutionary Ecology,* B. Sharrocks (ed.). Blackwell, Oxford. pp. 1–25. (10)

Bradshaw, A. D., and T. McNeilly. 1982. *Evolution and Pollution.* Arnold, London. (10)

Braun-Blanquet, J. 1964. *Pflanzensoziologie, Grundzüge der Vegetations Kunde.* 3 Aufl. Springer, Wien. (3)

Braverman, M. H., and R. G. Schrandt. 1966. Colony development of a polymorphic hydroid as a problem in pattern formation. In *The Cnidaria and Their Evolution,* W. J. Rees (ed.). *Symp. Zool. Soc. Lond.* 16, 168–198. Academic Press, London. (1,2)

Breese, E. L. 1966. Selection for differing degrees of out-breeding in *Nicotiana rustica. Ann. Bot. N.S.* 23, 331–334. (13)

Breese, E. L. 1966. Reproduction in ryegrass. In *Reproductive Biology and Taxonomy of Vascular Plants,* J. G. Hawkes (ed.). Pergamon, Oxford. pp. 51–58. (19)

Bremermann, H. J. 1980. Sex and polymorphism as strategies in host-pathogen interactions. *J. Theor. Biol.* 87, 671–702. (6)

Bridwell, J. C. 1918. Notes on the Bruchidae and their parasites in the Hawaiian Islands. *Proc. Hawaiian Entomol. Soc.* 3, 465–505. (4)

Briggs, D., and S. M. Walters. 1969. *Plant Variation and Evolution.* Weidenfeld and Nicolson, London. (1)

Brokaw, N. V. L. 1982. Treefalls: Frequency, timing and consequences. In *Ecology of a Tropical Forest: Season Rhythms and Long-Term Changes,* E. G. Leigh, A. S. Rand, and D. M. Wondsor (eds.). Smithsonian Institution Press, Washington. pp. 101–108. (4)

Broker, W. 1963. Genetisch-physiologische Untersuchungen über die zinkverträglichkeit von *Silene inflata* Sm. *Flora Jena B* 153, 122–156. (10)

Brown, A. H. D. 1979. Enzyme polymorphism in plant populations. *Theor. Popul. Biol.* 15, 1–42. (19)

Brown, A. H. D., and L. Albrecht. 1980. Variable outcrossing and the genetic structure of predominantly self-pollinated species. *J. Theor. Biol.* 82, 591–606. (13)

Brown, B. A., and M. T. Clegg. in press. Influence of flower color polymorphism on genetic transmission in a natural population of the common morning glory, *Ipomoea purpurea. Evolution.* (13)

Browning, J. A. 1981. The agroecosystem—natural ecosystem dichotomy and its impact on phytopathological concepts. In *Pests, Pathogens and Vegetation,* J. M. Thresh (ed.). Pitman, London. pp. 159–174. (18)

Bucholtz, K. P. 1971. The influence of allelopathy on mineral nutrition. In *Biochemical Interactions Amongst Plants,* U.S. Nat. Comm. for I.B.P. Natn. Acad. Sci., Washington. pp. 86–89. (18)

Bull, J. J. 1981. Sex ratio evolution when fitness varies. *Heredity* 46, 9–26. (13)

Bullock, S. H. 1980. Demography of an undergrowth palm in littoral Cameroon. *Biotropica* 12, 247–255. (4)
Bullock, S. H. 1982. Population structure and reproduction in the Neotropical dioecious tree *Compsoneura sprucei. Oecologia* 55, 238–242. (4)
Bullock, S. H., and K. S. Bawa. 1981. Sexual dimorphism and the annual flowering pattern in *Jacaratia dolichaula* (D. Smith) Woodson (Caricaceae) in a Costa Rican rain forest. *Ecology* 62, 1494–1504. (4)
Bulmer, M. G., and P. D. Taylor. 1980. Dispersal and the sex ratio. *Nature* 284, 448–449. (13)
Bunting, A. H. 1960. Some reflections on the ecology of weeds. In *The Biology of Weeds,* J. L. Harper (ed.). Blackwell, Oxford. pp. 11–26. (18)
Burdon, J. J. 1980. Intra-specific diversity in a natural population of *Trifolium repens. J. Ecol.* 68, 717–735. (5,8,13)
Burdon, J. J. 1983. Biological flora of the British Isles. *Trifolium repens* L. *J. Ecol.* 71, 307–330. (1,5)
Burdon, J. J., and J. L. Harper. 1980. Relative growth rates of individual members of a plant population. *J. Ecol.* 68, 953–957. (4,5)
Burley, N. in press. The meaning of assortative mating. *J. Ethol. Sociobiol. (13)*
Büsgen, M., and E. Münch. 1929. *The Structure and Life of Forest Trees.* Chapman and Hall, London. (1)
Buss, L. W. 1983. Evolution, development, and the units of selection. *Proc. Natl. Acad. Sci. USA* 80, 1387–1391. (1)
Cahn, M. G., and J. L. Harper. 1976a. The biology of the leaf mark polymorphism in *Trifolium repens* L. 1. Distribution of phenotypes at a local scale. *Heredity* 37, 309–325. (17)
Cahn, M. G., and J. L. Harper. 1976b. The biology of the leaf mark polymorphism in *Trifolium repens* L. 2. Evidence for the selection of leaf mark by rumen fistulated sheep. *Heredity* 37, 327–333. (17)
Caldwell, M. M., J. H. Richards, D. A. Johnson, R. S. Nowak, and R. S. Dzurec. 1981. Coping with herbivory: photosynthetic capacity and resource allocation in two semiarid *Agropyron* bunchgrasses. *Oecologia* 50, 14–24. (15)
Caligari, P. D. S., and M. J. Hanks. 1978. Genetical analysis of components of overall plant shape. *Theor. Appl. Gen.* 52, 65–72. (1)
Callaghan, T. V. 1976. Strategies of growth and population dynamics of plants. 3. Growth and population dynamics of *Carex bigelowii* in an alpine environment. *Oikos* 27, 402–413. (4)
Callaghan, T. V. 1980. Strategies of growth and population dynamics of tundra plants. 5. Age-related patterns of nutrient allocation in *Lycopodium annotinum* from Swedish Lapland. *Oikos* 35, 373–386. (1)
Callaghan, T. V., and N. J. Collins. 1981. Life cycles, population dynamics and the growth of tundra plants. In *Tundra Ecosystems: A Comparative Analysis,* L. C. Bliss, J. B. Cragg, D. W. Heal, and J. J. Moore (eds.). Cambridge University Press, Cambridge. pp. 257–284. (1)
Campbell, J. M., and R. J. Abbott. 1976. Variability of outcrossing frequency in *Senecio vulgaris* L. *Heredity* 36, 267–274. (13)

Candolle, A. P. de. 1813. *Théorie Élémentaire de la Botanique*. Paris. (1)
Candolle, A. P. de. 1827. *Organographie Végétale*. Paris. (1)
Cardwell, V. B. 1982. Fifty years of Minnesota corn production: Sources of yield increase. *Agron. Jour.* 74, 984–990. (19)
Cargill, S. M. 1981. *The Effects of Grazing by Lesser Snow Geese on the Vegetation of An Arctic Salt Marsh.* M.Sc. thesis, University of Toronto. (17)
Carmi, A., and D. Koller. 1979. Regulation of photosynthetic activity in the primary leaves of bean (*Phaseolus vulgaris* L.) by materials moving in the water conducting system. *Plant Physiol.* 64, 285–288. (15)
Carpenter, I. W., and A. T. Guard. 1950. Some effects of cross-pollination of seed production and hybrid vigor of tulip trees. *J. Forest.* 48, 852–855. (12)
Carpenter, S. R. 1980. Estimating shoot production by a hierarchical cohort method of herbaceous plant subject to high mortality. *Amer. Midl. Nat.* 104, 163–175. (1)
Caswell, H., and P. A. Werner. 1977. Transient behavior and life history analysis of Teasel (*Dipsacus sylvestris* Huds.). *Ecology* 59, 53–66. (18)
Cates, R. 1975. The interface between slugs and wild ginger: some evolutionary aspects. *Ecology* 56, 391–400. (7)
Caughley, G., and J. H. Lawton. 1981. Plant-herbivore systems. In *Theoretical Ecology, Principles and Applications.* (2nd Ed.), R. May (ed.). Blackwell, Oxford. pp. 132–166. (7)
Ceballos, G. J. 1982. *Experimental studies of grazing and its role in the balance of plant species.* M.Sc. thesis, University of Wales. (7)
Chabot, B. F., and D. J. Hicks. 1982. The ecology of leaf life spans. *Ann. Rev. Ecol. Syst.* 13, 229–259. (1)
Chacko, E. K., Y. T. N. Readdy, and T. V. Ananthanarayanan. 1982. Studies on the relationship between leaf number and area and fruit development in mango (*Mangifera indica* L.). *J. Hort. Sci.* 57, 483–492. (1)
Chandler, R. F. 1969. Plant morphology and stand geometry in relation to nitrogen. In *Physiological Aspects of Crop Yield,* J. D. Eastin, F. A. Haskins, C. Y. Sullivan, and C. H. M. van Bavel (eds.). Madison: American Society of Agronomy and Crop Science of America. pp. 265–285. (1)
Charles, A. H. 1961. Differential survival of cultivars of *Lolium, Dactylis* and *Phleum. J. Br. Grassl. Soc.* 16, 69–75. (19)
Charlesworth, B. 1980. The cost of meiosis with alternation of sexual and asexual generations. *J. Theoret. Biol.* 87, 517–528. (1)
Charlesworth, B., and M. A. Toro. 1982. Female-biased sex ratios. *Nature* 298, 494. (14)
Charlesworth, D., and B. Charlesworth. 1979. The evolutionary genetics of sexual systems in flowering plants. *Proc. Royal Soc. London B* 205, 513–530. (6)
Charlesworth, D., and B. Charlesworth. 1981. Allocation of resources to male and female functions in hermaphrodites. *Biol. J. Linn. Soc.* 15, 57–74. (13,14)
Charnov, E. L. 1979. Simultaneous hermaphroditism and sexual selection. *Proc. Nat. Acad. Sci.* 76, 2480–2484. (14)
Charnov, E. L. 1980. Sex allocation and local mate competition in barnacles. *Marine Biol. Letters* 1, 269–272. (14)
Charnov, E. L. 1982. *The Theory of Sex Allocation.* Princeton University Press, Princeton. (13,14)

Chaudonneret, J. 1979. La notion de métamère chez les invertébrés. *Bull. Soc. Zool. France* 104, 241–270. (1)
Cherrett, J. M. 1968. The foraging behavior of *Atta cephalotes* L. (Hymenoptera, Formicidae). I. Foraging pattern and plant species attacked in tropical rain forest. *J. Anim. Ecol.* 37, 387–403. (2)
Cideciyan, M. A., and A. J. C. Malloch. 1982. Effects of seed size on the germination, growth, and competitive ability of *Rumex crispus* and *Rumex obtusifolius. J. Ecol.* 70, 227–232. (4,9)
Cisne, J. L. 1979. Arthropoda. In *The Encyclopedia of Paleontology,* R. W. Fairbridge and D. Jablonski (eds.). Dowden, Hutchinson and Ross, Stroudsburg, Pa. pp. 51–60. (1)
Clark, A. B. 1978. Sex ratio and local resource competition in a prosimian primate. *Science* 201, 163–165. (14)
Clark, R. B. 1964. *Dynamics in Metazoan Evolution: The Origin of the Coelom and Segments.* Clarendon Press, Oxford. (1)
Clark, R. B. 1980. Natur und Entstehungen der metameren Segmentierung. *Zool. J. Anat.* 103, 169–195. (1)
Clark, S. C. 1980. Reproductive and vegetative performance in two winter annual grasses, *Catapodium rigidum* (L.) C. E. Hubbard and *C. marinum* (L.) C. E. Hubbard. I. The effects of soil and genotype on reproductive performance in the field and in a growth-room. *New Phytol.* 84, 59–78. (8)
Clausen, J. 1926. Genetical and cytological investigations on *Viola tricolor* L. and *V. arvensis* Murr. *Hereditas* 8, 1–156. (1)
Clausen, J., and W. M. Hiesey. 1958a. Experimental studies on the nature of species. IV. Genetic structure of ecological races. *Carnegie Inst. Washington Publ.* 615. (5,10,11)
Clausen, J., and W. M. Hiesey. 1958b. Phenotypic expression of genotypes in contrasting environments. *Rep. Scottish Plant Breeding Stat.* 1958, 41–51. (17)
Clausen, J., D. D. Keck, and W. M. Hiesey. 1940. Experimental studies on the nature of species. I. The effect of varied environments on western American plants. *Carnegie Inst. Washington Publ.* 520. (1,5,10,17)
Clausen, J., D. D. Keck, and W. M. Hiesey. 1948. Experimental studies on the nature of species. III. Environmental responses of climatic races of *Achillea. Carnegie Inst. Washington Publ.* 581. (5,10)
Clegg, M. T. 1980. Measuring plant mating systems. *BioScience* 30, 814–818. (13)
Clegg, M. T., A. L. Kahler, and R. W. Allard. 1978. Estimation of life cycle components of selection in an experimental plant population. *Genetics* 89, 765–792. (13)
Clements, F. E., and G. W. Goldsmith. 1924. *The Phytometer Method in Ecology. Carnegie Inst. Washington Publ.* 356. (11)
Clements, F. E., J. E. Weaver, and H. C. Hanson. 1929. Competition in cultivated crops. *Carnegie Inst. Washington Publ.* 398, 202–233. (9)
Cody, M. L. 1966. A general theory of clutch size. *Evolution* 20, 174–184. (4)
Cody, M. L. 1974a. *Competition and the Structure of Bird Communities.* Princeton University Press, Princeton. (5)

Cody, M. L. 1974b. Optimization in ecology. *Science* 183, 1156–1164. (8)
Cohen, D. 1966. Optimizing reproduction in a randomly varying environment. *J. Theor. Biol.* 12, 119–129. (15)
Cohen, D. 1971. Maximizing final yield when growth is limited by time or by limiting resources. *J. Theor. Biol.* 33, 299–307. (15)
Coleman, J. R., and M. A. Coleman. 1982. Reproductive biology of an andromonoecious *Solanum* (*S. palinacanthum* Dunal). *Biotropica* 14, 69–75. (14)
Coleman, W. 1964. *Georges Cuvier, Zoologist: A Study in The History of Evolution Theory.* Harvard University Press, Cambridge, Mass. (1)
Coleman, W. 1976. Morphology between the type concept and descent theory. *J. Hist. Med.* 31, 149–175. (1)
Coleman, W. 1980. Morphology in the evolutionary synthesis. In *The Evolutionary Synthesis,* E. Mayr and W. B. Provine (eds.). Harvard University Press, Cambridge, Mass. pp. 174–180. (1)
Coles, J. F., and D. P. Fowler. 1976. Inbreeding in neighboring trees in two white spruce populations. *Silvae. Genet.* 25, 29–34. (13)
Colwell, R. K. 1981. Group selection is implicated in the evolution of female-biased sex ratios. *Nature* 290, 401–404. (14)
Connell, J. H. 1980. Diversity and the coevolution of competitors, or the ghost of competition past. *Oikos* 35, 131–138. (5)
Connell, J. H., and R. O. Slatyer. 1977. Mechanisms of succession in natural communities and their role in community stability and organization. *Amer. Nat.* 11, 1119–1144. (3)
Conrad, G. G., and S. R. Radesovich. 1979. Ecological fitness of *Senecio vulgaris* and *Amaranthus retroflexus* biotypes susceptible or resistant to atrazine. *J. Appl. Ecol.* 16, 171–179. (18)
Conway, G. 1981. Man versus pests. In *Theoretical Ecology, Principles and Applications* (2nd Ed.), R. M. May (ed.). Blackwell, Oxford. pp. 356–386. (18)
Cook, C. A., C. Lefebre, and T. McNeilly. 1972. Competition between metal tolerant and normal plant populations on normal soil. *Evolution* 26, 366–372. (5)
Cook, R. E. 1979a. Asexual reproduction: A further consideration. *Amer. Nat.* 113, 769–772. (17)
Cook, R. E. 1979b. Patterns of juvenile mortality and recruitment in plants. In *Topics in Plant Population Biology,* O. T. Solbrig, S. Jain, G. B. Johnson and P. H. Raven (eds.). Columbia University Press, New York, pp. 207–231 (1,4,5)
Cook, R. E. 1980. Germination and size-dependent mortality in *Viola blanda. Oecologia* 47, 115–117. (4,18)
Cook, S. A., and M. P. Johnson. 1968. Adaptation to heterogeneous environments. I. Variation in heterophylly in *Ranunculus flammula* L. *Evolution* 22, 496–516. (10)
Cook, S. A., C. Lefebvre, and T. McNeilly. 1972. Competition between metal tolerant and normal plant populations on normal soil. *Evolution* 26, 366–372. (10,12)
Cooper, J. P. 1964. Climatic variation in forage grasses: leaf development in climatic races of *Lolium* and *Dactylis. J. Appl. Ecol.* 1, 45–61. (10)
Cooper, W. S., and R. H. Kaplan. 1982. Adaptive "coin-flipping": a decision-theoretic examination of natural selection for random individual variation.

J. Theor. Biol. 94, 135–151. (2)
Cormstock, V. E., and R. N. Anderson. 1968. An inheritance study of tolerance in flax (*Linum usitatissimum* L.) treated with MCPA. *Crop Sci.* 8, 423–427. (18)
Corner, E. J. H. 1958. Transference of function. *J. Linn. Soc. Bot. Zool.* 56, 33–40. (1)
Courtney, A. D. 1977. *Some Studies on The Growth and Development of Rhizomatous Grasses.* Ph.D. thesis, Queens Univ., Belfast. (18)
Coyne, D. P. 1980. Modification of plant architecture and crop yield by breeding. *Hort. Sci.* 15, 244–247. (1)
Cowan, I., and G. Farquhar. 1977. Stomatal function in relation to leaf metabolism and environment. *Symp. Soc. Exp. Biol.* 31, 471–505. (15)
Cramer, H. H. 1967. Plant protection and world crop production. *Pflanzenschutznachrichten* 20, 1–524. (18)
Crane, M. B., and K. Mather. 1943. The natural cross-pollination of crop plants with particular reference to radish. *J. Appl. Biol.* 30, 301–308. (12)
Crisp, M. D., and R. T. Lange. 1976. Age structure, distribution and survival under grazing of the arid zone shrub *Acacia burkitti. Oikos* 27, 86–92. (4)
Croft, B. A., J. L. Howes, and S. M. Welch. 1976. A computer based extension pest management delivery system. *Env. Ent.* 5, 20–34. (18)
Crow, J. F., and T. Maruyama. 1971. The number of neutral alleles maintained in a finite geographically structured population. *Theor. Pop. Biol.* 2, 437–453. (12)
Crowe, l. 1971. The polygenic control of outbreeding in *Borago officinalis. Heredity* 27, 111–118. (13)
Cruden, R. W. 1977. Pollen-ovule ratios: A conservative indicator of breeding systems in flowering plants. *Evolution* 31, 32–46. (14)
Cruden, R. W., and K. G. Jensen. 1979. Viscin threads, pollination efficiency and low pollen-ovule ratios. *Amer. J. Bot.* 66, 875–879. (14)
Cruden, R. W., and S. Miller-Ward. 1981. Pollen-ovule ratio, pollen size, and the ratio of stigmatic area to the pollen-bearing area of the pollinator: An hypothesis. *Evolution* 35, 964–974. (14)
Curtis, J. T., and R. P. McIntosh, 1951. An upland forest continuum in the prairie-forest border region of Wisconsin. *Ecology* 32, 476–496. (3)
Cussans, G. W. 1980. Strategic planning for weed control—a researcher's view. *Proc. Brit. Crop Prot. Conf. Weeds* 3, 823–832. (18)
Cussans, G. W., S. R. Moss, F. Pollard, and B. J. Wilson. 1979. Studies on the effects of tillage on annual weed populations. *Symp. EWRS: Influence of Different Factors on the Development and Control of Weeds.* pp. 383–390. (18)
Cusset, G. 1982. The conceptual bases of plant morphology. In *Axioms and Principles of Plant Construction,* R. Sattler (ed.). Junk, The Hague. pp. 8–86. (1)
Dainty, J. 1979. The ionic and water relations of plants which adjust to a fluctuating saline environment. In *Ecological Processes in Coastal Environments,* R. L. Jefferies and A. J. Davy (eds.). Blackwell, Oxford. pp. 201–209. (17)

Darwin, C. 1857. Productiveness of foreign seed. *Gardeners Chronicle and Agricultural Gazette* 46, 779. (18)

Darwin, C. 1859. *On the Origin of Species by Means of Natural Selection.* Facsimile of First Edition, Harvard University Press, Cambridge, Mass., 1964. (5,7,10)

Darwin, C. 1862. *The Various Contrivances by Which Orchids Are Fertilised by Insects.* 1st Ed. Murray, London. (14)

Darwin, C. 1868. *Variation of Animals and Plants under Domestication* (2 vols.). Orange Judd & Company, New York. (19)

Darwin, C. 1875. *Insectivorous Plants.* Murray, London. (2)

Darwin, C. 1876. *The Effects of Cross and Self Fertilisation in the Vegetable Kingdom.* Murray, London. (14)

Darwin, C. 1877. *The Different Forms of Flowers on Plants of the Same Species.* Murray, London. (2,14)

Darwin, C. 1885. *The Movements and Habits of Climbing Plants.* Murray, London. (2)

Darwinkel, A. 1978. Patterns of tillering and grain production of winter wheat at a wide range of plant densities. *Neth. J. Agric. Sci.* 26, 383–398. (1)

Davidson, B. R. 1965. Significance of differences between variety yields under experimental and farm conditions. *Nature* 207, 1009. (19)

Davies, M. S., and R. W. Snaydon. 1973a. Physiological differences among populations of *Anthoxanthum odoratum* L. collected from the Park Grass Experiment, Rothamsted. I. Response to calcium. *J. Appl. Ecol.* 10, 33–45. (5)

Davies, M. S., and R. W. Snaydon. 1973b. Physiological differences among populations of *Anthoxanthum odoratum* L. collected from the Park Grass Experiment, Rothamsted. II. Response to aluminium. *J. Appl. Ecol.* 10, 47–55. (5)

Davies, M. S., and R. W. Snaydon. 1973c. Physiological differences among populations of *Anthoxanthum odoratum* L. collected from the Park Grass Experiment, Rothamsted. III. Response to phosphate. *J. Appl. Ecol.* 11, 669–707. (5)

Davies, M. S., and R. W. Snaydon. 1976. Rapid population differentiation in a mosaic environment. III. Measures of selection pressures. *Heredity* 36, 59–66. (10,12)

Davis, M. B. 1981. Quaternary history and the stability of forest communities. In *Forest Succession: Concepts and Applications,* D. C. West, H. H. Shugart, and D. B. Botkin (eds.). Springer-Verlag, New York. pp. 132–153. (13)

Day, R. J. 1972. Stand structure, succession, and use of southern Alberta's Rocky Mountain forest. *Ecology* 53, 472–478. (4)

de Reffye, *see* Reffye

Deschamp, P. A., and T. J. Cooke. 1983. Leaf dimorphism in aquatic angiosperms: significance of turgor pressure and cell expansion. *Science* 219, 505–507. (17)

Detling, J. K., M. I. Dyer, and D. T. Winn. 1979. Net photosynthesis, root respiration, and regrowth of *Bouteloua gracilis* following simulated grazing. *Oecologia* 41, 127–134. (15)

Dew, D. A. 1972. An index of competition for estimating crop loss due to weeds. *Can. J. Plant Sci.* 52, 921–927. (18)

de Wit, *see* Wit
Diamond, J. M. 1978. Niche shifts and the rediscovery of interspecific competition. *Amer. Sci.* 66, 322–331. (5)
Dickman, D. I., and T. T. Kozlowski. 1968. Mobilization by *Pinus resinosa* cones and shoots of C14 photosynthate from needles of different ages. *Am. J. Bot.* 55, 900–906. (7)
Dingle, H. and J. P. Hegmann (eds.). 1982. *Evolution and Genetics of Life Histories.* Springer-Verlag, New York. (9)
Dirzo, R. 1980a. Experimental studies on slug-plant interactions. I. The acceptability of thirty plant species to the slug *Agriolimax caruanae. J. Ecol.* 68, 981–998. (7)
Dirzo, R. 1980b. *Studies on Plant-Animal Interactions: Terrestrial Molluscs and Their Food Plants.* Ph.D. thesis, University of Wales. (7)
Dirzo, R., and J. L. Harper. 1982a. Experimental studies on slug-plant interactions. III. Differences in the acceptability of individual plants of *Trifolium repens* to slugs and snails. *J. Ecol.* 70, 101–118. (4,7)
Dirzo, R., and J. L. Harper. 1982b. Experimental studies on slug-plant interaction. IV. The performance of cyanogenic and acyanogenic morphs of *Trifolium repens* in the field. *J. Ecol.* 70, 119–138. (4,7)
Dobzhansky, T. 1951. *Genetics and the Origin of Species* (3rd Ed.). Columbia University Press, New York. (11)
Donald, C. M. 1963. Competition among crop and pasture plants. *Adv. Agron.* 15, 1–118. (5,17,18)
Donald, C. M. 1968. The breeding of crop ideotypes. *Euphytica* 17, 385–403. (1)
Donald, C. M. 1981. Competitive plants, communal plants and yield in wheat crops. In *Wheat Science—Today and Tomorrow,* L. T. Evans, and W. J. Peacock (eds.). Cambridge University Press, Cambridge. pp. 223–247. (19)
Donald, C. M., and J. Hamblin. 1976. The biological yield and harvest index of cereals as agronomic and plant breeding criteria. *Adv. Agron.* 28, 361–405. (19)
Drayner, J. K. 1959. Self- and cross-fertility in field beans (*Vicia faba* Linn.). *J. Agric. Sci.* 53, 387–403. (13)
Drew, A. P., and F. A. Bazzaz. 1982. Effect of night temperature on daytime stomatal conductance in early and late successional plants. *Oecologia* 54, 76–79. (16)
Dunbier, M. W. 1972. Genetic variability in *Medicago lupulina* L. across a valley in the Weka Pass, New Zealand. *N.Z.J. Bot.* 10, 48–58. (5)
Duncan, W. G., R. S. Loomis, W. A. Williams, and R. Hanau. 1967. A model for simulating photosynthesis in plant communities. *Hilgardia* 38, 181–205. (15)
du Rietz, *see* Rietz
Eagles, H. A., and A. K. Hardacre. 1979. Genetic variation in maize (*Zea mays* L.) for germination and emergence at 10°C. *Euphytica* 28, 287–295. (4)
Edelin, C. 1977. *Images de l'architecture des conifères.* Doctoral thesis, Université des Sciences et Techniques du Langedoc, Montpellier. (1)
Edwards, C. A. 1975. Effect of direct drilling on the soil fauna. *Outlook on Agriculture* 8, 243–244. (18)

Egler, F. W. 1954. Philosophical and practical considerations of the Braun-Blanquet system of plant sociology. *Castanea* 19, 45–60. (3)
Ehleringer, J., and H. A. Mooney. 1978. Leaf hairs: effects on physiological activity and adaptive value to a desert shrub. *Oecologia* 37, 183–200. (15)
Ehleringer, J., O. Björkman, and H. A. Mooney. 1976. Leaf pubescence: Effects on absorptance and photosynthesis in a desert shrub. *Science* 192, 376–377. (15)
Ehrlich, P. R., and P. H. Raven. 1964. Butterflies and plants: A study in coevolution. *Evolution* 18, 586–608. (7)
Einarson, B., and G. K. K. Link. 1976. *Theophrastus: De Causis Plantarum.* Harvard University Press, Cambridge, Mass. (1)
Elias, C. O., and M. J. Chadwick. 1979. Growth characteristics of grass and legume cultivars and their potential for land reclamation. *J. Appl. Ecol.* 16, 537–544. (10)
Ellenberg, H. 1956. Grundlagen der Vegetationsgliederung. 1. Teil: Aufgaben und Methoden der Vegetationskunde. In *H. Walter. Einführung in die Phytologie 4(1),* Ulmer, Stuttgart. (3)
Elliott, J. G. 1980. The economic significance of weeds in the harvesting of grain. *Proc.* (1980) *Brit. Crop Prot. Conf. Weeds* 3, 787–797. (18)
Elliott, J. G. 1982. Weed control in cereals—strategy and tactics. *Proc.* (1982) *Brit. Crop Prot. Symp. Decision Making in the Practice of Crop Protection. Monog.* 25, 115–119. (18)
Elliott, J. G., and P. J. Boyle. 1963. Crop situations where cultivations for weed control may be eliminated by use of herbicides. A-agriculture. In *Crop Production in a Weed Free Environment,* E. K. Woodford (ed.). Symp. Brit. Weed Control Council, Blackwell, Oxford. 2, 4–13. (18)
Ellstrand, N. C., and D. A. Levin. 1982. Genotypic diversity in *Oenothera laciniata* (Onagraceae), a permanent translocation heterozygote. *Evolution* 36, 63–69. (5,13)
Ellstrand, N. C., A. M. Torres, and D. A. Levin. 1978. Density and the rate of apparent outcrossing in *Helianthus* (Asteraceae). *Syst. Bot.* 3, 403–407. (13)
Elmore, C. D. 1982. The paradox of no correlation between leaf photosynthesis rates and crop yield. In *Predicting Photosynthesis for Ecosystem Models,* II. J. D. Hesketh and J. W. Jones (eds.). C. R. C. Press, Florida. pp. 155–167. (19)
Emerson, A. E. 1960. The evolution of adaptation in population systems. In *Evolution after Darwin* (Vol. 1), S. Tox (ed.). University of Chicago Press, Chicago. (6)
Endler, J. A. 1977. *Geographic Variations, Speciation and Clines.* Princeton University Press, Princeton. (6)
Ennos, R. A. 1981. Detection of selection in populations of white clover (*Trifolium repens* L.) *Biol. J. Linn. Soc.* 15, 75–82. (12)
Ennos, R. A., and M. T. Clegg. 1981. Effect of population substructuring on estimates of outcrossing rate in plant populations. *Heredity* 48, 283–292. (6)
Ericsson, A., S. Larsson, and O. Tenow. 1980a. Effects of early and late season defoliation on growth and carbohydrate dynamics in Scots pine. *J. Appl. Ecol.* 17, 747–769. (1)

Ericsson, A., J. Hellkvist, K. Hillerdal-Hagstromer, S. Larsson, E. Mattson-Djos, and O. Tenow. 1980b. Consumption and pine growth—hypothesis on effects on growth processes by needle-eating insects. In *Structure and Function of Northern Coniferous Forests, an Ecosystem Study,* T. Persson (ed.). *Ecol. Bull. Stockholm.* pp. 1-9. (15)

Evans, L. T. 1980a. Response to challenge: William Farrer and the making of wheats. *J. Aust. Inst. Agric. Sci.* 46, 3-13. (19)

Evans, L. T. 1980b. The natural history of crop yield. *Amer. Sci.* 68, 388-397. (19)

Evans, L. T. 1981 Yield improvement in wheat: empirical or analytical? In *Wheat Science—Today and Tomorrow,* L. T. Evans and W. J. Peacock (eds.). Cambridge University Press, Cambridge. pp. 203-222. (19)

Evans, L. T., and I. F. Wardlaw. 1976. Aspects of the comparative physiology of grain yield in cereals. *Adv. Agron.* 28, 301-359. (19)

Evans, M. W. 1958. Growth and development in certain economic grasses. *Ohio Agric. Exp. Stn., Agron. Ser.* 147, 1-123. (1)

Eyde, R. H. 1975. The foliar theory of the flower. *Amer. Sci.* 63, 430-437. (1)

Faegri, K., and L. van der Pijl. 1979. *The Principles of Pollination Ecology* (3rd Ed.). Pergamon, Oxford. (14)

Fakorede, M. A. B., and D. K. Ojo. 1981. Variability for seedling vigour in maize. *Exp. Agric.* 17, 195-201. (4,8)

Falconer, D. S. 1960. *Introduction to Quantitative Genetics.* Oliver and Boyd, Edinburgh, (11,13)

Falconer, D. S. 1981. *Introduction to Quantitative Genetics* (2nd ed.). Longman, London. (8,10)

Faliński, J. B. 1978. Vegetation dynamics. *Proc. 3rd Symposium Working Group Succession Research on Permanent Plots. Phytocenosis.* Vol. 17. (3)

Faulkner, J. S. 1982. Breeding herbicide-tolerant crop cultivars by conventional methods. In *Herbicide Resistance in Plants*, H. M. Lebaron and J. Gressel (eds.). Wiley, New York. pp. 235-256. (18)

Fawzi, A. F. A., and M. M. El-fouly. 1979. Amylase and invertase activities and carbohydrate contents in relation to physiological sink in carnation. *Physiol. Plant.* 47, 245-249. (15)

Feeny, P. 1970. Seasonal changes in oak leaf tannins and nutrients as a cause of spring feeding by winter moth caterpillars. *Ecology* 51, 565-581. (7)

Feeny, P. 1976. Plant apparency and chemical defense. In *Recent Advances in Phytochemistry,* Vol. 10: *Interactions between Plants and Insects,* J. W. Wallace and R. L. Mansell (eds.). Plenum Press, New York. pp. 1-40. (7)

Fellows, R. J., D. B. Egli, and J. E. Leggett. 1979. Rapid changes in translocation patterns in soybeans following source-sink alterations. *Plant Physiol.* 64, 652-655. (15)

Felton, W. L. 1976. The influence of row spacing and plant population on the effect of weed competition in soybeans. *Austr. J. Expt. Agric. Anim. Husb.* 16, 926-931. (18)

Fick, G. W., W. A. Williams, and R. S. Loomis. 1973. Computer simulation of dry matter distribution during sugar beet growth. *Crop Sci.* 13, 413-417. (15)

Field, C. 1981. Leaf age effects on the carbon gain of individual leaves in relation to microsite. In *Components of Productivity of Mediterranean Regions—Basic and Applied Aspects,* N. S. Margaris and H. Mooney (eds.). Junk, The Hague. pp. 41–50. (15)

Field, C. 1983. Allocating leaf nitrogen for the maximization of carbon gain: leaf age as a control on the allocation program. *Oecologia* 56, 341–347. (15)

Finlay, K. W., and G. N. Wilkinson. 1963. The analysis of adaptation in a plant breeding programme. *Aust. J. Agric. Res.* 14, 742–754. (10)

Fischer, R. A. 1981. Developments in wheat agronomy. In *Wheat Science—Today and Tomorrow,* L. T. Evans and W. J. Peacock (eds.). Cambridge University Press, Cambridge. pp. 249–269. (19)

Fischer, R. A., and Z. Kertesz. 1976. Harvest index in spaced populations and grain weight in microplots as indicators of yielding ability in spring wheat. *Crop Sci.* 16, 56–59. (19)

Fisher, J. B., and H. Honda. 1979. Branch geometry and effective leaf area: A study of *Terminalia*—branching pattern. 1. Theoretical trees. 2. Survey of real trees. *Amer. J. Bot.* 66, 633–644, 645–655. (1)

Fisher, R. A. 1930. *The Genetical Theory of Natural Selection.* Oxford University Press, Oxford. (13,14)

Fisher, R. A. 1958. *The Genetical Theory of Natural Selection* (2nd Ed.). Dover, New York.

Flower-Ellis, J. G. K. 1980. Diurnal dry weight variation and dry matter allocation of some tundra plants. 1. *Andromeda polifolia* L. *Ecol. Bull. (Stockholm)* 30, 139–162. (1)

Flower-Ellis, J. G. K., A. Albrektsson, and L. Olsson. 1976. Structure and growth of some young Scots pine stands: (1) Dimensional and numerical relationships. *Swedish Coniferous Forest Project Technical Report 3.* (1)

Flower-Ellis, J. G. K., and H. Persson. 1980. Investigation of structural properties and dynamics of Scots pine stands. *Ecol. Bull. (Stockholm)* 32, 125–138. (1)

Fondy, B. R., and D. R. Geiger. 1980. Regulation of export by integration of sink and source activity. *What's New in Plant Physiology* 12, 33–36. (5)

Ford, E. B. 1964. *Ecological Genetics.* Methuen, London. (11)

Ford, E. D. 1974. Competition and stand structure in some even-aged plant monocultures. *J. Ecol.* 63, 311–333. (4)

Ford, H. 1981. Competitive relationships amongst apomictic dandelions. *Biol. J. Linn. Soc.* 15, 355–368. (5,12)

Ford, H. 1982. Leaf demography and the plastochron index. *Biol. J. Linn. Soc.* 17, 361–373. (1)

Fowden, L. 1982. Credibility of forecasting—agriculture and food. *Chem. Ind. L.* 16, 582–589. (18)

Fowler, N. L., and J. Antonovics. 1981a. Small-scale variability in the demography of transplants of two herbaceous species. *Ecology* 62, 1450–1457. (4,9)

Fowler, N. L., and J. Antonovics. 1981b. Competition and coexistence in a North Carolina grassland. I. Patterns in undisturbed vegetation. *J. Ecol.* 69, 825–841. (5,9)

Franco, M., and J. Sarukhán. 1981. Un modelo de simulación de la productividad forestal de un bosque de pino. *Ser. Premio Nal. Forestal* 1, 1–71. (4)

Frankie, G. W., P. A. Opler, and K. S. Bawa. 1976. Foraging behaviour of solitary bees: implications for outcrossing of a neotropical forest tree species. *J. Ecol.* 64, 1049–1057. (13)

Frey, K. J., J. A. Browning, and M. D. Simons. 1975. Multiline cultivars of autogamous crop plants. *SABRAO J.* 7, 113–123. (6)

Fryer, J. D. 1981. Weed control practices and changing weed problems. In *Pests, Pathogens and Vegetation,* J. M. Thresh (ed.). Pitman, London. pp. 403–414. (18)

Fryer, J. D., and R. J. Chancellor. 1970. Herbicides and our changing arable weeds. In *The Flora of a Changing Britain,* F. Perrin (ed.). Symp. Brit. Soc. Brit. Isles 1969. pp. 105–110. (18)

Fryxell, P. A. 1956. Effect of varietal mass on the percentage of outcrossing in *Gossypium hirsutum. J. Hered.* 57, 299–301. (12)

Gaastra, P. 1959. Photosynthesis of crop plants as influenced by light, carbon dioxide, temperature, and stomatal diffusion resistance. *Meded. Landbonwhogesch.* 59, 1–68. (15)

Gabriel, W. J. 1967. Reproductive behavior in sugar maple: Self-compatibility, cross-compatibility, agamospermy, and agamocarpy. *Silvae Genet.* 16, 165–168. (13)

Gadgil, M., and O. T. Solbrig. 1972. The concept of r- and K- selection: Evidence from wild flowers and some theoretical considerations. *Amer. Nat.* 106, 14–31. (18)

Gadgil, S., and M. Gadgil. 1975. Can a single resource support many consumer species? *J. Genet.* 62, 33–47. (15)

Gaines, M. S., J. H. Myers, and C. J. Krebs. 1971. Experimental analysis of relative fitness in transferrin genotypes of *Microtus ochrogaster. Evolution* 25, 443–450. (11)

Gartside, D. W., and T. McNeilly. 1974. The potential for evolution of heavy metal tolerance in plants. II. Copper tolerance in normal populations of different plant species. *Heredity* 32, 335–348. (10)

Gasc, J. P. 1979. La métamerie est-elle une clef fondamentale pour comprendre l'organization des vertébrés? *Bull. Soc. Zool. France* 104, 315–323. (1)

Gates, D. M. 1962. *Energy Exchange in the Biosphere.* Harper & Row, New York. (15)

Gates, D. M. 1965. Energy, plants and ecology. *Ecology* 46, 1–13. (17)

Gatsuk, L. E. 1974a. (Gemmaxillar plants and the system of co-ordinative units of their shoots.) *Biull. Mosk. Obshch. Isp. Prir. Otd. Biol.* 79, 100–113. (Russian) (1)

Gatsuk, L. E. 1974b. (On the methods of description and definition of life forms in seasonal climates.) *Biull. Mosk. Obschch. Isp. Prir. Otd. Biol.* 73, 84–100. (Russian) (1)

Gatsuk, L. E., O. V. Smirnova, L. I. Vorontzova, L. B. Zaugolnova, and L. A. Zhukova. 1980. Age states of plants of various growth forms: A review. *J. Ecol.* 68, 675–696. (1,4)

Gauch, H. G., Jr. 1982. *Multivariate Analysis in Community Ecology.* Cambridge University Press, Cambridge. (3)

Geier, P. W., and L. R. Clark. 1979. The nature and future of pest control: Production process or applied ecology? *Prot. Ecol.* 1, 79–101. (18)

Georghiou, G. P., and C. E. Taylor. 1977. Operational influences in the evolution of insecticide resistance. *J. Econ. Entomol.* 70, 653. (18)

Ghiselin, M. T. 1980. The failure of morphology to assimilate Darwinism. In *The Evolutionary Synthesis,* E. Mayr and W. B. Provine (eds.). Harvard University Press, Cambridge, Mass. pp. 180–193. (1)

Gibbs, A. and A. Harrison. 1976. *Plant Virology: The Principles.* E. Arnold, London. (4)

Gifford, R. M., and L. T. Evans. 1981. Photosynthesis, carbon partitioning and yield. *Ann. Rev. Plant Physiol.* 32, 485–509. (19)

Gifford, R. M., and C. L. Jenkins. 1981. Prospects of applying knowledge of photosynthesis towards improving crop production. In *Photosynthesis: CO_2 Assimilation and Plant Productivity,* II. Govindjee (ed.). Academic Press, New York. (19)

Gifford, R. M., and C. Marshall. 1973. Photosynthesis and assimilate distribution in *Lolium multiflorum* Lam. following differential tiller defoliation. *Aust. J. Biol. Sci.* 26, 517–526. (15)

Gilbert, L. E. 1977. Development of theory in the analysis of insect-plant interactions. In *Analysis of Ecological Systems,* D. H. Horn, R. D. Mitchell and G. R. Stairs (eds.). Ohio State University Press, pp. 117–154. (7)

Gilbert, L. E., and P. H. Raven (eds.). 1975. *Coevolution of Animals and Plants.* University of Texas Press, Austin. (5,7)

Gill, D. E. 1974. Intrinsic rate of increase, saturation density and competitive ability. II. The evolution of competitive ability. *Amer. Nat.* 108, 103–116. (5)

Gill, F. B., and L. L. Wolf. 1975. Economics of feeding territoriality in the golden-winged sunbird. *Ecology* 56, 333–345. (13)

Gillespie, J. H. 1974. Natural selection for within-generation variance in offspring number. *Genetics* 76, 601–606. (13)

Givnish, T. J. 1979. On the adaptive significance of leaf form. In *Topics in Plant Population Biology,* O. T. Solbrig, S. Jain, G. B. Johnson and P. H. Raven (eds.). Columbia University Press, New York. pp. 351–380. (17)

Givnish, T. J. 1980. Ecological constraints on the evolution of breeding systems in seed plants: dioecy and dispersal in gymnosperms. *Evolution* 34, 959–972. (6,14)

Gleason, H. A. 1926. The individualistic concept of the plant association. *Bull. Torrey Bot. Club* 53, 7–26. (3)

Glesener, R. R., and D. Tilman. 1978. Sexuality and the components of environmental uncertainty: Clues from geographic parthenogenesis in terrestrial animals. *Amer. Nat.* 112, 659–673. (5)

Gluch, W. 1967. Wuchsformstudien an zentraleuropäischen Fabaceen. I. Die Stauden der Gattung *Medicago* L. und *Trifolium* L. *Fedd. Repert.* 76, 221–265. (1)

Godwin, H. A. 1960. The history of weeds in Britain. In *The Biology of Weeds,* J. L. Harper (ed.). Symp. Brit. Ecol. Soc. Blackwell, London. pp. 1–10. (18)

Goloff, A. A. 1973. *A Germination Model for Natural Seed Populations.* Ph.D. thesis, University of Illinois, Urbana. (16)

Gönen, Y., and G. Wricke. 1978. Untersuchungen zur Vererbung der Internodienzahl und Internodienlange bei determinierten Freilandgurken. *Z. Pflanzenzüchtg.* 81, 258–270. (1)

Goodwin, B. C. 1982. Development and evolution. *J. Theoret. Biol.* 97, 43–55. (1)

Gottlieb, L. D. 1977. Genotypic similarity of large and small individuals in a natural population of the annual plant *Stephanomeria exigua* ssp. *coronaria* (Compositae). *J. Ecol.* 65, 127–134. (4,9,13)

Gottschlich, D. E., and A. P. Smith. 1982. Convective heat transfer characteristics of toothed leaves. *Oecologia* 53, 418–420. (15)

Gould, S. J. 1970. Evolutionary paleontology and the science of form. *Earth Sci. Rev.* 6, 77–119. (2)

Gould, S. J., and R. C. Lewontin. 1979. The spandrels of San Marco and the Panglossian paradigm: A critique of the adaptationist programme. *Proc. Roy. Soc. London B* 205, 581–598. (1,8,10)

Grant, M. C., and J. Antonovics. 1978. Biology of ecologically marginal populations of *Anthoxanthum odoratum.* I. Phenetics and dynamics. *Evolution* 32, 822–838. (5,8,12,17,18)

Grant, V. 1963. *The Origin of Adaptations.* Columbia University Press, New York. (17)

Grant, V. 1981. *Plant speciation* (2nd Ed.). Columbia University Press. New York. (1)

Green, J. M., and M. D. Jones. 1953. Isolation of cotton for seed increase. *Agron. J.* 45, 366–368. (12)

Gressel, J., H. V. Ammon, H. Fogelfors, J. Gasquez, Q. O. N. Kay, and H. Kees. 1982. Discovery and distribution of herbicide-resistant weeds outside North America. In *Herbicide Resistance in Plants,* H. M. Lebaron and J. Gressel (eds.). Wiley, New York. pp. 31–55. (10,18)

Gressel, J., and L. A. Segel. 1978. The paucity of plants evolving genetic resistance to herbicides: possible reasons and implications. *J. Theor. Biol.* 75, 349–361. (18)

Gressel, J., and L. A. Segel. 1982. Interrelating factors controlling the rate of appearance of resistance: The outlook for the future. In *Herbicide Resistance in Plants,* H. M. Lebaron and J. Gessel (eds.). Wiley, New York. pp. 325–348. (18)

Griffing, B. 1977. Selection for populations of interacting genotypes. In *Proc. Intern. Conf. Quant. Genetics,* E. Pollak, O. Kempthorne and T. B. Bailey Jr. (eds.). Iowa State University Press, Ames. pp. 413–434. (6)

Grime, J. P. 1977. Evidence for the existence of three primary strategies in plants and its relevance to ecological and evolutionary theory. *Amer. Nat.* 111, 1169–1194. (3,18)

Grime, J. P. 1979a. *Plant Strategies and Vegetation Processes.* Wiley, New York. (3,8)

Grime, J. P. 1979b. Competition and struggle for existence. In *Population Dynamics,* R. M. Anderson, B. D. Turner and L. R. Taylor (eds.). Blackwell, Oxford. pp. 123–139. (3)

Gross, K. L. 1981. Predictions of fate from rosette size in four "biennial" plant species: *Verbascum thapsus, Oenothera biennis, Daucus carota,* and *Tragopogon dubius. Oecologia* 48, 209–213. (4)
Grubb, P. J. 1977. The maintenance of species richness in plant communities: The importance of the regeneration niche. *Biol. Rev.* 52, 107–145. (3,5)
Guédès, M. 1979. *Morphology of Seed-Plants.* J. Cramer, Vaduz. (1)
Gulmon, S. L., and C. C. Chu. 1981. The effects of light and nitrogen on photosynthesis, leaf characteristics, and dry matter allocation in the chaparral shrub, *Diplacus aurantiacus. Oecologia* 49, 207–212. (15)
Gunn, J. S. 1977. The role of herbicides in highly mechanised cash root crop production. *Proc. (1976) Brit. Crop Prot. Conf. Weeds* 3, 831–858. (18)
Guries, R. P., and F. T. Ledig. 1982. Genetic diversity and population structure in Pitch Pine (*Pinus rigida* Mill.). *Evolution* 36, 387–402. (5)
Gutiérrez, A. P., Y. Wang, and R. E. Jones. 1979. Systems analysis applied to crop protection. *EPPO Bull.* 9, 133–148. (7)
Haas, H., and J. C. Streibig. 1982. Changing patterns of weed distribution as a result of herbicide use and other agronomic factors. In *Herbicide Resistance in Plants,* H. M. Lebaron and J. Gressel (eds.). Wiley, New York. pp. 57–80. (18)
Haeckel, E. H. 1866. *Generelle Morphologie der Organismen: Allegemeine Grundzüge der organischen Formen-Wissenschaft, mechanisch begründet durch die von Charles Darwin reformie Descendenz-Theorie* (2 vols.). Georg Reimer, Berlin. (1)
Hagemann, I. 1983. Wuchformenuntersuchungen an zentraleuropäischen *Hypericum*-Arten. *Flora* 173, 97–142. (1)
Hageman, R. H., D. Flesher, and A. Gitter. 1961. Diurnal variation and other light effects influencing the activity of nitrate reductase and nitrogen metabolism in corn. *Crop Sci.* 1, 201–204. (19)
Hagman, M. 1971. On self- and cross-incompatibility shown by *Betula verrucosa* Ehrh. and *Betula pubescens* Ehrh. *Commun. Inst. Forest. Fenn.* 73, 1–125. (13)
Håkansson, S. 1967. The growth of *Agropyron repens* (L.) Beauv. I. Development and growth and the response to burial at different stages. *Lantbrukshogst. Ann.* 33, 823–873. (18)
Håkansson, S. 1982. Multiplication, growth and persistence of perennial weeds. In *Biology and Ecology of Weeds,* W. Holzner and N. Numata (eds.). Junk, The Hague. pp. 125–135. (1)
Hall, R. L. 1974. Analysis of the nature of interference between plants of different species. I. Concepts and extension of the de Wit analysis to examine effects. *Austr. J. Agric. Res.* 25, 739–747. (18)
Hallé, F., and R. A. A. Oldeman. 1970. *Essai sur l'architecture et la Dynamique de Croissance des Arbres Tropicaux.* Masson, Paris. (1)
Hallé, F., R. A. A. Oldeman, and P. B. Tomlinson. 1978. *Tropical Trees and Forests: An Architectural Analysis.* Springer-Verlag, New York. (1,2)
Halloran, G. M., and J. W. Lee. 1979. Plant nitrogen distribution in wheat cultivars. *Aust. J. Agric. Res.* 30, 779–789. (19)
Hamblin, J., and J. G. Rowell. 1975. Breeding implications of the relationship between competitive ability and pure culture yield in self-pollinated grain crops. *Euphytica* 24, 221–228. (19)

Hamid, Z. A., and J. E. Grafius. 1978. Developmental allometry and its implications to grain yield in barley. *Crop Sci.* 18, 83–86. (1)

Hamilton, W. D. 1967. Extraordinary sex ratios. *Science* 156, 477–488. (14)

Hamilton, W. D. 1980. Sex versus non-sex versus parasite. *Oikos* 35, 282–290. (11)

Hamrick, J. L. 1979. Genetic variation and longevity. In *Topics in Plant Population Biology,* O. T. Solbrig, S. Jain, G. B. Johnson and P. H. Raven (eds.). Columbia University Press, New York. pp. 84–113. (5,11)

Hamrick, J. L. 1982. Plant population genetics and evolution. *Amer. J. Bot.* 69, 1685–1693. (5)

Hamrick, J. L., and R. W. Allard. 1972. Microgeographical variation in allozyme frequencies in *Avena barbata. Proc. Nat. Acad. Sci.* USA 69, 2100–2104. (5)

Hamrick, J. L., and L. R. Holden. 1979. Influence of microhabitat heterogeneity on gene frequency distribution and gametic phase disequilibrium in *Avena barbata. Evolution* 33, 521–533. (5,11)

Hamrick, J. L., Y. B. Linhart, and J. B. Mitton. 1979. Relationship between life-history characteristics and electrophoretically detectable genetic variation in plants. *Ann. Rev. Ecol. Syst.* 10, 173–200. (11)

Hancock, J. F., and R. S. Bringhurst. 1978. Inter-populational differentiation and adaptation in the perennial, diploid species *Fragaria vesca* L. *Amer. J. Bot.* 65, 795–803. (8)

Hancock, J. F., and R. E. Wilson. 1976. Biotype selection in *Erigeron annuus* during old field succession. *Bull. Torr. Bot. Club* 103, 122–125. (5)

Hansen, W. R., and R. Shibles. 1978. Seasonal log of the flowering and podding activity of field-grown soybeans. *Agron. J.* 70, 47–50. (2)

Harberd, D. J. 1957. The within population variance in genecological trials. *New Phytol.* 56, 269–280. (17)

Harborne, J. B. (ed.). 1978. *Biochemical Aspects of Plant and Animal Coevolution.* Academic Press, London. (7)

Harding, J. A., R. W. Allard, and D. G. Smeltzer. 1966. Population studies in predominantly self-pollinated species. IX. Frequency dependent selection in *Phaseolus lunatus. Proc. Nat. Acad. Sci. USA* 56, 99–104. (6)

Harding, J. A., and K. Barnes. 1977. Genetics of *Lupinus.* X. Genetic variability, heterozygosity and outcrossing in colonial populations of *Lupinus succulentus. Evolution* 31, 247–255. (13)

Harding, J. A., and C. L. Tucker. 1964. Quantitative studies on mating systems. I. Evidence for the non-randomness of outcrossing in *Phaseolus lunatus. Heredity* 19, 369–381. (13)

Hare, J. D. 1980. Variation in fruit size and susceptibility to seed predation among and within populations of the cocklebur, *Xanthium strumarium* L. *Oecologia* 46, 217–222. (4)

Hare, J. D., and D. J. Futuyma. 1978. Different effects of variation in *Xanthium strumarium* L. (Compositae) on two insect seed predators. *Oecologia* 37, 109–120. (4)

Harlan, H. V., and M. L. Martini. 1938. The effect of natural selection in a mixture of barley varieties. *J. Agric. Res.* 57, 189–199. (10)

Harper, J. L. 1957. Ecological aspects of weed control. *Outlook on Agriculture* 1, 197–205. (18)
Harper, J. L. 1961. Approaches to the study of plant competition. In *Mechanisms in Biological Competition,* F. L. Milthorpe (ed.). *Symp. Soc. Exp. Biol.* 15, 1–39. (5)
Harper, J. L. 1964. The nature and consequence of interference amongst plants. In *Genetics Today. Proc. XI. Int. Cong. Genetics* 2, 465–482. (5)
Harper, J. L. 1967. A Darwinian approach to plant ecology. *J. Ecol.* 55, 247–270. (4,7)
Harper, J. L. 1969. The role of predation in vegetational diversity. In *Diversity and Stability in Ecological Systems, Brookhaven Symp. in Biol.* 22, 48–62. (7)
Harper, J. L. 1977. *Population Biology of Plants.* Academic Press, London. (1,3,4,5,6,7,9,16,17,18,19)
Harper, J. L. 1978. Plant relations in pastures: A keynote address. In *Plant Relations in Pastures,* J. R. Wilson (ed.). CSIRO, Melbourne. pp. 3–16. (6,7)
Harper, J. L. 1981. The concept of population in modular organisms. In *Theoretical Ecology: Principles and Applications* (2nd Ed.), R. M. May (ed.). Blackwell, Oxford. pp. 57–77. (1)
Harper, J. L. 1982. After description. In *The Plant Community as a Working Mechanism,* E. I. Newman (ed.). Spec. Publ. No. 1, Brit. Ecol. Soc., Blackwell, Oxford. pp. 11–25. (3,5,10)
Harper, J. L., and A. D. Bell. 1979. The population dynamics of growth form in organisms with modular construction. In *Population Dynamics, the 20th Symposium of the British Ecological Society,* R. M. Anderson et al. (eds.). Blackwell, Oxford. pp. 29–52. (1,2,18)
Harper, J. L., P. H. Lovell, and K. G. Moore. 1970. The shapes and sizes of seeds. *Ann. Rev. Ecol. Syst.* 1, 327–356. (4,8,9,18)
Harper, J. L., and M. Obeid. 1967. Influence of seed size and depth of sowing on the establishment and growth of varieties of fiber and oil seed flax. *Crop Sci.* 7, 527–532. (4)
Harper, J. L., and J. White. 1974. The demography of plants. *Ann. Rev. Ecol. Syst.* 5, 419–463. (1,4)
Harris, P. M. 1980. Agronomic research and potato production practice. In *Opportunities for Increasing Crop Yields,* R. G. Hurd, P. U. Biscoe and C. Dennis (eds.). Pitman, London. pp. 205–213. (19)
Harris, W. 1978. Defoliation as a determinant of the growth persistence and composition of pasture. In *Plant Relations in Pastures,* J. R. Wilson (ed.). CSIRO. pp. 67–85. (18)
Hartnett, D. C. and F. A. Bazzaz. 1983. Physiological integration among interclonal ramets in *Solidago canadensis. Ecology* 64, 779–788. (16)
Hartshorn, G. S. 1975. A matrix model of tree population dynamics. In *Tropical Ecological Systems.* II. *Trends in Terrestrial and Aquatic Research,* F. B. Golley and E. Medina (eds.). Springer-Verlag, New York. pp. 41–51. (4)
Hartshorn, G. S. 1978. Tree falls and tropical dynamics. In *Tropical Trees as Living Systems,* P. B. Tomlinson and M. H. Zimmermann (eds.). Cambridge University Press, London. pp. 617–638. (7)
Hartung, R. C., J. E. Specht, and J. H. Williams. 1981. Modification of soy-

bean plant architecture by genes for stem growth habit and maturity. *Crop Sci.* 21, 51–56. (1)

Harvey, H. J. 1979. *The Regulation of Vegetative Reproduction.* Ph.D. thesis, University of Wales. (1)

Hawton, D. 1980. Yield effects of herbicides on competition between crop and weed communities. *Aust. J. Agric. Res.* 31, 1075–1081. (18)

Hayward, M. D. 1967. The genetic organisation of natural populations of *Lolium perenne.* II. Inflorescence production. *Heredity* 22, 105–116. (19)

Heath, D. J. 1977. Simultaneous hermaphroditism: Cost and benefit. *J. Theor. Biol.* 64, 363–373. (14)

Heath, D. J. 1979. Brooding and the evolution of hermaphroditism. *J. Theor. Biol.* 81, 151–155. (14)

Hedley, C. L., and M. J. Ambrose. 1981. Designing "leafless" plants for improving yields of the dried pea crop. *Adv. Agron.* 34, 225–277. (1)

Hedrick, P. W. 1983. *Genetics of Populations.* Science Books International, New York. (6)

Hedrick, P. W., M. E. Ginevan, and E. P. Ewing. 1976. Genetic polymorphism in heterogeneous environments. *Ann. Rev. Ecol. Syst.* 7, 1–32. (6)

Heinrich, B. 1979. Resource heterogeneity and patterns of movement in foraging bumblebees. *Oecologia* 40, 235–245. (13)

Herd, E. M., and G. R. Squire. 1976. Observations on the winter dormancy of tea (*Camellia sinensis* L.) in Malawi. *J. Hort. Sci.* 51, 267–279. (1)

Herrera, C. M. 1982a. Breeding systems and dispersal-related maternal reproductive effort of southern Spanish bird-dispersed plants. *Evolution* 36, 1299–1314. (14)

Herrera, C. M. 1982b. Grasses, grazers, mutualism and coevolution: A comment. *Oikos* 38, 254–257. (17)

Heslop-Harrison, J. 1964. Forty years of genecology. *Adv. Ecol. Res.* 2, 159–247. (5,8,10,17)

Heslop-Harrison, J. 1979. Pollen-stigma interaction in grasses: A brief review. *New Zeal. J. Bot.* 17, 537–546. (13)

Hett, J. M., and O. L. Loucks. 1976. Age structure models of balsam fir and Eastern hemlock. *J. Ecol.* 59, 507–520. (4)

Hickey, D. A., and T. McNeilly. 1975. Competition between metal tolerant and normal plant populations: A field experiment on normal soil. *Evolution* 29, 458–464. (10,12)

Hickman, J. D. 1979. The basic biology of plant numbers. In *Topics in Plant Population Biology,* O. T. Solbrig, S. K. Jain, G. B. Johnson and P. H. Raven (eds.). Columbia University Press, New York. pp. 232–263. (8)

Highsmith, R. C. 1982. Reproduction by fragmentation in corals. *Mar. Ecol. Prog. Ser.* 7, 207–226. (1)

Hill, J. 1975. Genotype-environment interactions—a challenge to plant breeding. *J. Agric. Res.* 85, 477–494. (19)

Hill, J., and Y. Shimamoto. 1973. Methods of analysing competition with special reference to herbage plants. *J. Agric. Sci.* 81, 77–89. (5)

Hilu, K. W., and J. M. J. de Wet. 1980. Effect of artificial selection on grain dormancy in *Eleusine* (Gramineae). *Syst. Bot.* 5, 54–60. (8)

Hiroi, T., and M. Monsi. 1966. Dry matter economy of *Helianthus annuus* communities grown at varying densities and light intensities. *J. Fac. Sci. Univ. of Tokyo* 9, 241–285. (4)

Hirose, Y., and N. Kachi. 1982. Critical plant size for flowering in biennials with special reference to their distribution in a sand dune system. *Oecologia* 55, 281–284. (15)

Hockett, E. A., and P. F. Knowles. 1970. Inheritance of branching in sunflowers, *Helianthus annuus* L. *Crop Sci.* 10, 432–436. (1)

Hodges, C. M., and R. B. Miller. 1981. Pollinator flight directionality and the assessment of pollen returns. *Oecologia* 50, 376–379. (13)

Hodgkin, T. 1981. The inheritance of node number and rate of node production in Brussels sprouts. *Theor. Appl. Gen.* 59, 79–82. (1)

Hodgkinson, K. C. 1974. Influence of partial defoliation on photosynthesis, photorespiration and transpiration by lucerne leaves of different ages. *Aust. J. Plant Physiol.* 1, 561–578. (15)

Hoekstra, F. A., and J. Bruinsma. 1975. Respiration and vitality of binucleate and trinucleate pollen. *Physiol. Plant.* 34, 221–225. (13)

Hoff, V. J. 1962. An analysis of outcrossing in certain complex-heterozygous *Euoenotheras.* I. Frequency of outcrossing. *Amer. J. Bot.* 49, 715–721. (13)

Holland, J. H. 1975. *Adaptation in Natural and Artificial Systems.* The University of Michigan Press, Ann Arbor. (2)

Hölldobler, B., and M. Moglich. 1980. The foraging system of *Pheidole militicida* (Hymenoptera: Formicidae). *Insectes Sociaux* 27, 237–264. (2)

Holzner, W. 1982. Concepts, categories and characteristics of weeds. In *Biology and Ecology of Weeds,* W. Holzner and M. Numata (eds.). Series GeoBotany, Junk Publishers, Amsterdam. (18)

Holzner, W., and M. Numata. 1982. *Biology and Ecology of Weeds.* Series GeoBotany, Junk Publishers, Amsterdam. (18)

Honda, H., P. B. Tomlinson, and J. B. Fisher. 1981. Computer simulation of branch interaction and regulation by unequal flow rates in botanical trees. *Amer. J. Bot.* 68, 569–585. (1,2)

Hoopkingson, J. M. 1964. Studies on the expansion of the leaf surface. IV. The carbon and phosphorus economy of a leaf. *J. Exp. Bot.* 15, 125–137. (7)

Horowitz, A., and J. Harding. 1972a. Genetics of *Lupinus.* 5: Intraspecific variability for reproductive traits in *Lupinus nanus. Bot. Gaz.* 133, 155–165. (13)

Horowitz, A., and J. Harding. 1972b. The concept of male outcrossing in hermaphrodite higher plants. *Heredity* 29, 223–236. (13)

Hounshell, D. A. 1980. Edison and the pure science ideal in 19th-century America. *Science* 207, 612–617. (19)

Houssard, C., J. Escarré, and F. Romane. 1980. Development of species diversity in some Mediterranean plant communities. *Vegetatio* 43, 59–72. (3)

Howe, H. F., and G. F. Estabrook. 1977. On intraspecific competition for avian dispersers in tropical trees. *Amer. Nat.* 111, 817–832. (4)

Howe, H. F., and W. M. Richter. 1982. Effects of seed size on seedling size in *Virola surinamensis;* a within and between tree analysis. *Oecologia* 53, 347–351. (4)

Hubbell, S. P. 1979. Tree dispersion, abundance and diversity in a tropical dry forest. *Science* 203, 1299–1309. (3)

Hubby, J. L., and R. C. Lewontin. 1966. A molecular approach to the study of genic heterozygosity in natural populations. I. The number of alleles at different loci in *Drosophila pseudoobscura. Genetics* 54, 595–609. (11)

Huff, D. 1954. *How to Lie with Statistics.* W. W. Norton, New York. (8)

Huffaker, C. B. 1957. Fundamentals of biological control of weeds. *Hilgardia* 27, 101–157. (18)

Hughes, T. P., and J. B. C. Jackson. 1980. Do corals lie about their age? Some demographic consequences of partial mortality, fission and fusion. *Science* 209, 713–715. (1)

Hull, D. L. 1980. Individuality and selection. *Ann. Rev. Ecol. Syst.* 11, 311–332. (6)

Hume, L., and P. B. Cavers. 1982. Adaptation in widespread populations of *Rumex crispus* as determined using composite diagrams. *Can. J. Bot.* 60, 2637–2651. (1)

Hunt, R., and F. A. Bazzaz. 1980. The biology of *Ambrosia trifida* L. V. Response to fertilizer, with growth analysis at the organismal and suborganismal levels. *New Phytol.* 84, 113–121. (16)

Huxley, J. 1942. *Evolution: The Modern Synthesis.* Allen and Unwin, London. (11)

Hyder, D. N. 1972. Defoliation in relation to vegetative growth. In *The Biology and Utilization of Grasses*, V. B. Youngner and C. M. McKell (eds.). Academic Press, New York. pp. 304–317. (1)

Imam, A. G., and R. W. Allard. 1965. Population studies in predominantly self-pollinated species. VI. Genetic variability between and within natural populations of wild oats from different habitats in California. *Genetics* 51, 49–62. (5,8)

Ismail, A. M. A., and G. R. Sagar. 1981. The reciprocal transfer of radiocarbon between a lateral branch and its parent shoot under normal and stress conditions in plants of *Vicia faba* L. *J. Hort. Sci.* 56, 155–159. (1)

Istock, C. A. 1982. Some theoretical considerations concerning life history evolution. In *Evolution and Genetics of Life Histories,* H. Dingle and J. Hegmann (eds.). Springer-Verlag, New York. pp. 21–29. (9)

Jackson, J. B. C., L. W. Buss, and R. E. Cook (eds.). 1983. *Population Biology and Evolution of Clonal Organisms.* Yale University Press, New Haven. (6)

Jacobs, D. L. 1946. Shoot segmentation in *Anacharis densa. Amer. Midl. Nat.* 35, 282–286. (1)

Jain, S. K. 1969. Comparative ecogenetics of two *Avena* species occurring in central California. *Evol. Biol.* 3, 73–118. (5)

Jain, S. K. 1975. Patterns of survival and microevolution in plant populations. In *Population Genetics and Ecology,* S. Karlin and E. Nevo (eds.). Academic Press, New York. pp. 49–89. (6,13)

Jain, S. K. 1978. Inheritance of phenotypic plasticity in soft chess, *Bromus mollis* L. (Gramineae). *Experientia* 34, 835–836. (8)

Jain, S. K. and A. D. Bradshaw. 1966. Evolutionary divergence among adjacent plant populations. I. Evidence and its theoretical analysis. *Heredity* 21, 407–441. (5,10,12)

Jain, S. K., and K. B. L. Jain. 1969. Polymorphisms in an inbreeding population under models involving underdominance. *Science* 166, 1294–1296. (6)
Jain, S. K., C. O. Qualset, J. C. Williams, H. E. Vogt, and P. Kulakow. 1981. Population dynamics of composite crosses: Heterozygote disadvantage and frequency dependent selection at locus *b*. *Proc. IV Intern. Barley Genetics Symp.*, Edinburgh. (6)
Jain, S. K., and C. A. Suneson. 1966. Increased recombination and selection in barley populations carrying a male sterility factor. *Genetics* 54, 1215–1224. (6)
Jana, S., B. L. Harvey, and E. T. Thomas. 1973. Cyclic heterozygote advantage at the R/r locus. *Barley Genetics Newsletter* 3, 20–22. (6)
Jana, S., and J. M. Naylor. 1980. Dormancy studies in seed of *Avena fatua*. II. Heritability for seed dormancy. *Can. J. Bot.* 58, 91–93. (8)
Janzen, D. H. 1968. Host plants as islands in evolutionary and contemporary time. *Amer. Nat.* 102, 592–595. (7)
Janzen, D. H. 1969. Seed eaters versus seed size, number, toxicity and dispersal. *Evolution* 23, 1–27. (4)
Janzen, D. H. 1971. Seed predation by animals. *Ann. Rev. Ecol. Syst.* 2, 465–492. (7)
Janzen, D. H. 1976. Reduction of *Mucuna andreana* (Leguminosae) seedling fitness by artificial seed damage. *Ecology* 57, 826–828. (4,7)
Janzen, D. H. 1977. What are dandelions and aphids? *Amer. Nat.* 111, 586–589. (1)
Janzen, D. H. 1980. Specificity of seed-attacking beetles in a Costa Rican deciduous forest. *J. Ecol.* 68, 929–952. (7)
Jefferies, R. L., A. J. Davy, and T. Rudnick. 1981. Population biology of the salt marsh annual *Salicornia europaea* agg. *J. Ecol.* 69, 17–31. (4,9)
Jefferies, R. L., and L. D. Gottlieb. 1982. Genetic differentiation of the microspecies. *Salicornia europaea* L. (*sensu stricto*) and *S. ramosissima* J. Woods. *New Phytol.* 92, 123–129. (17)
Jefferies, R. L., and L. D. Gottlieb. 1983. Genetic variation within and between populations of the asexual plant *Puccinellia* x *phryganodes* (Trin.) Scribner and Merr. *Can. J. Bot.* 61, 774–779. (17)
Jefferies, R. L., A. Jensen, and D. Bazely. 1983. The biology of the annual, *Salicornia europaea* agg., at the limits of its range in Hudson Bay. *Can. J. Bot.* 61, 762–773. (17)
Jennings, D. L., and A. Dale. 1982. Variation in the growth habit of red raspberries with particular reference to cane height and node production. *J. Hort. Sci.* 57, 197–204. (1)
Jennings, P. R., and J. de Jesus. 1968. Studies on competition in rice I. Competition in mixtures of varieties. *Evolution* 22, 119–124. (10)
Jones, D. A. 1973. Coevolution and cyanogenesis. In *Taxonomy and Ecology*, V. H. Heywood (ed.). Academic Press, London. pp. 213–242. (7)
Jones, J. S., B. H. Leith, and P. Rawlings. 1977. Polymorphism in *Cepaea:* A problem with too many solutions? *Ann. Rev. Ecol. Syst.* 8, 109–143. (11)
Jones, M. D., and J. S. Brooks. 1950. Effectiveness of distance and border rows in preventing outcrossing in corn. *Oklahoma Agric. Res. Sta. Tech. Bull.* No. T-45. (12)
Jones, M. E. 1971. The population genetics of *Arabidopsis thaliana*. II. Population structure. *Heredity* 27, 51–58. (8)

Joy, P., and A. Laitinen, 1980. Breeding for coadaptation between red clover and timothy. *Hankkija's Seed Publ.* No. 13, Hankkija Plant Breeding Institute, Finland. (5)

Kannenberg, L. W., and R. W. Allard. 1967. Population studies in predominantly self-pollinated species. VIII. Genetic variability in the *Festuca microstachys* complex. *Evolution* 21, 227–240. (8)

Kasperbauer, M. J., R. C. Bruckner, and W. D. Springer. 1980. Haploid plants by anther-panicle culture of tall fescue. *Crop Sci.* 20, 103–106. (11)

Kästner, A. 1981. Beitrage zur Wuchsformanalyse und systematischen Gliederung von *Teucrium* L. III. Wuchsformen und Verbreitung von Arten der Sektionen *Teucropsis* und *Teucrium. Flora* 171, 466–519. (1)

Kates, R. W. 1970. Natural hazard in human ecological perspective: Hypotheses and models. Institute of Behavioral Science, Natural Hazards Working. Paper No. 14. University of Colorado, Boulder. (18)

Kawano, S. 1975. The productive and reproductive biology of flowering plants: II. The concept of life history strategy in plants. *J. Coll. Lib. Arts Toyama Univ. Japan* 8, 51–86. (4)

Kays, S., and J. L. Harper. 1974. The regulation of plant and tiller density in a grass sward. *J. Ecol.* 62, 97–105. (4,19)

Keddy, P. A. 1980. Population ecology in an environmental mosaic: *Cakile edentula* on a gravel bar. *Can. J. Bot.* 58, 1095–1100. (5)

Keddy, P. A. 1981. Experimental demography of the sand-dune annual, *Cakile edentula*, growing along an environmental gradient in Nova Scotia. *J. Ecol.* 69, 615–630. (8,9)

Kemp, W. B. 1937. Natural selection within plant species. *J. Heredity* 28, 329–333. (5)

Kerner von Marilaun, A. 1891 *Pflanzenleben* Vol. 2. Leipzig. (10)

Kernick, M. D. 1961. Seed production in specific crops. In *Agricultural and Horticultural Seeds*, FAO Agricultural Studies No. 55, pp. 181–547. (12)

Kershaw, K. 1964. *Quantitative and Dynamic Plant Ecology*. Edward Arnold, London. (6)

Kettlewell, H. B. D. 1956. Further selection experiments on industrial melanism in the Lepidoptera. *Heredity* 10, 287–301. (11)

Kiang, Y. T. 1972. Pollination study in a natural population of *Mimulus guttatus. Evolution* 26, 308–310. (13)

Kiang, Y. T. 1982. Local differentiation of *Anthoxanthum odoratum* L. populations on roadsides. *Amer. Midl. Nat.* 107, 340–350. (12)

Kikuzawa, K. 1983. Leaf survival of woody plants in deciduous broad-leaved forests. 1. Tall trees. *Can. J. Bot.* 61, 2133–2139. (1)

Kimura, M., and T. Maruyama. 1971. Pattern of neutral allele polymorphism in a geographically structured population. *Genet. Res.* 18, 125–131. (12)

Kimura, M., and G. H. Weiss. 1964. The stepping stone model of population structure and the decrease of genetic correlation with distance. *Genetics* 49, 561–576. (12)

Kimura, M., Y. Yokoi, and K. Hogetsu. 1978. Quantitative relationships between growth and respiration in growing *Helianthus tuberosus* leaves. *Bot. Mag.* (Tokyo) 91, 43–56. (15)

Kira, T., H. Ogawa, and K. Shinozaki. 1953. Intraspecific competition among higher plants. 1. Competition-density-yield inter-relationships in regularly dispersed populations. *J. Inst. Polytech. Osaka Cy. Univ. D.* 4, 1–16. (4)

Klemow, K. M., and D. J. Raynal. 1981. Population ecology of *Melilotus alba* in a limestone quarry. *J. Ecol.* 69, 33–44. (4)

Knowles, R. P., and H. Baenziger. 1962. Fertility indices in cross-pollinated grasses. *Can. J. Plant Sci.* 42, 460–471. (12)

Knowles, R. P., and A. W. Ghosh. 1968. Isolation requirements for smooth bromegrass *Bromus inermis,* as determined by a genetic marker. *Crop Sci.* 3, 571–574. (12)

Knowles, P., and M. C. Grant. 1983. Age and size structure analyses of Engelmann spruce, Lodgepole pine, and Limber pine in Colorado. *Ecology* 64, 1–9. (4)

Kobayashi, S. 1975. Growth analysis of plants as an assemblage of internodal segments—a case of sunflower plants in pure stand. *Japn. J. Ecol.* 25, 61–70. (1)

Koblet, R. 1979. Entwicklung, jahreszeitlicher Verlauf des Stoffzuwaches und Wettberwerbsverhalten von Wiesenpflanzen im Alpenraum. *Z. Acker-und Pflanzenbau* 148, 23–53. (1)

Koehn, R. K. 1978. Physiology and biochemistry of enzyme variation: The interface of ecology and population genetics. In *Ecological Genetics: The Interface,* P. Brussard (ed.). Springer, New York. pp. 51–72. (11)

Koehn, R. K., R. J. E. Newell, and F. Immerman. 1980. Maintenance of an aminopeptidase allele frequency cline by natural selection. *Proc. Nat. Acad. Sci. USA* 77, 5385–5389. (11)

Kohyama, T. 1981. Studies on the *Abies* populations of Mt. Shimagare. II. Reproductive and life history traits. *Bot. Mag. Tokyo.* 95, 167–181. (4)

Koyama, H., and T. Kira. 1956. Intraspecific competition among higher plants. VIII. Frequency distribution of individual plant weight as affected by the interaction between plants. *J. Inst. Polytech. Osaka Cy. Univ.* 7, 73–94. (4)

Kozlowski, T. T. 1973. Extent and significance of shedding of plant parts. In *Shedding of Plant Parts,* T. T. Kozlowski (ed.). Academic Press, New York. pp. 1–44. (7)

Krause, R. A., L. B. Massie, and R. A. Hyre. 1975. Blitecast: A computerised forecast of potato late blight. *Pl. Dis. Reptr.* 59, 95–98. (18)

Krückeberg, A. R. 1951. Response of plants to serpentine soils. *Amer. J. Bot.* 38, 408–418. (5)

Kujala, V. 1926. Untersuchungen über die Waldvegetation in Sud-und MittlelFinnland. I. Zur Kenntnis des ökologisch-biologischen Charakters der Pflanzenarten unter spezieller Berücksichtigung der Bildung von Pflanzenvereinen. A. Gefässpflanzen. *Metsätiet. Koelait. Julk.* 10, 1–154. (1)

Kumar, D., and P. F. Wareing. 1972. Factors controlling stolon development in the potato plant. *New Phytol.* 71, 639–648. (17)

Kurihara, H., T. Kuroda, and O. Kinoshita. 1978. Morphological bases of shoot growth to estimate tuber yields with special reference to phytometer concept in potato plant. *Japn. J. Crop Sci.* 47, 690–698. (1)

Lammerink, J. 1968. Genetic variability in commencement of flowering in *Medicago lupulina* L. in South Island of New Zealand. *N. Z. J. Bot.* 6, 33–42. (5)

Lande, R. 1976. Natural selection and random genetic drift in phenotypic evolution. *Evolution* 30, 314-334. (13)
Lande, R. 1982a. Elements of a quantitative genetic model of life history evolution. In *Evolution and Genetics of Life Histories*, H. Dingle and J. Hegman (eds.). Springer-Verlag, New York. pp. 21-29. (9)
Lande, R. 1982b. A quantitative genetic theory of life history evolution. *Ecology* 62, 607-615. (8)
Lande, R., and S. J. Arnold. in press. The measurement of selection on correlated characters. *Evolution.* (8)
Lang, A. L., J. W. Pendleton, and G. H. Dungan. 1956. Influence of population and nitrogen levels on yield and protein and oil content of nine corn hybrids. *Agron. J.* 48, 284-289. (19)
Langer, R. H. M., and F. K. Y. Liew. 1973. Effects of varying nitrogen supply at different stages of the reproductive phase on spikelet and grain production and on grain nitrogen in wheat. *Aust. J. Agric. Res.* 24, 646-656. (19)
Langer, R. H. M., S. M. Ryle, and O. R. Jewiss. 1964. The changing plant and tiller populations of timothy and meadow fescue swards. I. Plant survival and the pattern of tillering. *J. Appl. Ecol.* 1, 197-208. (19)
Langlet, O. 1959. A cline or not a cline—a question of Scots Pine. *Sylvae Genet.* 8, 15-22. (10)
Langlet, O. 1971. Two hundred years of genecology. *Taxon.* 20, 653-722. (5)
Lankester, E. R. 1904. The structure and classification of the Arthropoda. *Q. J. Microscop. Sci.* 47, 523-582. (1)
Lanner, R. M. 1966. Needed: A new approach to the study of pollen dispersion. *Silvae. Genet.* 15, 50-52. (13)
Larsson, S., and O. Tenow. 1980. Needle-eating insects and grazing dynamics in a mature Scots pine forest in central Sweden. *Ecol. Bull. (Stockholm)* 32, 296-306. (1)
Latter, B. D. H. 1973. The island model of population differentiation: A general solution. *Genetics* 73, 147-157. (12)
Latter, B. D. H., and J. A. Sved. 1981. Migration and mutation in stochastic models of gene frequency change. II. Stochastic migration with a finite number of islands. *J. Math. Biol.* 13, 95-104. (12)
Lauder, G. V. 1981. Form and function: Structural analysis in evolutionary morphology. *Paleobiol.* 7, 430-442. (1)
Law, R. 1975. *Colonization and the Evolution of Life Histories in* Poa annua. Ph.D. thesis, University of Liverpool. (5,8)
Law, R. 1979. The cost of reproduction in annual meadow grass. *Amer. Nat.* 113, 3-16. (4,8)
Law, R., A. D. Bradshaw, and P. D. Putwain. 1977. Life history variations in *Poa annua. Evolution* 31, 233-246. (4,5,8,9,10,13,18)
Lawlor, L. R., and J. Maynard Smith. 1976. The coevolution and stability of competing species. *Amer. Nat.* 110, 79-99. (5)
Leak, W. B. 1964. An expression of diameter distribution for unbalanced, uneven aged stands and forests. *Forest Sci.* 10, 39-50. (4)
Leakey, R. R. B. 1981. Adaptive biology of vegetatively regenerating weeds. *Adv. Appl. Biol.* 6, 57-90. (1)

Lebaron, H. M., and J. Gressel. 1982. *Herbicide Resistance in Plants.* Wiley, New York. (18)

Ledig, F. T., F. H. Bormann, and K. F. Wenger. 1970. The distribution of dry matter growth between shoot and roots in loblolly pine. *Bot. Gaz.* 131, 349–359. (15)

Ledig, F. T., and J. H. Fryer. 1972. A pocket of variability in *Pinus rigida. Evolution* 26, 259–266. (13)

Lee, T. D., and F. A. Bazzaz. 1982. Regulation of fruit maturation pattern in an annual legume, *Cassia fasciculata. Ecology* 63, 1374–1388. (9)

Lenoir, T. 1978. Generational factors in the origin of Romantische Naturphilosophie. *J. Hist. Biol.* 11, 57–100. (1)

Lenski, R. E., and P. M. Service. 1982. The statistical analysis of population growth rates calculated from schedules of survivorship and fecundity. *Ecology* 63, 655–662. (11)

Leon, J. A. 1974. Selection in contexts of interspecific competition. *Amer. Nat.* 108, 739–757. (5)

Lercari, B. 1982. The promoting effects of far-red light on bulb formation in the long day plant *Allium cepa* L. *Plant Sci. Lett.* 27, 243–254. (15)

Lerner, I. M., and E. R. Dempster. 1962. Indeterminism in interspecific competition. *Proc. Nat. Acad. Sci. USA* 48, 821–826. (6)

Leverich, W. J., and D. A. Levin. 1979. Age-specific survivorship and fecundity in *Phlox drummondii* Hook. *Amer. Nat.* 113, 881–903. (4,9,13)

Levin, D. A. 1972. Low frequency disadvantage in the exploitation of pollinators by corolla variants in Phlox. *Amer. Nat.* 106, 453–460. (13)

Levin, D. A. 1975a. Interspecific hybridization, heterozygosity and gene exchange in *Phlox. Evolution* 29, 37–51. (13)

Levin, D. A. 1975b. Pest pressure and recombination systems in plants. *Amer. Nat.* 190, 437–451. (11)

Levin, D. A. 1976. Consequences of long-term artificial selection, inbreeding, and isolation in *Phlox.* I. The evolution of cross-incompatibility. *Evolution* 30, 335–344. (13)

Levin, D. A. 1978a. Some genetic consequences of being a plant. In *Ecological Genetics: The Interface,* P. F. Brussard (ed.). Springer-Verlag, New York. pp. 189–212. (13)

Levin, D. A. 1978b. The origin of isolating mechanisms in flowering plants. *Evol. Biol.* 11, 185–317. (12)

Levin, D. A. 1979. The nature of plant species. *Science* 204, 381–384. (14)

Levin, D. A. 1981. Dispersal versus gene flow in plants. *Ann. Mo. Bot. Gard.* 68, 233–253. (12,13)

Levin, D. A. 1983a. Plant parentage: An alternate view of the breeding structure of populations. In *Population Biology: Retrospect and Prospect,* C. E. King and P. S. Dawson (eds.). Columbia University Press, New York. pp. 171–188. (13)

Levin, D. A. 1983b. An immigration-hybridization episode in *Phlox. Evolution* 37:575–582 (12)

Levin, D. A., and H. W. Kerster. 1969. The dependence of bee-mediated pollen and gene dispersal upon plant density. *Evolution* 23, 560–571. (13)

Levin, D. A., and H. W. Kerster. 1974. Gene flow in seed plants. *Evol. Biol.* 7, 139–220. (12,13)

Levin, D. A., and H. W. Kerster. 1975. The effect of gene dispersal on the statics and dynamics of gene substitution in plants. *Heredity* 35, 317–336. (12)

Levins, R. 1966. The strategy of model building in population biology. *Amer. Sci.* 54, 421–431. (8)

Levins, R. 1968. *Evolution in Changing Environments.* Princeton University Press, Princeton. (5)

Levins, R. 1970. Extinction. In *Some Mathematical Problems in Biology*, M. Gerstenhaber (ed.). Amer. Math. Soc., Providence. pp. 77–107. (6)

Levins, R., and R. H. MacArthur. 1966. The maintenance of genetic polymorphism in a spatially heterogeneous environment: Variations on a theme by Howard Levine. *Amer. Nat.* 100, 585–589. (5)

Lewis, W. H. 1976. Temporal adaptation correlated with ploidy in *Claytonia virginica. Syst. Bot.* 1, 340–347. (5)

Lewontin, R. C. 1974. *The Genetic Basis of Evolutionary Change.* Columbia University Press, New York. (11)

Libby, W. J., R. F. Stettler, and F. W. Seitz. 1969. Forest genetics and forest-tree breeding. *Annu. Rev. Genet.* 3, 469–494. (13)

Linhart, Y. B. 1974. Intra-population differentiation in annual plants. I. *Veronica peregrina* L. raised under non-competitive conditions. *Evolution* 28, 232–243. (5,8,11)

Linhart, Y. B., and I. Baker. 1973. Intra-population differentiation in reponse to flooding in a population of *Veronica peregrina* L. *Nature* 242, 275–276. (5)

Linhart, Y. B., and P. Feinsinger. 1980. Plant-hummingbird interactions: Effects of island size and degree of specialization on pollination. *J. Ecol.* 68, 745–760. (13)

Linkola, K. 1935. Über die Dauer und Jahresklassenverhältnisse des Jungenstadiums bei einigen Wiesenstauden. *Acta For. Fenn.* 42, 1–56. (1)

Lint, P. J. A. L. de, and G. Heij. 1982. Night temperature and the number of nodes and flowering of the main stem of glasshouse cucumber. (*Cucumis sativus* L.). *Neth. J. Agric. Sci.* 30, 149–159. (1)

Lloyd, D. G. 1972. Breeding systems in *Cotula* L. (Compositae, Anthenudae). II. Monoecious populations. *New Phytol.* 71, 1195–1202. (13)

Lloyd, D. G. 1980a. Benefits and handicaps of sexual reproduction. *Evol. Biol.* 13, 69–111. (11)

Lloyd, D. G. 1980b. Sexual strategies in plants. I. An hypothesis of serial adjustment of maternal investment during one reproductive session. *New Phytol.* 86, 69–79. (14)

Lloyd, D. G. 1980c. Demographic factors and mating patterns in angiosperms. In *Demography and Evolution in Plant Populations,* O. T. Solbrig (ed.). Blackwell, Oxford. pp. 67–87. (4)

Lloyd, D. G. 1982. Selection of combined versus separate sexes in seed plants. *Amer. Nat.* 120, 571–585. (13)

Lloyd, D. G. 1983. Evolutionarily stable sex ratios and sex allocations. *J. Theoret. Biol.* (14)

Lloyd, D. G. in press. Variation strategies of plants in heterogeneous environments. *Biol. J. Linn Soc.* (14)

Lloyd, D. G., and K. S. Bawa. in press. Modification of the gender of seed plants in varying conditions. *Evol. Biol.* (14)

Lloyd, D. G., and J. M. A. Yates. 1982. Intrasexual selection and the segregation of pollen and stigmas in hermaphrodite plants, exemplified by *Wahlenbergia albomarginata* (Campanulaceae). *Evolution* 36, 903–913. (14)

Lloyd, D. G., and C. S. Webb. 1977. Secondary sex characters in seed plants. *Bot. Rev.* 43, 177–216. (4)

Lovett Doust, J., and G. W. Eaton. 1982. Demographic aspects of flower and fruit production in bean plants, *Phaseolus vulgaris* L. *Amer. J. Bot.* 69, 1156–1164. (1)

Lovett Doust, J., and J. L. Harper. 1980. The resource costs of gender and material support in an andromonoecious umbellifer, *Smyrnium olusatrum* L. *New Phytol.* 85, 251–264. (14)

Lovett Doust, L. 1981a. Intraclonal variation and competition in *Ranunculus repens*. *New Phytol.* 89, 495–502. (5)

Lovett Doust, L. 1981b. Population dynamics and local specialization in a clonal perennial (*Ranunculus repens*). I. The dynamics of ramets in contrasting habitats. *J. Ecol.* 69, 743–755. (2)

Lovett Doust, L. 1981c. Population dynamics and local specialization in a clonal perennial (*Ranunculus repens*). II. The dynamics of leaves, and a reciprocal transplant-replant experiment. *J. Ecol.* 69, 757–768. (8,10)

Lubchenco, J. 1978. Plant species diversity in a marine intertidal community: Importance of herbivore food preference and algal competitive abilities. *Amer. Nat.* 112, 23–39. (7)

Lubchenco, J., and J. Cubit. 1980. Heteromorphic life histories of certain marine algae as adaptations to variations in herbivory. *Ecology* 61, 676–687. (7)

Lubchenco, J., and S. D. Gaines. 1981. A unified approach to marine plant-herbivore interactions. I. Populations and communities. *Ann. Rev. Ecol. Syst.* 12, 405–437. (1,7)

Lupton, F. G. H., and F. M. Pushman. 1975. The crop, the environment and the genotype. In *Bread*, A. Spicer (ed.). Applied Science Publishers, London. pp. 67–86. (19)

Lynch, M., and W. J. Gabriel. in press. Phenotypic evolution and parthenogenesis. *Amer. Nat.* (13)

Maarel, E. van der. 1969. On the use of ordination models in phytosociology. *Vegetatio* 19, 21–46. (3)

Maarel, E. van der. 1976. On the establishment of plant communities boundaries. *Ber. Deutsch. Bot. Ges.* 89, 415–433. (3)

Maarel, E. van der (ed.). 1980a. *Succession: Advances in Vegetation Science.* Vol. 3, Junk, The Hague. (3)

Maarel, E. van der. 1980b. Vegetation development in a former orchard under different treatments: A preliminary report. *Vegetatio* 43, 95–102. (3)

Maarel, E. van der, N. de Cook, and E. de Wildt. in press. Population dynamics of some major woody species in relation to long-term succession in the dunes of Voorne. *Vegetatio.* (3,4)

Maarel, E. van der, L. Orloci, and S. Pignatti (eds.). 1980. *Data-processing in Phytosociology*. Junk, The Hague. (3)
Maarel, E. van der, and V. Westhoff. 1964. The vegetation of the dunes near Oostvoorne, Netherlands. *Wentia* 12, 1-61. (3)
MacArthur, R. H., and R. Levins, 1964. Competition, habitat selection, and character displacement in a patchy environment. *Proc. Nat. Acad. Sci. USA* 51, 1207-1210. (5)
MacArthur, R. H., and R. Levins. 1967. The limiting similarity, convergence and divergence of coexisting species. *Amer. Nat.* 101, 377-385. (5)
MacArthur, R. H., and E. O. Wilson. 1967. *The Theory of Island Biogeography,* Princeton University Press, Princeton. (18)
MacDougal, D. T. 1936. Studies in tree growth by the dendrographic method. *Carnegie Inst. Wash. Publ.* 462. (1)
MacDougal, D. T. 1938. *Tree Growth.* Chronica Botanica, Leiden. (1)
Mackie, G. O. 1963. Siphonophores, bud colonies, and superorganisms. In *The Lower Metazoa: Comparative Biology and Phylogeny.* California University Press. pp. 329-337. (1)
MacMahon, J. A., D. L. Phillips, J. V. Robinson, and D. J. Schimpf. 1978. Levels of biological organization: an organism-centered approach. *BioScience* 28, 700-704. (5)
MacMahon, J. A., D. J. Schimpf, D. C. Anderson, K. G. Smith, and R. L. Bayn Jr. 1981. An organism centered approach to some community and ecosystem concepts. *J. Theor. Biol.* 88, 287-307. (5)
Mahmoud, A., J. P. Grime, and S. B. Furness. 1975. Polymorphism in *Arrhenatherum elatius* (L.). Beauv. ex J. and C. Presl. *New Phytol.* 75, 269-276. (1)
Maillette, L. 1981. Needle demography and growth pattern of Corsican pine. *Can. J. Bot.* 60, 105-116. (1)
Maillette, L. 1982a. Structural dynamics of silver birch. I. The fates of buds. *J. App. Ecol.* 19, 203-218. (1,2)
Maillette, L. 1982b. Structural dynamics of silver birch. II. A matrix model of the bud population. *J. Appl. Ecol.* 19, 219-238. (1,2)
Mariani, A., and E. Avieli. 1973. Heterosis during the early phases of growth in intraspecific and interspecific crosses of cotton. *Crop Sci.* 13, 15-18. (2)
Martin, M. M., and J. Harding. 1981. Evidence for the evolution of competition between two species of annual plants. *Evolution* 35, 975-987. (5,6)
Maruyama, T., and M. Kimura. 1980. Genetic variability and effective population size when local extinction and recolonization of subpopulations are frequent. *Proc. Nat. Acad. Sci. USA* 77, 6710-6714. (12)
Masselink, A. K. 1980. Germination and seed population dynamics in *Melampyrum pratense. Acta Bot. Neerl.* 29, 451-468. (18)
Masle-Meynard, J., and M. Sebillotte. 1981. Etúde de l'hétérogénéité d'un peuplement de blé d'hiver. I. Notion de structure du peuplement. II. Origine des différentes catégories d'individus du peuplement; éléments de description de sa structure. *Agronomie* 1, 207-216; 217-224. (1)
Mather, K. 1949. *Biometrical Genetics.* Methuen, London. (11)
Mather, K. 1969. Selection through competition. *Heredity* 25, 529-540. (6)

Mather, K., and J. L. Jinks. 1977. *Introduction to Biometrical Genetics.* Chapman and Hall, London. (8)

Matzinger, D. F., and E. A. Wernsman. 1967. Genetic diversity and heterosis in *Nicotiana.* L. Interspecific cross. *Der Züchter* 37, 186–191. (12)

May, R. H. 1975. Stability in ecosystems: Some comments. In *Unifying Concepts in Ecology,* W. H. van Dobben and R. H. Lowe-McConnell (eds.). Junk, The Hague. pp. 161–168. (18)

Maynard Smith, J. 1971. The origin and maintenance of sex. In *Group Selection,* G. C. Williams (ed.). Aldine Atherton, Chicago. pp. 163–175. (14)

Maynard Smith, J. 1974. The theory of games and the evolution of animal conflicts. *J. Theor. Biol.* 47, 209–221. (14)

Maynard Smith, J. 1978a. *The Evolution of Sex.* Cambridge University Press, Cambridge. (11,13,14)

Maynard Smith, J. 1978b. Optimization theory in evolution. *Ann. Rev. Ecol. Syst.* 9, 31–56. (8,11,13)

Mayr, E. 1982. *The Growth of Biological Thought.* The Belknap Press of Harvard University Press, Cambridge, Mass. (1)

Mayr, E., and W. B. Provine. 1980. *The Evolutionary Synthesis: Perspectives on the Unification of Biology.* Harvard University Press, Cambridge, Mass. (11)

McCarl, B. A. 1981. *Economics of Integrated Pest Management.* Special Report 636, International Plant Protection Centre and Department of Agriculture and Resource Economics, Oregon State University, Corvallis. (18)

McClure, F. A. 1966. *The Bamboos—a Fresh Perspective.* Harvard University Press, Cambridge, Mass. (2)

McDermitt, D. K., and R. S. Loomis. 1981. Elemental composition of biomass and its relation to energy content, growth efficiency, and growth yield. *Ann. Bot.* 48, 245–290. (15)

McIntosh, R. P. 1967. The continuum concept of vegetation. *Bot. Rev.* 33, 131–187. (3)

McIntosh, R. P. 1970. Community, competition and adaptation. *Quart. Rev. Biol.* 45, 259–280. (5)

McIntosh, R. P. 1976. Ecology since 1900. In *Issues and Ideas in America,* B. J. Taylor and T. J. White (eds.). University of Oklahoma Press, Norman. (3)

McIntosh, R. P., and H. A. Gleason. 1976. "Individualist Ecologist" 1882–1975. His contributions to ecological theory. *Bull. Torr. Bot. Club* 102, 253–273. (3)

McIntyre, G. I. 1967. Apical dominance in the rhizomes of *Agropyron repens:* the influence of water stress on bud activity. *Can. J. Bot.* 54, 2744–2754. (18)

McKey, D. 1974. Adaptive patterns in alkaloid physiology. *Amer. Nat.* 108, 305–320. (7)

McKey, D. 1979. The distribution of secondary compounds within plants. In *Herbivores,* G. A. Rosenthal and D. H. Janzen (eds.). Academic Press, New York. pp. 56–133. (7)

McMahon, D. J. 1982. *The Population Dynamics of Agropyron repens* (L.) Beauv. Ph.D. thesis, University of Liverpool. (18)

McMillan, C. 1960. Ecotypes and community function. *Amer. Nat.* 94, 245–257. (5)

McNaughton, S. J. 1979. Grazing as an optimization process: Grass-ungulate relationships in the Serengeti. *Amer. Nat.* 113, 691–703. (15)

McNaughton, S. J., T. C. Folsom, T. Lee, F. Park, C. Price, D. Roeder, J. Schmitz, and C. Stockwell. 1974. Heavy metal tolerance in *Typha latifolia* without the evolution of tolerant races. *Ecology* 55, 1163–1165. (5)

McNaughton, S. J., and L. L. Wolf. 1970. Dominance and the niche in ecological systems. *Science* 167, 131–139. (5)

McNeilly, T. 1968. Evolution in closely adjacent plant populations. III. *Agrostis tenuis* on a small copper mine. *Heredity* 23, 99–108. (12)

McNeilly, T. 1981. Ecotypic differentiation in *Poa annua:* Interpopulation differences in response to competition and cutting. *New Phytol.* 88, 539–547. (5)

McNeilly, T., and J. Antonovics. 1968. Evolution in closely adjacent plant populations. IV. Barriers to gene flow. *Heredity* 23, 205–218. (5,12)

McRill, M., and G. R. Sagar. 1973. Earthworms and seeds. *Nature* 243, 482. (18)

McWhorter, C. G., and W. C. Shaw. 1982. Research needs for integrated weed management systems. *Weed Sci.* 30 (Suppl), 40–45. (18)

Melin, D. 1977. Corrélations de croissance chez une plante grimpante à rameaux polymorphes: *Periploca graeca* L. *Ann. Sci. Nat. Bot., Ser.* 12, 18, 251–274. (1)

Mendoza, A. E. 1981. *Modificaciones del equilibrio foliar y sus efectos en el comportamiento reproductivo y* vegetativo en *Astrocaryum mexicanum.* B.Sc. thesis, Facultad de Ciencias, UNAM, Mexico. (7)

Meusel, H. 1970. Wuchsformenreihen mediterran-mitteleuropäischer Angiospermen-Taxa. *Fedd. Repert.* 81, 41–59. (1)

Meusel, H., and G. Mörchen. 1977. Zur ökogeographischen und morphologischen Differenzierung einiger *Scrophularia*-Arten. *Flora* 166, 1–20. (1)

Michener, G. R. 1980. Differential reproduction among female Richardson's ground squirrels and its relation to sex ratio. *Behav. Ecol. Sociobiol.* 7, 173–178. (13)

Michod, R. E. 1982. The theory of kin selection. *Ann. Rev. Ecol. Syst.* 13, 23–55. (6)

Miles, J. 1979. *Vegetation Dynamics.* Chapman and Hall, London. (3)

Miller, P. C. 1967. Leaf orientation and energy exchange in quaking aspen (*Populus tremuloides*) and gambell's oak (*Quercus gambellii*) in central Colorado. *Oecol. Plant.* 2, 241–270. (2,15)

Miller, R. S. 1967. Pattern and process in competition. *Adv. Ecol. Res.* 4, 1–81. (2,15)

Milton, K., D. M. Windsor, and D. W. Morrison. 1982. Fruiting phenologies of two neotropical *Ficus* species. Ecology 63, 752–762. (4)

Mohler, C. L., P. L. Marks, and D. G. Sprugel. 1978. Stand structure and allometry of trees during self-thinning of pure stands. *J. Ecol.* 66, 599–614. (4)

Moll, R. H., J. H. Lonnquist, V. Fortuno, and E. C. Johnson. 1965. The relationship of heterosis and genetic divergence in maize. *Genetics* 52, 139–144. (12)

Monsi, M. 1968. Mathematical models of plant communities. In *Functioning of Terrestrial Ecosystems at the Primary Production Level,* F. Eckardt (ed.). UNESCO, Paris. pp. 131–149. (15)

Monsi, M., and T. Saeki. 1953. Über den Lichtfactor in den Pflanzengesellschaften und seine Bedeutung für die Stoffproduktion. *Jap. J. Bot.* 14, 22–52. (15)

Moody, K., and S. K. De Datta. 1980. Economics of weed control in tropical and subtropical rice growing regions with emphasis on reduced tillage *Proc. (1980) Brit. Crop Prot. Conf. Weeds* 3, 931–940. (18)

Mooney, H. A., J. Ehleringer, and O. Björkman. 1977. The energy balance of leaves of the evergreen desert shrub, *Atriplex hymenelytra. Oecologia* 29, 301–310. (15)

Mooney, H. A., and S. L. Gulmon. 1979. Environmental and evolutionary constraints on the photosynthetic characteristics of higher plants. In *Topics in Plant Population Biology,* O. T. Solbrig, S. Jain, G. B. Johnson, and P. H. Raven (eds.). Columbia University Press, New York. pp. 316–337. (5)

Mooney, H. A., and S. L. Gulmon. 1982. Constraints on leaf structure and function in reference to herbivory. *BioScience* 32, 198–206. (15,16)

Moore, D. M., and H. Lewis. 1965. The evolution of self-pollination in *Clarkia xantiana. Evolution* 19, 104–117. (13)

Moore, P. D. 1976. How far does pollen travel? *Nature* 260, 388–389. (13)

Morishima, H. 1978. Breeding systems as conditioned by adaptive strategies in wild rice species. *US-Japan Seminar on the Dynamics of Speciation in Plants and Animals,* Tokyo. pp. 42–47. (18)

Morley, F. H. W. 1958. The inheritance and ecological significance of seed dormancy in subterranean clover (*Trifolium subterraneum* L.). *Aust. J. Biol. Sci.* 11, 261–274. (8)

Morrison, D. A., and P. J. Myercough. 1982. Genecological differentiation in *Leptospermum flavescens* Sm. in the Sydney region. *Aust. J. Bot.* 30, 461–475. (5)

Morrison, J. 1979. Botanical change in agricultural grasslands in Britain. In *Changes in Sward Composition and Productivity,* A. H. Charles and R. J. Haggar (eds.). British Grassland Society, Hurley. pp. 5–10. (19)

Mortimer, A. M. 1983. On weed demography. In *Recent Advances in Weed Research,* W. W. Fletcher (ed.). Commonwealth Agriculture Bureaux, Farnham Royal. pp. 3–40. (18)

Mortimer, A. M., and D. J. McMahon. 1982. Time, space and the growth of couch grass. *Proc. (1982) Brit. Crop Prot. Conf.—Weeds* 2, 771–778. (18)

Moss, S. 1980. A study of populations of black grass *(Alopecurus myosuroides)* in winter wheat as influenced by seed shed in the previous crop, cultivation system and straw disposal method. *Ann. Appl. Biol.* 94, 121–126. (18)

Muchow, R. C. 1979. Effects of plant population and season on kenaf *(Hibiscus cannabinus* L.) grown under irrigation in tropical Australia. I. Influence on the components of yield. II. Influence on growth parameters and yield prediction. *Field Crops Res.* 2, 55–66; 67–76. (1)

Mueller-Dombois, D., and H. Ellenberg. 1974. *Aims and Methods of Vegetation Ecology,* Wiley, New York. (3)

Munro, J. M and H. G. Farbrother. 1969. Composite plant diagrams in cotton. *Cotton Grow. Rev.* 46, 261–282. (1)

Murphy, G. I. 1968. Pattern in life history and the environment. *Amer. Nat.* 102, 391–403. (8)
Murray, J., M. S. Johnson, and B. Clarke. 1982. Microhabitat differences among genetically similar species of *Partula. Evolution* 36, 316–325. (5)
Nagylaki, T. 1979. The island model with stochastic migration. *Genetics* 91, 163–176. (12)
Nei, M., A. Chakravarti, and Y. Tateno. 1977. Mean and variance of F_{ST} in a finite number of incompletely isolated populations. *Theor. Pop. Biol.* 11, 291–306. (12)
Neilson-Jones, W. 1969. *Plant Chimeras.* Methuen, London. (1)
Nelson, L. R. 1980. Recurrent selection for improved rate of germination in ryegrass. *Crop Sci.* 20, 219–221. (4)
New, J. 1961. Biological flora of the British Isles: *Spergula arvensis. J. Ecol.* 49, 205–215. (1)
Newell, S. J. 1982. Translocation of 14C-photoassimilate in two stoloniferous *Viola* species. *Bull. Torrey Bot. Club* 109, 306–317. (1)
Ng, F. S. P. 1977. Strategies of establishment in Malayan forest trees. In *Tropical Trees as Living Systems,* P. B. Tomlinson and M. H. Zimmerman (eds.). Cambridge University Press, Cambridge. pp. 129–162. (4)
N. I. A. B. 1982. *Recommended Varieties of Grasses.* Nat. Inst. Agric. Bot., Cambridge. (19)
Nielsen, D. B. 1978. The economic impact of poisonous plants on the range of livestock injury in the 17 western states of USA. *J. Range Manage.* 31, 325–328. (18)
Nielsen, E. L. 1968. Intraplant morphological variation in grasses. *Amer. J. Bot.* 55, 116–122. (1)
Niemann, P. 1980. Evaluation of long term trends for control of weeds in spring barley (1970–9). *Gesunde Pflanzen* 32, 239–240. (18)
Nieto, J. H., M. Anaya Brondó, and J. T. González. 1968. Critical periods of the crop growth cycle for competition from weeds. *PANS* 14, 159–166. (18)
Niklas, K. J. 1978. Morphometric relationships and rates of evolution among Paleozoic vascular plants. *Evol. Biol.* 11, 509–543. (1)
Niklas, K. J. 1982. Computer simulations of early land plant branching morphologies: canalization of patterns during evolution? *Paleobiol.* 8, 196–210. (1)
Niklas, K. J., and T. D. O'Rourke. 1982. Growth patterns of plants that maximize vertical growth and minimize internal stresses. *Amer. J. Bot.* 69, 1367–1374. (1)
Nilsson, B. D. 1974. Heterosis in an intraspecific hybridization experiment in Norway spruce. *Proc. Joint IUFRO meeting.* S.02.041-3. (12)
Nix, J. 1978. The economics of growing wheat and barley 1977–1980 and recent trends. 1987. I. *Farm Business Unit Occasional Paper No. 1.* Wye College University, London. pp. 24. (18)
Noble, I. R., and R. O. Slatyer. 1980. The use of vital attributes to predict successional changes in plant communities subject to recurrent disturbances. *Vegetatio* 43, 5–21. (3)

Noble, J. C., A. D. Bell, and J. L. Harper. 1979. The population biology of plants with clonal growth. I. The demography and structural demography of *Carex arenaria*. *J. Ecol.* 67, 983–1008. (2)
Noodén, L. D. 1980. Senescence in the whole plant. In *Senescence in Plants,* K. Thimann (ed.). CRC Press Inc. Boca Raton. pp. 219–258. (15)
North, J. J. 1981. Cereal growing: The changing scene. *AAB Grass Weeds in Cereals in the U.K. Conf.* 1–4. (18)
Norton, G. A., and G. R. Conway. 1977. The economic and social context of pest disease and weed problems. In *Origin of Pest, Parasite, Disease and Weed Problems,* J. M. Cherret and G. R. Sagar (eds.). Blackwell, Oxford. pp. 205–226. (18)
Noy-Meir, I. 1975. Stability of grazing systems: An application of predator-prey graphs. *J. Ecol.* 63, 459–481. (7,18)
Nozeran, R. 1978a. Réflexions sur les enchâinements de fonctionnement au cours du cycle des végétaux supérieurs. *Bull. Soc. Bot. France* 125, 263–280. (1)
Nozeran, R. 1978b. Multiple growth correlations in phanerogams. In *Tropical Trees as Living Systems,* P. B. Tomlinson and M. H. Zimmerman (eds.). Cambridge University Press, Cambridge. pp. 423–443. (1)
Nozeran, R., L. Bancilhon, and P. Neville. 1971. Intervention of internal correlations in the morphogenesis of higher plants. *Adv. Morphogen.* 9, 1–66. (1)
Obeid, M., M. Machin, and J. L. Harper. 1967. Influence of density on plant to plant variation in fiber flax, *Linnum usitatissimum. Crop Sci.* 7, 471–473. (4)
Ogden, J. 1970. Plant population structure and productivity. *Proc. N.Z. Ecol. Soc.* 17, 1–9. (4)
Oka, H. I. 1976. Mortality and adaptive mechanisms of *Oryza perennis* strains. *Evolution* 30, 380–392. (8)
O'Keefe, M. G. 1980. The control of *Agropyron repens* and broad-leaved weeds preharvest of wheat and barley with the isopropylamine salt of glyphosphate. *Proc. (1980) Brit. Crop Prot. Conf.—Weeds* 3, 53–60. (18)
Olsen, S. R. 1982. Removing barriers to crop productivity. *Agron. J.* 74, 1–4. (18)
Olufajo, O. O., R. W. Daniels, and D. H. Scarisbrick. 1982. The effect of pod removal on the translocation of 14C photosynthate from leaves of *Phaseolus vulgaris* L. cv. Lochness. *J. Hort. Sci.* 57, 333–338. (1)
Orlóci, L. 1978. *Multivariate Analysis in Vegetation Research* (2nd Ed.). Junk, The Hague. (3)
Orr-Ewing, A. L. 1969. Racial crossing in Douglas fir. *Proc. Working Group on Quantitative Genetics Section 22 IUFRO* (North Carolina State University, Raleigh). (12)
Ospovat, D. 1981. *The Development of Darwin's Theory: Natural History, Natural Theology and Natural Selection, 1838–1859.* Cambridge University Press, Cambridge. (1)
Owen, R. 1848. *On the Archetype and Homologies of the Vertebrate Skeleton.* London. (1)
Owen, R. 1849. *On Parthenogenesis, or the Successive Production of Procreating Individuals from a Single Ovum: A Discourse.* John van Voorst, London. (1)

Owen, D. F., and R. G. Wiegert. 1981. Mutualism between grasses and grazers: An evolutionary hypothesis. *Oikos* 36, 376–378. (15)

Owino, S., and B. Zobel. 1977. Genotype × environment interactions and genotypic stability in loblolly pine. III. Heterosis and heterosis × environment interaction. *Silvae. Genet.* 26, 114–116. (12)

Paine, R. T. 1966. Food web complexity and species diversity. *Amer. Nat.* 100, 65–75. (7)

Paine, R. T. 1969. A note on trophic complexity and community stability. *Amer. Nat.* 103, 91–93. (16)

Painter, E. I., and J. K. Detling. 1981. Effects of defoliation on net photosynthesis and regrowth of western wheatgrass. *J. Range Mgmt.* 34, 68–71. (15)

Palmblad, I. G. 1968. Competition in experimental populations of weeds with emphasis on the regulation of population size. *Ecology* 49, 26–34. (13)

Palmer, T. P. 1972. Variation in flowering time among and within populations of *Trifolium arvense* L. in New Zealand. *N.Z.J. Bot.* 10, 59–68. (5)

Parker, C., and J. D. Fryer. 1975. Weed control problems causing major reductions in world food supplies. *FAO Plant Protection Bulletin* 23, 83–95. (18)

Parker, E. D. 1979. Phenotypic consequences of parthenogenesis in *Cnemidophorus* lizards. I. Variability in parthenogenetic and sexual populations. *Evolution* 33, 1150–1166. (13)

Parker, M. A. 1982. Association with mature plants protects seedlings from predation in an arid grassland shrub, *Gutierrezia microcephala. Oecologia* 53, 276–280. (4)

Passioura, J. B. 1979. Accountability, philosophy and plant physiology. *Search* 10, 347–350. (19)

Passioura, J. B. 1981. The interaction between the physiology and breeding of wheat. In *Wheat Science-Today and Tomorrow*, L. T. Evans and W. J. Peacock (eds.). Cambridge University Press, Cambridge. pp. 191–201. (19)

Paterson, J. G. 1977. Interaction between herbicides, time of application and genotype of wild oats (*Avena fatua* L.). *Aust. J. Agric. Res.* 28, 671–680. (18)

Pearson, K. 1903. Mathematical contributions to the theory of evolution. XI. On the influence of natural selection on the variability and correlation of organs. *Phil. Trans. Roy. Soc. Lond.* A. 200, 1–66. (8)

Pedersen, P. N., R. L. Hurst, M. D. Levin, and G. L. Stoker. 1961. Computer analysis of the genetic contamination of alfalfa seed. *Crop Sci.* 9, 1–4. (12)

Peet, M. M., and P. J. Kramer. 1980. Effects of decreasing source/sink ratio in soybeans on photosynthesis, photorespiration, transpiration and yield. *Plant, Cell, Environ.* 3, 201–206. (15)

Penning de Vries, F. W. T., A. H. M. Brunsting, and H. H. van Laar. 1974. Products, requirements and efficiency of biosynthesis: A quantitative approach. *J. Theor. Biol.* 45, 339–377. (15)

Peters, C. M. 1983. *Reproduction, Growth and the Population Dynamics of* Brosimum alicastrum *Sw. in a Moist Tropical Forest of Central Veracruz, Mexico.* Ph.D. dissertation, Yale University. (4)

Peters, N. C. B. 1978. *Competition Studies with* Avena fatua. Ph.D. thesis, University of Reading. (18)

Pfahler, P. L. 1965. Fertilization ability of maize pollen grains. I. Pollen sources. *Genetics* 52, 513–520. (13)
Pfahler, P. L. 1967. Fertilization ability of maize pollen grains. II. Pollen genotype, female sporophyte and pollen storage interaction. *Genetics* 57, 513–521. (13)
Pfirsch, E. 1972. Déterminisme de la croissance en stolon chez l'*Ajuga reptans* L. Influence des noeuds de base sur l'action corrélative émanant de l'inflorescence. *C. R. Acad. Sci. Paris* 274, 2499–2502. (1)
Pfirsch, E. 1978. Induction de la croissance plagiotrope chez le *Stachys silvatica* L. Rôle de l'acide abscissique dans le mécanisme autorépétitif. *Bull. Soc. Bot. France* 125, 231–242. (1)
Pianka, E. R. 1970. On r and K selection. *Amer. Nat.* 104, 592–597. (8)
Pianka, E. R. 1978. *Evolutionary Ecology* (2nd Ed.). Harper & Row, New York. (5)
Pigott, C. D. 1982. The experimental study of vegetation. *New Phytol.* 90, 389–404. (17)
Pimentel, D. 1961. Species diversity and insect population outbreaks. *Ann. Entomol. Soc. Amer.* 54, 76–86. (5)
Pimentel, D. 1964. Population ecology and the genetic feedback mechanism. In *Genetics Today, Proc. XI Intern. Congr. Genet.* Pergamon Press, Oxford. pp. 483–488. (6)
Pimentel, D. 1968. Population regulation and genetic feedback. *Science* 159, 1432–1437. (5)
Pimentel, D. 1976. World food crisis: Energy and pests. *Bull. Entomol. Soc. Amer.* 22, 20–26. (18)
Pimentel, D. 1979. Environmental risks associated with biological controls. In *Environmental Protection and Biological Forms of Control of Pest Organisms,* B. Lundholm and E. Stackerud (eds.). Proc. Int. Workshop Stockholm Swedish Natural Science Research Council, pp. 11–24. (18)
Pimentel, D., D. Andow, R. Dyson Hudson, D. Gallahan, S. Jacobson, M. Irish, S. Kroop, A. Moss, I. Schreinr, M. Shepard, T. Thompson and B. Vinzant. 1980. Environmental and social costs of pesticides: A preliminary assessment. *Oikos* 34, 126–160. (18)
Pimentel, D., E. H. Feinburg, P. W. Wood, and J. T. Hayes. 1965. Selection, spatial distribution and the coexistence of competing fly species. *Amer. Nat.* 99, 97–109. (5)
Pimentel, D., W. P. Nagel, and J. L. Madden. 1963. Space-time structure of the environment and the survival of parasite-host systems. *Amer. Nat.* 97,
Piñero, D., J. Sarukhán, and E. González. 1977. Estudios demográficos en plantas. *Astrocaryum mexicanum.* I. Estructura de las poblaciones. *Bol. Soc. Bot. Méx.* 37, 69–118. (4)
Piñero, D., and J. Sarukhán. 1982. Reproductive behaviour and its individual variability in a tropical palm, *Astrocaryum mexicanum. J. Ecol.* 70, 461–472. (4)
Platt, W. J., and I. M. Weis. 1977. Resource partitioning and competition within a guild of fugitive prairie plants. *Amer. Nat.* 111, 479–513. (5)
Pleasants, J. M., and M. Zimmerman. 1979. Patchiness in the dispersion of nectar resources: Evidence for hot and cold spots. *Oecologia* 41, 283–288. (13)

Plummer, G. L., and C. Keever. 1963. Autumnal daylight weather and camphorwood dispersal in the Georgia piedmont region. *Bot. Gaz.* 124, 283–289. (12)

Pohl, F. 1937. Die Pollenerzeugung der Windbluter. *Beih. Bot. Centralbl.* 56, 365–470. (14)

Poissonet, P., F. Romane, M. P. Austin, E. van der Maarel, and W. Schmidt (eds.). 1981. *Vegetation Dynamics in Grasslands, Heathlands and Mediterranean Lignous Formations.* Junk, The Hague. (3)

Policansky, D. 1982. Sex change in plants and animals. *Ann. Rev. Ecol. Syst.* 13, 471–495. (17)

Pontin, A. J. 1982. *Competition and Coexistence of Species.* Pitman, London. (5)

Porter, J. R. 1983. A modular approach to analysis of plant growth. II. Methods and results. *New Phytol.* 94, 191–200. (1)

Powell, G. R. 1977. Biennial strobilus production in balsam fir: A review of its morphogenesis and a discussion of its apparent physiological basis. *Can. J. For. Res.* 7, 547–555. (1)

Prat, S. 1934. Die Erblichkeit der Resistenz gegen Kupfer. *Ber. Deut. Bot. Ges.* 102, 65–67. (10)

Prévost, M. F. 1967. Architecture de quelques Apocynacées ligneuses. *Mém. Soc. Bot. France* 114, 24–36. (1)

Prévost, M. F. 1978. Modular construction and its distribution in tropical woody plants. In *Tropical Trees as Living Systems*, P. B. Tomlinson and M. H. Zimmerman (eds.). Cambridge University Press, Cambridge. pp. 223–231. (1)

Price, M. V., and N. M. Waser. 1979. Pollen dispersal and optimal outcrossing in *Delphinium nelsoni. Nature* 277, 294–297. (13)

Price, M. V., and N. M. Waser. 1982. Population structure, frequency dependent selection, and the maintenance of sexual reproduction. *Evolution* 36, 35–43. (11)

Primack, R. B. 1980. Phenotypic variation of rare and widespread species of *Plantago. Rhodora* 82, 87–95. (13)

Primack, R. B., and J. Antonovics. 1981. Experimental ecological genetics in *Plantago.* Components of seed yield in the ribwort plantain *Plantago lanceolata* L. *Evolution* 35, 1069–1079. (4,8)

Primack, R. B., and J. Antonovics. 1982. Experimental ecological genetics in *Plantago.* VII. Reproductive effort in populations of *P. lanceolata* L. *Evolution* 36, 742–752. (4,8)

Proctor, M., and P. Yeo. 1973. *The Pollination of Flowers.* Collins, London. (13)

Putwain, P. D. 1982. Herbicide resistance in weeds—an inevitable consequence of herbicide use? *Proc (1982) Brit. Crop Prot. Conf.-Weeds* 3, 719–728. (18)

Putwain, P. D., K. R. Scott, and R. J. Holliday. 1982. The nature of resistance to triazine herbicides: Case studies of phenology and population studies. In *Herbicide Resistance in Plants,* H. M. Lebaron and J. Gressel (eds.). Wiley, New York. pp. 99–116. (18)

Pyke, G. H. 1978. Optimal foraging in bumblebees and coevolution with their plants. *Oecologia* 36, 181–193. (13)
Rabotnov, T. A. 1950. The life cycle of perennial herbaceous plants in meadow coenoses. *Trudy Bot. Inst. Akad. Nauk SSSR,* Ser. III, 6, 7–204. (Russian) (1)
Rabotnov, T. A. 1960. (Methods of age and longevity determination in herbage.) *Field Geobotany. Akad. Nauk SSSR,* Moscow. (Russian) (4)
Rabotnov, T. A. 1969. On coenopopulations of perennial herbaceous plants in natural coenoses. *Vegetatio* 19, 87–95. (4)
Rabotnov, T. A. 1978a. Structure and method of studying coenotic populations of perennial herbaceous plants. *Sov. J. Ecol.* 9, 99–105. (4)
Rabotnov, T. A. 1978b. On coenopopulations of plants reproducing by seeds. *Verh. Kon. Nederl. Akad. Wetensch., Afd. Natuurk.,* Tweede Reeks 70, 1–26. (1)
Rabotnov, T. A. 1980, 1981. Bibliography of papers on the problems of coenopopulations published in the U.S.S.R. Parts I, II, III. *Excerpta Bot.* B20, 71–96, 171–206; B21, 91–119. (1)
Rai, K. N., and S. K. Jain. 1982. Population biology of *Avena.* IX. Gene flow and neighborhood size in relation to microgeographic variation in *Avena barbata. Oecologia* 53, 399–405. (6,13)
Raschke, K. 1956. Über die physikalischen Beziehungen zwischen Wärmeübergangszahl, Strahlungsanstausch, Temperatur und Transpiration eines Blattes. *Planta* 40, 200–237. (15)
Raschke, K. 1960. Heat transfer between the plant and the environment. *Ann. Rev. Plant Physiol.* 11, 111–126. (17)
Raup, D. M. 1972. Approaches to morphologic analysis. In *Models in Paleobiology,* T. J. M. Schopf (ed.). Freeman, Cooper and Co., San Francisco. pp. 28–44. (1)
Raup, D. M., and A. Seilacher. 1969. Fossil foraging behaviour: Computer simulation. *Science* 166, 994–995. (2)
Raven, P. H. 1976. Systematics and plant population biology. *Syst. Bot.* 1, 284–317. (5)
Raven, P. H. 1979. Future directions in plant population biology. In *Topics in Plant Population Biology,* O. T. Solbrig, S. Jain, G. B. Johnson, and P. H. Raven (eds.). Columbia University Press, New York. pp. 461–481. (5)
Raynal, D. J., and F. A. Bazzaz. 1975. Interference of winter annuals with *Ambrosia artemisiifolia* in early successional fields. *Ecology* 56, 35–49. (16)
Reed, K. F. M. 1972. The performance of sheep grazing different pasture types. In *Plants for Sheep in Australia,* J. Leigh, I. Noble (eds.). Angus & Robertson, Sidney. (18)
Reffye, P. de. 1981–1982. Modèle mathématique aléatoire et simulation de la croissance et de l'architecture du caféier Robusta. 1. Étude du fonctionnement des méristèmes et de la croissance des axes végétatifs. 2. Étude de la mortalité des méristèmes plagiotropes. 3. Étude de la ramification sylleptique des rameaux primaires et de la ramification proleptique des rameaux secondaires. *Café Cacao Thé* 25, 83–104, 219–230; 26, 77–96. (1)
Regehr, D. L., and F. A. Bazzaz. 1976. Low temperature photosynthesis in successional winter annuals. *Ecology* 57, 1297–1303. (16)
Regehr, D. L., and F. A. Bazzaz. 1979. The population dynamics of *Erigeron canadensis,* a succesional winter annual. *J. Ecol.* 67, 923–933. (16)

Renard, C. 1971. Quelques caracteres des auxiblastes chez le hêtre en Haut-Ardenne. *Lejeunia* 59. (1)

Reznick, D. 1981. "Grandfather effects": The genetics of interpopulation differences in offspring size in the mosquito fish. *Evolution* 35, 941–953. (9)

Rhoades, D. F. 1979. Evolution of plant chemical defense against herbivores. In *Herbivores,* G. A. Rosenthal and D. H. Janzen (eds.). Academic Press, New York. pp. 4–54. (7)

Rhoades, D. F., and R. G. Cates. 1976. A general theory of plant anti-herbivore chemistry. In *Recent Advances in Phytochemistry,* Vol. 10: *Interactions between Plants and Insects,* J. W. Wallace and R. L. Mansell (eds.). Plenum Press, New York and London. pp. 168–213. (7)

Rhodes, I., and W. Harris. 1979. The nature and basis of differences in sward composition and yield in ryegrass—white clover mixtures. In *Changes in Sward Composition and Productivity,* A. H. Charles and R. J. Haggar (eds.). British Grassland Society, Hurley. pp. 55–60. (19)

Riedl, R. 1977. A systems-analytical approach to macro-evolutionary phenomena. *Q. Rev. Biol.* 52, 351–370. (1)

Riedl, R. 1978. *Order in Living Organisms: A Systems Analysis of Evolution.* Wiley, Chichester. (1)

Rietz, G. E. du. 1931. Life-forms of terrestrial flowering plants. 1. *Acta Phytogeog. Suec.* 3, 1–95. (1)

Riggs, T. J., P. R. Hanson, N. D. Start, D. M. Miles, C. L. Morgan, and M. A. Ford. 1981. Comparison of spring barley varieties grown in England and Wales between 1880 and 1980. *J. Agric. Sci.* 97, 599–610. (19)

Riley, R., and V. Chapman. 1957. Haploids and polyhaploids in *Aegilops* and *Triticum. Heredity* 11, 195–207. (1)

Rinard, R. G. 1981. The problem of the organic individual: Ernst Haeckel and the development of the biogenetic law. *J. Hist. Biol.* 14, 249–275. (1)

Risser, P. G. 1969. Competitive relationships among herbaceous grassland plants. *Bot. Rev.* 35, 251–284. (5)

Ritland, K. in press. Consanguineous systems of matings in inbred plant populations: The effective proportion of self-fertilization. *Genetics.* (6)

Ritterbusch, A. 1977. Homolog-und Analog-Modell einer Spermatophyten und einer terrestren Pflanze. *Ber. Deutsch. Bot. Ges.* 90, 363–368. (1)

Roberts, H. A. 1976. Weed competition in vegetable crops. *Ann. Appl. Biol.* 83, 321–324. (18)

Roberts, H. A. 1982. *Weed Control Handbook: Principles.* British Crop Protection Council, Blackwell, Oxford. (18)

Roberts, H. A., and M. E. Ricketts. 1979. Quantitative relationships between the weed flora after cultivation and the seed population in the soil. *Weed Res.* 19, 269–275. (18)

Robertson, A. 1982. Credibility of forecasting—the environment. *Chem. Ind. L.* 16, 589–593. (18)

Robinson, D. W. 1978. The challenge of the next generation of weed problems. *Proc. (1978) Brit. Crop Prot. Conf.-Weeds* 3, 799–821. (18)

Rockwood, L. L. 1973. The effect of defoliation on seed production of six Costa Rican tree species. *Ecology* 54, 1363–1369. (7)

Romberger, J. A. 1963. Meristems, growth and development in woody plants. *U.S. Dept. Agric. For. Serv. Tech. Bull.* 1293. (1)
Rosen, B. R. 1979. Modules, members and communes: A postscript introduction to social organisms. In *Biology and Systematics of Colonial Organisms,* G. Larwood and B. R. Rosen (eds.). Academic Press, London and New York. pp. xiii–xxxv. (1)
Rosenzweig, M. L. 1973. Evolution of the predator isocline. *Evolution* 27, 84–94. (5)
Ross, M. D. 1982. Five evolutionary pathways to subdioecy. *Amer. Nat.* 119, 296–318. (6)
Ross, M. D., and H. R. Gregorius. 1983. Outcrossing and sex function in hermaphrodites: A resource allocation model. *Amer. Nat.* 121, 204–222. (14)
Ross, M. D., and B. S. Weir. 1975. Maintenance of male sterility in plant populations. III. Mixed selfing and random mating. *Heredity* 35, 21–29. (6)
Roughgarden, J. 1979 *Theory of Population Genetics Evolutionary Ecology.* Macmillan, New York. (6)
Rudmik, T. 1983. *Morphological, Anatomical, and Physiological Changes in* Triglochin maritima *in Response to Changes in Salinity and Nitrogen.* MSc Thesis, University of Toronto. (17)
Russell, E. S. 1916. *Form and Function: A Contribution to the History of Animal Morphology.* John Murray, London. (1)
Sagar, G. R. 1959. *The Biology of Some Sympatric Species of Grassland.* D. Phil. thesis, University of Oxford. (4)
Sagar, G. R. 1974. On the ecology of weed control. In *Biology in Disease and Pest Control, Brit. Ecol. Symp. 13.,* D. Price Jones and M. E. Solomon (eds.). Blackwell, Oxford. pp. 42–58. (18)
Sagar, G. R., and A. M. Mortimer. 1976. The population biology of plants with special reference to weeds. *Appl. Biol.* 1, 1–43. (18)
Salisbury, E. J. 1942. *The Reproductive Capacity of Plants.* Bell and Sons Ltd., London. (4,8,9)
Salisbury, E. 1961. *Weeds and Aliens.* Collins, London. (12)
Sari Gorla, M., E. Ottaviano, and D. Faini. 1975. Genetic variability of gametophytic growth rate in maize. *Theor. Appl. Genet.* 46, 289–294. (13)
Sarukhán, J. 1978. Studies on the demography of tropical trees. In *Tropical Trees as Living Systems,* P. B. Tomlinson and M. H. Zimmerman (eds.). Cambridge University Press. pp. 163–184. (4)
Sarukhán, J. 1980. Demographic problems in tropical systems. In *Demography and Evolution in Plant Populations,* O. T. Solbrig (ed.). Blackwell, Oxford. pp. 161–188. (4)
Sarukhán, J., and M. Gadgil. 1974. Studies on plant demography: *Ranunculus repens* L., *R. bulbosus* L. and *R. acris* L. III. A mathematical model incorporating multiple modes of reproduction. *J. Ecol.* 62, 921–936. (4,18)
Sarukhán, J., and J. L. Harper. 1973. Studies on plant demography: *Ranunculus repens* L., *R. bulbosus* L. and *R. acris* L. I. Population flux and survivorship. *J. Ecol.* 61, 675–716. (1,4,7)
Sattler, R. 1974. A new conception of the shoot of higher plants. *J. Theor. Biol.* 47, 367–382. (1)
Schaaf, H. M., and R. R. Hill. 1979. Cross-fertility differentials in birdsfoot trefoil. *Crop Sci.* 19, 451–454. (13)

Schaal, B. A. 1975. Population structure and local differentiation in *Liatris cylindracea. Amer. Nat.* 109, 511–528. (5,11,13)
Schaal, B. A. 1978. Density dependent foraging on *Liatris pycnostachya. Evolution* 32, 452–454. (13)
Schaal, B. A. 1980a. Measurement of gene flow in *Lupinus texensis. Nature* 284, 450–451. (13)
Schaal, B. A. 1980b. Reproductive capacity and seed size in *Lupinus texensis Amer. J. Bot.* 67, 703–709. (4,9)
Schaal, B. A., and D. A. Levin. 1976. The demographic genetics of *Liatris cylindracea. Amer. Nat.* 110, 191–206. (1)
Schaal, B. A., and D. A. Levin. 1978. Morphological differentiation and neighbourhood size in *Liatris cylindracea. Amer. J. Bot.* 65, 923–928. (5)
Schaffer, W. M. 1974. Optimal reproductive effort in fluctuating environments. *Amer. Nat.* 108, 783–790. (8)
Schaffer, W. M. 1977. Some observations on the evolution of reproductive rate and competitive ability in flowering plants. *Theor. Pop. Biol.* 11, 90–104. (15)
Schaffer, W. M., and M. V. Schaffer. 1977. The adaptative significance of variations in reproductive habit in the Agavaceae. In *Evolutionary Ecology*, B. Stonehouse and C. M. Perrins (eds.). Macmillan, London. pp. 261–276. (8)
Schaffer, W. M., and M. V. Schaffer. 1979. The adaptative significance of variations in reproductive habit in the Agavaceae. II. Pollinator foraging behavior and selection for increased reproductive expenditure. *Ecology* 60, 1051–1069. (8)
Schellner, R. A., S. J. Newell, and O. T. Solbrig. 1982. Studies on the population biology of the genus *Viola.* IV. Spatial pattern of ramets and seedlings in three stoloniferous species. *J. Ecol.* 70, 273–290. (2)
Schemske, D. W. 1978. Evolution of reproductive characteristics in *Impatiens* (Balsaminaceae): The significance of cleistogamy and chasmogamy. *Ecology* 59, 596–613. (14)
Schemske, D. W. 1980. Floral ecology and hummingbird pollination of *Combretum farinosum* in Costa Rica. *Biotropica* 12, 169–181. (13)
Schemske, D. W. 1983. Limits to specialization and coevolution in plant-animal mutualisms. In *Coevolution,* M. Nitecki (ed.). University of Chicago Press, Chicago. pp. 67–109. (13)
Schemske, D. W. in press. Population structure and local selection in *Impatiens pallida* (Balsaminaceae), a selfing annual. *Evolution.* (8)
Schmalhansen, I. I. 1949. *Factors of Evolution.* McGraw-Hill, New York. (17)
Schmid, G. 1930. Goethes Metamorphose der Pflanzen. In *Goethe als Seher und Erforscher der Natur,* J. Walther (ed.). Kaiserlich-Leopodinische Deutsche Akademie der Naturforscher, Halle. pp. 205–319. (1)
Schmidt, K. P. 1982. *The Comparative Demography of Reciprocally Sown Populations of* Phlox drummondii *Hook.* Ph.D. thesis, University of Texas, Austin. (8)
Schmitt, J. 1980. Pollinator foraging behavior and gene dispersal in *Senecio* (Compositae). *Evolution* 34, 934–943. (13)

Schoen, D. J. 1982a. Male reproductive effort and breeding system in an hermaphroditic plant. *Oecologia* 53, 255–257. (14)
Schoen, D. J. 1982b. The breeding system of *Gilia achilleifolia:* Variation in floral characteristics and outcrossing rate. *Evolution* 36, 352–360. (13)
Schooler, A. B., A. R. Bell, and J. D. Nalewaja. 1972. Inheritance of siduron tolerance in foxtail barley. *Weed Sci.* 20, 167–172. (18)
Schutz, W. M., and S. A. Usanis. 1969. Intergenotypic interactions in plant populations. II. Maintenance of allelic polymorphisms with frequency-dependent selection and mixed selfing and random mating. *Genetics* 61, 875–891. (6)
Scott, K. R., and P. D. Putwain. 1981. Maternal inheritance of simazine resistance in a population of *Senecio vulgaris. Weed Res.* 21 137–143. (18)
Scott, K. R., and S. J. Wilcockson. 1976. Weed biology and the growth of sugar beet. *Ann. Appl. Biol.* 83, 331–335. (18)
Sculthorpe, C. D. 1967. *The Biology of Aquatic Vascular Plants.* Edward Arnold, London. (1)
Seaton, A. J. P., and J. Antonovics. 1967. Population interrelationships. I. Evolution in mixtures of *Drosophila* mutants. *Heredity* 22, 19–33. (5)
Service, P. M., and R. E. Lenski. 1982. Aphid genotypes, plant phenotypes, and genetic diversity: A demographic analysis of experimental data. *Evolution* 36, 1276–1282. (11)
Shafranova, L. M. 1980. (On metamerism and metameres in plants.) *Zur. Obshch. Biol.* 41, 437–447. (Russian) (1)
Sharitz, R. R., and J. F. McCormick. 1975. Population dynamics of two competing annual plant species. *Ecology* 54, 723–740. (4)
Shaver, G. R., F. S. Chapin, and W. D. Billings. 1979. Ecotypic differentiation in *Carex aquatilis* on ice-wedge polygons in the Alaskan coastal tundra. *J. Ecol.* 67, 1025–1046. (5)
Shaw, R. F., and J. D. Mohler. 1953. The selective significance of the sex ratio. *Amer. Nat.* 87, 337–342. (14)
Shaw, W. C., and K. Loustalot. 1963. Revolution in weed science. *Agric. Sci. Rev.* 30–47. (18)
Sheppard, P. M. 1975. *Natural Selection and Heredity.* Hutchinson University Lib., London. (5)
Shugart, J. R., and D. C. West. 1980. Forest succession models. *BioScience* 30, 308–313. (3)
Sibma, L., and J. Kort, and C. T. de Wit. 1964. Experiments on competition as a means of detecting possible damage by nematodes. *Jaarb. I.B.S.* 1964, 119–124. (7)
Silander, J. A., and J. Antonovics. 1979. The genetic basis of the ecological amplitude of *Spartina patens.* I. Morphometric and physiological traits. *Evolution* 33, 1114–1127. (8)
Silvertown, J. W. 1982a. No evolved mutualism between grasses and grazers. *Oikos* 38, 253–254. (4,17)
Silvertown, J. W. 1982b. *Introduction to Plant Population Ecology.* Longman, London. (4,9)
Simon, H. A. 1962. The architecture of complexity. *Proc. Amer. Phil. Soc.* 106, 467–482. (1)
Simmonds, N. W. 1964. The genetics of seed and tuber dormancy in the cultivated potatoes. *Heredity* 19, 489–504. (8)

Singh, R. S. 1972. *Genetic Variability and Selective Forces in Two Bulk Hybrid Populations of Barley.* Ph.D. thesis, University of California, Davis. (6)

Slatkin, M. 1977. Gene flow and genetic drift in a species subject to frequent local extinctions. *Theor. Pop. Biol.* 12, 253–263. (12)

Slatkin, M. 1981a. Estimating levels of gene flow in natural populations. *Genetics* 99, 323–335. (12)

Slatkin, M. 1981b. Populational heritability. *Evolution* 35, 859–871. (6)

Slatkin, M., and M. J. Wade. 1978. Group selection a quantitative character. *Proc. Nat. Acad. Sci. USA* 75, 3531–3534. (12)

Slobodkin, L. B. 1961. *Growth and Regulation of Animal Populations.* Holt, Rinehart and Winston, New York. (5)

Slocum, J. C. 1980. Differential susceptibility to grazers in two phases of an intertidal alga: advantages of heteromorphic generations. *J. Exp. Mar. Biol. Ecol.* 46, 99–110. (7)

Smirnova, E. S. 1970. (Morphological classification of flowering plants according to vegetative characteristics.) *Dokl. Akad. Nauk. SSSR* 190, 1243–1245. (Russian) (1)

Smirnova, O. V. 1967. (Ontogeny and age states of *Carex pilosa* Scop. and *Aegopodium podagraria* L. In *Ontogeny and Age State Composition of Flowering Plant Populations*), A. A. Uranov (ed.). Nauka, Moscow. pp. 100–113. (Russian) (4)

Smirnova, O. V. 1968. (Some peculiarities of vegetatively-movable plant life cycles.) *Uchoenyje Zapiski Permskogo Pedagogicheskogo Instituta* 64, 153–158. (Russian) (4)

Smith, B. H. 1983. Demography of *Floerkea proserpinacoides,* a forest-floor annual. 1. Density-dependent growth and mortality. *J. Ecol.* 71, 391–404. (1)

Smith, C. C. 1981. The facultative adjustment of sex ratio in lodgepole pine. *Amer. Nat.* 118, 297–305. (13)

Smith, D. L., and P. G. Rogan. 1980. Correlative inhibition in the shoot of *Agropyron repens* (L.) Beauv. *Ann. Bot.* 46, 285–296. (18)

Smith, R. A. H., and A. D. Bradshaw. 1979. The use of metal tolerant plant populations for the reclamation of metalliferous wastes. *J. Appl. Ecol.* 16, 595–612. (10)

Smith, W. K. 1978. Temperatures of desert plants: Another perspective on the adaptability of leaf size. *Science* 201, 614–616. (15)

Snaydon, R. W. 1962. The growth and competitive ability of contrasting natural populations of *Trifolium repens* L. on calcareous and acid soils. *J. Ecol.* 50, 439–447. (10)

Snaydon, R. W. 1963. The diversity and complexity of ecotypic differentiation within plant species in response to soil factors. *Proc. XI Inter. Genet. Cong. The Hague* 1, 1–43. (5)

Snaydon, R. W. 1970. Rapid population differentiation in a mosaic environment I. The response of *Anthoxanthum odoratum* populations to soils. *Evolution* 24, 257–269. (5)

Snaydon, R. W. 1978a. Indigenous species in perspective. *Proc. (1978) Brit. Crop Prot. Conf.—Weeds* 3, 905–914. (14)

Snaydon, R. W. 1978b. Genetic change in pasture populations. In *Plant Relations in Pastures,* J. R. Wilson (ed.). CSIRO, East Melbourne, Australia. pp. 253–269. (5,14)

Snaydon, R. W. 1980a. Plant demography in agricultural systems. In *Demography and Evolution of Plant Populations,* O. T. Solbrig (ed.). Blackwell, Oxford. pp. 131–160. (18,19)

Snaydon, R. W. 1980b. Ecological aspects of management—a perspective. In *Amenity Grassland: An Ecological Perspective,* I. H. Rorison and R. Hunt (eds.). Wiley, New York. pp. 219–231. (19)

Snaydon, R. W. 1982. Weeds and crop yield. *Proc. (1982) Brit. Crop Prot. Conf. —Weeds* 3, 729–739. (18)

Snaydon, R. W., and A. D. Bradshaw. 1961. Differential responses to calcium within the species *Festuca ovina* L. *New Phytol.* 60, 219–234. (10)

Snaydon, R. W., and A. D. Bradshaw. 1962. Differences between natural populations of *Trifolium repens* L. in response to mineral nutrients. I. Phosphate. *Jour. Exp. Bot.* 13, 422–424. (10)

Snaydon, R. W., and M. S. Davies. 1976. Rapid population differentiation in a mosaic environment. IV. Populations of *Anthoxanthum odoratum* at sharp boundaires. *Heredity* 37, 9–25. (5)

Snaydon, R. W., and T. M. Davies. 1982. Rapid divergence of plant populations in response to recent changes in soil conditions. *Evolution* 36, 289–297. (5)

Snoad, B. 1981. Plant form, growth rate and relative growth rate compared in conventional, semi-leafless and leafless peas. *Sci. Hort.* 14, 9–18. (1)

Sohn, J. J., and D. Policansky. 1977. The costs of reproduction in the mayapple *Podophyllum peltatum* (Berberidaceae). *Ecology* 58, 1366–1374. (2,8)

Solbrig, O. T. 1980. *Demography and Evolution in Plant Populations.* Blackwell, Oxford. (9)

Solbrig, O. T. 1981. Studies on the population biology of the genus *Viola.* II. The effect of plant size on fitness in *Viola sororia. Evolution* 35, 1080–1093. (4,13)

Solbrig, O. T., and B. B. Simpson. 1974. Components of regulation of a population of dandelions in Michigan. *J. Ecol.* 62, 473–486. (5,8)

Solbrig, O. T., and B. B. Simpson. 1977. A garden experiment on competition between biotypes of the common dandelion (*Taraxacum officinale*). *J. Ecol.* 65, 427–430. (8)

Solbrig, O. T., S. J. Newell, and D. T. Kincaid. 1980. The population biology of the genus *Viola.* The demography of *Viola sororia. J. Ecol.* 68, 521–546. (4)

Sørensen, T. 1954. Adaptation of small plants to deficient nutrition and a short growing season illustrated by cultivation experiments with *Capsella bursa-pastoris* (L.). *Botaniska Tidsskrift* 51, 339–361. (10)

Southward, E. C. 1980. Regionation and metamerisation in Pogonophora. *Zool. Jb. Anat.* 103, 264–275. (1)

Southwood, T. R. E. 1962. The numbers of species of insects associated with various trees. *J. Anim. Ecol.* 30, 1–8. (7)

Southwood, T. R. E. 1981. The rise and fall of ecology. *New Sci.* 92, 512–514. (18)

Southwood, T. R. E., and G. A. Norton. 1973. Economic aspects of pest management strategies and decisions. In *Insects: Studies in Pest Management,* P. Geier et al. (eds.). Mem. Ecological Soc. Aust., Canberra. pp. 168–184. (18)
Sparnaaij, L. D., Y. O. Kho, and J. Baer. 1968. Investigations on seed production in tetraploid freesias. *Euphytica* 17, 289-297. (13)
Spieth, P. T. 1974. Gene flow and genetic differentiation. *Genetics* 78, 961-965. (12)
Spitters, C. J. T. 1979. Competition and its consequences for selection in barley breeding. *Center for Agric. Publ. Docum.,* Wageningen. (6)
Squillace, A. E., and R. J. Bingham. 1958. Selective fertilization in *Pinus monticola* L. *Silvae. Genet.* 7, 188–196. (13)
Staniforth, R. J., and P. B. Cavers. 1976. An experimental study of water dispersal in *Polygonum* spp. *Can. J. Bot.* 54, 2587–2596. (13)
Stanley, S. M. 1973. An ecological theory for the sudden origin of multicellular life in the late Precambrian. *Proc. Nat. Acad. Sci. USA* 70, 1486–1489. (7)
Stearns, S. C. 1976. Life-history tactics: A review of the ideas. *Quart. Rev. Ecol.* 51, 3–47. (7)
Stearns, S. C. 1977. The evolution of life history traits. *Ann. Rev. Ecol. Syst.* 8, 145–172. (9)
Stearns, S. C. 1980. A new view of life-history evolution. *Oikos* 35, 266–281. (8)
Stearns, S. C. 1982. Evolution of life histories. In *Evolution and Development,* J. T. Bonner (ed.). Springer-Verlag, Heidelberg and New York. pp. 238–258. (17)
Stebbins, G. L. 1950. *Variation and Evolution in Plants.* Columbia University Press, New York. (1)
Stebbins, G. L. 1970. Transference of function as a factor in the evolution of seeds and their accessory structures. *Israel J. Bot.* 19, 59–70. (1)
Stebbins, G. L. 1980. Botany and the synthetic theory of evolution. In *The Evolutionary Synthesis,* E. Mayr and W. B. Provine (eds.). Harvard University Press, Cambridge, Mass. pp. 139–152. (1)
Stebbins, G. L., and K. Daly. 1961. Changes in the variation pattern of a hybrid population of *Helianthus* over an eight-year period. *Evolution* 15, 60–71. (5)
Steenbergh, W. F., and C. H. Lowe. 1977. *Ecology of the Saguaro:* II. *Reproduction, Germination, Growth, and Survival of the Young Plant.* National Park Service Scientific Monograph Series. No. 8. (4)
Stevens, P. S. 1974. *Patterns in Nature.* Atlantic-Little, Brown, Boston. (2)
Stinson, H. T., and D. N. Moss. 1960. Some effects of shade on corn hybrids tolerant and intolerant of dense planting. *Agron. Jour.* 52, 482–484. (19)
Stobbs, T. H. 1973. The effect of plant structure on the intake of tropical pastures. II. Difference in sward structure, nutritive value and bite size of animals grazing *Setaria anceps* and *Chloris gayana* at various stages of growth. *Austr. J. Agric. Res.* 24, 821–829. (18)
Stringam, G. R., and R. K. Downey. 1978. Effectiveness of isolation distance in turnip rape. *Can. J. Plant Sci.* 58, 427–434. (12)

Sulzbach, D. S. 1980. Selection for competitive ability: Negative effects in *Drosophila. Evolution* 34, 431–436. (5)

Suneson, C. A. 1969. Evolutionary plant breeding. *Crop Sci.* 9, 119–121. (11)

Suneson, C. A., and G. A. Weibe. 1942. Survival of barley and wheat varieties in mixtures. *J. Amer. Soc. Agron.* 34, 1052–1056. (10)

Sylven, N. 1937. The influence of climatic conditions on type composition. *Imp. Bur. Plant Gen. Herb. Bull.* 21, 8. (10)

Syme, J. R. 1972. Single-plant characters as a measure of field plot performance of wheat cultivars. *Aust. J. Agric. Res.* 23, 753–760. (19)

Symonides, E. 1977. Mortality of seedlings in natural psammophyte populations. *Ekol. Pol.* 25, 635–651. (4)

Tachenouchi, Y. 1926. On the rhizome of Japanese bamboos. *Trans. Nat. Hist. Soc. Formosa* 16, 37–46. (2)

Tamm, C. O. 1956. Further observations on the survival and flowering of some perennial herbs: 1. *Oikos* 7, 274–292. (4)

Tanton, T. W. 1981. Growth and yield of the tea bush. *Expl. Agric.* 17, 323–331. (1)

Tansley, A. G., and R. S. Adamson. 1925. Studies on the vegetation of the English chalk. III. The chalk grasslands of the Hampshire-Sussex border. *J. Ecol.* 13, 177–223. (7)

Taylor, P. D. 1981. Intra-sex and inter-sex sibling interactions as sex ratio determinants. *Nature* 291, 64–66. (14)

Taylor, P. D., and M. G. Bulmer. 1980. Local mate competition and the sex ratio. *J. Theor. Biol.* 86, 409–419. (13)

Teeri, J. A. 1972. *Microenvironmental Adaptations of* Saxifraga oppositifolia *in the High Arctic.* Ph.D. thesis, Duke University, Durham, N.C. (5)

Templeton, A. R. 1982. The prophecies of parthenogenesis. In *Evolution and Genetics of Life Histories,* H. Dingle and J. P. Hegmann (eds.). Springer-Verlag, New York. pp. 75–101. (13)

Templeton, A. R., and D. A. Levin. 1979. Evolutionary consequences of seed pools. *Amer. Nat.* 114, 232–249. (13)

Teramura, A. H., and B. R. Strain. 1979. Localized populational differences in the photosynthetic response to temperature and irradiance in *Plantago lanceolata. Can. J. Bot.* 57, 2559–2563. (5)

Ter Borg, S. J. 1979. Some topics in plant population biology. In *The Study of Vegetation,* M. J. A. Werger (ed.). Dr. W. Junk, London. pp. 13–55. (12)

Thompson, D. W. 1917. *On Growth and Form.* Cambridge University Press, Cambridge. (1)

Thompson, K., and J. P. Grime. 1979. Seasonal variation in the seed banks of herbaceous species in ten contrasting habitats. *J. Ecol.* 67, 893–922. (18)

Thomson, J. D. 1981. The spatial and temporal components of resource assessment by flower feeding insects. *J. Anim. Ecol.* 50, 49–59. (2)

Thomson, J. D., and S. C. H. Barrett. 1981. Selection for outcrossing, sexual selection, and the evolution of dioecy in plants. *Amer. Nat.* 118, 443–449. (6)

Thomson, J. D., W. P. Maddison, and R. C. Plowright. 1982. Behavior of bumblebee pollinators of *Aralia hispida* Vent. (Araliaceae). *Oecologia* 54, 326–336. (13,14)

Thorne, G. 1974. Physiology of grain yield in wheat and barley. *Rep. Rothamsted Exp. Stat.* 1973, Part 2, 5–25. (19)

Thorne, J. H., and H. R. Koller. 1974. Influence of assimilate demand on photosynthesis, diffusive resistances, translocation, and carbohydrate levels of soybean leaves. *Plant Physiol.* 54, 201–207. (15)

Thornley, J. H. M. 1972. A balanced quantitative model for root:shoot ratios in vegetative plants. *Ann. Bot.* 36, 431–441. (15)

Throughton, A. 1956. Studies on the growth of young grass plants with special reference to the relationship between the root and shoot systems. *J. Brit. Grassld. Soc.* 11, 56–65. (15)

Thurston, J. M. 1957. Morphological and physiological variation in wild oats (*Avena fatua* L. and *A. ludoviciana* Dur.) and in hybrids between wild and cultivated oats. *J. Agric. Sci. Camb.* 49, 260–274. (17)

Tietema, T., and J. Vroman. 1978. Ecophysiology of the sand sedge, *Carex arenaria* L. I. Growth and dry matter distribution. *Acta Bot. Neerl.* 27, 161–173. (2)

Tomlinson, P. B. 1970. Monocotyledons—toward an understanding of their morphology and anatomy. *Adv. Bot. Res.* 3, 207–292. (1)

Tomlinson, P. B. 1974. Vegetative morphology and meristem dependence—the foundation of productivity in seagrasses. *Aquaculture* 4, 107–130. (1,2)

Tomlinson, P. B. 1982. Chance and design in the construction of plants. In *Axioms and Principles of Plant Construction,* R. Sattler (ed.). Martinus Nijhoff and Dr. W. Junk Publ, The Hague. pp. 162–183. (1)

Tomlinson, P. B., and P. Soderholm. 1975. *Corypha elata* in flower in South Florida. *Principes* 19, 83–99. (2)

Tomlinson, P. B., and M. H. Zimmermann. 1978. *Tropical Trees as Living Systems.* Cambridge University Press, Cambridge. (1)

Trewavas, A. 1981. How do plant growth substances work? *Plant, Cell Envir.* 4, 203–228. (1)

Tso, T. C. 1972. *Physiology and Biochemistry of Tobacco Plants.* Dowden, Hutchinson and Ross, Stroudsburg, Pa. (15)

Tuomi, J., P. Niemela, and R. Mannila. 1982. Resource allocation on dwarf shoots of birch (*Betula pendula*): Reproduction and leaf growth. *New Phytol.* 91, 483–487. (1)

Tuomi, J., J. Salo, E. Haukioja, P. Niemela, T. Hakala, and R. Mannila. in press. The existential game of individual self-maintaining units: Selection and defense tactics of trees. *Oikos.* (1)

Turesson, G. 1922. The genotypical response of the plant species to the habitat. *Hereditas* 3, 211–350. (10)

Turesson, G. 1925. The plant species in relation to habitat and climate. *Hereditas* 6, 147–236. (10)

Turkington, R. 1979. Neighbour relationships in grass-legume communities. IV. Fine scale biotic differentiation. *Can. J. Bot.* 57, 2711–2716. (5)

Turkington, R. 1983. Leaf and flower demography of *Trifolium repens* L. 1. Growth in mixture with grasses. *New Phytol.* 93, 599–616. (1)

Turkington, R., and J. J. Burdon. 1983. The biology of Canadian weeds. *Trifolium repens* L. *Can. Jour. Pl. Sci.* 63, 243–266. (5)

Turkington, R., and J. L. Harper. 1979a. The growth, distribution and neighbour relationships of *Trifolium repens* in a permanent pasture. I. Ordination, pattern and contact. *J. Ecol.* 67, 201–218. (4,5)

Turkington, R., and J. L. Harper. 1979b. The growth, distribution and neighbour relationships of *Trifolium repens* in a permanent pasture. IV. Finescale biotic differentiation. *J. Ecol.* 67, 245–254. (4,5,10,11,13)

Turner, M. E., J. C. Stephens, and W. W. Anderson. 1982. Homozygosity and patch structure in plant populations as a result of nearest-neighbor pollination. *Proc. Nat. Acad. Sci. USA* 79, 203–207. (6)

Ulam, S. 1966. Patterns of growth of figures: Mathematical aspects. In *Module, Symmetry, Rhythm, k 64.74,* G. Kepes (ed.). Braziller, New York. (2)

Uranov, A. A. 1960. (Living conditions of species in plant associations.) *Bulletin Moscovscogo obteshestva ispytatelej prirody otdel biologicheskij* 64, 77–92. (Russian) (4)

Uranov, A. A. 1975. (Age spectrum of the phytocoenopopulation as a function of time and energetic wave processes.) *Biologicheskie Nauki* 2, 7–34. (Russian) (4)

van den Bergh, *see* Bergh

Vanderlip, R. L., and G. F. Arkin. 1977. Simulating accumulation of dry matter in grain sorghum. *Agron. J.* 69, 917–923. (15)

van der Maarel, *see* Maarel

VanderWall, S. B., and R. P. Balda. 1977. Coadaptations of the Clark's nutcracker and the pinyon pine for efficient seed harvest and dispersal. *Ecol. Monogr.* 47, 89–111. (13)

Van Valen, L. 1975. Life, death and energy of a tree. *Biotropica* 7, 260–269. (4)

Varley, G. C., and G. B. Gradwell. 1960. Key factors in population studies. *J. Anim. Ecol.* 29, 399–401. (18)

Vasek, F. C. 1967. Outcrossing in natural populations. III. The Deer Creek population of *Clarkia exilis. Evolution* 21, 241–248. (13)

Vasek, F. C. 1980. Creosote bush-long-lived clones in the Mojave desert. *Amer. J. Bot.* 67, 246–255. (1)

Vázquez-Yanes, C. 1981. Notas sobre la autoecología de los árboles pioneros de la selva tropical lluviosa. *Trop. Ecol.* 21, 103–112. (4)

Venable, D. L., and L. Lawlor. 1980. Delayed germination and dispersal in desert annuals: Escape in space and time. *Oecologia* 46, 272–282. (8)

Venable, D. L. and D. A. Levin. 1983. Morphological dispersal structures in relation to growth habit in the Compositae. *Plant Syst. Evol.* 134, 1–16. (8)

Venable, D. L. and D. A. Levin. in press a. Evolutionary ecology of achene dimorphism in *Heterotheca latifolia.* I. Achene structure, germination and dispersal. *J. Ecol.* (8)

Venable, D. L. and D. A. Levin. in press b. Evolutionary ecology of achene dimorphism in *Heterotheca latifolia.* II. Intrapopulational demographic variation. *J. Ecol.* (8)

Vere, D. T., and B. A. Auld. 1982. The cost of weeds. *Prot. Ecol.* 4, 29–42. (18)

Verheij, E. W. M. 1970. Spacing experiments with Brussels sprouts grown for single-pick harvests. *Neth. J. Agric. Sci.* 18, 89–104. (1)

Vickery, R. K. 1978. Case studies in the evolution of species complexes in *Mimulus. Evol. Biol.* 11, 405–507. (12)

Vogel, S. 1970. Convective cooling at low airspeeds and the shapes of broad leaves. *J. Exp. Bot.* 21, 91–101. (15)

Voight, P. W., and H. W. Brown. 1969. Phenotypic recurrent selection for seedling vigor in side-oats gramma *Bouteloua curtipendula. Crop Sci.* 9, 664–666. (4)

Waddington, K. D. 1981. Factors influencing pollen flow in bumblebee-pollinated *Delphinium virescens. Oikos* 37, 153–159. (13)

Wade, M. J., and D. E. McCauley. 1980. Group selection: The phenotypic and genotypic differentiation of small populations. *Evolution* 34, 799–812. (6)

Walker, D. I. T., and N. W. Simmonds. 1981. Comparisons of the performance of sugarcane varieties in trials and in agriculture. *Exp. Agric.* 17, 137–144. (19)

Wall, G., G. McBoyle, and T. Mock. 1977. The nature of weeds. *Bull. Conserv. Coun. Ontario* 24, 13–19. (18)

Wall, G., G. McBoyle, and T. Mock. 1979. Perception of weeds as an agricultural hazard. *Ontario Geography* 14, 5–19. (18)

Wallace, L. L., S. J. McNaughton, and M. B. Coughenour. 1982. The effects of clipping and fertilization on nitrogen nutrition and allocation by mycorrhizal and nonmycorrhizal *Panicum coloratum* L., a C4 grass. *Oecologia* 54, 68–71. (15)

Waller, D. M. 1980. Environmental determinants of outcrossing in *Impatiens capensis* (Balsaminaceae). *Evolution* 34, 747–761. (6)

Waller, D. M., and D. A. Steingraeber. in press. Branching and modular growth: Theoretical models and empirical patterns. In *The Biology of Clonal Organisms,* J. Jackson, R. Cook, and L. Buss (eds.). Yale University Press. (1,2)

Walley, K. A., M. S. I. Khan, and A. D. Bradshaw. 1974. The potential for evolution of heavy metal tolerance in plants. I. Copper and zinc tolerance in *Agrostis tenuis. Heredity* 32, 309–319. (10)

Wallgren, B. 1980. Weed control in spring cereals. *21st Swedish Weed Conf.* 18–23. (18)

Wang, C. W., T. O. Perry, and A. G. Johnson. 1960. Pollen dispersion of slash pine (*Pinus elliottii* Engelm.) with special reference to seed orchard management. *Sylvae. Genet.* 9, 78–86. (13)

Wann, M., and C. D. Raper, Jr. 1979. A dynamic model for plant growth: Adaptation for vegetative growth of soybeans. *Crop Sci.* 19, 461–467. (15)

Wann, M., C. D. Raper, Jr., and H. L. Lucas Jr. 1978. A dynamic model for plant growth: A simulation of dry matter accumulation in tobacco. *Photosynthetica* 12, 121–136. (15)

Wapshere, A. J. 1982. Priorities in the selection of agents for the biological control of weeds. *Proc. (1982) Brit. Crop Prot. Conf.—Weeds* 3, 779–785. (18)

Wardlaw, C. W. 1965. *Organization and Evolution in Plants.* Longman and Green, London. (17)

Wardlaw, C. W. 1968. *Morphogenesis in Plants: A Contemporary Study.* Methuen, London. (1)

Wareing, P. F. 1977. Growth substances and integration in the whole plant. In *Integration of Activity in the Higher Plant,* D. H. Jennings (ed.). Cambridge University Press, Cambridge. pp. 337–365. (17)

Wareing, P. F., and J. Patrick. 1975. Source-sink relations and the partition of assimilates in the plant. In *Photosynthesis and Productivity in Different Environments,* J. P. Cooper (ed.). Cambridge University Press, Cambridge. pp. 481–499. (15)

Warwick, S. I. 1980. The genecology of lawn weeds. VII. The response of different growth forms of *Plantago major* L. and *Poa annua* L. to simulated trampling. *New Phytol.* 85, 461–469. (5)

Warwick, S. I., and L. Black. 1980. Uniparental inheritance of atrazine in *Chenopodium album* L. *Can. J. Plant. Sci.* 60, 751–756. (18)

Warwick, S. I., and D. Briggs. 1978a. The genecology of lawn weeds. I. Evidence for disruptive selection in *Poa annua* L. in a mosaic environment of bowling green lawns and flower beds. *New Phytol.* 81, 711–723. (8)

Warwick, S. I., and D. Briggs. 1978b. The genecology of lawn weeds. II. Evidence for disruptive selection in *Poa annua* L. in a mosaic environment of bowling green lawns and flower beds. *New Phytol.* 81, 725–737. (5,8)

Warwick, S. I., and D. Briggs. 1979. The genecology of lawn weeds. III. Cultivation experiments with *Achillea millefolium* L., *Bellis perennis* L., *Plantago lanceolata* L., *Plantago major* L., and *Prunella vulgaris* L., collected from lawns and contrasting grassland habitats. *New Phytol.* 83, 509–536. (1,5,12)

Warwick, S. I., and D. Briggs. 1980a. The genecology of lawn weeds. V. The adaptive significance of different growth habits in lawn and roadside populations of *Plantago major* L. *New Phytol.* 85, 289–300. (12)

Warwick, S. I., and D. Briggs. 1980b. The genecology of lawn weeds. VI. The adaptive significance of variation in *Achillea millefolium* L. as investigated by transplant experiments. *New Phytol.* 85, 451–460. (15)

Warwick, S. I., and P. B. Marriage. 1982a. Geographical variation in populations of *Chenopodium album* resistant and susceptible to atrazine. I. Between- and within-population variation in growth and response to atrazine. *Can. J. Bot.* 60, 483–493. (12)

Warwick, S. I., and P. B. Marriage. 1982b. Geographical variation in populations of *Chenopodium album* resistant and susceptible to atrazine. II. Photoperiod and reciprocal transplant studies. *Can. J. Bot.* 60, 494–504. (12)

Waser, N. M., and M. V. Price. in press. Optimal and actual outcrossing in plants, and the nature of plant-pollinator interaction. In *Handbook of Experimental Pollination Ecology,* C. E. Jones, and R. J. Little (eds.). Van Nostrand, New York. (6)

Waser, N. M., R. K. Vickery, and M. V. Price. 1982. Patterns of seed dispersal and population differentiation in *Mimulus guttatus. Evolution* 36, 753-761. (5,13)

Watkinson, A. R. 1980. Density dependence in single species populations of plants. *J. Theor. Biol.* 83, 345–357. (18)

Watkinson, A. R. 1981. Interference in pure and mixed populations of *Agrostemma githago. J. Appl. Ecol.* 18, 967–976. (15,18)

Watson, M. A. in press. Developmental constraints: Effect on population growth of resource allocation in a clonal plant. *Amer. Nat.* (1)

Watson, M., and B. Casper. in press. Branch autonomy, patterns of vascular transport, and the question of resource allocation in plants. *Ann. Rev. Ecol. Syst.* (1)

Watson, P. J. 1969. Evolution in closely adjacent plant populations. VI. An entomophilous species, *Potentilla erecta,* in two contrasting habitats. *Heredity* 24, 407–422. (5)

Watt, A. S. 1947. Pattern and process in the plant community. *J. Ecol.* 35, 1–22. (17)

Way, M., and M. E. Cammell. 1981. Effects of weeds and weed control in invertebrate pest ecology. In *Pests, Pathogens and Vegetation,* J. M. Thresh (ed.). Pitman, London. pp. 443–458. (18)

Webb, S. L. 1982. Long-distance dispersal: The Holocene extension of the range of beech *(Fagus grandifolia)* into Wisconsin. *Abst. Amer. Quater. Assoc. Program* 7, 177. (13)

Welbank, P. J. 1961. A study of the nitrogen and water factor in competition with *Agropyron repens* (L.) Beauv. *Ann. Bot.* 25, 116–137. (18)

Werner, P. A. 1975. Predictions of fate from rosette size in teasel (*Dipsacus fullonum* L.). *Oecologia* 20, 197–201. (4,15)

Werner, P. A., and H. Caswell. 1977. Population growth rates and age vs. stage-distribution models for teasel (*Dipsacus sylvestris* Huds.). *Ecology* 58, 1103–1111. (4)

Werner, P. A., and W. J. Platt. 1976. Ecological relationships of co-occurring golden rods (*Solidago,* Compositae). *Amer. Nat.* 110, 959–971. (3)

Werner, P. A., and R. Rioux. 1977. The biology of Canadian weeds 24. *Agropyron repens* (L.) Beauv. *Can. J. Plant Sci.* 57, 905–919. (18)

West, D. C., H. H. Shugart, and D. B. Botkin (eds.). 1981. *Forest Succession: Concepts and Applications.* Springer, New York. (3)

West, N. E., K. H. Rea, and R. O. Harviss. 1979. Plant demographic studies in sagebrushgrass communities of southeastern Idaho. *Ecology* 60, 376–388. (4)

Westhoff, V., and E. van de Maarel. 1978. The Braun-Blanquet approach. In *Classification of Plant Communities,* R. H. Whittaker (ed.). Junk, The Hague. pp. 287–399. (3)

Westoby, M. 1981. How diversified seed germination behavior is selected. *Amer. Nat.* 118, 882–885. (5)

Westoby, M., and B. Rice. 1982. Evolution of the seed plants and inclusive fitness of plant tissues. *Evolution* 36, 713–724. (4)

White, J. 1979. The plant as a metapopulation. *Ann. Rev. Ecol. Syst.* 10, 109–145. (1)

White, J. 1980. Demographic factors in populations of plants. In *Demography and Evolution in Plant Populations,* O. T. Solbrig (ed.). Blackwell, Oxford. pp. 21–48. (1,4)

White, J. 1981. The allometric interpretation of the self-thinning rule. *J. Theor. Biol.* 89, 475–500. (4)

White, J., and J. L. Harper. 1970. Correlated changes in plant size and number in plant populations. *J. Ecol.* 58, 467–485. (4)

White, P. S. 1979. Pattern, process, and natural disturbance in vegetation. *Bot. Rev.* 45, 229–299. (2,7)

Whitehead, W. L., and R. L. Davis. 1954. Self- and cross-compatibility in alfalfa, *Medicago sativa. Agron. J.* 46, 452–456. (13)

Whitham, T. G. 1983. Host manipulation of parasites: Within-plant variation as a defense against rapidly evolving pests. In *Variable Plants and Herbivores in Natural and Managed Systems,* R. F. Denno and M. S. McClure (eds.). Academic Press, London and New York. pp. 15–41. (1)

Whitham, T. G., and C. N. Slobodchikoff. 1981. Evolution by individuals, plant-herbivore interactions, and mosaics of genetic variability: The adaptive significance of somatic mutations in plants. *Oecologia* 49, 287–292. (1,13,17)

Whitney, G. G. 1982. A demographic analysis of the leaves of open and shade grown *Pinus strobus* L. and *Tsuga canadensis* (L.) Carr. *New Phytol.* 90, 447–453. (1)

Whittaker, J. B. 1979. Invertebrate grazing: Competition and plant dynamics. In *Population Dynamics,* R. M. Anderson, B. D. Turner and L. R. Taylor (eds.). Blackwell, Oxford. pp. 207–222. (7)

Whittaker, R. H. 1960. Vegetation of the Siskiyou Mountains, Oregon and California. *Ecol. Monogr.* 30, 279–338. (3)

Whittaker, R. H. 1962. Classification of natural communities. *Bot. Rev.* 28, 1–239. (3)

Whittaker, R. H. 1965. Dominance and diversity in land plant communities. *Science* 147, 250–260. (3)

Whittaker, R. H. 1967. Gradient analysis of vegetation. *Biol. Rev.* 42, 207–264. (3)

Whittaker, R. H. 1969. Evolution of diversity in plant communities. *Biology* 22, 178–196. (3)

Whittaker, R. H. (ed.). 1978a. *Classification of Plant Communities.* Junk, The Hague. (3)

Whittaker, R. H. (ed.). 1978b. *Ordination of Plant Communities.* Junk, The Hague. (3)

Whittaker, R. H. 1978c. Direct gradient analysis. In *Ordination of Plant Communities,* R. H. Whittaker (ed.). Junk, The Hague. pp. 5–50. (3)

Wiens, J. A. 1977. On competition and variable environments. *Am. Sci.* 65, 590–597. (5)

Wilbur, H. M. 1976. Life history evolution in seven milkweeds of the genus *Asclepias. J. Ecol.* 64, 223–240. (8)

Wilbur, H. M. 1977. Propagule size, number, and dispersion pattern in *Ambystoma* and *Asclepias. Amer. Nat.* 111, 43–68. (8)

Wilcox, J. R., and K. A. Taft. 1969. Genetics of yellow poplar *(Liriodendron). USDA For. Serv. Res. Pap.* WO-6. (13)

Wilkins, D. A. 1959. Sampling for genecology. *Rep. Scottish Plant Breed. Stat.* 1959, 92–96. (17)

Willey, R. W. 1982. Plant population and crop yield. In *Handbook of Agricultural Productivity,* I. M. Rechcigl (ed.). C. R. C. Press, Florida. pp. 201–207. (19)

Willey, R. W., and S. B. Heath. 1969. The quantitative relationships between plant population and crop yield. *Adv. Agron.* 21, 281–321. (18,19)

Williams, E. D. 1970. Effects of decreasing light intensity on the growth of *Agropyron repens* (L.) Beauv. in the field. *Weed Res.* 10, 360–366. (18)

Williams, G. C. 1975. *Sex and Evolution.* Princeton Unviersity Press, Princeton. (6,11)

Williams, J. T., and J. L. Harper. 1965. Seed polymorphism and germination. I. The influence of nitrates and low temperatures on the germination of *Chenopodium album. Weed Res.* 5, 141–150. (17)
Williams, R. D., and G. Evans. 1935. The efficiency of spatial isolation in maintaining the purity of red clover. *Welsh J. Agric.* 11, 164–171. (12)
Williams, W. E. 1983. Optimal water-use efficiency in a California shrub. *Plant, Cell, Environ.* 6, 145–151. (15)
Williamson, G. B. 1982. Plant mimicry: Evolutionary constraints. *Biol. J. Linn. Soc.* 18, 49–58. (1)
Williamson, M. H. 1958. Selection, controlling factors, and polymorphism. *Amer. Nat.* 92, 329–335. (11)
Willson, M. F. 1979. Sexual selection in plants. *Amer. Nat.* 113, 777–790. (14)
Willson, M. F. 1983. *Plant Reproductive Ecology.* Wiley-Interscience, New York. (13)
Willson, M. F. 1984. *Vertebrate Natural History.* Saunders, Philadelphia. (13)
Willson, M. F., and N. Burley. 1983. *Mate Choice in Plants: Mechanisms, Tactics and Consequences.* Princeton University Press, Princeton, N.J. (13)
Willson, M. F., K. P. Ruppel. in press. Resource allocation and floral sex ratios in *Zizania aquatica* L. *Can. J. Bot.* (13)
Wilman, D., and A. A. Mohamed. 1980. Response to nitrogen application and interval between harvests in five grasses. I. Dry matter yield, nitrogen content, number and weight of tillers, and proportion of crop fractions. *Fert. Res.* 1, 245–263. (19)
Wilson, B. J. 1978. The long term decline of a population of *Avena fatua* L. with different cultivations associated with spring barley cropping. *Weed Res.* 18, 25–31. (18)
Wilson, B. J., and G. W. Cussans. 1975. A study of the population dynamics of *Avena fatua* L. as influenced by straw burning, seed shedding, and cultivations. *Weed Res.* 15, 249–258. (18)
Wilson, E. O. 1971. *The Insect Societies.* The Belknap Press of Harvard University Press, Cambridge, Mass. (2)
Wilson, E. O., and W. H. Bossert. 1971. *A Primer of Population Biology.* Sinauer Associates, Sunderland, Mass. (16)
Wilson, J. R. 1978. *Plant Relations in Pastures.* CSIRO, Melbourne. (5,11)
Wit, C. T. de. 1960. On competition. *Versl. Landbouwk. Onderz.* 66, 1–82. (5,6)
Witcombe, J. R., and W. J. Whittington. 1972. The effects of selection for reduced dormancy in charlock *(Sinapis arvensis). Heredity* 29, 37–49. (8)
Woessner, R. A. 1972. Crossing among loblolly pines indigenous to different areas as a means of genetic improvement. *Silvae. Genet.* 21, 35–39. (12)
Woldenberg, M. J. 1968. Energy flow and spatial order—mixed hexagonal hierarchies of central places. *Geogr. Rev.* 58, 552–574. (2)
Wolfenbarger, D. O. 1946. Dispersion of small organisms, distance dispersion rates of bacteria, spores, seeds, pollen, and insects: Incidence rates of diseases and injuries. *Am. Midl. Nat.* 35, 1–152. (13)

Woodell, S. J. 1978. Directionality in bumblebees in relation to environmental factors. *Linn. Soc. Symp. Ser.* 6, 31–39. (13)

Wool, D., and S. Mendlinger. 1981. Genetical and ecological consequences of subdivision in *Tribolium* populations. *J. Anim. Ecol.* 50, 421–433. (13)

Woolley, D. J., and P. F. Wareing. 1972. The role of roots, cytokinins and apical dominance in the control of lateral shoot form in *Solanum andigena. Planta* 195, 33–42. (17)

Wright, J. W. 1952. Pollen dispersion of some forest trees. *Northeast Forest Exper. Station Pap.* 46. (13)

Wright, J. W. 1962. *Genetics of Forest Tree Improvement.* F. A. O., Rome. (10)

Wright, S. 1931. Evolution in Mendelian populations. *Genetics* 16, 97–159. (12,17)

Wright, S. 1940. Breeding structure of populations in relation to speciation. *Amer. Nat.* 74, 232–248. (13)

Wright, S. 1943. Isolation by distance. *Genetics* 28, 114–138. (12)

Wright, S. 1948. On the roles of directed and random changes in gene frequency in the gametes of populations. *Evolution* 2, 279–295. (12)

Wright, S. 1969. *Evolution and Genetics of Populations.* Vol. 2. University of Chicago Press, Chicago. (6)

Wright, S. 1970. Random drift and the shifting balance theory of evolution. In *Mathematical Topics in Population Genetics,* K. Kojima (ed.). Springer-Verlag, New York. pp. 1–30. (6)

Wu, K. K., and S. K. Jain. 1979. Population regulation in *Bromus rubens* and *B. mollis:* Life cycle components and competition. *Oecologia* 39, 337–357. (6)

Wu, L., and J. Antonovics. 1976. Experimental ecological genetics in *Plantago.* II. Lead tolerance in *Plantago lanceolata* and *Cynodon dactylon* from a roadside. *Ecology* 57, 205–208. (5)

Yabuno, T. 1966. Biosystematic study of the genus *Echinochloa. Jap. J. Bot.* 19, 277–323. (18)

Yadav, A. S., and R. S. Tripathi. 1981. Population dynamics of the ruderal weed *Eupatorium odoratum* and its natural regulation. *Oikos* 36, 355–361. (4)

Yazdi-Samadi, B., K. K. Wu, and S. K. Jain. 1978. The role of genetic polymorphisms in the outcome of interspecific competition in *Avena. Genetics* 48, 151–159. (6)

Yoda, K., T. Kira, H. Ogawa, and K. Hozumi. 1963. Self-thinning in overcrowded pure stands under cultivated and natural conditions. *J. Biol. Osaka Cy. Univ.* 14, 107–129. (4)

Zadoks, J. C. 1971. Systems analysis and the dynamics of epidemics. *Phytopathol.* 61, 600–610. (18)

Zhukova, L. A. 1961. (The peculiarities of age changes in tussock hairgrass.) *Morphogenesis of Plants* 2, 635–638. (Russian) (4)

Zimdahl, R. L. 1980. *Weed-Crop Competition—A Review.* IPCC Oregon State University, Oregon. (18)

Zimmerman, M. 1982. The effect of nectar production on neighborhood size. *Oecologia* 52, 104–108. (13)

Zimmermann, M. H. 1978. Hydraulic architecture of diffuse-porous trees. *Can. J. Bot.* 56, 2286–2295. (1)

INDEX

Abies balsamea, 34, 84
Abies veichtii, 103–104
Abutilon theophrasti, 332, 333
Acer saccharum, 226, 271
achenes, germination strategies and, 182–186
Achillea millefolium, 108, 124, 215, 216, 222–224
Agriolimax caruanae, 146, 158, 160
agronomy, theoretical plant demography and, 27–28
Agropyron cristatum, 400
Agropyron repens, 376
Agrostis stolonifera, 215, 249
Agrostis tenuis
  differentiation in, 215, 217, 224, 225
  displacement in flowering time, 111
  immigration experiments, 248, 249
  metal tolerance in, 124, 219
algae, heteromorphic life-history, 162, 163
alkaloids, 153
allele identity, 259
Allium cepa, 376
allozymes, reproductive fitness and, 191
Alnus glutinosa, 76
Alopecurus pratensis, 225
Alpinia speciosa, 57, 60–61
Alstonia, 57
"altruistic behavior," 42
Amaranthus retroflexus, 331
Ambrosia artemisiifolia, 333, 336
Ambrosia trifida, 330–341
Ammophila arenaria, 76
Anacharis densa, 26
Andromeda polifolia, 44
andromonoecy, 300
animals
  plant metamers and, 45
  as pollinators, 263–264
  seed traits and, 88–89
Anser caerulescens caerulescens, 356
Anthoxanthum odoratum
  breeding systems, 123–124
  competition experiments with, 219
  copper tolerance in, 225
  differentiation in, 121, 123–124
  displacement of flowering time, 111
  genetic variance experiments, 234–240
  immigration experiments, 249
  intraspecific variation in, 215
  life-history studies, 170
  transplant experiments, 220, 221
Anthriscus sylvestris, 37
ants, foraging systems, 52, 53
apical dominance, mineral deficiencies and, 353
applied studies, nature of, 390–392
Arabidopsis thaliana, 170
Aralia hispida, 267
Aralia nudicaulis, 104
Arrhenatherum elatius, 73–75, 225
arthropods, metamerism and, 22–24
artificial competition experiments, 217–219
Asclepias speciosa, 271
assortative mating, 268
Aster pilosus, 345
aster yellows, 330
Astrocaryum mexicanum
  age-specific survivorship, 92–93
  defoliation studies, 151–153
  environment and reproduction, 101–103
  growth and reproductive variability, 95–103
  leaf area and survivorship, 93, 94
  seedling survival, 89–90
ATP, 308
auxin, 358
Avena barbata, 121, 135–136
Avena fatua
  herbicide experiments, 380
  life-history studies, 170–172
  microdifferentiation in, 135–136
  polymorphisms of, 110, 121, 123, 135–136
  succession study, 71
  weed control studies, 386
average selection coefficients, 232

bamboos, *see* Gramineae
Bambusa, 58–59
barley
  Avena fatua infestation, 386
  competition experiments with, 218
  gynodioecy models and, 130
  harvest index, 393–394
  heterogeneity of, 115–116
  heterozygosity experiments, 134–135, 136
  predation by nematode, 158–159
  see also Hordeum vulgare

bees, 263, 267
Bellis perennis, 249
Beta vulgaris, 365
Betula, 76, 265, 267, 271
bilateral costs, of reproduction, 278–279
Bildung, plant morphology and, 31
bioassay techniques, 349
biomass, 44
  herbivory and, 154
  intraplant competition and, 41–42
  range of, 46
Borago officinalis, 271
Brachypodium phoenicoides, 80
branching patterns
  clonal, 51-56, 60-65
  "clumps" and "runners," 60–64
  computer simulation of, 56–65
  evolution of, 65
Brassica oleracea, *see* Brussels sprouts
breeding systems
  Darwin on, 277
  effects on fitness, 200–205
  modification by growth substances, 353
  population differentiation and, 123–124
  see also mating patterns
Bromus, 137, 138, 225
Bromus mollis, 137, 138, 400
Bromus tectorum, 269
Brosimum alicastrum, 103–104
Brussels sprouts, 27, 35, 38

cabbage, 395, 400
Callitriche heterophylla, 354
Camelina sativa, 215
Campsis radicans, 271, 272
canopy structure, 333–334
Capsella bursa-pastoris, 146, 147, 155, 158, 160, 215, 269
carbamates (herbicides), 382
carbon
  interaction with nitrogen, 317–320
  intraplant competition and, 40–42
  productivity and, 44
carbon balance, 313–316
carbon dioxide fixation, 308
Carex aquatilis, 110, 121, 123
Carex arenaria, 47
Carex subspathacea, 356–357
cassava, 395
Cassia fasciculata, 201
Catapodium marinum, 171
Catapodium rigidum, 171
Cepaea nemoralis, 231
cereal crops, 392–399
  weed damage functions, 368
Chenopodium album, 249, 333
chi-squared comparisons, 155–156
Claoxylon sandwicense, 322
Clarkia exilis, 272
Clarkia xantiana, 276
Claytonia virginica, 112
cleistogamy, 262
clonal plants
  branching patterns, 51–56, 60–65
  Elymus repens, 378–379
  growth and reproduction, 26
  Solidago canadensis, 341–346
clover, *see Trifolium*
clumps, 56, 60–64
coevolution, 126, 162, 325
Coffea arabica, 29, 33, 40
cohort life table, 339
cold, effects on plant physiology, 325
coloniality, fitness and metamerism, 25–27
colonization
  defined, 242–243
  differentiation and, 255–259
community dynamics, classical theory of, 113
competition
  density and, 38–40
  environmental factors and, 108–109
  interspecific, 137–138, 139
  intraplant, 40–42
  intraspecific differentiation and, 107–109
  models of, 320–322
  niche differentiation and, 115–118
  physiological correlates, 321
  weed control and, 375, 376
component yield ratio, 116, 118–120
Compsoneura sprucei, 103–104
computer simulation, of clonal morphology, 56–65
convergence, of physiological traits, 324
Conyza canadensis, 269
copper, tolerance to, 225, 248
  see also metal tolerance
corn, *see Zea mays*
cost-effectiveness, weed control and, 364, 369
cotton, 27
Cotula, 275
couch grass, *see Elymus repens*
Crataegus monogyna, 76–78
"critical period of competition," 375, 376
crop(s)
  as weed control agent, 367, 375–380
  cereal, 368, 392–399
  metamers and growth dynamics, 27–28
crop density, 395–396, 397, 403
crop-weed communities, 375–381
cropping cycle, 367, 368
cross compatibility, 247
cucumbers, 27, 35, 152
Cucumis sativus, *see* cucumbers

cyanogenesis, in *Trifolium repens*, 147, 160–161, 163–164
Cynosurus cristatus, 225
CYR, *see* component yield ratio
cytokinin, 353, 358
cytotypes, 112

Dactylis glomerata, 37, 116, 225, 402
damage functions, weed control and, 367–368
Danthonia semiannularis, 70
Darwin, C.
  on adventive species, 387–388
  on animal predation, 141–142, 146
  on applied studies, 389–390
  on breeding systems, 227
  on ecological amplitude, 227–228
  on grass, 234
  on intraspecific variation, 213, 229
  on plant morphology, 15–17, 48–49
  on variability, 83–84, 139–140
Datura stramonium, 331
Deevey type survivorship, 179
defoliation
  herbivory and, 148–150
  photosynthesis and, 310–313
  reproductive success and, 340–341
Delphinium nelsonii, 264
demography
  agronomy and, 27–28
  of different genotypes, 111–112
density
  crop-weed communities and, 372, 375
  demographic patterns and, 111–112
  effect on canopy structure, 333–334
  interplant competition and, 38–40
  life-history studies and, 177–178
Deschampsia flexuosa, 225
desert plants, 325
development
  heteroblastic, 354
  meristems and, 33–34
  turgor pressure and, 354
de Wit replacement series, 116–121, 129
  see also replacement series
dichogamy, 292
Dictyodora, 50
differential mating ability, 270–272
differentiation
  adaptive significance of, 215–221
  cumulative nature of, 221
  intraspecific, 125–127
  limits to, 224–227
  see also variability
Diplacus aurantiacus, 317, 318
Dipsacus fullonum, 91
disassortative mating, 268
disk achenes, 182–184
diversity analysis, 79–80
DNA, nonfunctional sequences and variation, 231
domestication, Darwin on, 389–390
dominance–diversity analysis, 80
Douglas fir, 247
Drosophila, evolutionary experiments on, 115, 122
dunes, succession study, 73, 75–77

Eciton, 52
ecological amplitude, *see* ecological range
ecological combining ability, *see* niche differentiation
ecological range, evolutionary significance of, 221–224, 225
effective population size, 232
Eichhornia crassipes, 34
electrophoresis, 230, 231, 350–351
Eleusine coracana, 172
Eleusine indica, 172, 383
Elymus repens, 376–379
  comparative control regimes, 384
Empoasca kraemeri, 383
energy exchange, 306
environment
  competition and, 108–109
  effects on reproduction, 102
  experimental perturbation of, 350, 353
  fluctuation and polymorphism, 122–123
  phenotype plasticity and, 349–350
environmental grain, 122
equilibrium gene frequency variance, 252, 255–257
Erigeron, 84
Erigeron annuus, 112, 326–330, 331
Erigeron canadensis, 326–330, 331
Erodium, 116
ESS, 280–298
Euoenothera, 271, 272
Euphorbia drummondii, 71
Euphorbia forbesii, 322
evolutionarily stable strategies, 280–298
Evolutionary Synthesis, 229–230
extinction, recolonization and, 255–259

F_{ST}, 251–258
Fagus sylvatica, 34, 41, 266
Faramea occidentalis, 150

fecundity, plant size and, 268–270
feeding sites, clonal organization and, 53, 60
fertilizer, nitrogen, 396–398
Festuca microstachys, 171
Festuca ovina, 214, 215, 225
Festuca pratensis, 113, 402, 403, 404
Festuca rubra, 222–223, 225
Ficus, 95
fitness
  components of, 86, 136
  environment vs. genotype, 105–106
  metamerism and, 24–26
fitness limits, 296–300
fitness set analysis, 281
flowering, seasonal patterns, 329–331
flysch, 50
foraging systems, summarized, 51
Fragaria vesca, 171
Freesia, 271
fruits
  photosynthetic capacity of, 340
  production by *Astrocaryum mexicanum*, 99
fugitive species guild, 112

GA_3, 353, 358
Galium aparine, 381
gamodemes, 128
gas exchange, 306
  defoliation and, 312–313
geitonogamous selfing, 262, 300
"gemmation," 20–22
gender allocations
  fitness and, 283–285, 287, 292, 295
  paternal, 279–280
gene flow, 243–245, 267
gene frequency variance (F_{ST}), 251–258
genet(s)
  intraplant competition and, 40–42
  metamerism and, 25–27
  morphological study of, 55
genet mortality, 346
genetic drift, 251–252, 254, 273
genetic variation, *see* variability
genostasis, 226–227
genotypes
  biotic specialization and, 126
  demography of, 111
Geranium sanguineum, 215
germination, soil temperature and, 331, 332
"germination syndromes," 111
germination time, 179–186
Gestalt, plant morphology and, 31
gibberellin, 358
Gilia achilleifolia, 275, 279–280
gingers, *see* Zingiberales
Glycine, see soybeans
goldenrod, *see Solidago canadensis*
Gossypium, 247
gradient analysis, 67–68
Gramineae, clonal branching patterns, 53–55
grassland, succession study, 70–72
grazing pressure, phenotypic response to, 356–358
  see also herbivory
growth rate
  light and, 322
  relative, 111
  sex type and, 104
  variability in, 94–97, 104
growth substances, 353, 358
  see also specific substance
guerrillas, 53, 58–59
Gutierrezia microcephala, 91
gynodioecy, maintenance of, 130

habitat expansion, post-Pleistocene, 266
harvest, community composition and, 381
harvest index, 393–394, 396, 398, 402
Hedypnois cretica, 71
Helianthus, 35, 40, 84, 197, 400
Heracleum sphondylium, 73, 75
herbicides
  evolution of resistance to, 226, 381–383
  weed control and, 365–366, 379–380
herbivory
  interspecific competition and, 158–159
  intraspecific competition and, 154–158
  tissue damage, 143, 151–153
  ultimate (evolutionary) interactions and, 160–165
  variables affecting plant fitness, 143–144
heritability, of life-history traits, 173
herkogamy, 292
Heterodera avenae, 158–159
Heteromeles arbutifolia, 309
heteromorphology, in algae, 162, 163
heterosis, 247–248
Heterosperma pinnatum, 183–185, 186
heterostyly, 292
Heterotheca latifolia, 180, 182–183, 184
Hibiscus cannabinus, 38
Hippophae rhamnoides, 76
Holcus lanatus
  component yield ratios, 118–120
  copper tolerance in, 225
  niche differentiation in, 116–121
  polymorphisms of, 114
  succession studies, 74
  variability in, 116–118

Hordeum vulgare, 376, 380
hybrids, fitness of, 247–251
hybrids, of cultivated pasture species, 401–402
hybrid seed, pollination and, 245–246
hydrostatic potential, 354
Hypochaeris radicata, 71

IAA, 358
identity, of genes and alleles, 258, 259
immigration
  defined, 242
  patterns and pollination types, 245–246
immigration rate, 243–245
Impatiens, 133
Impatiens biflora, 280
Impatiens pallida, 179, 280
inbreeding depression, 201–205
infrared gas analysis, 325
insects, weed control and, 383
  see also pollinators
interdeme selection, 129
interference
  between reproductive functions, 291–292
  in crop/weed communities, 375–380
intraclass correlation coefficients, 173
intraspecific comparisons, life-history traits and, 168–172
intraspecific differentiation, 125–127
  herbivory and, 154–158
  individual fitness and, 234–241
  in life-history traits, 169–176
  range of, 214–215
Ipomoea purpurea, 271, 272
island model, 251, 252
Isoetopsis graminifolia, 71
isolation distances, 244
Isotoma petraea, 248

Juniperus osteosperma, 77, 78

K-selection, 26, 113, 177–178
"keystone" species, 331
kin-structured migration, 252–255, 256
Koala bears, herbivory by, 152

Lactuca sativa, 146, 155
Lamium maculatum, 74, 75
"laws of metamerism", 22–24
lead, tolerance to, 217
  see also metal tolerance
leaf area, survivorship and, 93, 94
leaf area index, 334
leafhoppers, 383
leaves
  age-dependent fitness contribution, 151–153
  biophysical models, 306
  birth and death rates, 337–339
  carbon/nitrogen interactions, 314–320
  functions of, 336
  gas and energy exchange in, 306
  herbivory and, 148–150
  heteroblastic development of, 354
  as metameric units, 21
  shade and reproductive characteristics, 309
  sun- and shade-adapted, 339-340
Leptochloa filiformis, 383
Liatris cylindracea, 110, 123, 191
life-histories
  of algae, 162, 163
  demography and, 85
  intraspecific variation in traits, 168–172, 189
  literature survey, 169–172
  methods of studying, 167–168
life-history attributes, 79
light
  adjustments to, 308–310
  growth rates and, 322
  reproduction and, 102
Ligustrum vulgare, 76
Limnanthes douglasii, 130–131
Limonium vulgare, 80
Linum usitatissimum, 84
Liriodendron tulipifera, 247, 271
loblolly pine, 247
Lolium multiflorum, 171
  cultivation of, 401–406
  gas exchange in, 312–313
Lolium perenne
  component yield ratios, 118–120
  cultivation of, 400–406
  differentiation in, 113, 114, 215
  niche differentiation, 116–121
  polymorphisms of, 110, 121
  transplant experiments, 223–225
Lotus corniculatus, 146, 155, 271
Lupinus nanus, 271
Lupinus succulentus, 271
Lupinus texensis, 87, 190–205
Lycopodium annotinum, 30

maize, *see Zea mays*
male reproductive success, 270
male sterility, selection and, 131–132
marginal habitats, 350, 352
Marsilea, 354
maternal allocation, *see* gender allocations
maternal effects, life-history traits and, 194–200
mathematical models
  for gender allocation, 281–297
  life-history studies and, 167
mating patterns
  effect on evolutionary survival, 273–274
  effect on population structure, 273
  random and nonrandom, 261–262
  rare events, 264–266
Medeola virginiana, 33
Medicago lupulina, 124
Medicago polymorpha, 71
Medicago sativa, 110, 121, 270, 271
Melilotus alba, 400
meristems, plant morphology and, 28–30, 33–34
metal tolerance, 217, 218, 219, 221, 224–226
  see also specific metal
metamer(s)
  crop growth dynamics and, 27–28
  defined, 29–31
  intraplant competition and, 40–42
metamerism
  adaptive value of, 24–27
  Darwin on, 16, 46–47
  defined, 15
  historical background, 18–24
  laws of, 22–24
  plant-animal interactions and, 45
  productivity and, 42–45
microdifferentiation, 135–137
microenvironments
  selection and, 122
  variability and, 90, 92
Microlaena polynoda, 279, 280
migrants, relatedness of, 252–255
migration, colonization and extinction, 255–259
  see also immigration
migration rate, 251–252
Mimulus, 247
Mimulus guttatus, 262
minority-advantage experiment, 238–240
model(s)
  gender allocation, 281–297
  growth and carbon balance, 313–316
  morphological, 56–65
  optimization, 314–316
  physiological, 306
  Watkinson's two-species competition, 321
  see also mathematical models
module(s)
  defined, 28–30
  herbivory and, 143, 147–150
monoecy, 292
morphology
  of clonal organisms, 26, 49–51
  "clumps" and "runners," 60–64
  computer simulation of, 56–65
  Darwin on, 15–17, 48–49
  evolutionary fitness and, 46–47
  genetics and, 34–38
  idealistic (transcendental), 19–21
  synthesis with ecology and evolution, 17
mortality
  age-specific, 92–93
  juvenile, 111
  size-dependent, 87
mutations, migration and, 258

NADPH, 308
Narcissus pseudonarcissus, 55, 56
Nectandra ambigens, 89, 150
nectar levels, pollinator behavior and, 263
nematodes, predation by, 158–159
net reproductive rate, 235, 236
  equation for, 194
  inbreeding depression and, 202–204
  mean seed weight and, 198
niche differentiation, 115–118
Nicotiana, 247
Nicotiana rustica, 35, 271–272
nitrate supply, plant growth and, 318–319
nitrogen
  carbon balance and, 314, 317–320
  net photosynthesis and, 317
nitrogen fertilizer, 396–398
  effect on *Lolium*, 404–405, 406
nonlinear fitness curves, 285–287
Norway spruce, 247

oats, predation by nematode, 158–159
  see also Avena spp.
Obelia, 57
Oenothera laciniata, 122, 123
Omphalea oleifera, 145, 150, 156–157
Onagraceae, 299
ontogenesis, *see* development
"optimal outcrossing," 264–265
optimization models, 314–316
orchard, succession study, 72–73, 74–75
Orchidaceae, 299
ordination, 68
"organizer" species, 331
Oryza, 381
Oryza perennis, 169
osmotic potential, 354

Pagiocerus frontalis, 89
Papaver dubium, 146, 147, 155, 172, 180
Paraenis, 50
Park Grass Experiment, 220–221
parthenogenesis, 261
  Owen's theory of, 19–22
Partula, 124
pasture age, competitive ability and, 116–117
pasture crops, 399–406
paternal allocation, *see* gender allocations
peas, 395
periodicity, agriculture and, 365, 367
permanent plots, succession studies and, 69
phalanx, 53, 58–59
Phaseolus, 400
Phaseolus lunatus, 271
Phaseolus vulgaris, 40, 383
Pheidole militicide, 52, 53
phenostage, effect of herbivory on, 143, 144–147
phenotype
  classification of, 348–350
  grazing pressure and, 356–358
  plasticity of, 37–38, 249, 347–358
phenoxy herbicides, 382
Philodendron selloum, 28, 29
Phleum pratense, 116, 225, 402, 403, 404
Phlox, 137
Phlox cuspidata, 247
Phlox drummondii
  habitat range, 222
  immigration experiments, 249–251
  life-history studies, 169, 174
  "optimal outcrossing" in, 264
  preferential mating in, 271
  reciprocal transplant experiments, 221
  survivorship of, 92
Phlox pilosa, 267
photosynthesis
  defoliation and, 310–313
  by flowers and fruit, 340
  optimum temperature, 327–329
  rate of, 310, 343–344
  rate and leaf age, 336–340
Phyllostachys, 58–59
physiological ecology, 305–307, 322–326
phytocoenose, 68
phytomer, 29, 30
phyton, 29
phytosociology, 66–67
Picea abies, 34
Picea glauca, 264
Pinus, 265
Pinus edulis, 266
Pinus monticola, 271
Pinus radiata, 43
Pinus resinosa, 151
Pinus rigida, 124
Pinus sylvestris, 43, 215, 224
Pisum sativum, 35
Plantago lanceolata
  breeding system and differentiation, 123
  competition experiments with, 219
  density-dependent variation in, 111
  differential survivorship of, 90
  effect of seedling predation, 146, 155
  fecundity vs. plant size, 269
  growth rate, 95
  immigration experiments, 249
  intraspecific variation in, 215
  life-history studies, 170, 174, 175
  physiological differentiation in, 110–111
  seed survivorship, 374
Plantago major, 249, 269
Plantago maritima, 80, 215
Plantago varia, 71
plasticity
  genetic control of, 175
  life-history traits and, 168, 172–173
  phenotypic, 37–38, 347–358
Poa annua
  competitive ability of, 113
  effect of predation on seedlings, 146, 155, 158, 160
  immigration experiment, 249
  intraspecific variation in, 215
  life-history studies, 170, 174, 175, 177, 178, 191
  polymorphisms of, 110, 111
  variability in, 95
Poa compressa, 118–120
Poa pratensis, 225, 345
Poa trivialis, 225
Podocoryne, 49
Podophyllum peltatum, 179
pollen
  directional dispersal of, 266–267
  fitness limits and, 298–300
  local competition, 287–291
  long-distance movement of, 265–266
  movement by animal vectors, 263–264
pollen shadows, 262
pollination
  immigration patterns and, 245–246
pollinators
  behavior of, 263–264
  competition for, 330

pollution, evolution and, 227
Polygonum lapathifolium, 376, 381
Polygonum pensylvanicum, 331, 333
polymorphisms
  categories of, 110–114
  for defensive traits, 163–165
  interspecies competition and, 136–138, 139
  temporal, 112
population genetics, 230–234
Populus, 265
Populus tremula, 76
Potamogeton natans, 348
potato, metameric organization of, 30
Potentilla erecta, 110, 215
Potentilla glandulosa, 36, 108, 216
-3/2 power law, *see* self-thinning rule
predation, seed, 89
primary succession, 73
Prunella vulgaris, 215, 224
"pseudogenes," 231
Pseudotsuga, 265
Psychotria chiapensis, 148–150
Pterocarpus hayesii, 150
Pteticus cyanifrons, 89
Puccinellia phryganodes, 356–357

Quercus, 266

radiation (light)
  adjustments to level, 308–310
  growth rates and, 322
radish, 316, 400
Ranunculus friesianus, 37
Ranunculus repens, 113, 123, 171, 215, 220–221
raspberries, 27
ray achenes, 182–184
recessive genes, immigration patterns and, 246
reciprocal transplants, 216–224, 350–352
reciprocity, selection and, 125
relative growth rate, 111
relativity hypothesis, of niche differentiation, 118–120
replacement series, 139, 158–159
  see also de Wit replacement series
reproduction
  age-specific, 192–194, 203–204
  asexual vs. sexual, 235–237
  carbon balance and, 314
  costs of, 293–296
  environmental effects on, 102
  seed vs. vegetative, 179
  variability in, 95–105
  see also breeding systems
reproductive allocation
  in cereal crops, 394
  crop density and, 396
  nitrogen and, 397–398
  see also gender allocations
reproductive barriers, 137
reproductive rate, *see* net reproductive rate
reproductive value, equation for, 338
resins, 153
resource availability, phenotype plasticity and, 349
  variability and, 92
Rhabdopterus, 179
Rhamnus catharticus, 76, 78
RHIZOM, 57
ribulose bisphosphate, 308, 309, 339
ribulose bisphosphate carboxylase, 308, 309
rice, 35, 395
rosette size, 327–330
r-selection, 26, 113, 177–178
RuBP, *see* ribulose bisphosphate
Rumex, 88, 95
Rumex acetosa, 249
runners, 56, 60–64
rye, 400

safflower, 400
Salicornia europaea, 350–352
Salicornia ramosissima, 350–352
salinity, phenotype plasticity and, 355–356
Salvia lyrata, 90, 111, 123
Sambucus nigra, 76
Saxifraga oppositifolia, 110
Scabiosa, 32, 36
secondary succession, 72–73
sediment feeders, 50
seeds
  biotic stresses and, 111
  directional dispersal of, 266
  effects of breeding system on, 201–202
  hybrid, 245–246
  local competition, 287–291
  long-distance movement of, 265–266
  maternal investment in, 87–89, 195–200
  predation on, 145, 262–263, 336
  selective forces and size, 87–89, 195–200
seed bed preparation, for weed control, 372–374
seed predation, 262–263
  plant height and, 336
seed shadows, 262
seed survivorship, in weeds, 373–374
SEED3, 57

seedlings
  predation on, 146-147, 155-158
  size and survivorship, 89-92, 195-200
selection
  group vs. individual, 129
  in gynodioecious populations, 130-133
  niche differentiation and, 115-118
self-fertilization, 262, 270-271
  see also breeding systems
self-incompatibility (SI) system, 270-271
self-thinning rule, 87
Senecio viscosus, 269
Senecio vulgaris, 271, 382
Setaria faberii, 331, 333
sex expression, environmental effects on, 335-336
sex ratio (male:female), 275, 278
sex type (male/female)
  differential dispersal of, 274-276
  growth rate and, 104
sexual reproduction, fitness and, 130-133
shade, effect on clonal plants, 342-344
Shaw-Mohler equation, 281
Shibataea, 58-59
Silene anglica, 269
Silene vulgaris, 224
simazine, 383
Simmondsia chinensis, 91
simulation, *see* computer simulation
Sinapsis arvensis, 172
size
  differential survivorship and, 87
  of seedlings, 89-92, 195-200
  of seeds, 87-89, 195-200
size hierarchies, as population vector, 84
slash-and-burn agriculture, 365
slugs, seedling predation by, 146-147, 158, 160
Smyrnium olusatrum, 279, 280
snow geese, 356
soil microenvironment, variability and, 90, 92
soil temperature, germination and, 331, 332
Solanum andigena, 358
Solanum tuberosum, 172
Solidago canadensis, 338, 341-346
Solidago virgaurea, 215
Soviet Union, plant morphology studies in, 31
soybeans, 27, 35, 197, 395
spaced-plant experiments, 216-217
spacers, 53, 55, 60
Spartina patens, 171
species coexistence, 126-127
species richness, in orchard succession study, 75
species types, succession modeling and, 79
Spergula arvensis, 38
spider monkeys, herbivory by, 152
Stephanomeria exigua, 95, 191, 245
stomatal resistance, 332
straw burning, 373
Streptanthus glandulosus, 215
Stylidium, 264
succession modeling, 77, 79
succession studies
  abandoned orchard, 72-73, 74-75
  calcareous dunes, 73, 75-77
  grassland, 70-72
  population dynamics and, 69-77
sugar beets, 395, 400
sugar cane, 395
sunflower, *see Helianthus*
suvivorship
  plant vs. leaf, 334, 335
  seed size and, 87-89
survivorship curves, for *L. texensis*, 192, 193, 199, 203
sweet vernal grass, *see Anthoxanthum odoratum*
synecology
  correspondence with population biology, 80-81
  defined, 66

Tagetes patula, 84
tagmata, 23, 29-30
tannins, 153
Taraxacum (agamospecies), 249
Taraxacum officinale, 37, 112, 124, 169, 177
tea plantations, 27
Texas bluebonnet, *see Lupinus texensis*
tiller density, 404
tobacco, 319
tonoplast, 354
trees, intraspecific variation in, 110
triazene herbicides, 382
Tribolium, 137
Trifolium arvense, 124
Trifolium fragiferum, 400
Trifolium pratense, 116
Trifolium repens
  competition experiments with, 219
  competitive ability of, 113
  component yield ratio experiments, 118-120

cyanogenesis in, 92, 147, 160–161, 163–164
density and, 40
differential survivorship, 92
effect of predation on seedlings, 146, 155
genetic basis of morphology, 36, 88
grazing pressure and, 357
growth rate, 95, 111
immigration experiment, 249
intraplant competition in, 41
intraspecific variation in, 214, 215
life-history studies, 172, 173
metamerism and, 25–26
morphological variation, 110
niche differentiation in, 116–121, 125
polymorphism of, 113–114
resource allocation, 400–401
soil microenvironment and, 90
Trifolium subterraneum, 172, 400
Triglochin maritima, 355–356
Tsuga, 266
turgor pressure, 354
Turnera ulmifolia, 353
"twins," 240
two-species competition model, 321
Typha latifolia, 124

Ulam's packing game, 59
Ulmus, 265
uracils (herbicides), 382
Urtica dioica, 73–75
urticating hairs, herbivory and, 142–143

variability
Dawin on, 83–84, 139–140
vs. genetic uniformity, 235–237
in growth and reproduction, 93–105
gynodioecy and, 130–133
herbicide resistance and, 382
individual fitness and, 234–241
microdifferentiation and, 135–137
migration and, 251–259
patterns and demography, 105–106
planting design experiments, 235
resource availability and, 92
of seed size, 87–88
variance, *see* variability

vegetation analysis, 67–69
Verbascum thapsus, 91
Veronica peregrina, 111, 122, 169
vegetative reproduction, 261–262
Vicia faba, 271, 272
Viola, 41, 56, 103
Viola sororia, 89–91
Viola surinamensis, 88
Viola tricolor, 36
vital attributes, 79

Wahlund variance, 254
water-use efficiency index, 312
weed(s)
cost-benefit function, 364–366
economic costs of, 363–364
herbicide resistance in, 381–383
primitive agriculture and, 363
seed mortality in, 373–374
word derivation, 363
weed control
biological, 370
chemical, 370
crop timing, 372
cultural, 370
herbicides, 365–366, 379–380
weed-free cropping environment, 365
wheat, 27, 39, 134, 371, 393, 395
winter moth caterpillars, herbivory by, 152
winter wheat, 39
cost of production, 371

Xanthium strumarium, 87–88

Yucca whiplei, 95
Yushania, 58–59

Zea mays, 37, 172, 247, 271, 395
zinc, tolerance to, 217
Zingiberales, clonal branching patterns, 53–55
Zizania aquatica, 265
Zizania palustris, 280